35th International DAAAM Symposium on Intelligent Manufacturing and Automation 2024

Vienna, Austria
24-25 October 2024

Editor:

Branko Katalinic

ISBN: 979-8-3313-1393-7

Printed from e-media with permission by:

Curran Associates, Inc.
57 Morehouse Lane
Red Hook, NY 12571

DAAAM International Vienna
TU Wien, Getreidemarkt 9
1060 Wien
Austria, EU

Phone: +436646424398

http://www.daaam.info

Additional copies of this publication are available from:

Curran Associates, Inc.
57 Morehouse Lane
Red Hook, NY 12571 USA
Phone: 845-758-0400
Fax: 845-758-2633
Email: curran@proceedings.com
Web: www.proceedings.com

ISSN 1726-9679

DANUBE ADRIA ASSOCIATION FOR AUTOMATION & MANUFACTURING

DAAAM INTERNATIONAL VIENNA

www.daaam.info

PROCEEDINGS

OF THE 35TH INTERNATIONAL DAAAM SYMPOSIUM
"INTELLIGENT MANUFACTURING & AUTOMATION"
24-25TH OCTOBER 2024
HOSTED FROM VIENNA UNIVERSITY OF TECHNOLOGY, VIENNA, AUSTRIA, EU

ORGANIZED BY:
DAAAM INTERNATIONAL VIENNA & UNIVERSITY OF MOSTAR

IN COOPERATION WITH:
ÖIAV 1848, VIENNA UNIVERSITY OF TECHNOLOGY, INTERNATIONAL ACADEMY OF ENGINEERING
AND UNIVERSITY OF APPLIED SCIENCES – TECHNIKUM WIEN

UNDER THE AUSPICES OF THE DANUBE RECTORS' CONFERENCE &
RECTORS' AND PRESIDENTS' HONOR COMMITTEE OF DAAAM INTERNATIONAL

EDITOR: B.[RANKO] KATALINIC

IMPRESSUM 2024

Proceedings of the 35th International DAAAM Symposium "Intelligent Manufacturing & Automation"
24-25th October 2024, hosted from Vienna University of Technology, Vienna, Austria, EU, Volume 35, No.1,
Abbreviated key title: ***Ann. DAAAM 2024***

ISSN 1726-9679

Ed. B. Katalinic,

Published by DAAAM International, Vienna, Austria, EU, 2024
Layout & Design: B[ranko] Katalinic
Technical Editors: I[lija] Zec / L[uka] Zec

Publisher:

DAAAM International Vienna
TU Wien, Getreidemarkt 9, 1060 Wien, Austria
E-mails: president@daaam.info / branko.katalinic@tuwien.ac.at
DAAAM homepage: http://www.daaam.info

CONTENTS

AUTHOR INDEX

ORGANIZING COMMITTEE FOR 2024

KATALINIC, BRANKO – PRESIDENT OF DAAAM INTERNATIONAL
COVIC, DRAGAN – HONOR PRESIDENT OF ORGANIZING COMMITTEE
GUSEV, BORIS – PRESIDENT OF INTERNATIONAL ACADEMY OF ENGINEERING
ZEC, ILIJA - SECRETARY

ABURIA, MOHAMED
BEVANDA, IVAN
BIELIC, TONI
BOSNJAK, IGOR
BUCHMEISTER, BORUT
COLAK, IVO
COLAK, STANKO
COSIC, ILIJA
CVETKOVSKA, ROBERTA
DURAKBASA, NUMAN
FUERST, LAURENZ
GODEC, DAMIR
HOELLER, LIANE
KALENIC, MIODRAG
KNEZOVIC, NIKOLA
KOSTOLANI, DAVID
KUKUSHKIN, ILYA
MARIC, PETAR
NAUMBURGER, JOERG
PANFILOV, PETER
PARK, HONG-SEOK
PRYANICHNIKOV, VALENTIN
RAKIC, KRESIMIR
SARAVANJA, LUKA
SCHMOELLEBECK, FRITZ
SEREMET, ZELJKO
STAZHKOV, SERGEY
STAROVESKI, TOMISLAV
STOJKIC, INJA
STOJKIC, ZELJKO
STUJA, KEMAJL
TEKIC, ZELJKO
TOMIC, ZORAN
UDILJAK, TOMA
UGLESIC, ANTE
VLADOV, MAREK
ZEC, ANTE
ZEC, LUKA

RECTORS' AND PRESIDENTS' HONOR COMMITTEE

ABDRASILOV, BOLATBEK	RECTOR, KARAGANDA STATE INDUSTRIAL UNIVERSITY
ABRUDAN, IOAN VASILE	RECTOR, TRANSILVANIA UNIVERSITY OF BRASOV
ALEXANDROV, ANATOLIY	RECTOR, BAUMAN MOSCOW STATE TECHNICAL UNIVERSITY
BAAIJENS, FRANK	RECTOR, EINDHOVEN UNIVERSITY OF TECHNOLOGY
BIAŁKIEWICZ, ANDRZEJ ANTONI	RECTOR, CRACOW UNIVERSITY OF TECHNOLOGY
BONDREA, IOAN	PRESIDENT, LUCIAN BLAGA UNIVERSITY OF SIBIU
BORAS, DAMIR	RECTOR, UNIVERSITY OF ZAGREB
BOROVIKOV, YURY	RECTOR, THE YAROSLAV-THE-WISE NOVGOROD STATE UNIVERSITY
CHEON, YEON	PRESIDENT, UNIVERSITY OF ULSAN
CISNEROS PÉREZ, GUILLERMO	RECTOR, UNIVERSIDAD POLITECNICA DE MADRID
COSTOIU, MIHNEA	RECTOR, UNIVERSITY POLITEHNICA OF BUCHAREST
CZIGANY, TIBOR	RECTOR, BUDAPEST UNIV. OF TECHNOLOGY AND ECONOMICS
DAM, HENRIK	DIRECTOR, UNIVERSITY OF SOUTHERN DENMARK
DANIELS, MITCHELL	PRESIDENT, PURDUE UNIVERSITY
DI SCIASCIO, EUGENIO	RECTOR, POLYTECHNIC OF BARI
DRĂGAN, FLORIN	RECTOR, POLYTECHNIC UNIVERSITY OF TIMISOARA
G. BOUDOUVIS, ANDREAS	RECTOR, NATIONAL TECHNICAL UNIVERSITY OF ATHENS
GALIJEV, ILJHAM	PRESIDENT, OMSK STATE TRANSPORT UNIVERSITY
GARCÍA MARÍN, JUAN FRANCISCO	RECTOR, UNIVERSITY OF LEON
GARCIA, EVA FERREIRA	RECTOR, UNIVERSITY OF THE BASQUE COUNTRY
GELENCSER, ANDRAS	RECTOR, UNIVERSITY OF PANNONIA
GEYER, SYLVIA	RECTOR, UNIV. OF APPLIED SCIENCES TECHNIKUM, VIENNA
GONOKAMI, MAKOTO	PRESIDENT EMERITUS, UNIVERSITY OF TOKYO
GUBERAC, VLADO	RECTOR, STROSSMAYER UNIVERSITY OF OSIJEK

INTERNATIONAL SCIENTIFIC COMMITTEE

ALANDER, JARMO TAPANI (FIN)
ARAI, TAMIO (JPN)
ARES, ENRIQUE (SPA)
ARNDT, GUENTER (AUS)
ATTARD, MARIO M. (AUS)
BACHYNSKI, MORREL (CAN)
BELAJEV, GENNADY J. (BLR)
BILBERG, ARNE (DEN)
BLIEDTNER, JENS (DEU)
BOHEZ, ERIK L. J. (THA)
BRIGHT, GLEN (RSA)
BROWNE, J. (IRL)
CELAR, STIPE (HRV)
COELHO, REGINALDO T. (BRA)
COVIC, DRAGAN (BIH)
CRUZ, FERNANDO M. M. (POR)
DAR-EL, E. (ISR)
DE FILIPPI, A. (ITA)
DEVAHASTIN, SAKAMON (THA)
DINCMEN, M. (TUR)
DINIZ, ANSELMO EDUARDO (BRA)
DOHDA, KUNIAKI (JPN)
DONG, JIANHUA (CHN)
DORNFELD, DAVID (USA)
DOVE, RICK (USA)
DUDAS, ILLES (HUN)
EGOROV, SERGEY B. (RUS)
ENDTER, R. (DEU)
EVERSHEIM, W. (DEU)
GALANTUCCI, LUIGI MARIA (ITA)
GIACAGLIA, G. E. (BRA)
GIBSON, I. (AUS)
GIROT, FRANK (FRA)
GOLDFARB, VENYAMIN Y. (RUS)
GRABENWEGER, JOHANN (BRA)
GUSEV, BORIS V. (RUS)
HABCHI, GEORGES (FRA)
HAN, QING YAO (CHN)
HOCK, SEO SWEE (SNG)
HOFMANN, PETER (DEU)
HRYU, JEE-HWAN (KOR)
IBISHI, ISMET MALIQ (KWT)
ILTSCHEV, VELKO (BGR)
IRFAN, MOHAMMAD A. (PAK)
ITOH, KIYOSHI (JAP)
IVAN, NICOLAE-VALENTIN (ROM)
KALS, H. J. J. (NDL)
KATALINIC, BRANKO (AUT)
KIMURA, FUMIHIKO (JPN)
KINNANDER, ANDERS (SWE)
KNOSALA, RYCZARD (POL)
KOCH, JAN (POL)
KONDRATENKO, YURIY (RUS)
KUOSMANEN, PETRI (FIN)
KYTTNER, REIN (EST)
LE, CUNG (VTN)
LIN, CHYI-YEU JERRY (TWN)
LLORENTE, J. I. (ESP)
LOMBARDI, F. (ITA)
LOW, KIN HUAT (MAL)
LU CHANG, TAO (CHN)
LU, S. S. (TWN)
LUONG, LEE HONG SON (AUS)
MARCOS, MARIANO (ESP)
MEGHDARI, ALI (IRI)
MERTINS, KAI (DEU)
MICHELINI, R. C. (ITA)
MIKHAYLOV, YURI O. (RUS)
MUNTEANU RADU (ROM)
NANASI, JOZEF (SVK)
NEUMANN, H. J. (DEU)
NICOLETTO, G. (ITA)
NOVAKOVIC, BRANKO (HRV)
OCZOS, KAZIMIERZ E. (POL)
OTTO, TAUNO (EST)
PARK, HONG-SEOK (KOR)
PEGDEN, DENIS (USA)
PETERKA, JOZEF (SVK)
PORTER, BRIAN (HKG)
PRITSCHOW, GUENTER (DEU)
PRYANICHNIKOV, VALENTIN (RUS)
PRZYBYLSKI, W. (POL)
RADERMACHER, F. J. (DEU)
RADHAKRISHNAN, V. (IND)
RAMAYAH T. (MLA)
RUNGRUANG, CHANNARONG (THA)
SCHMID, STEVEN (USA)
SEGUI LLINARES, V. JESUS (ESP)
SLADEK, JERZY (POL)
STAZHKOV, SERGEY (RUS)
STEFANESCU, D. MICHAEL (USA)
TABUCANON, MARIO T. (THA)
TAKAKUWA, SOEMON (JAP)
TAKALA, J. (FIN)
TIKKA, H. (FIN)
TOPALOVA, IRINA (BGR)
TORRES-LEZA, F. (ESP)
TORVINEN, S. (FIN)
TSAI, B. S. Y. (TWN)
UDILJAK, TOMA (HRV)
UGLESIC, ANTE (CRO)
UHLMANN, ECKART (DEU)
UYAR, EROL (TUR)
VAN BRUSSEL, HENDRIK (BEL)
VAN OUDHEUSDEN, DIRK L. (BEL)
VARGA, GYULA (HUN)
VAVRIN, PETR (CZE)
VENKATESH, V. C. (SGP)
VETTURI, DAVID (ITA)
VITECEK, ANTONIN (CZE)
WALDIR, P. O. (BRA)
WERTHEIM, RAFI (ISR)
WESSELY, EMIL (SVK)
WESTKAEMPER, ENGELBERT (DEU)
WIENDAHL, HANS-PETER (DEU)
WILLIAMS, D. J. (GBR)
WINSTON, PATRICK H. (USA)
ZALZALA, ALI M. S. (GBR)
ZHAO, DAOZHI (CHN)

INTERNATIONAL PROGRAM COMMITTEE

ACKO, BOJAN
AHUJA, HARISH
ALBERS, ALBERT
ALEKSEYEV, WLADIMIR A.
AMBERG, MICHAEL
BANYAI, AGOTA
BANYAI, TAMAS
BARITZ, MIHAELA IOANA
BARREIRO, JOAQUIN
BEGIC, DERZIJA
BIJELONJA, IZET
BLEICHER, FRIEDRICH
LEVIN, ALEXANDER
LI, GUIHUA
LICA, DUMITRU
LIPA, ZDENKO
LIU, HUAMING
LO VALVO, ERNESTO
LOGHIN, CARMEN MARIA
LOPEZ DE L., LUIS NORBERTO
LOW, KIN HUAT
LUCA MOTOC, DANA I.
LUDOVICO, ANTONIO D.
LUJIC, ROBERTO
PERNIU, DANA
PERUTKA, KAREL
PETERS, JACQUES
PETKOVIC, DARKO
PETRE, CRISTIAN
PETRE, MARIAN
PETRESCU, VALENTIN
PINCA-BRETOTEAN, CAMELIA
PINTILIE, GHEORGHE
PIRTEA, GABRIEL MARILEN
PIVONKA, PETR
PJERO, ELENICA

BLEY, HELMUT
BOTEZATU, CORNELIA PAULINA
BOTEZATU, MIHAI
BRAJLIH, TOMAZ
BREZAK, DANKO
BREZOCNIK, MIRAN
BUCHMEISTER, BORUT
BUCURESTEANU, ANCA
CAMPBELL, R. IAN
CEKIC, AHMET
CELAR, STIPE
CHAN, CHEE SENG
CHAUS, ALEXANDER
CHRAMCOV, BRONISLAV
COHODAR, MAIDA
COLEMAN, J. J.
COSIC, ILIJA PETAR
COSMA, SMARANDA
COTET, COSTEL EMIL
CUBONOVA, NADEZDA
CURTU, IOAN
DAMIC, VJEKOSLAV
DENKENA, BEREND
DEVAHASTIN, SAKAMON
DOLGA, LIA
DOLGA, VALER
DOLINAY, JAN
DOLINAY, VILIAM
DOSTAL, PETR
DOSTALEK, PETR
DRAGCEVIC, ZVONKO
DRAGICEVIC, MARIJA
DRSTVENSEK, IGOR
DUHOVNIK, JOZE
EJUB DZAFEROVIC,
FIRST ROGALE, SNJEZANA
FUJII, NORIO
GHEORGHIU, HORIA
GODEC, DAMIR
GOMEZ, ANTONIO
GORDIN, DOINA MARGARETA
HLAVA, JAROSLAV
HOFREITER, MILAN
HRUBINA, KAMIL
HRUSKA, FRANTISEK
HRUSKOVA, ERIKA
IBISHI, ISMET M.
JANACOVA, DAGMAR
JANSSENS, THIERRY
JARVIS, CHRISTINE W.
JAVOROVA, ANGELA
JERBIC, BOJAN
JURAGA, IVAN
JURUM-KIPKE, JASNA
KARAULOVA, TATYANA
KATALINIC, BRANKO
KAWALEC, ANDRZEJ
KHANNA, ASHUTOSH
KHOMCHENKO, VASILY
KIRK, CHRIS. S.
KISS, IMRE
KITTL, BURKHARD
KLIMENKO, S. A.
KNEGO, NIKOLA
KNEZEVIC, PETAR
KOHLERT, HELMUT
LUKIC, IVANA
LUPEA, IULIAN
LUPU, CIPRIAN
LUPULESCU, NOURAS BARBU
MACHEDON PISU, TEODOR
MACKU, LUBOMIR
MACUROVA, ANNA
MAGHETI, IOAN
MAJETIC, DUBRAVKO
MALISA, VIKTORIO
MALOPOLSKI, WALDEMAR
MAN, MARIANA
MANDIC, LIDIJA
MANDRU, DAN
MANESCU, TIBERIU STEFAN
MARASCU KLEIN, VLADIMIR
MARCOS, MARIANO
MARCU, MARIUS
MAREK, KUBALCIK
MARIAN TOLNAY
MARIN, DUMITRU
MARIN, MIHNEA
MARINESCU, NICULAE ION
MARN, FERDINAND
MARONEK, MILAN
MARSLAND, DAVID
MARTA, CONSTANTIN
MARTINEZ, SUSANA
MASTACAN, LUCIAN
MATUSOVA, MIRIAM
MATUSU, RADEK
MEHTA, NARINDER KUMAR
MEIER, HORST
MENZ, PETER
MICHAL, BLIZNAK
MICIETA, BRANISLAV
MICLOSI, VIOREL
MOHORA, CRISTINA
MOK, H.
MORARU, SORIN AUREL
MORIWAKI, T.
MORTOIU, DOINA
MOTOMANCEA, ADRIAN
MUNTEAN, MIHAELA
MUNTEAN, SEBASTIAN
MUNTEANU, ANCA
MUSCA, GAVRIL
MUSTAFIZUL, KARIM
MYSHLYAVTSEV, ALEXANDR
NAGI, MIHAI
NAGLA, K. S.
NAKAMOTO, NOBUYUKI
NEUMANN, PETR
NICA COTET, GABRIELA BEATRICE
NIKOLIC, GOJKO
NKUMBWA, RODGER
NOH, SANG DO
NOVAC, OVIDIU
NOVOSELOVIC, DANIEL
NURUL AMIN A.K.M.
OANCEA, GHEORGHE I.
OKAMOTO, J.
OLARU, ADRIAN, DUMITRU
OLARU, ONISIFOR
OLEKSIK, VALENTIN
OLTEANU, CIPRIAN
PLESA, ANGELA
PODURAEV, JURY
POKORNY, PAVEL
POMAZAN, VALENTINA
POPA, CICERONE LAURENTIU
POPPEOVA, VIERA
PRODAN, DAN
PROKOP, ROMAN
PRYANICHNIKOV, VALENTIN
PYTKO, STANISLAW
QUINN, ROGER D.
RAMAJ, VEHBI XH.
RAVAS, RUDOLF
RAY, T. S.
RIKALOVIC, ALKSANDAR
RIZESCU, SABIN VASILE
ROGALE, DUBRAVKO
ROHDE, ULRICH L.
ROMERO, FERNANDO
ROS, OLIMPIA
ROSADO, V.O.G.
RUDAS, IMRE J.
RUGESCU, DRAGOS RADU DAN
RUJA, IOAN
RUSU ANGHEL, NICOLAE
RUSU, LIGIA
RUSU-CASANDRA, AURELIA
RUTEMI, GANI BRAHIM
RUZIC, DRAGO
SAKULSKI, DUSAN
SALIHU, AVDI HAJDIN
SAMEK, DAVID
SCHREIBER, PETER
SELVAKUMAR, ARUL LAWRENCE
SEONG, POONG HYUN
SERCER, MLADEN
SHIBASAKA, TOSHIROH
SHIH, MING CHANG
SIEMIATKOWSKI, M.
SIHN, WILFRIED
SINAY, JURAJ
SISEK, BORIS
SITUM, ZELJKO
SKOCZYNSKI, WACLAW
SKOLUD, BOZENA
SLUGA, ALOJZ
SMIRNOV, ALEXANDER
STANKOVSKI, STEVAN
STAROVEŠKI, TOMISLAV
STEFANIC, NEDELJKO
STOPPER, MARKUS
STRGAR KURECIC, MAJA
SUBRAMANIAN, ARUNKUMAR
SUGAR, PETER
SYROVA, LIVIA
SYSALA, TOMAS
TADAMURA, KATSUMI
TAKALA, JOSU
TAMASILA, MATEI
TAN, KOK KIONG
THIERHEIMER, WALTER WILHELM
ULRICH, GUENTHER
VAARIO, JARI
VASEK, VLADIMIR
VELISEK, KAROL
VEZA, IVICA

REVIEWING COMMITTEE

CULJAK, EVA
CUS, FRANCI
DAR-EL, E.
DE FILIPPI, A.
DEVAHASTIN, SAKAMON
DINCMEN, M.
DINIZ, ANSELMO EDUARDO
DOHDA, KUNIAKI
DOINA MARGARETA
DONG, JIANHUA
DOVE, RICK
DURAKBASA, NUMAN
EGOROV, SERGEY
ENDTER, R.
EVERSHEIM, W.
GALANTUCCI, LUIGI MARIA
GAWLIK, JOZEF
GIACAGLIA, G. E.
GIBSON, I.
GIROT, FRANK
GOLDFARB, VENYAMIN Y.
GOMEZ, ANTONIO
GRABENWEGER, JOHANN
GUSEV, BORIS V.
HABCHI, GEORGES
HAN, QING YAO
HOBLE, DOREL ANTON
HOCK, SEO SWEE
HOFMANN, PETER
HRYU, JEE-HWAN
ILTSCHEV, VELKO
MACUTA, SILVIU-DANUTI
MARCOS, MARIANO
MARCOS, MARIANO
MARIC, PETAR
MATUSU, RADEK
MEGHDARI, ALI
MERTINS, KAI
MICHELINI, R. C.
MILOS, LAURA RAISA
MIRONEASA, SILVIA
MOMETE, DANIELA CRISTINA
MUNTEANU RADU
NASTAC, SILVIU
NEUMANN, H. J.
NICOLETTO, G.
NOVAC, CORNELIA MIHAELA
NOVAC, OVIDIU CONSTANTIN
OCZOS, KAZIMIERZ E.
OTTO, TAUNO
PALCIC, IZTOK
PANDA, ANTON
PANFILOV, PETER
PARK, HONG-SEOK
PEGDEN, DENIS
PERAKOVIC, DRAGAN
PERUTKA, KAREL
PETERKA, JOZEF
PIVONKA, PETR
PODURAEV, JURY
POP, PETRU ADRIAN
POPESCU, DANIEL
PORTER, BRIAN
TOPALOVA, IRINA
TOPLICEANU, LILIANA
TORRES-LEZA, F.
TORVINEN, S.
TSAI, B. S. Y.
UDILJAK, TOMA
UHLMANN, ECKART
UYAR, EROL
VALMOHAMMADI, CHANGIZ
VAN BRUSSEL, HENDRIK
VAN OUDHEUSDEN, DIRK L.
VARGA, GYULA
VAVRIN, PETR
VENKATESH, V. C.
VETTURI, DAVID
VITECEK, ANTONIN
VOSNIAKOS, G. CHRISTOPHER
VOUK, RUDOLF
WALDIR, P. O.
WERTHEIM, RAFI
WESSELY, EMIL
WESTKAEMPER, ENGELBERT
WIENDAHL, HANS-PETER
WILLIAMS, D. J.
WINSTON, PATRICK H.
WOZNIAK, ADAM
ZALZALA, ALI M. S.
ZEC, ILIJA
ZEC, LUKA
ZHAO, DAOZHI
ZLOTO, TADEUSZ

35TH DAAAM INTERNATIONAL SYMPOSIUM ON INTELLIGENT MANUFACTURING AND AUTOMATION

Preface

The 35th DAAAM International Symposium on Intelligent Manufacturing and Automation was organised as virtual online conference hosted by the Vienna University of Technology, Vienna, Austria, between the 24th and 25th October 2024, during the DAAAM International Week. The Symposium was organized by DAAAM International Vienna and University of Mostar in cooperation with ÖIAV 1848, Vienna University of Technology, International Academy of Engineering and University of Applied Sciences – Technikum Wien and Under the Auspices of the Danube Rectors' Conference & Rectors' and Presidents' Honor Committee of DAAAM International for 2024. The Symposium was hosted from Vienna, Austria. This year's symposium aimed at continuing the success of the previous years, focusing on the five-fold traditional objectives of the symposium: the presentation of the most recent high-quality results, support of development of young scientists and researchers, organization of international (summer) doctoral school, inauguration of new members of Central European Branch of International Academy of Engineering and the provision of the necessary setting for stimulating discussions, brainstorming and networking among European and international researchers coming both from the academia government agencies and industry

DAAAM International Symposium has been established as the leading Central European conference devoted to the cutting-edge advancements in the field of modern production and automation. Symposium provides an international forum for engineers, scientists and industrial managers to present and discuss the current status and impact of advanced manufacturing and automation. The international series of Symposium was initiated by the first meeting in Vienna (1990) and continued in Štrbske Pleso (1991), Budapest (1992), Brno (1993), Maribor (1994), Cracow (1995), Vienna (1996), Dubrovnik (1997), Cluj - Napoca (1998), Vienna (1999), Opatija (2000), Jena (2001), Vienna (2002), Sarajevo (2003), Vienna (2004), Opatija (2005), Vienna (2006), Zadar (2007), Trnava (2008), Vienna (2009), Zadar (2010), Vienna (2011), Zadar (2012), Zadar (2013), Vienna (2014), Zadar (2015), Mostar (2016), Zadar (2017), Zadar (2018), Zadar (2019), Mostar (2020), Vienna (2021), Vienna (2022), Vienna (2023), Vienna (2024).

Regular organization of the annual Symposium is one of the central activities of the DAAAM International association. The association was founded on 5. November 1990 in Vienna during a meeting organised by Professor B. Katalinic in order to celebrate the 175 Years of Vienna University of Technology. The main idea of the association was to establish a network for a new kind of international scientific and academic cooperation in the fields of intelligent manufacturing and automation. Our primary focus is on: all aspects of production of technical products made out of parts and corresponding automation and knowledge, trans-disciplinary application of methods and all kinds of education.

DAAAM International activities, during the last thirty-five years, have been supported by scientists and experts from over fifty countries all over the world. In the course of the last decade, DAAAM International has focused its activities on the regional and global long-term co-operation between (technical) universities. The result of this initiative is the founding of the University Network of DAAAM International Vienna and DAAAM International Network for Advanced Technologies which was founded on 2002-10-25 in Vienna.

The main activities of DAAAM are:

- Organisation of annual DAAAM International Symposium.
- To maintain and improve the University Network of DAAAM International Vienna as a strategic alliance and permanent international open platform for long-term co-operation and partnership in all fields of technology, research and education.
- Organisation of long-term projects, initially known as Inter-University Doctoral Studies.
- Organisation of long-term projects, initially known as the Virtual University.
- Publishing of international journals, books, and proceedings.
- Promotion of the exchange of people, scientific, and technical information.
- Promotion of international co-operation among scientists and engineers.
- Promotion of R & D, education, and industrial application.
- To co-operate with other national and international organisations in matters of common interest.
- Organisation of International Doctoral School as a platform for the support of the next generation of engineers at the highest level.

This proceeding contains 52 refereed papers from 154 authors presented at the 35th DAAAM 2024. In DAAAM proceedings, since the first symposium in 1990, we published papers from 9811 authors.

All the papers included in these proceedings are reviewed by experts in double blind peer review way, and finally revised by the authors for the improvement according to referee's reports and advices.

The papers discuss many aspects of modern manufacturing and automation such as: Algorithms, Artificial Intelligence, CAX, Computer Integration, Control, Cutting Tools, Design, FEM, Invited Lectures, Knowledge, Management, Manufacturing System, Mechatronics, Methodology, Methods, Modelling, Optimization, Robotics, Simulation, Technical Solutions, Technology and Trends.

Appreciation is forwarded to reviewers who contributed to improve the quality of papers reviewing the papers, providing constructive critical comments, improvements, and corrections to the authors. We also recognize the support of reviewers, who were in charge of acceptance/rejection process. We wish to thank the Scientific, Program and Organizing Committees, Session Chairs, student helpers and administrative assistants, for contributing to the success of this Conference. They did an excellent job.

Finally, we wish to thank all the authors who submitted papers, making this conference possible and the authors of accepted papers for making appropriate modifications and corrections with care and patience and for preparing the final manuscript. It is the quality of their presentations and their passion to communicate with other participants that really makes this conference series a success.

We would like to thank all members of the International Academy of Engineering – Central European Branch, Rector of the Vienna University of Technology Professor Dr Sabine Seidler, and Rector of the University of Mostar Professor Dr. Sc. Zoran Tomić and his team, management and all those nice colleagues for their excellent support in the organization of 35th DAAAM International Symposium.

Vienna, 2024-12-09
Editor
Chair of the 35th DAAAM 2024

Branko Katalinic

Professor rtd Dr.sc. Dr.mult.h.c.Prof.h.c. Branko Katalinic
President of DAAAM International
TU Wien, IFLT-IMS, Getreidemarkt 9
1060 Wien, Austria, Europe
URL: http://www.daaam.info
E-mail: president@daaam.info
branko.katalinic@tuwien.ac.at

DOI: 10.2507/35th.daaam.proceedings.001

ANALYZING THE ROLE OF BRAND IDENTITY ELEMENTS IN SHAPING CONSUMER PERCEPTION: A REVIEW STUDY OF CULTURAL HERITAGE BRANDING IN TOURISM DESTINATION

Berislav Andrlić, Orhan Uludag & Mehmet Saydam

This Publication has to be referred as: Andrlic, B[erislav]; Uludag, O[rhan] & Saydam, M[ehmet] (2024). Analyzing the Role of Brand Identity Elements in Shaping Consumer Perception: A Review Study of Cultural Heritage Branding in Tourism Destination, Proceedings of the 35th DAAAM International Symposium, pp.0004-0009, B. Katalinic (Ed.), Published by DAAAM International, ISBN 978-3-902734-44-0, ISSN 1726-9679, Vienna, Austria
DOI: 10.2507/35th.daaam.proceedings.001

Abstract

This research aims to analyze how brand identity elements—such as name, logo, characters, slogans, and packaging—contribute to building a recognizable and authentic brand identity. The introduction defines the term "brand," examines the key elements of brand identity with examples of best practices, and discusses the importance of brand identity in today's business and marketing landscape. The study explores visual elements and language used in communication, as well as the concept of brand personality and its impact on consumer perception. The second part focuses on creating a new brand for a tourist destination.

Keywords: marketing; internet; tourism destination; branding elements

1. Introduction

Brand consists of a name and/or logo, as well as other elements and activities assigned by producers to a product, service, or idea to communicate its uniqueness to the market. A brand includes the name, symbols, and all sensory stimuli linked to the product's features and its functional or psychological benefits. [12]

Pavlek states that brand identity is a relatively new concept in marketing theory and practice. It refers to the creation of distinctive features that differentiate an individual or group from others or indicate belonging to a group with shared traits, important for building relationships with the environment. [7] The quality a brand reflects is essential for a product's survival in the market. Companies understand the value of branding, evident in the growing number of new brands created by domestic producers as markets open and new sales and distribution channels emerge. Managers must find innovative and effective ways to compete and build strong brands.

From the definitions of a brand, it is clear that various authors describe it in their own way, but common elements are present in all. Brands consist of both tangible and intangible characteristics, which help communicate with consumers and make it easier for them to decide on a purchase. When customers recognize a brand, their loyalty to its products or services increases the value of both the brand and the company itself. In the context of global growth, authors mention that the "new normal" requires a shift from old business models to sustainable ideas and new development patterns, representing a transition to a new societal and economic structure. [8]

A brand must align with the company's vision and mission, and employees should believe in the brand for customers to do the same. [12] Communicating the brand to the public is crucial, especially in field of tourism what is focus of research inj tourism industry.

2. Brand identity prism in tourism

A prerequisite for building a successful brand (assuming appropriate product and service quality) is a suitable brand identity that must realistically and reliably reflect the brand's value to customers. Brand identity can be represented through a "brand prism". The brand identity prism is a model used to visually represent a brand's identity, illustrating the key components that define how a brand is perceived by consumers. It consists of six elements that shape the brand's internal and external image: [4]

1. Physical characteristics (Physique) – The tangible aspects of the brand, including the product design, logo, and packaging, that consumers can immediately identify.
2. Personality – The human characteristics or traits associated with the brand, often communicated through tone of voice, marketing style, or brand characters.
3. Culture – The values and principles that the brand embodies, often rooted in its country of origin or corporate culture.
4. Relationship – The type of relationship the brand seeks to establish with its customers, whether it's built on trust, loyalty, or emotional connection.
5. Reflection – The perceived image of the target consumer, or how the brand makes its customers feel or see themselves.
6. Self-image – How the brand sees itself and the internal view it holds regarding its mission and purpose.

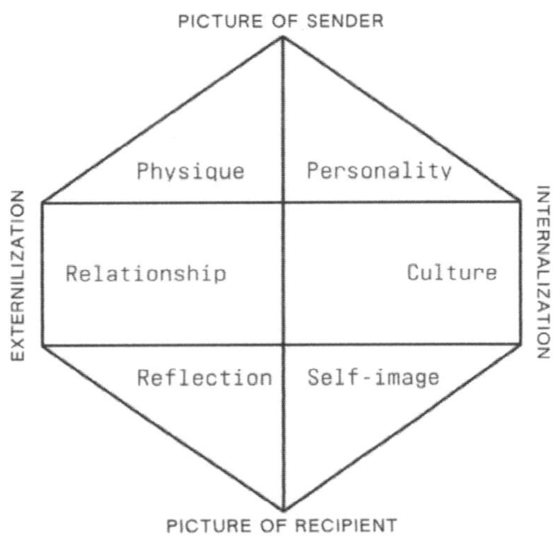

Fig. 1. The brand identity prism elements [4]

This prism helps create a cohesive and consistent brand identity, guiding how the brand communicates with the market. In the context of tourism, the brand identity prism plays a vital role in shaping how a destination is perceived by visitors. The physical characteristics might include iconic landmarks, cultural symbols, or even unique local experiences that differentiate the destination from others. The personality of a tourist destination could be reflected in the tone of its promotional materials—whether adventurous, relaxing, or culturally enriching. Culture is especially important in tourism branding, as it encompasses the traditions, heritage, and lifestyle of the destination. The relationship between a destination and its visitors is often based on providing memorable, authentic experiences, fostering emotional connections and encouraging repeat visits. The prism also helps in portraying the reflection of the ideal tourist, influencing how travelers perceive themselves while experiencing the destination, while the self-image of a tourist brand defines its commitment to sustainability, hospitality, or adventure, shaping its overall identity in the tourism market. Physical characteristic sare crucial for understanding and recognizing the brand. Packaging can influence brand value, especially if the design is distinctive compared to others and offers an additional functional advantage.

Also, each brand should build the concept of sustainable tourist development which indicates a necessity of optimal use of environment resources that are the foundation of tourist development and by doing that keeping the important ecological processes intact and helping the protection and preservation of natural heritage and biodiversity. [9]

3. Focus on Tourism Destination: Brand Identity Strategy

The identity and image of a brand are not synonymous, and why it is necessary to distinguish them is explained by this way: "When talking about image, it refers to the current state determined by research, penetrating into consumer perception, and if it ends there, it is a passive relationship, a statement, but—what next? When identity is mentioned, it implies branding as an active relationship towards brand creation, which is at the core of brand management.". [13]

In his work, author states that the identity of a product brand "arises from the need to determine what existing product brands actually are, or what makes them special and unique. [4] In the communication process, especially in advertising, a specific copy strategy is adopted, which changes from campaign to campaign, and in reality, only a few product or service brands have a clear image or a firmly defined long-term identity and uniqueness.". Creating the identity of a product brand is the starting point and fundamental aspect of the entire brand management process. By developing a strategy and following the process of building a strong and unique brand identity, greater market value of the brand is achieved, and clear recognition of that product brand among other product categories is accomplished, with the potential for brand extension into other markets. "When choosing brand elements, it is necessary to adhere to six criteria: memorability, meaningfulness, likability, transferability, adaptability, and protectability" [2]

To create a strong tourism destination brand, it is important to align the key elements of brand identity, as outlined by scientific paper according to Keller: [2]

- Brand name
- Brand mark (logo, symbol)
- Characters
- Slogans
- Jingles
- Packaging
- Color.

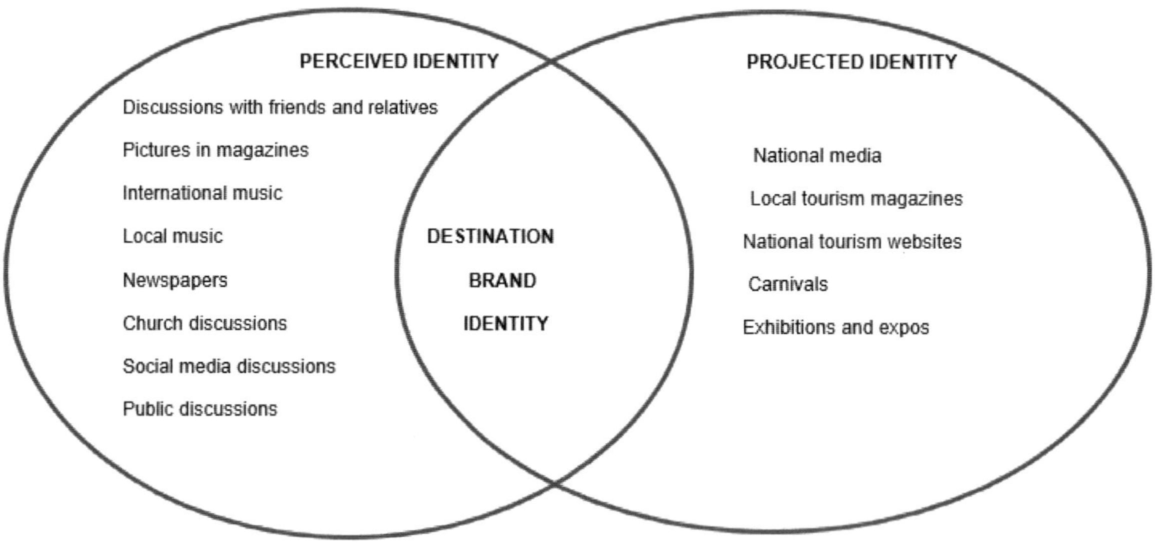

Fig. 2. Brand communication process in tourism industry [13]

Vranešević emphasizes that the criteria related to brand building, such as memorability, meaning, and likability, can also be applied to tourism. These elements are crucial for creating recognition for tourist destinations or services. Memorability refers to how easily visitors can recognize and recall a particular destination or tourist product. (Fig.2.). [12] The criterion of meaning highlights how a specific tourism offering can communicate its unique advantages, such as natural beauty, cultural heritage, or unique experiences, playing a key role in building an authentic brand identity. Likability is also important as it positively influences potential tourists; the destination must be interesting, appealing, and stimulate the imagination. Furthermore, the criteria of transferability, adaptability and protectability are particularly relevant to tourism. Transferability means that brand elements can extend to different tourism services or various regions. Adaptability requires flexibility in marketing and offerings, allowing destinations to modernize and adjust to changes in the market. Protectability pertains to those brand elements that can be legally protected to distinguish them from competitors.

As consumer acceptance of a brand increases, so does their loyalty to it, making the brand stronger. This emotional connection is particularly significant in tourism, where experiences and feelings greatly impact consumer choices. A tourism product lacking these elements (brand acceptance, loyalty, and emotional connection) is considered a generic product, which fundamentally differs from a branded tourism experience.

4. Brand identity prism: focus on tourism destination

Since humanity became aware of its existence, there has been a persistent desire to explore the surrounding environment and beyond. According to the definition by the World Tourism Organization (WTO), a tourist is "a person who travels to countries or places outside their usual environment for personal, professional, or business reasons" [10]

In today's highly competitive tourism industry, destinations are feeling increasing pressure. [5] This dynamic competitive environment necessitates the creation of a clear identity or "brand" based on the destination's "personality" and its unique core values, which will differentiate it from others. Creating a brand for a destination is increasingly becoming a crucial strategic issue in marketing and business. The first and essential step in creating a new brand is developing a strategy, followed by an action plan and its implementation. Given the rapid advancement of technology and the complexity of systems,

AI can analyze data about visitor behavior, traveler preferences, reviews, and other information to provide personalized travel recommendations, activities, accommodations, and experiences. It can also assess market dynamics, seasonal changes, and other factors to adjust accommodation prices or tourism packages, enhancing competitiveness and profitability. Demand forecasting, capacity management, and overbooking can all be monitored using AI, which is vital for improving customer experiences and reviews.

Authors note that with the emergence of a new array of technological achievements worldwide, "technological tools" are the most powerful means of communication, creating a lasting impact on people's minds. [3]

The accelerated technological development has certainly improved some business processes but has also necessitated modifications to marketing research techniques in tourism. While tourism marketing emerged more frequently only after the 1960s, the global theoretical knowledge on this discipline is now well developed, providing a foundation for managing businesses in tourism.

In paper of Foley and Fahy, authors state that "the marketing strategy of a destination is a key factor in managing its image." It must be based on the core qualities of the destination to produce a consistent and rich image that forms the basis of the brand. The image of the destination should be grounded in real qualities—compelling, simple, attractive, and distinct from others. Continuous research into the opinions of visitors and the local population is essential for strategic destination marketing. [1]

When positioning a tourist destination, it is vital to know the competitive advantages, select the most relevant ones, and effectively communicate them to the target audience or visitors. Diversity of tourist needs and numerous travel motives require the development of specific forms of tourism.

In terms of rural tourism development, rural areas are structurally and functionally diverse. This diversity arises from their geographic, demographic, socioeconomic, and natural characteristics, which must be considered when planning tourism development and destinations.

The process of creating a brand for a tourist destination consists of several phases, as outlined Morgan & Pritchard:

1. Market research, analysis, and strategic recommendations.
2. Creation of brand identity.
3. Launching the brand into the market and communicating its vision (both internal and external).
4. Brand implementation.
5. Brand monitoring, evaluation, and critical review. [6]

Numerous other variables contribute to brand value, such as quality, innovation, personality, and symbolism. Gathering valid and reliable information involves a series of activities collectively referred to as the market research process. This process consists of a logical sequence of procedures that must be followed to ultimately obtain reliable and valid information that aids in decision-making in market operations. A destination brand is a name, symbol, logo, word, label, or other graphic sign that simultaneously identifies a destination and distinguishes it from others; furthermore, the brand conveys the promise of an unforgettable travel experience uniquely tied to the destination; it also reinforces and enhances the recollection of pleasant memories associated with the destination's experiences.

Creating a new tourist destination brand requires a strategic approach and collaboration with all relevant stakeholders, including the local community, entrepreneurs, tourism service providers, agencies, and marketing experts, all aimed at attracting new visitors and retaining existing ones. Theory and tourism practice notes that alongside the primary motivations for tourists visiting a destination (such as swimming, relaxation, or culture), more complex travel motivation packages arise, including intense enjoyment, outdoor activities, a strong desire for experiences, emotions, recreation, and health. Tourists seek to experience new adventures that distinctly differ from their everyday surroundings and life circumstances. Increasing tourist expectations are supported by media, advanced telecommunications, greater education, and prior personal experiences. The same factors influence the faster transmission of information and the growing significance of word-of-mouth promotion.

5. Brand identity elements research review- case study of Trogir town

On November 25, 2019, the presentation of the new brand strategy and visual identity of Trogir will was held at the Square of Ivan Pavlo II in Trogir. The city of Trogir is the first in Dalmatia to embark on a destination branding project. The aim of the project was to emphasize the authentic values and beauties of the city and to profile itself as a unique destination on the Croatian coast as well as on the Mediterranean. After a year of thorough research, work and production of new visuals, slogans, brochures and video materials, the time had come to present what was highlighted on the map of Croatian coastal cities and beyond.

The Historic City of Trogir was inscribed on the World Heritage List in 1997. The city presents a remarkable example of urban continuity. The orthogonal street plan of this island settlement dates back to the Hellenistic period and it was embellished by successive rulers with many fine public and domestic buildings and fortifications. Its beautiful Romanesque churches are complemented by the outstanding Renaissance and Baroque buildings from the Venetian period.

This brand project outlines the comprehensive rebranding strategy developed for the city of Trogir, Croatia, whose historic core is recognized as a UNESCO World Heritage site. Over the course of a year, our team crafted a brand strategy that included a new slogan and visual identity aimed at revitalizing Trogir's appeal as a tourist destination while instilling a sense of pride among its residents. Our approach sought to harmonize Trogir's rich historical heritage with contemporary marketing techniques to foster increased tourism and economic growth. By leveraging the city's historical depth and cultural assets, town employees aimed to enhance its visibility as a destination, encouraging longer stays and greater spending by visitors. Methodology involved a multi-faceted approach which included: [10]

1. Comparative snalysis- successful branding strategies of European cities were examined such as Eindhoven, Glasgow, and Leeuwarden. These case studies provided insights into the impact of effective branding on tourism and economic growth.
2. Surveys and interviews- extensive surveys targeting both residents and tourists to gauge their perceptions of Trogir. Additionally, they engaged in interviews with key stakeholders, including representatives from the City Council, the Tourist Board, and local cultural institutions. Insights were gathered from experts such as art historians, tourism professionals, and local craftsmen.
3. Archival research- a thorough investigation of historical documents and expert literature was undertaken to identify unique aspects of Trogir's history, geography, and culture that could be woven into the branding narrative.

The research revealed several unique elements of Trogir that distinguish it from other destinations, including its well-preserved medieval architecture, vibrant local traditions, and rich cultural events. The sentiment expressed by both residents and tourists highlighted a desire for greater community engagement and year-round activities that celebrate Trogir's heritage. The resultant branding strategy includes:

- New slogan: a memorable and engaging slogan that encapsulates the essence of Trogir.
- Visual identity: a modernized visual identity that reflects Trogir's historical significance while appealing to contemporary audiences.
- Storytelling approach-a narrative that integrates Trogir's past with its present, aiming to evoke local pride and attract visitors through immersive experiences.

Fig. 3. Brand identity rebranded elemets of Trogir town[11]

The rebranding of Trogir represents a strategic initiative designed to foster economic growth through enhanced tourism. By connecting the city's historical narrative with modern marketing practices, we aim to create a dynamic identity that resonates with both visitors and residents.

This comprehensive approach not only seeks to increase visitor numbers but also aims to reinvigorate community pride and engagement throughout the year. Ongoing evaluation of the branding strategy's effectiveness will be essential, including tracking tourism metrics and resident feedback to ensure that the new identity continues to meet the needs of both visitors and the local community.

6. Conclusion

In the concluding discussion of this paper on brand identity elements, important insights are reached regarding the impact of visual, verbal, and emotional elements on consumer perception. By presenting examples of successful brands, it is concluded that creating an authentic identity is crucial for market differentiation and building long-term consumer loyalty. Visual elements such as logos, packaging, and colors are often the first interactions with consumers and can strongly influence the impressions formed about a brand. This paper emphasizes the importance of brand identity elements, particularly visual identity, in establishing a recognizable and trustworthy brand. Limitations of paper arise from situation that research is focused on specific tourist destinations, and findings may not be generalizable to other locations.

Slogans, jingles, and other verbal components help deepen the emotional connection between the brand and its consumers. Strong verbal identification can foster emotional bonds and create stories that consumers want to share. Additionally, associating a brand with characters or personalities can contribute to the emotional feelings that cultivate loyalty to a particular brand. From this, it can be concluded that brand personality is a key aspect of identity. Consumers often perceive brands as entities with human characteristics, making it essential to strategically define a personality that resonates with the target audience. Ultimately, successful brand identities are not static; they evolve as society, technology, and consumer preferences change. Tourism companies must continuously monitor trends, conduct market research, and adapt their image to remain relevant and appealing. The use of new technologies and innovations is desirable, as is artificial intelligence, which can be a significant support in the strategic decision-making process.

In the context of creating a tourist destination brand of Trogir town, it can be conclude that brand identity elements also play a very important role in marketing. Destinations need to view their identity from a holistic perspective, considering various cultural activities in the broadest sense, and reflect on the long-term impact of tourism on the local community and natural environment. They should present their resources and stories to potential visitors in a way that is attractive, authentic, and sustainable. Future plans for scientific research will include surveys in field of tourist satisfaction with quality of brand identity elements design on sample of Mediterranean tourist destinations.

7. References

[1] Foley, A., & Fahy, J. (2004). Incongruity Between Expression and Experience: The Role of Imagery in Supporting the Positioning of a Tourism Destination Brand. Journal of Brand Management, 11, 209-217.

[2] Kotler, P., & Keller, K. (2008). Marketing Management, Zagreb: Mate.

[3] Kurian, S., Andrlić, B., & Ramanathan, N. H. (2020). Gauging the Usage of Internal Marketing by Teachers for Effective Teaching and Communication*. URL: https://files.eric.ed.gov/fulltext/ED621957.pdf

[4] Lijović, B. (2012). Brand Identity of Products–A Factor of Successful Market Positioning of Brands*. URL: https://hrcak.srce.hr/file/139707

[5] Mansur, S., Saragih, N., Susilawati, S., Yusiatie, U. D., & Endri, E. (2021). Consumer Brand Engagement and Brand Communications on Destination Brand Equity in Maritime Tourism in Indonesia. *Journal of Environmental Management and Tourism, 12(4), 1032-1042.

[6] Morgan, N., & Pritchard, A. (2005). On Souvenirs and Metonymy: Narratives of Memory, Metaphor, and Materiality. *Tourist Studies, 5(1), 29-53.

[7] Pavlek, Z. (2008). Branding: How to Build the Best Brand*, Zagreb: M.E.P. Consult, p.153.

[8] Ristanović, V., & Andrlić, B. (2023). The Concept of Global Growth and Development With the New Normal, Sustainable Growth and Global Social Development in Competitive Economies. URL: https://repozitorij.ftrr.hr/islandora/object/ftrr%3A142

[9] Rados, B. (2014). Development of a new tourist destination by building a new railway infrastructure. *DAAAM International Scientific Book, Vienna

[10] Theobald, W. F. (2005). The Meaning, Scope, and Measurement of Travel and Tourism. Global Tourism, 3, 23-48.

[11] Tourist Board Trogir Research Data.

[12] Vranešević, T. (2007).Brand Management, Zagreb: Accent, p.10.

[13] Winata, K. S. K., & Adiwaty, M. R. (2024). Kapferer's Brand Identity Prism Analysis on Brand Stord to Increase Brand Awareness. East Asian Journal of Multidisciplinary Research, 3(7), 3079-3088.

DOI: 10.2507/35th.daaam.proceedings.002

MANUFACTURING METHOD AND CHARACTERIZATION OF A NUMBER OF THERMOPLASTIC COMPOSITE MATERIALS WITH NATURAL FIBRES

Cornel Ciupan, Emilia Ciupan & Mihai Ciupan

This Publication has to be referred as: Ciupan, C[ornel]; Ciupan, E[milia] & Ciupan, M[ihai] (2024). Manufacturing Method and Characterization of a Number of Thermoplastic Composite Materials with Natural Fibres, Proceedings of the 35th DAAAM International Symposium, pp.0010-0017, B. Katalinic (Ed.), Published by DAAAM International, ISBN 978-3-902734-44-0, ISSN 1726-9679, Vienna, Austria
DOI: 10.2507/35th.daaam.proceedings.002

Abstract

The furniture industry is one of the world's most important sectors, but due to the scarcity of wood and its high-cost companies have had to look for alternatives. Natural fibre composites represent one alternative because many fibrous plants are easy to cultivate and generate high yields. The authors have also developed a simple and cost-efficient way of blending natural fibres and polypropylene and forming them into boards needed for samples in laboratory testing. The paper studies the mechanical properties of several natural fibre composites which incorporate hemp, willow or jute. The authors developed and tested several formulations consisting of one type of natural fibre (hemp, willow and jute) with varying percentages of polypropylene as thermoplastic matrix. The tests show that below a certain percentage of matrix material, the composite loses its mechanical properties quickly. However, certain natural fibre and polypropylene blends resulted in good mechanical properties and indicated excellent potential for use in the structural elements of furniture.

Keywords: natural fibre composites; thermoplastic matrix; sample manufacturing method.

1. Introduction

The furniture industry is one of Romania's strategic branches. In 2023, it brought a contribution of 1.23 billion euros to Romania's trade balance. It represents the second economic branch as added value. From the amount of furniture produced in Romania, 90% is exported, Romania ranking as the 5th country in Europe, and the 12th in the world, from this point of view [1].

However, furniture manufacturing in Romania has decreased significantly for the last two years. One of the important factors that contributed to this is the steady increase, for the last eight years, of the price of wood on the local market, which has led to decreased competitiveness on the international market. The decrease of furniture manufacturing resulted, in 2023, in a reduction by 19% of the number of employees hired in the furniture sector, as compared to 2022.

Finding advantageous alternative solutions to the replacement of the main raw material—wood—can contribute to a revival of the steady growth of this economic sector. A solution in this sense is the use of composite materials made of natural fibre (flax, hemp, willow, poplar, coconut, etc.), mixed with other matrices.

1.1. Advantages of using thermoplastic composite materials with natural fibres

The industry of upholstered furniture, as a branch of the furniture industry, uses, to a large extent, wood, chipboard, multi-layered and other wooden boards, polyurethane, non-woven materials, textile materials, and leather and substitutes of leather. Natural resources are limited. In the current situation in which the world is facing environmental degradation and global warming, rethinking the design process of products or services is of major interest in all fields of activity. Some authors propose the adoption of an eco-design process that reconsiders the entire life cycle of products or services [2]. At the same time, there are concerns about the classification, evaluation and ranking of the negative effects that an organization can produce on the environment, in accordance with the ISO 14001 standard [3].

Under these circumstances, the partial replacement of wood and wood-based products in the structure of an item upholstered with composite materials with matrices from polypropylene, or adhesives and natural fibres, as fibres of enforcement, has a significant number of advantages. Among them count the preservation of the current forests, the reduction of pollution and of the greenhouse emissions, the preservation of flora and fauna, increased productivity of the fibre-producing cultures, and reduced costs [4], [5], [6]. The use of the new materials contributes to the development of a more competitive upholstered industry, and to an increased number of jobs.

For a cubic meter of wood utilized in the upholstered industry, around six cubic meters of raw wood must be cut down from the forest. While for a hectare of forest, the quantity of wood grows by 3-5t/year, the production of natural fibres per hectare is larger, namely 10-14 t/ha, for hemp, 50-60 t/ha, for willow and poplar, 16-20 t/ha, for miscanthus.

The "environmental friendliness" of the composite material is also given by the biodegradability of the natural fibres, their origin from renewable sources, and the chances of recycling the material [7], [8], [9]. Furniture factories must comply with the Forest Stewardship Council (FSC) requirements, which impose awareness of wood traceability and responsible management of forests worldwide.

The interest and necessity of replacing certain wooden parts in the structure of upholstered furniture, without affecting the external appearance of the product and its functional requirements, were presented in more detail in papers [4] and [10].

Criteria	Wood/Chipboard	Composite materials
Excessive use of wood	Yes	No
Raw material availability	Lengthy harvesting cycle (80-100 years) and little annual growth (3-5 t/ha/year)	Yearly harvesting and renewal, and significant yearly growth (50-60 t/ha/year, in the cases of willow and poplar)
Material usage yield	Low	Significant
Complex-shaped items	Involves a wide range of frames and types of raw materials. Example: an armchair side uses 21 elements and 3 types of raw material (wood, chipboard, cardboard)	Reduced number of frames and the possibility of using a single raw material. Example: an armchair side can be made of 2 elements of composite material
Technological complexity	It is difficult to realize complex shapes, with many manual operations for the realization of the frames and assembly	It is easy to realize complex spatial shapes through thermoforming or injection in moulds
Dimension of the technological line	Large surface both at the producer of raw material and the one of the frames, as it requires many operations of cutting, processing, assembling	A compact line and a short technological flow, which can be automated
Content of formaldehyde (carcinogenic substance)	Yes (Chipboard- 15%)	No
Purchase price of raw material	Price in a continuous growth	Slight variations, price stability
Recycling / reusage	Reduced	Excellent

Table 1. Advantages of using composite material

When traditional technologies are used, the manufacturing of an upholstered item involves the use of several wooden frames [4], higher manufacturing costs and a lengthier manufacturing process. The replacement of wood by natural fibre composite to obtain complex-shaped frames comes along with a number of advantages (Table 1).

1.2. Problem statement

The authors have specific achievements in the development of thermoplastic composite materials and their usage in the field of upholstered furniture, at TAPARO SA from Targu Lapus [4], [10], [11], [12]. A wide range of frames (sides of sofas, armchairs, chairs, etc.) were manufactured of composite materials made of a fibrous thermoplastic (polypropylene, with a length of 40-60 mm, and a linear density of 7-16 DEN, in percentages that ranged from 40% to 50% of the total weight of the mixture) and of natural fibrous component (fibres of hemp, jute, sisal, coconut, etc.), or a mixture of such fibres with a linear mass density of 70-80 DEN, and a length of the fibre between 5 and 100 mm, in a percentage between 50% and 60% of the total weight of the mixture. These materials are obtained as fibrous layers mechanically reinforced by interweaving. The transformation of the fibrous layer-shaped composite into frames is achieved through the following operations: cutting the material in accordance with the dimensions of the item, heating the material in a hot plate press at a temperature higher than the melting temperature of the matrix, transferring it into the thermoforming mould, and consolidating it through cooling.

From the research conducted [4], [13], the authors noticed that for the items to be competitive, the time necessary for the consolidation of the composite material into items should be reduced. For instance, the time necessary for the thermoforming of a piece with the dimension of a shell for an armchair is greater than 15 min. This cannot be reduced due to the extensive amount of time required in the mould, which is due, in its turn, to the reduced thermal transfer coefficient of the natural fibres.

A significant reduction of composite transformation time could be obtained by replacing the thermoforming technology with injection, which is used to obtain plastic products. Composite mixtures with long fibres cannot be used for injection, as this requires the use of short fibres, or granules.

For a new composite to be used industrially, it should be characterized, to determine its properties. The characterization of the material involves blending the raw materials and thermoforming several samples that should be submitted to mechanical trials on specialized machines [11], [14], [15], [16]. The authors developed a simple and effective method of obtaining new materials by using simple and cost-effective technological equipment.

2. Method and Equipment Presentation

To develop a simple and effective method of making and improving the natural fibre composite materials, the authors turned to using technological equipment of reduced complexity for both the obtaining of the fibrous mixture and the thermal consolidation of the samples. The characterization of the material is made on special machines, based on industry specific standards.

The method of making and characterizing the mixture formulations for the thermoplastic natural fibres composite materials involves the following stages:

A. Elaborating the material formulation:
- choosing the constituents (natural fibres, thermoplastic matrix and their parameters)
- elaborating the material formulations (proportion of the constituents)

B. Making the composite material and thermally consolidating it:
- preparing the constituents
- realizing the fibrous mixture
- thermally consolidating it in the shape of boards using different process parameters

C. Characterizing the composite material:
- cutting the samples out of the consolidated composite boards
- characterizing the material by determining the tensile strength, Young's modulus, and elongation
- interpreting the results and selecting the formulations that offer the most appropriate physical and mechanical properties.

2.1. Developing the material formulation

The goal of this stage is to establish the optimum percentage of thermoplastic fibre in the mixture, depending on the usage of the composite material. The percentage of polypropylene used should be as little as possible, given its higher cost, as compared to the natural fibre, and the negative impact on the environment.

Composite formulations can be made using virgin or recycled polypropylene to optimize the costs and use resources sustainably. However, the validation of a formulation can be made solely based on the characterization of the material. The technology of obtaining and consolidating the fibrous mixture should also be analysed when elaborating a formulation for use in series production.

Composite materials were made using different types of natural fibres. Three formulations were designed for each fibre material where the percentage of natural fibre and polypropylene differed. Similarly, formulations made of three elements (for example, hemp, willow and polypropylene) could be made, but the current paper focused on formulations containing just one type of natural fibre.

2.2. Making and thermal consolidation of the composite material

The first stage in the realization of the composite material consists in the preparation of the natural and polypropylene fibres at the required parameters. The second one involves weighing the right percentage of the constituents based on the formulation. Subsequently, the fibrous mixture is made whilst taking care to obtain good homogeneity.

In most of the applications of natural fibre composites, the fibrous mixture must be consolidated as a piece of textile, so that it can be easily cut and placed in the thermoforming mould. To simplify the method, a mould of reduced complexity was conceived, in which the material is placed directly, with no need of previous consolidation as a fibrous layer. To thermally consolidate the fibres as a board, a mould was made consisting of the following parts: core 1, handle 2, pressing screws 3, cavity 4 and centring pins 5 (Fig. 1).

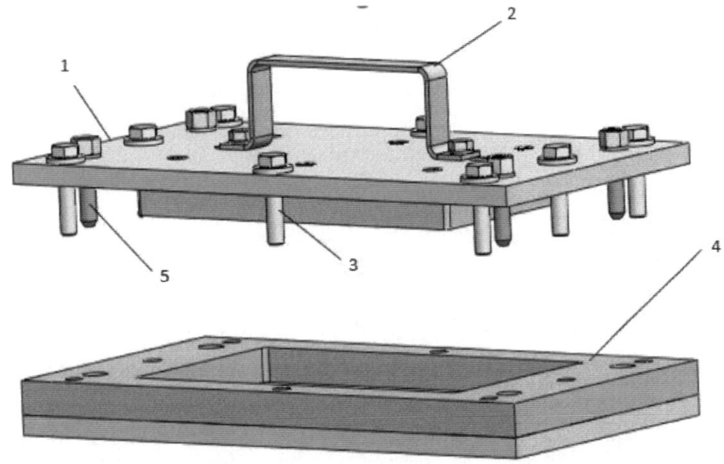

Fig. 1. Thermoforming mould

The composite mixture is placed in cavity 4 (Fig. 2). Afterwards, the mould is closed, and screws 3 are tightened to a torque of 5 Nm, which ensures a pressure of 10 MPa on the surface of the composite material. Thermoforming involves the heating of the mixture above the melting temperature of the polypropylene. The mould was placed in a convection oven (Fig. 2). The oven offers the possibility of regulating temperature (up to 240° C) and heating time (15-60 minutes). The oven is equipped with a thermometer for temperature control.

Fig. 2. Oven for thermoforming

The temperature of the oven and the heating time were established through experimental research and depended on the type of natural fibres and the thickness of the mixed layer.

2.3. Characterization of the composite material

The samples were cut from the thermally consolidated board and tested in terms of tensile strength through traction. The tensile strength, the specific elongation and Young's modulus were determined. The samples and tests were in accordance with standard ASTM D3039, "Test Method for Tensile Properties of Polymer Matrix Composite Materials".

3. Results

3.1. Composite material formulations

Nine formulations of material marked FS1-FS9 (Table 1) were elaborated using natural fibres of jute, willow and hemp.

No.	**Vegetable fibre**	**Polypropylene [%]**		
		50 %	**40%**	**30%**
1.	Jute	FS1	FS2	FS3
2.	Willow	FS4	FS5	FS6
3.	Hemp	FS7	FS8	FS9

Table 2. formulations of composite materials of short fibre

3.2. Making of the thermally consolidated boards

Natural and polypropylene fibres were prepared during the first stage. The natural fibres were acquired from several providers and were cut so that their length was between 1 and 5 mm (Fig. 3, Fig. 4).

willow coconut hemp jute cane

Fig. 3. Vegetable fibres

Virgin polypropylene 70% willow-30% polypropylene Cutting waste (hemp and polypropylene) 50% hemp-50% recycled polypropylene

Fig. 4. Polypropylene fibre and mixture

The blends were prepared based on the formulations during the second stage. A scale with a precision of one gram was used to weigh the fibres and the blend was made manually (Fig. 5).

After some trials, the oven was set at a temperature of 220 °C as it was observed that at this temperature the blend reaches the temperature of 180 °C after a time of 45 minutes. At this temperature, the polypropylene is melted, and the natural fibre is not affected. After being taken from the oven, the bolts of the mould are torqued again to 5 Nm, and then the mould is left to cool. After the temperature of the mould drops below 60 °C, the thermoformed board is taken out (Fig. 6, 7).

Fig. 5. Making the fibrous mixture and placing it in the mould

Fig. 6. Mould cavity with thermoformed board

Fig.7. Thermoformed boards of hemp and polypropylene

3.3. Characterization of new materials

Five samples of each type of consolidated material were subjected to tensile testing. The machine Zwick/Roell Z150 from the CERTETA laboratory of the Technical University of Cluj-Napoca was used. Figure 8 shows the graphs from the Test report for material FS1.

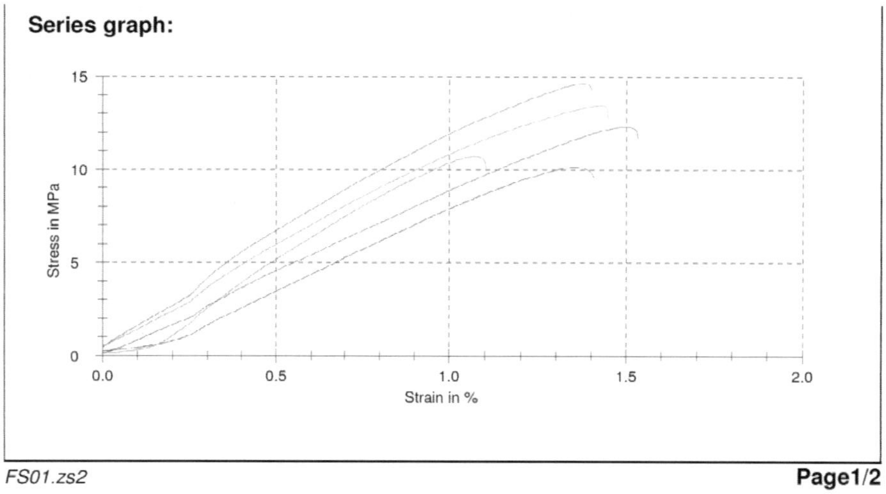

Fig. 8. Test report FS1

The following parameters were evaluated for each material formulation: tensile strength, specific elongation and Young's modulus. The results are shown in Table 3. The optimization regards the modification of the parameters of thermoforming, with the goal of reducing the percentage of polypropylene from the mixture, correlated with the physical and mechanical characteristics of the consolidated material and the decrease of the production cost per unit.

Nr. crt.	Formulation	Medium density ρ[g/cm³]	Ultimate tensile strength UTS [MPa]	Young's modulus E [MPa]	Elongation %
1.	FS1 (jute 50%, polypropylene 50%)	0.66	12.30	845.0	1.30
2.	FS2 (jute 60%, polypropylene 40%)	0.51	5.80	423.0	1.30
3.	FS3 (jute 70%, polypropylene 30%)	0.47	5.69	362.0	1.30
4.	FS4 (willow 50%, polypropylene 50%)	0.51	5.17	446.0	1.10
5.	FS5 (willow 60%, polypropylene 40%)	0.52	5.03	4.23	1.10
6.	FS6 (willow 70%, polypropylene 30%)	0.51	4.86	417.0	1.20
7.	FS7 (hemp 50%, polypropylene 50%)	0.76	24.60	867.0	2.80
8.	FS8 (hemp 60%, polypropylene 40%)	0.74	17.90	942.0	2.20
9.	FS9 (hemp 70%, polypropylene 30%)	0.69	13.60	637.0	1.90

Table 3. Results obtained

4. Conclusion

The authors identified the need to replace wood in the structure of upholstered furniture with thermoplastic composite materials with plant fibres. To create new composite materials, it is necessary to try different formulations, and this cannot be done economically on the industrial production lines of the material.

To solve the problem, the paper presents a simple and effective method of elaborating and characterizing new formulations of composites made of natural fibres and a thermoplastic matrix, by using economic, reliable and easy-to-use technological equipment. The analysis of the newly elaborated and consolidated materials by using the above-mentioned method leads to the following conclusions:

- The best results were obtained in the case of the composite mixtures that had hemp fibres. Material FS7 used short fibres with a blend of 50% hemp and 50% polypropylene and had mechanical properties like the material obtained from fibrous layers consolidated through interweaving and having the same amount of polypropylene.
- The materials of jute fibres and polypropylene have a tensile strength below 50%, as compared to hemp. The materials with willow and polypropylene have significantly reduced mechanical properties, as compared to those made of hemp and jute. This can be explained by the deterioration of the natural fibre when heated, and/or a much weaker bond between the consolidating fibres and the thermoplastic matrix. The strength of the composites with willow fibres is inferior to the other ones, since the fibres are very short and have a large number of particles.

With the proposed method, the authors will develop new composite materials using combinations of plant fibres (hemp with willow, flax with poplar etc.) and varying proportions with polypropylene fibres.

5. Acknowledgments

The paper resulted from research conducted in the project "The development of a research centre in the field of composite material at SC TAPARO SA", project code SMIS: 121434.

6. References

[1] https://www.zf.ro/business-construct/dumitru-blaga-presedintele-apmr-industria-mobilei-doilea-sector-22490378, (2024). Ziarul Financiar, Accessed on: 2024-09-20

[2] Bodur, O.; Walcher, E. M.; Durakbasa, N. & Sterca, D. A. (2021). Implementation of an Eco-Design Process to Minimize Negative Environmental Impacts – Case Study, Chapter 10 in DAAAM International Scientific Book 2021, pp.117-130, B. Katalinic (Ed.), Published by DAAAM International, ISBN 978-3-902734-31-0, ISSN 1726-9687, Vienna, Austria, DOI: 10.2507/daaam.scibook.2021.10]

[3] Tomaskova, T. & Bicova, K. (2021). Impact of Machining on the Environment, Proceedings of the 32nd DAAAM International Symposium, pp.0623-0630, B. Katalinic (Ed.), Published by DAAAM International, ISBN 978-3-902734-33-4, ISSN 1726-9679, Vienna, Austria, DOI: 10.2507/32nd.daaam.proceedings.088

[4] Ciupan, E., Ciupan, C., Câmpean, E. M., Stelea, L., Policsek, C. E., Lungu, F., & Jucan, D. C. Opportunities of Sustainable Development of the Industry of Upholstered Furniture in Romania. A Case Study, Sustainability, Vol. 10, Issue 9, (SEP 2018), eISSN 2071-1050

[5] Barton-Pudlik, J. & Czaja, K. Fast-growing willow (Salix viminalis) as a filler in polyethylene composites, Composites Part B: Engineering. Vol. 143, (JUN 2018), pp. 68-74, ISSN 1359-8368, eISSN 1879-1069

[6] Niemczyk, M., Kaliszewski, A., Jewiarz, M., Wrobel, M. & Mudryk, K. Productivity and biomass characteristics of selected poplar (Populus spp.) cultivars under the climatic conditions of northern Poland. Biomass & Bioenergy, Vol. 111, (APR 2018), pp. 46–51, ISSN 0961-9534, eISSN 1873-2909

[7] Ichim, M., Filip, I., Stelea, L., Lisa, G. & Muresan, E. I. Recycling of Nonwoven Waste Resulting from the Manufacturing Process of Hemp Fiber-Reinforced Recycled Polypropylene Composites for Upholstered Furniture Products, Sustainability, Vol. 15, Issue 4, (FEB 2023), eISSN 2071-1050

[8] Branciforti, M. C., Marinelli, A. L., Marcio, K. & Nobre, A., 2009. Wood Polymer Composites Technology Supporting the Recovery and Protection of Tropical Forests: The Amazonian Phoenix Project, Sustainability, Vol. 1, Issue 4, (DEC 2009), pp. 1431-1443

[9] Renouard, N., Merotte, J., Kervoelen, A., Behlouli, K., Baley, C. & Bourmaud, A. Exploring two innovative recycling ways for poly-(propylene)-flax non wovens wastes, Polymer Degradation and Stability, Vol. 142, (AUG 2017), pp. 89–10, ISSN 0141-3910, eISSN 1873-2321

[10] Ciupan, C., Pop, E., Filip, I., Ciupan, E., Câmpean, E., Cionca, I. & Hereş, V. A new approach of the design process for replacing wooden parts of furniture, Proceedings of Modern Technologies in Manufacturing (MTEM 2017 - AMATUC), OCT 12-13, 2017, Cluj-Napoca, ISSN 2261-236X, ISBN 978-2-7598-9027-9, MATEC Web of Conferences, Art. No. 06002, E D P Sciences, F-91944 CEDEX A, FRANCE, DOI: 10.1051/matecconf/201713706002

[11] Ciupan, E., Lăzărescu, L., Filip, I., Ciupan, C., Câmpean, E., Cionca, I. & Pop, E. Characterization of a thermoforming composite material made from hemp fibers and polypropylene, Proceedings of Modern Technologies in Manufacturing (MTEM 2017 - AMATUC), OCT 12-13, 2017, Cluj-Napoca, ISSN 2261-236X, ISBN 978-2-7598-9027-9, MATEC Web of Conferences, Art. No. 08003, E D P Sciences, F-91944 CEDEX A, FRANCE, DOI: 10.1051/matecconf/201713708003

[12] Ciupan, C., Steopan, M., Pop, E., Campean, E., Filip, I. & Ciupan, E. Comparative analysis of different ribs used to rigidize the resistance structure of a sofa side made of composite materials based on vegetable fibers, Acta Technica Napocensis Series-Applied Mathematics, Mechanics and Engineering, Vol. 61, Issue I, (MAR 2018), pp. 39-44, ISSN 1221-5872

[13] Ciupan, E., Cionca, I., Ciupan, M. & Câmpean., E. The Optimization of Costs for the Thermoformed Frames from Natural Fiber Composites, Acta Technica Napocensis Series-Applied Mathematics, Mechanics and Engineering, Vol. 65, Issue 1, (FEB 2022), pp. 57-64, ISSN 1221-5872

[14] Ichim, M., Stelea, L., Filip, I., Lisa, G. & Muresan, E. I. Thermal and Mechanical Characterization of Coir Fibre–Reinforced Polypropylene Biocomposites, Crystals, Vol. 12, Issue 9, (SEP 2022), Art. No. 1249, eISSN 2073-4352

[15] Ezenwankwo, J., Zak, J. & Petrikova, I. (2022). Comparative Analysis of Polymeric Composite Reinforced with Glass Fibre and Viscose Rayon, Proceedings of the 33rd DAAAM International Symposium, pp.0496-0502, B. Katalinic (Ed.), Published by DAAAM International, ISBN 978-3-902734- 36-5, ISSN 1726-9679, Vienna, Austria, DOI: 10.2507/33rd.daaam.proceedings.069

[16] Cionca, I., Ciupan, E., Ciupan, M. & Ciupan, C. New Ecological Composite Materials for the Upholstered Furniture Industry, Acta Technica Napocensis Series-Applied Mathematics, Mechanics and Engineering, Vol. 66, Special Issue SI, (SEP 2023), ISSN 1221-5872, eISSN 2393-2988

DOI: 10.2507/35th.daaam.proceedings.003

ROBOTIC SPINE FOR VERTICALLY ORIENTED MOBILE ROBOTS BALANCING BY CONTROLLING THE GYROSCOPIC MOMENTS

Valentin Pryanichnikov & Viatcheslav Iastrebov

This Publication has to be referred as: Prianichnikov, V[alentin] & Iastrebov, V[iatcheslav] (2024). Robotic Spine for Vertically Oriented Mobile Robots Balancing By Controlling The Gyroscopic Moments, Proceedings of the 35th DAAAM International Symposium, pp.0018-0025, B. Katalinic (Ed.), Published by DAAAM International, ISBN 978-3-902734-44-0, ISSN 1726-9679, Vienna, Austria
DOI: 10.2507/35th.daaam.proceedings.003

Abstract

This paper introduces a novel self-balancing robotic spine, based on the principle of active control of the gyroscope momentum. Mobile and walking robots, especially vertically oriented machines, are able to achieve better mobility and rotation in small spaces and move faster if their vertical balance could be independently supported by a kind of automatically applied force, originated from the spine of the robot. Such effect can be reached by the real-time controlling of the momentums of the fast-spinning flywheels inside the gimbals, based on the data from the tilt sensor. Corrective force is applied opposite to the direction of the spine tilt. Depending on the angular speed of the flywheels their gyroscopic momentums could be strong, thus requiring the greater gear rate of the servomotors to overcome and control these momentums. This is what we call CMG (Control Moment Gyroscope). Usually when an unexpected disturbance affects the robot, it maintains its balance with movements of the whole body and tilting of the body. Such balancing of the humanoid robot is reached by the software. Prevention of the falling is reached by applying a large rotation of the joints which repositions the CMG of the robot. Such software balancing often takes extra computing and power resources. If the disturbance exceeds the response capability of the robot, the robot may lose stability and safety of the robot may be put at risk by movements to maintain balance. By using a pair or two pairs of counteract-rotating CMGs in our spine, the robot can maintain balance with just small movements of its body, creating at the same time a double balancing force. Smaller flywheels diameter, smaller mass and smaller gimbals can still produce a large compensating balancing force due to the higher flywheel angular rotation and high enough gear rate of the servo motors, overcoming the gyro precession torque, applied to the gimbals. Improved performance and stability were verified with experiments.

Keywords: Balancing; control moment gyroscope; flywheel; precession; tilt sensor; variable speed control.

1. Introduction

Robots are being deployed for industrial and civil applications due to their accuracy, strength, repeatability and intelligence. Common approach in the design of mobile robot ensures its stability by keeping the gravity vector through the centre of mass. Such stability fails when the robot moves at high speed as the inertial forces become significantly compared to the static gravitational force.

However, the momentum of the moving structure can be exploited to enhance stability if it is dynamically controlled. This principle was used in building a gyroscopically stabilized single-wheeled robot "Gyrover" by the researchers at Carnegie Melon University [1], [2]. Another application of this principle had been implemented in a wearable balancing aid, consisting of a set of control moment gyroscopes (CMGs) contained into a backpack-like orthopedic corset, and was designed at the Department of Biomechanical Engineering, Delft University of Technology of Netherlands [5], [6]. Our design is assuming a similar principle for stability but uses a different construction and is aimed to become a universal part for different applications.

2. Motivation

Every time looking at the balancing of the different vertically-oriented robots, same question came to our mind again and again: why not design a universal self-balancing spine to free the robot from complicated attempts of balancing using the entire body? Indeed, why the self-balancing ball exists, self-balancing plate exists, but self-balancing spine for robot we have not met. We felt motivated to filling this gap.

2.1. Control Momentum Gyroscope - State of the Art

A gyroscopically stabilized single-wheeled mobile platform (Gyro Bot) was designed at the Department of Mechanical and Production Engineering of National University of Singapore according to the same principle used to design Gyrover, but included some modifications in the actuation mechanism to generate forward and backward motion [3]. A wearable balancing aid, consisting of a set of control moment gyroscopes (CMGs) contained into a backpack-like orthopedic corset, was designed at the Department of Biomechanical Engineering, Delft University of Technology of Netherlands [5], [6].

A bipedal robotic ball-shaped chassis, namely "Any Walker" with the unique kinematic scheme which has the possibility to locomote in complicated multi-level environment has been designed at the Kuban State University (Russia), AnyWalker is equipped with the system of compensation of external impacts with motor-wheels which can stabilize the robotic system in 3 dimensions [7].

An approach to control the orientation of the robot using CMG has been conducted at the Robotics Research Group from the University of Texas of Austin USA. They first designed a backpack mount for the gyroscope gimbal and conducted experiments using human subjects to estimate the performance of the system that could potentially be used to turn bipedal humanoid robots [8]. A study of the novel femto-satellite design that uses a micro-electromechanical system Control Moment Gyroscope is described in [9]. The focus was on the principal design, modelling, and discussion of the proposed Control Moment Gyroscope while detailing a controllable femto-satellite design that can make use of attitude control for simple sensing missions, as most femto-satellite designs have no attitude control capability due to the power and size constraints on attitude control actuators.

2.2. Idea of the Gyroscopic Spine Construction and Steering

A drawing of the idea of one section of the self-balancing robotic spine, based on the principle of active control of the gyroscope momentum is shown in Figure.1.

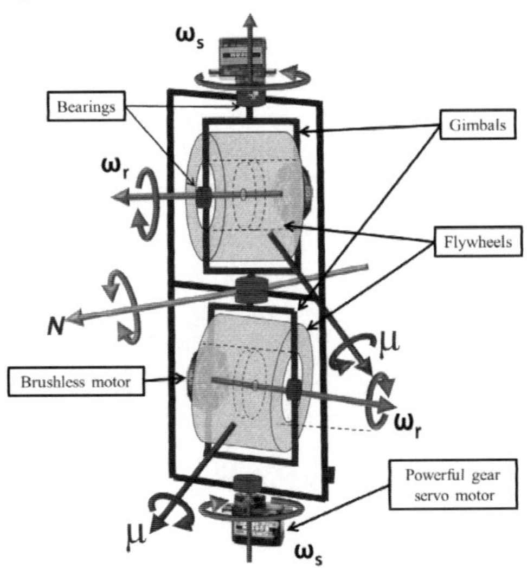

Fig. 1. A section of the robotic spine

Idea is as follows. Two actuators are capable of modifying their angular momentum to impact a torque on a body of the Spine. They consist of 2 self-rotating flywheels mounted inside the gimbals, which, in turn, are swivelled by the two servomotors. Usually by changing the angular velocity ω_r of the flywheels, an angular moment μ is generated and affects the gimbal. But here, the rotation of an outer gimbal (by swivelling angular velocity ω_S reorients the flywheel and produces a significantly larger gyroscopic moment denoted by μ.
Because this momentum is orthogonal to the gimbal and flywheel spin axes, it is applied directly to the gimbal bearings. This allows to apply a relatively smaller gimbal motor torque and power and considerably larger output momentum N is possible.

Fig.2 shows a scheme for parrying disturbances acting on a precessing flywheel. A sector (quadrant) of the rotating flywheel is highlighted, which is installed inside the gimbal, which, in turn, is attached to the housing using bearings. For calculations, we visualize the spinning of the flywheel quadrant by quadrant.
Symbols (Fig.3.):

- Flywheel rotation – spinning;
- Body tilt around the Y axis – roll;
- Body tilt around the X–axis - pitch;
- Rotation of the gimbal around the vertical Z axis – yaw, turn.

All parts of the highlighted flywheel quadrant move towards the rotation around the vertical Z axis. The flywheel rotates at a constant speed. Each time the mass of the quadrant approaches the Z axis of rotation, it tends to "outpace" the movement in a circle. The red arrows in the drawing reflect this trend. Conversely, when the mass moves away from the Z axis of rotation, it begins to lag behind. Collectively, in all four quadrants, the flywheel will have a strong tendency to roll. If the flywheel mounted inside the gimbal has a random roll, then the roll axis will tend to become co-axial with the axis of rotation around the vertical Z axis. The pitch DOF is mechanically fixed. Our goal is to prevent roll by limiting the roll freedom, and use this torque to prevent the entire spine body from falling. This is reached via the forced swivel control of the gimbal by the gear servomotor.

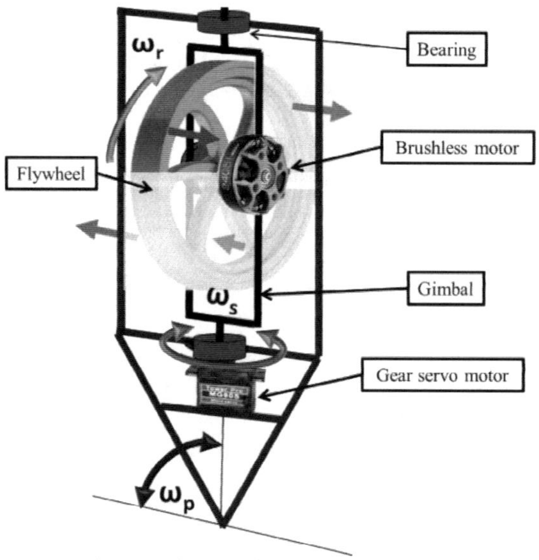

Fig. 2. Scheme of parrying disturbances

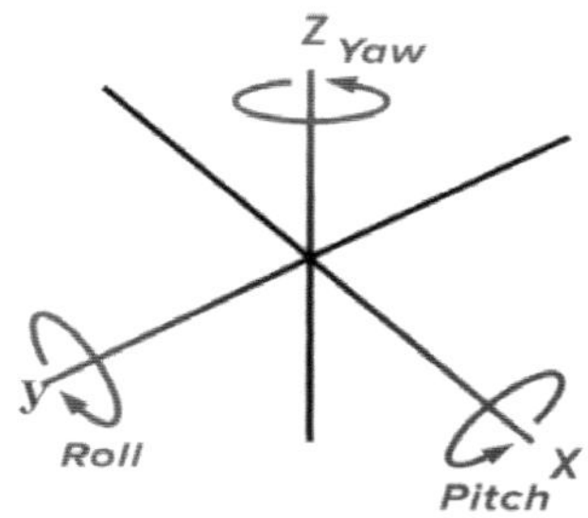

Fig. 3. Axis of degrees of freedom

3. CMG Torque estimation

It was necessary to evaluate the following:

- Given a certain rate of rotation and roll, estimate what torque is required to prevent the entire spine structure from rolling (falling)?

We performed an estimated calculation for a simple case: a brass flywheel with a radius of R and a mass of M. At the same time, we noted several basic concepts:

- The moment of inertia when controlling the gyroscopic moment (CMG) can be calculated by multiplying the mass of the flywheel by the square of its radius of rotation.

- The radius of rotation is an indicator of how much mass is distributed relative to the axis of rotation. The units of measurement of the moment of inertia are usually kg.m^2. It is a unit of inertia of rotation, which is a measure of the resistance of an object to changes in rotation.
- The efficiency of the CMG is directly related to its torque. A higher efficiency means that more torque can be generated at a given angular velocity. This allows control position of the spine more accurately.
- The moment of inertia of the CMG can usually be adjusted by changing the mass or radius of rotation of the flywheel. This can be done by adding or removing mass, as well as changing the shape or material of the flywheel.
- In our case, gyroscopic moments are controlled by synchronous rotation of gimbals with flywheels using servomotors with powerful gearboxes based on data from spinal tilt sensors, which reorients the flywheels.

Then angular moment or torque enough to compensate the falling is:

$$\mu = \omega_s \omega_r M R^2 \tag{1}$$

Where:

ω_r - Angular velocity of the Rolling flywheel,
ω_s - Angular velocity of the Swivelling servo motor,
M - Mass of the flywheel,
R - Diameter of the flywheel.
In our case:
R = 3cm = 0,03m
ω_r = 7000 rpm = 733 rad/sec (lower value was selected; angular speed can be doubled)
ω_s = 0.68 rad/sec (a slow rotation was selected; this servo can rotate 10 times faster)
M = 330g = 0,33kg
μ = = 0.68 rad/sec x 733rad/sec x 0.33kg x 0.0009m^2 = **0.15 kg-m = 1.47 N-m**

This means, that the torque obtained from the single control momentum gyroscope (CMG) module consisting of a single flywheel diameter of 6cm, mass 330g, spinning at constant speed of 7000rpm, and swivelled by the single gear servo motor at maximal angular speed of 0.68 rad/sec, can produce a pitch-compensating torque **μ = 1.47 N-m** (Fig. 4a). By using a pair of counteract-rotating and counteract-swivelling CMGs in our spine, this torque doubles. Counteraction prevents from the slow rotation around the yaw axis of the entire construction (Fig. 4b). Adding another pair of CMGs along with the pitch-balancing pair, adds a roll-balancing option. Then the spine, consisting of four interlaced CMGs for pitch & roll balancing should be able to maintain balance with just small movements of the swivelling servos, creating at the same time a double balancing torque N=**2.94** N-m around pitch and roll axes (Fig.4c). Final torque N is counter-directed to the pitch and roll directions of the spine, which is measured by the dual-axes tilt sensor +/-90°.

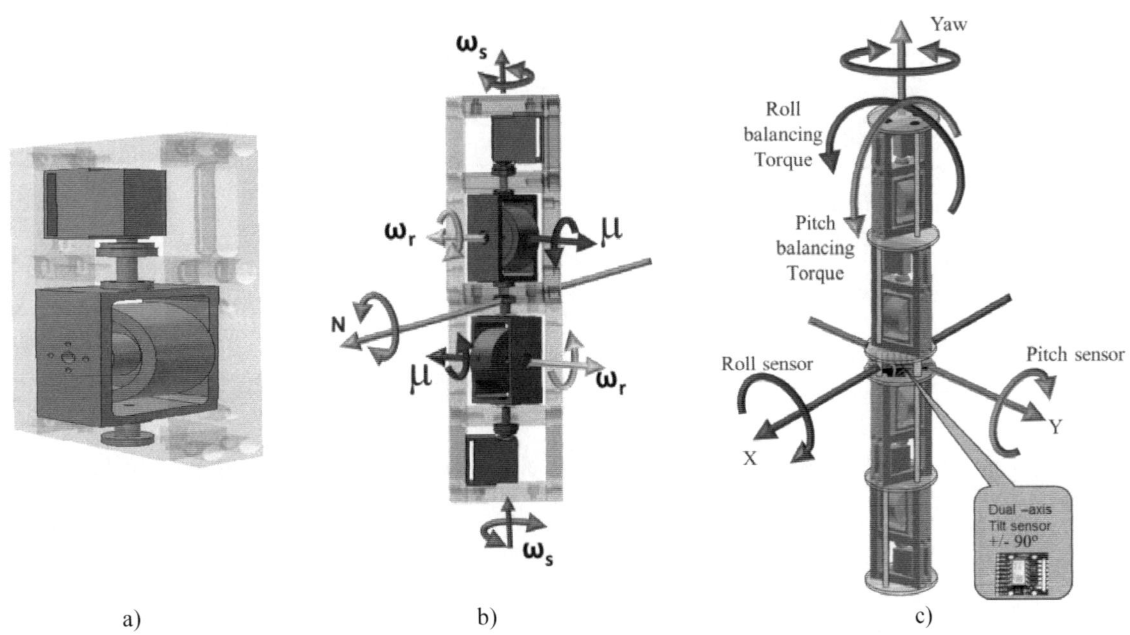

Fig.4. 3D model of the CMG self-balancing robotic spine

3.1. Self-balancing Spine – The Construction Requirements

From the equation (1) we could see, that CMG moment is proportional to its mass, angular velocity of the flywheel, angular velocity of the swivelling servomotor and to the square of the flywheel radius. In other words:

- Doubling the mass, doubles the momentum,
- Doubling any of the angular velocities, doubles the momentum,
- Doubling the radius, quadruples the momentum.

Usually the easiest solution would be selection of a flywheel of bigger radius and mass. But we are creating a spine and it should be as thin as possible and also not too heavy. We have prototyped multiple different construction combinations of flywheels shape, mass diameter, spinning motors and gimbals. Few of them are shown in Fig.5.

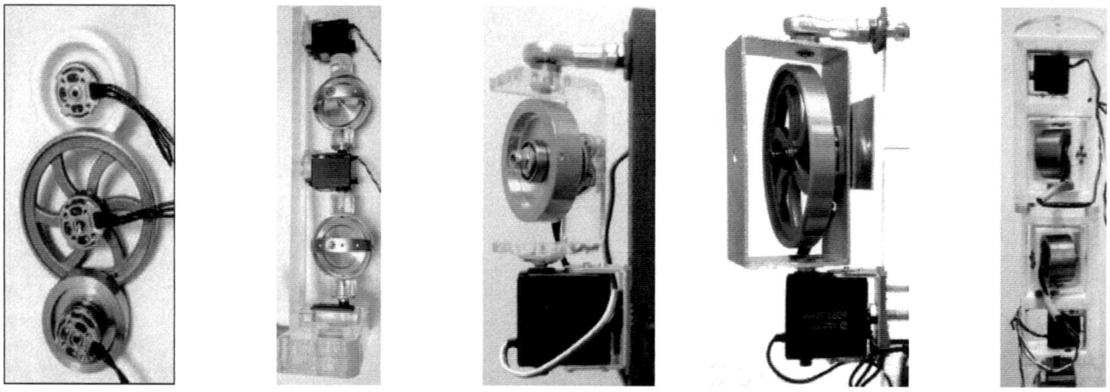

Fig. 5. Evolution of the attempts of CMG prototyping

After conducting experimental tests with the flywheels of different mass, diameter and material, we have come to implement a brass flywheel of 3cm radius, 3cm depth and 300g mass. This flywheel should be self-rotating. In our meaning it should be a kind of brushless "Hub motor", shaft of which with coil inductors around is stator, and outer heavy wheel with permanent magnets spinning around those coils, is rotor. This would be a self-spinning flywheel. Our requirements for the flywheels were a high speed of angular rotation at minimal level of vibration. Our search for a small size of such hub motor brought us to a few available products, maximal angular velocity of which did not exceed 3000rpm. Hence we began prototyping a flywheel, driven by the separate motor, which would provide angular velocity of at least 7000rpm, have enough torque to start rotation of heavy flywheel, and to be mounted inside this flywheel. Such motor was found. We used brushless motor M2205S 2600KV for Quadcopters (45g).

Description & Features of the brushless motor:

- Model: M2205S;
- Shaft: M5;
- Weight: 45g;
- Max Angular velocity - 24500rpm at 12V 30A control.
- High Strength N52 Neodymium Magnets;
- Active Cooling Fins Greatly Reduce Motor Temperatures;
- Race Spec Performances;
- Active Cooling Fins Greatly Reduce Motor Temperatures;
- Light Weight Design;
- Japanese Bearing For Performances.

3.2. Tilt sensor

To measure the tilt angle and angular velocity of the robot, an inertia measurement unit (IMU) was used. CJMCU-100 is a dual-axis tilt sensor chip measuring 90 °, based on 3D-MEMS technology. The sensing axes of the internal sensing elements are parallel to the mounting plane and are orthogonal to each other. Low temperature drift, high resolution, low noise and robust design made the SCA100T an ideal choice for horizontal instruments. Murata's tilt sensor is better able to withstand vibrations and can withstand up to 20,000 shocks. The IMU consists of an inclinometer) and a gyro sensor. Even though the inclinometer can accurately measure the tilt angle of the body, it cannot measure the correct angle during fast rotation because of centrifugal force. On the other hand, the gyro sensor used to measure the angular velocity of the body can estimate the angle of the body by integrating angular velocity.

3.3. CMG Module

The module consists of two CMGs. Each CMG has two actuators. One actuator rotates the flywheel and the other rotates the gimbal in which the flywheel is spinning. If both flywheels are controlled to rotate with the same magnitude and in opposite directions, then both gimbals are also to be controlled in opposite direction. Then the final torque doubles. The CMG generates torque in a direction perpendicular to the rotational axes of the flywheel and gimbal. The magnitude of the torque is the product of the angular velocity of the gimbal and the momentum of the flywheel. In order to generate enough torque, it is necessary to have a bigger moment of inertia and a faster flywheel speed. To ensure that the torque of the CMG is accurate, the two flywheels need to rotate at the same constant speed. Thus the PWM controller is sending the same PWM pulses to both flywheel motor controllers. To reach doubling the torque the second CMG was mounted so, that it was mirrored vertically and horizontally (Fig.6).

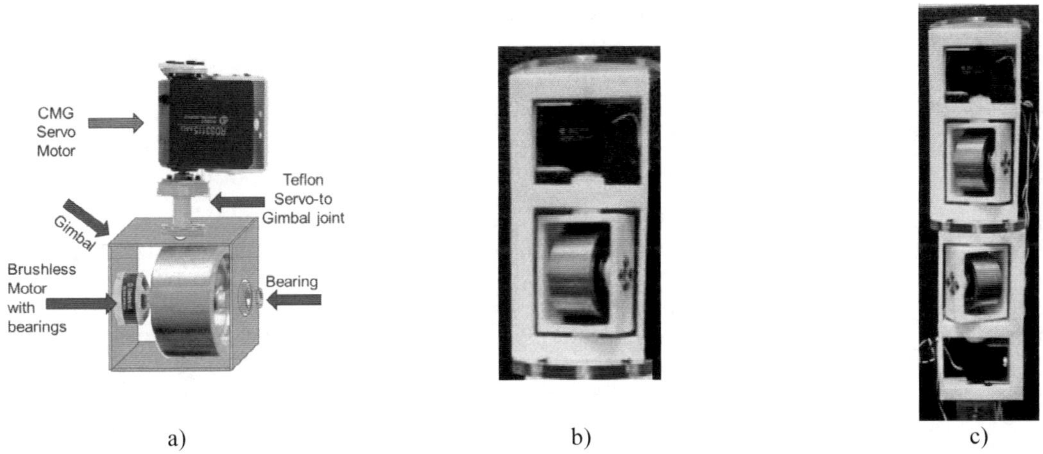

Fig. 6. Brushless motor and CMG module

The gimbal motors used in the CMG do not need high speed, but just sufficient torque. The Robot Digital Servo RDS3115MG was selected. Torque varies between **1.3 - 1.6 N-m**, which is commensurate with the required torque estimation. The outer frame and the gimbal were made of the white Teflon material, which is a synthetic fluoropolymer that can take temperature of +/-200 degrees Celsius and has good resistance against chemical and corrosion. Another reason, why Teflon has been selected, is its exceptional low frictional coefficient, making it one of the prime choice for sliding plate applications. This feature of Teflon allowed us to avoid using bearings. As a result, the frame construction obtained a capability to dampen the vibration from the extremely high speed (up to 20,000rpm) flywheel motors (Fig.7).

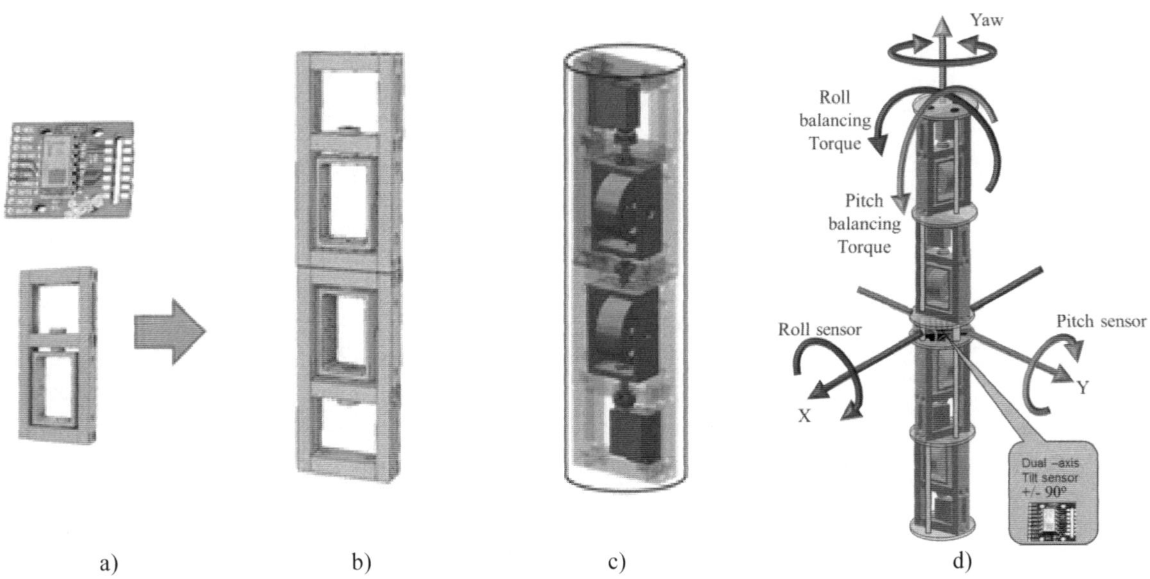

Fig. 7. Dual-Axis tilt sensor + evolution of Gimbals & Frame + future construction

Fig.7b shows a design for balancing both tilt and pitch, which we plan to assemble and test as the next step.

3.4. Alternative analog tilt sensor approach for CMG control

Initially, the control of the brushless flywheels motors and gimbal servomotors, as well as the reading of tilt sensors and the generation of control signals for the gimbal servomotors was carried out by an Arduino microprocessor. It seemed to us that in some cases the control was somewhat delayed, as a result of which the spine twitched excessively when returning to balance. Alternative approach was to use the analog outputs of the tilt sensor to control the CMG due to their higher speed of operation. The CJMCU-100 tilt sensor has three analog voltage outputs for each axis. We have implemented such analog control for the RDS3115MG servo motor, intended for gimbal swivelling. The idea is that we generated a triangular voltage at the PWM frequency of the servo control and used a comparator to compare the triangular pulses with signals from the tilt sensor to generate pulses in the range of 1-2 milliseconds.
The functional and electrical circuits are shown in fig.8 and Fig.9, respectively.

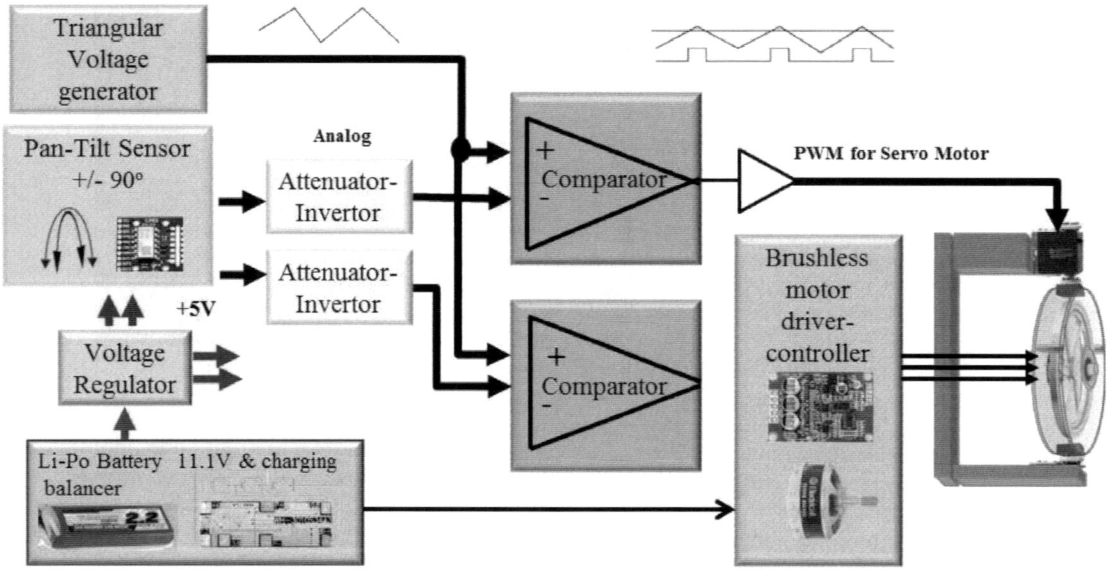

Fig. 8. Functional Diagram of the CMG Analog control

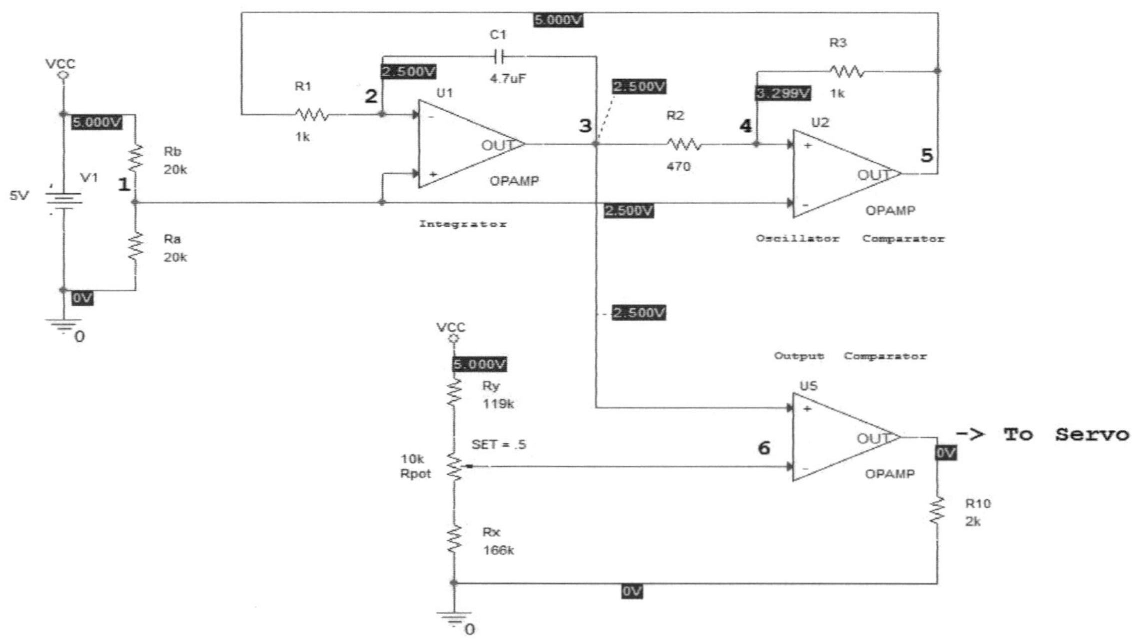

Fig. 9. Electrical diagram of the CMG Analog control

4. Acknowledgement

This research had been initially supported by the BeingTogether Centre (BTC) and the Robotics Research Centre (RRC) at the Nanyang Technological University (NTU) Singapore). Special gratitude is expressed to Associate Professor Seet Gim Lee Gerald and Professor Nadia Magnenat Thalmann. Later and till now the initiative was taken up by the Keldysh Institute of Applied Mathematics (KIAM, Moscow, Russia).

5. Conclusion

Initially a supposed to walk humanoid robot of child size at NTU did not meet a walking ability and often fell down. Balancing using joints was not satisfactory. This was a start point of an idea to design a self-balancing spine for it. By prototyping, experimenting, research and development, a prototype of a self-balancing robotic spine has been designed and tested. But when it was ready, the child robot was modified so that its legs were taken away and replaced by a hard skirt, attaches to a smart vacuum cleaner to roll on the floor. The prototype worked well enough. So we decided to give it a life in other different applications. The prototype meets our expectations to be implemented in many other vertically oriented robots, besides humanoids. We consider this device as a kind of an instrumentation tool for implementation in different mobile vertical oriented robots. As a first step we are planning to implement this spine in a retractable unobtrusive vertical self-balancing mobile robot for reconnaissance on the ground from a height of telescopic mast.

6. References

[1] H. B. Brown and Y. Xu, "A single wheel, gyroscopically stabilized robot", Proceedings of the JEEE International Conference on Robotics and Automation, Minneapolis, Minnesota, 1996;

[2] H. B. Brown and Y. Xu, "A single-wheel, gyroscopically stabilized robot", in IEEE Robotics and Automation Magazine, vol. 4, pp. 3944;

[3] Tanveer Saleh, Yap Haw Hann, Zhu Zhen, A. Al Mamun. Design of a Gyroscopically Stabilized Single-Wheeled Robot. Proceedings of the 2004 IEEE Conference on Robotics, Automation and Mechatronics Singapore, 1-3 December, 2004;

[4] Ji-Hyun Park and Baek-Kyu Cho. Development of a self-balancing robot with a control moment gyroscope. Proceedings of the International Journal of Advanced Robotic Systems March-April 2018: (1–11).

[5] Daniel Lemus, Jan van Frankenhuyzen, Heike Vallery. Design and Evaluation of a Balance Assistance Control Moment Gyroscope. Journal of Mechanisms and Robotics Oct 2017, Vol. 9 / 051007-9.

[6] Andrew Berry, Daniel Lemus, Robert Babuska, Directional Singularity-Robust Torque Control for Gyroscopic Actuators. IEEE/ASME Transactions on Mechatronics doi: 10.1109/TMECH.2016.2603601

[7] Igor Ryadchikov, Semyon Sechenev, Alexander Svidlov & Oth. AnyWalker: all-terrain robotic chassis. Proceedings of 47th International Symposium on Robotics ISR 2016, At Munich, Germany.

[8] A. Boddiford, C. Manion, Kwan Suk Kim, Luis Sentis. Experiments to validate the use of a Control Momentum Gyroscope (CMG) to Turn Robots. Proceedings of the ASME 2013 Portland, Oregon, USA.

[9] Mark Post, Ralf Bauer, Junquan Li, Regina Lee. Study for Femto Satellites Using Micro Control Moment Gyroscope. 978-1-4799-5380-6/15/$31:00 c 2015 IEEE.

[10] O. J. Woodman: An introduction to inertial navigation. Technical Report UCAM-CLTR-696, University of Cambridge, Computer Laboratory, Aug.2007

DOI: 10.2507/35th.daaam.proceedings.004

COMPARISON OF CREATING TECHNOLOGICAL OPERATIONS BASED ON SIMILARITY IN CAM SOFTWARE

Jozef Peterka & Jozef Hipp

This Publication has to be referred as: Peterka, J[ozef] & Hipp, J[ozef] (2024). Comparison of Creating Technological Operations Based on Similarity in CAM Software, Proceedings of the 35th DAAAM International Symposium, pp.0026-0032, B. Katalinic (Ed.), Published by DAAAM International, ISBN 978-3-902734-44-0, ISSN 1726-9679, Vienna, Austria
DOI: 10.2507/35th.daaam.proceedings.004

Abstract

The article will examine the comparison of creating simple operations based on the similarity of two parts in CAM software, specifically SolidCAM, MasterCAM, and Fusion 360. The introduction presents two similar parts. In the following sections, the article summarises and explains how operations can be created in each CAM software based on similarity. The principle of creating an operation for a similar part in SolidCAM and Fusion 360 is done through templates. The template is created from the operation for the first part. Then, when inserted into the program for a similar part, the necessary elements are edited, and the operation is created. In MasterCAM, a similar operation must be manually modified, where the principle lies in editing the elements directly within the created operation. The next step, of the research will involve comparing with more complex and intricate parts.

Keywords: CAM; part similarity; technological operation; CAM software

1. Introduction

Computer-Aided Manufacturing (CAM) software converts Computer-Aided Design (CAD) models into information that can be used by shop floor machines. This type of software helps engineering companies optimize the process of transforming raw materials and components into finished products [1]. CAM involves the use of engineers and programmers in manufacturing or prototyping components and products, either with or without the aid of CAD [2]. CAM refers to the use of software to control machine tools and automate specific manufacturing processes. It is a type of manufacturing automation process known as programmable automation. CAM is one of a series of steps that include modelling the design in CAD software. Next, Computer-Aided Engineering (CAE) tools analyse and validate the product design, and only then is the product ready for the manufacturing phase [3], [10]. The most common technological method in CAM is milling, which is divided into 3- to 5-axis milling [4].

The manufacturing phase begins when the 3D model of the product is imported into CAM software. This model contains vector graphic information defining the product's geometry and properties. However, the 3D model alone is not sufficient for manufacturing. Accordingly, CAM software requires the user to provide additional information, such as the machine, available tools, materials, post-processor, maximum supported feed rate, supported acceleration and deceleration rates, and the machine's work cycle [3].

By combining all this information, the software determines the required cutting/tool path, feed rate, and cutting speed. It can also simulate the machining/manufacturing process depending on the software's capabilities. This allows the user to identify errors in aspects such as the generated tool path [3]. Finally, the post-processor generates CNC programs. These include lines of code documenting all these steps, tool paths, and other manufacturing information, such as tool definition and height above the workpiece. This NC code is compliant with the machine's conventions, meaning it cannot be universally deployed across all machines. Post-processors are not agnostic; rather, they are customized and fine-tuned for use with specific machines. Therefore, a custom post-processor for your device needs to be installed whenever CAM software is set up [3].

Once generated, the programs are transferred to the machine's control unit, which then directs its operations and tool movements. Simply put, the machine relies on these programs to create the geometric features on the part [3]. The integration of CAM software in the manufacturing process offers many advantages that streamline operations, improve accuracy, and increase overall productivity. CAM software is designed to enhance machining efficiency. It also helps eliminate human errors, thus reducing time spent on adjustments. The level of precision and consistency that can be achieved is difficult to compare with manual machine programming. It allows the replication of parts with complex geometries and tight tolerances, ensuring consistently high-quality output [8], [11].

In the industry, most new product designs are derived by partially modifying existing products. It is very rare that completely new products need to be designed from scratch. If everything were new, it would be time- and cost-intensive. Therefore, in many cases, designers aim to use existing product data during the design process of a new product. Today, most mechanical designers use feature-based 3D modelling systems. In this system, new models can be easily created by modifying some constraints of existing models instead of completely redesigning them [9].

Based on this idea, it can be argued that such an approach could also be applied to CAM programming, thus partially automating the creation of certain technological operations when creating a new CAM program for similar parts. This would increase the productivity of creating CAM programs and reduce the time required to create a new CAM program for a similar part. The current state in the field of creating technological procedures for CNC machines is oriented either to the creation of CAM paths using various techniques [11], [12] or to the interconnection and integration of CAD/CAM/CAPP systems [13], [14], [15].

The article will compare the creation of technological operations based on certain similarities between two parts and then compare them in the CAM software SolidCAM, MasterCAM, and Fusion 360. The pairs of parts are created with dimensional similarities, meaning the features are the same but have different dimensions. Creating specific operations based on similarity can help accelerate the process of generating CAM programs for complex parts, meaning efficient creation of new programs in CAM software. This article aims to explain the principle and establish the foundation for further potential research in the area of automated CAM program generation. The research will be limited to a simple part, basic operations, and three CAM software programs. In Fig. 1, you can see the dimensional differences between Part 1 (orange) and Part 2 (grey). To compare, two operations will be created: face milling and rough pocketing.

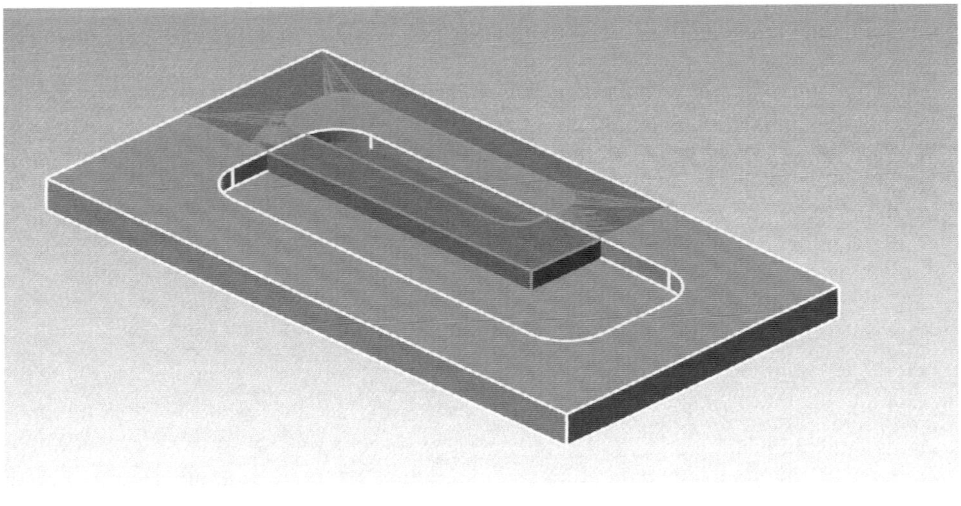

Fig. 1. Part 1 (orange) and Part 2 (grey)

The basic method of our research is analysis and similarity. In this contribution, we want to point out the possibility of applying the well-known technological statement: "Similar parts are produced by a similar technological procedure", which was presented by prominent personalities such as Prof. A.P. Sokolovskij and further developed by Prof. J. Békés. This rule was stated for the so-called conventional machining technologies. Our hypothesis is to apply this rule when designing technological procedures using CAM software. In the first step of our research (and in this post), we focused on the analysis of several CAM software, as offered by software tools for creating NC paths and how they fulfil the mentioned technological rule.

2. SolidCAM

The principle of creating operations based on similarity in SolidCAM is as follows. After creating the CAM program, it is necessary to create templates from the operations. When creating a new CAM program for a new part that is similar, the operations are created from these templates, where various parameters of the operations are already defined, and adjustments are made, such as the machining boundaries, machining depth, and so on. For Part 1, operations were created, from which templates were generated, as shown in Fig. 2, where the operations 'Face Roughing' and 'Pocket Roughing' are displayed. In the face milling operation, all parameters remain the same, with only the face being adjusted. After all the preparatory steps, such as defining the zero point and the stock material, the CAM program creation follows. The process will differ from that of a new part because a template has been created, which will be used to generate the face milling operation.

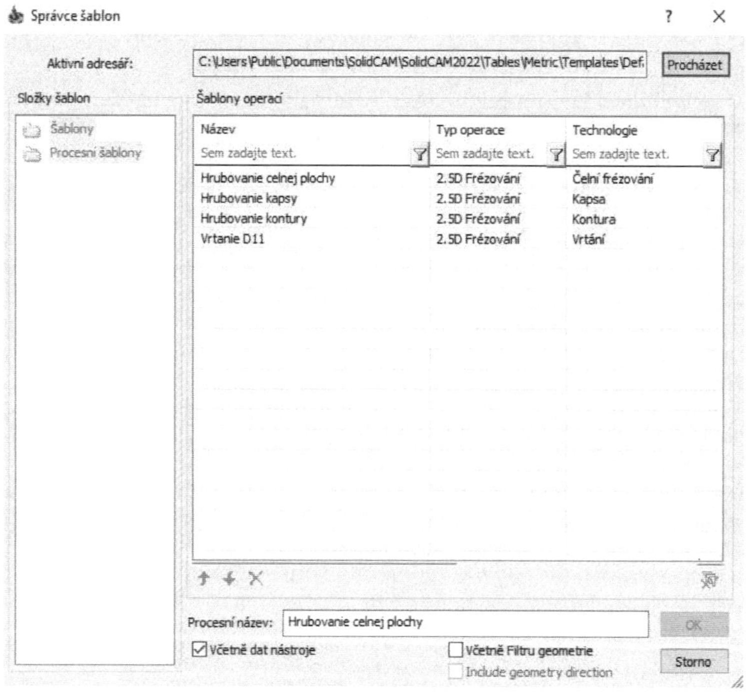

Fig. 2. Creation of a Template in SolidCAM

In Fig. 3, there is a dialogue box that appears after adding the template. In this window, as seen in Fig. 3, you can fill in parameters such as the machining curve or the Z coordinates for the surface and machining depth. For this operation, no additional input is needed, as the face milling operation takes the largest dimension of the part, meaning it is automatically calculated. The operation needs to be recalculated, and then the toolpaths are generated.

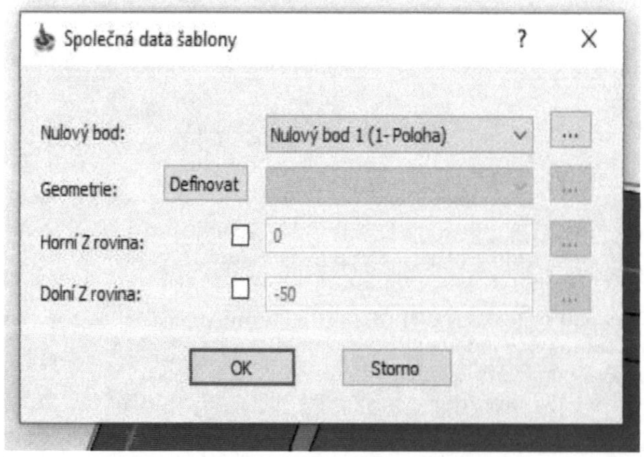

Fig. 3. Adding a Template in SolidCAM

For the pocket roughing operation, the procedure is the same. The change occurs when selecting a new curve, as shown in Fig. 4. In this operation, it is necessary to define a new machining curve in the dialogue box shown in Fig. 3.

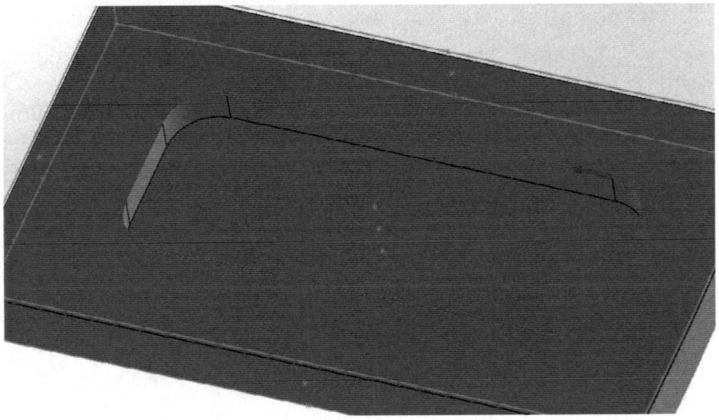

Fig. 4. Selecting the Machining Curve in SolidCAM

The surface coordinate remains the same, meaning that the thickness of Part 2 is the same as that of Part 1. The other difference is in the depth of this pocket. In the dialogue box, it is necessary to check the depth box, which will then highlight the box to the right with three dots. You need to click on this box and then select the pocket depth on the 3D model (red dot in Fig. 5), which represents the value Z-30 mm, as shown in Fig. 5.

Fig. 5. Pocket Depth in SolidCAM

The next steps are the same as for the previous operation, which means recalculating the operation and then generating the toolpaths.

3. MasterCAM

The principle of creating operations based on similarity in MasterCAM differs from that in SolidCAM and Fusion 360. The principle is as follows: For simple 2D operations, most parameters remain the same for similar parts. In cases of dimensional similarity, the boundary curve, which defines the machining limits or machining depth, often changes. After creating both operations for Part 1, Part 2 is inserted into the program. First, it is necessary to align Part 2 to one of the 4 corners, as shown in Fig. 1. Then, the stock for Part 2 is adjusted. The operations are still created for Part 1.

The face milling operation will be adjusted first. The tool remains the same, so there is no need to edit it. For the face milling operation, only the curve defining the area to be milled will be adjusted. Fig. 6 shows the selection of the new curve. The other parameters remain as defined for in Part 1. This operation is then regenerated, and the toolpaths are generated for Part 2.

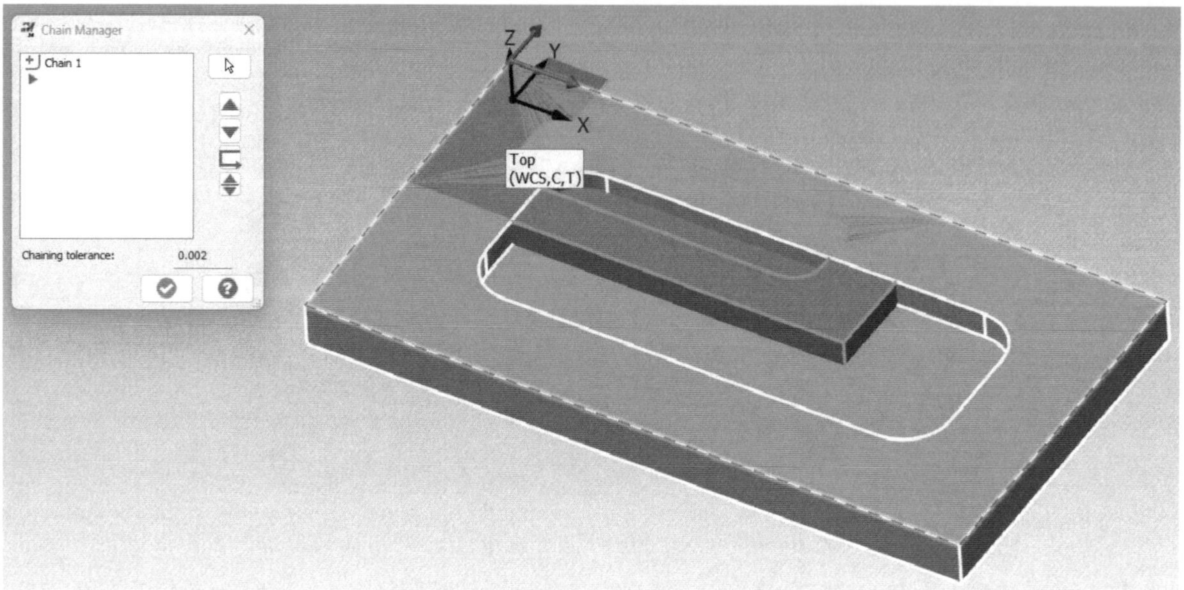

Fig. 6. Selecting the Curve in MasterCAM

The next operation to be adjusted is pocket milling. As with the previous operation, the tool and other parameters will remain the same as for Part 1, so no changes are necessary. A new machining boundary will be defined, similar to the previous operation, meaning a new curve will be marked. Another parameter that differs is the pocket depth. Fig. 7 shows the selection of the machining depth for Part 2. The pocket in Part 1 was milled to a depth of 20 mm. In the highlighted window, it can be seen that the pocket has a depth of 30 mm.

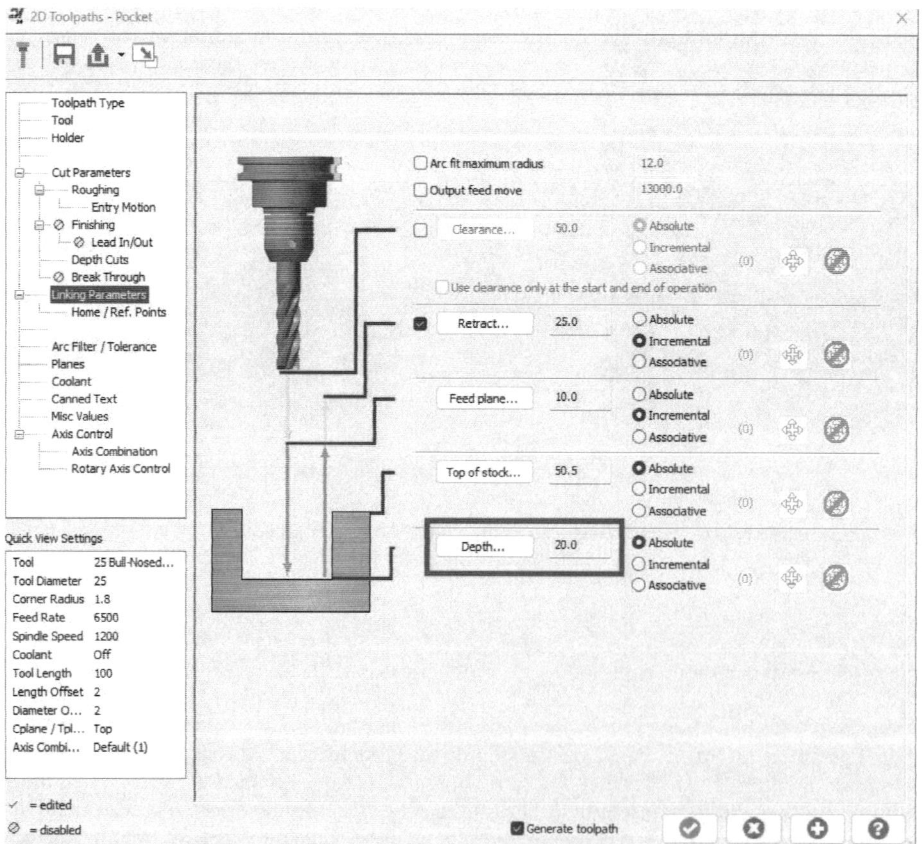

Fig. 7. Selecting the Machining Depth in MasterCAM

4. Fusion 360

The principle of creating operations based on similarity in Fusion 360 is similar to that in SolidCAM. This means that, just like in SolidCAM, after creating an operation for a representative part, it is necessary to create the operation as a template. Subsequently, when creating a new CAM program, the operation from the template is used. The same parameters of the operation remain unchanged, while some need to be adjusted. Fig. 8 a) shows an example of creating a template from an already created face milling operation.

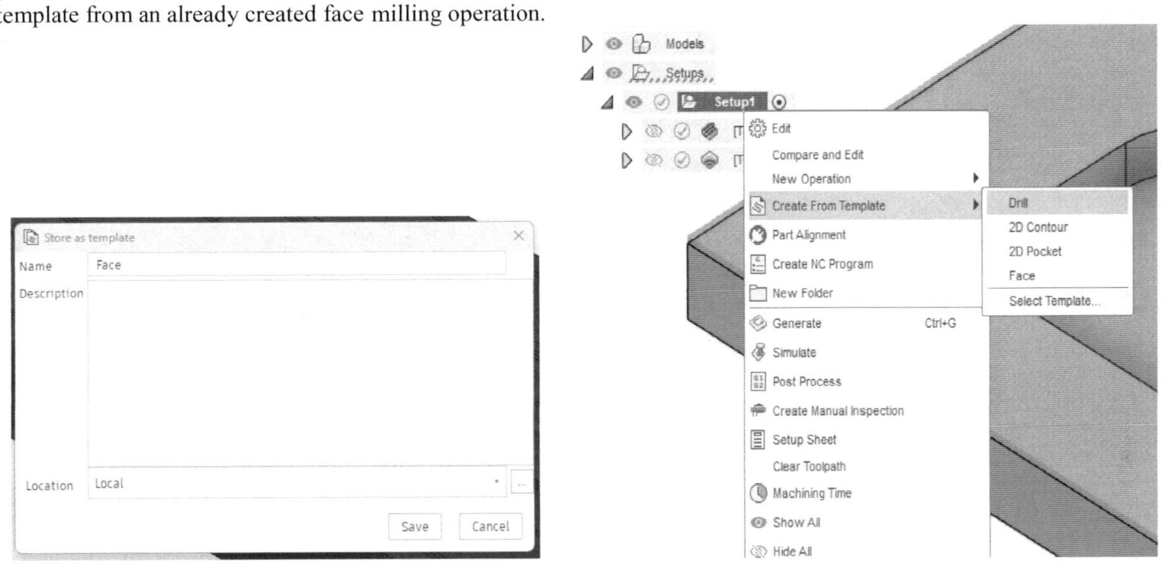

a) b)

Fig. 8. Creating a Template and an Operation from a Template in Fusion 360
a) Creating a Template, b) Creating an Operation

Fig. 8 b) shows the path to access the created template and subsequently use it when creating a new CAM program. The face milling operation is then automatically recalculated, and the toolpaths are generated. After adding the template for the pocket roughing operation, it is necessary to select a new machining curve, as shown in Fig. 9. In the next step, it is necessary to directly adjust the milling depth according to the 3D model.

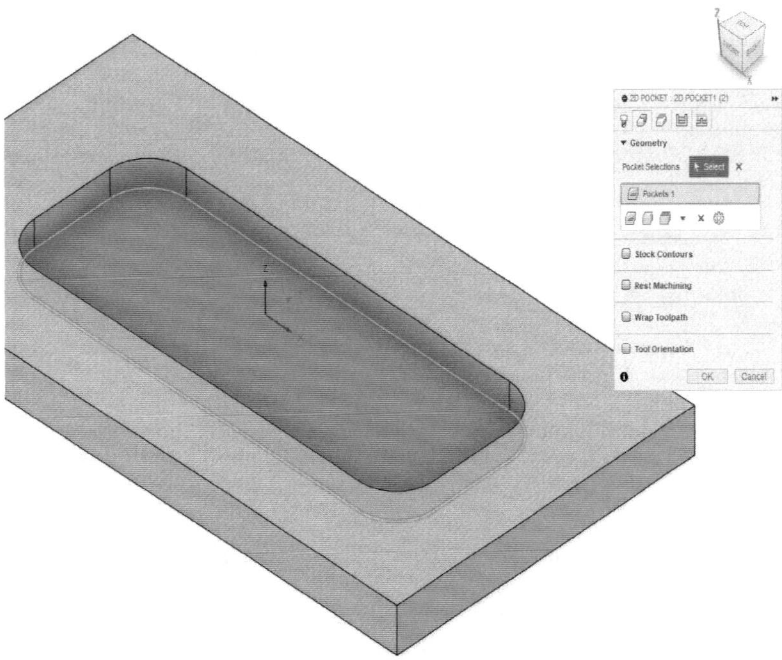

Fig. 9. Selecting the Curve in Fusion 360

5. Conclusion

The article summarizes the comparison of creating simple operations in CAM software. It can be stated that the process and principle of creating operations based on similarity are similar in SolidCAM and Fusion 360. In SolidCAM, an operation is first created for the initial part. This operation is then saved as a template. This template is used to create the operation for the second part, thus reducing the time required to create a similar operation. This means that the principle involves creating templates. In MasterCAM, the procedure and principle are different. The process is somewhat more labour-intensive and manual, and the principle involves adjusting certain elements of the operations directly during operation editing. The research identified the process that can be used for creating operations for a simple part based on similarity. Further research would be beneficial to continue comparing with more complex operations or to focus on a more complex part of one CAM software.

6. Acknowledgments

The project is supported by grants KEGA 001STU-4/2022 and VEGA 1/0391/247.

7. References

[1] Sengupta, A. (2024). Best Computer-Aided Manufacturing (CAM) Software. What is Computer-Aided Manufacturing (CAM) software. Available from: https://www.g2.com/categories/computer-aided-manufacturing-cam, Accessed on: 2024-07-10

[2] Sanders, N. (2017). A possible first use of CAM/CAD. HAL open science, Available from: https://inria.hal.science/hal-01526813/document, Accessed on: 2024-07-10

[3] Kevin, A. (2023). What is CAM (Computer Aided Manufacturing)? Computer-Aided Manufacturing and its role in manufacturing. Scan2CAD. Available from: https://www.scan2cad.com/blog/cad/cam-explained/, (2023), Accessed on: 2024-07-10

[4] Peterka, J; Pokorny, P and Vaclav, S. (2008). CAM strategies and surfaces accuracy. 19th International Symposium of the Danube-Adria-Association-for-Automation-and-Manufacturing. 2008. Annals of DAAAM for 2008 & Proceedings of the 19th International DAAAM Symposium , pp. 1061-1062

[5] Beno, M; Zvoncan, M; Kovac, M and Peterka, J. (2013). Circular interpolation and positioning accuracy deviation measurement on five axis machine tools with different structures. Jun 2013. Tehnicki Vjesnik - Technical Gazette 20 (3), pp. 479-484

[6] Pokorny, P; Peterka, J and Václav, S. (2012). The task of 5-axis milling. Mar 2012. Tehnicki Vjesnik - Technical Gazette 19 (1), pp. 147-150

[7] Kovác, M and Peterka, J. (2014). Selected 5-axis Strategies for High-speed milling of Thin-walled Parts. International Conference on Materials Science and Mechanical Engineering (ICMSME 2013). Materials science and mechanical engineering 467, pp. 466-469

[8] Sarah, A. (2023). The role of CAM software in modern manufacturing. Available from: https://proleantech.com/the-role-of-cam-software-in-modern-manufacturing/, Accessed on: 2024-07-10

[9] Taesik Hong; Kunwoo Lee; Sungchan Kim; Chongnam Chu and Hyunchan Lee. (2005). Similarity comparison of mechanical parts. Computer-Aide3d Design & Applications, Vol. 2, No. 6, 2005, pp. 759-768. Available from: https://cad-journal.net/files/vol_2/CAD_2(6)_2005_759-768.pdf#:~:text=In%20this%20research,%20the%20solid%20parts%20are%20represented%20by%20B-rep Accessed on: 2024-07-10

[10] Pobożniak, J. (2000). Product modelling for CAD/CAM integration, Available from: https://komunikacie.uniza.sk/artkey/csl-200001-0001_product-modelling-for-aac-capp-integration.php. DOI: 10.26552/com.C.2000.1.4-9, Accesed: 2024-07-10

[11] Zhang, X.; Afsharizand, B.; Essink, W.; Newman, S. T and Nassehi A. (2014). A STEP-compliant method for manufacturing knowledge capture, CIRP 20, 2nd Intern. Conf. On Ramp-Up Manag. 2014 (ICRP), pp. 103-108. Available from: https://www.sciencedirect.com/science/article/pii/S2212827114007136. Accessed on: 2024-07-10

[12] Kamran, L; Anbia, A; Yusri, Y and Abdul, K.A.Z. (2021). A review of G code, STEP, STEP-NC, and open architecture control technologies based embedded CNC systems. International Journal of Advanced Manufacturing Technology Volume 114, Issue 9-10, Pages 2549 – 2566. DOI 10.1007/s00170-021-06741-z

[13] Salomons, W; Kuipers, J.M; Graaff, J; Slooten, F; Houte, F.J.A.M and Kals, H.J.J. (1996). Collaborative product development in CAD and CAPP. An approach based on communication due to constraint conflicts and user initiative. IFIT Intern. Federat. For Information Processing 1996. Available from: https://link.springer.com/content/pdf/10.1007/978-0-387-34930-5_5.pdf#:~:text=This%20paper%20addresses%20collaborative%20product%20development,%20focusing%20on%20computer%20support. Accessed on: 2024-07-10

[14] Gerken, H and Weyrich, M. (2001). Management of experiences with an assistance system for the CAD/CAM process chain. VDI Berichte, Issue 1614, pp. 233 – 251, ISSN 00835560

[15] Deepanshu, S and Venkateswara Rao, K. (2023). STEP-NC AP238 – an excellent paradigm for smart manufacturing. International Journal on Interactive Design and Manufacturing Volume 17, Issue 4, pp. 1475 – 1487. DOI 10.1007/s12008-023-01289-6

DOI: 10.2507/35th.daaam.proceedings.005

SPATIAL ANALYSIS FOR OPTIMAL LOCATION OF NEW CLUSTERS OF MEDICAL CENTERS IN A DEFINED ADMINISTRATIVE REGION

Irina Topalova & Pavlinka Radoyska

This Publication has to be referred as: Topalova, I[rina] & Radoyska, P[avlinka] (2024). Spatial Analysis for Optimal Location of New Clusters of Medical Centers in a Defined Administrative Region, Proceedings of the 35th DAAAM International Symposium, pp.0033-0039, B. Katalinic (Ed.), Published by DAAAM International, ISBN 978-3-902734-44-0, ISSN 1726-9679, Vienna, Austria
DOI: 10.2507/35th.daaam.proceedings.005

Abstract

Administrative populated areas, such as territorial districts, correspond to demographic and infrastructure databases, as well as data on the utilities provided, health care facilities available, and more. Determining the location and the required number of health centers in a given administrative region depends on all these parameters. The article presents the developed cluster method for determining the location and number of necessary new health centers for a certain populated region in Bulgaria. A basic requirement for access to these centers through the existing transport network and within a certain time interval has been met. Spatial data has been prepared for the integration of the transport infrastructure with the geographical coordinates of the settlements, as well as of the existing health facilities. The number of clusters has been empirically optimized to achieve maximum health care coverage of all settlements in the municipality, within the set time interval for movement. The results for determining the optimal number of clusters representing the new health centers are presented and analysed. The possibilities for continuing the research were discussed.

Keywords: spatial databases; health regions; geographic information systems; cluster analysis

1. Introduction

Currently, there is a lot of research on the need to create efficient health networks in Europe and appropriate analyses on the necessary complement to the available health facilities that provide timely care in the administrative regions [1], [2]. Together with the analysis of the perspectives and the necessary measures to improve health care, the different approaches to implementing these measures have been analysed [3], [4]. Health regions are not just territorial areas, but represent a set of demographic, infrastructural, transport network parameters, as well as health facilities/centres for qualified health care. For this reason, determining the appropriate location of needed additional new health care centers by region is a function of this set of parameters.

In this type of research, it is most appropriate to use Geographic Information Systems (GIS) methods to optimally determine the geographic location of the necessary health centers, due to the possibility they offer to integrate the above parameters/databases with the geographic coordinates of the settlements. The researchers in [5] apply five different methods for determination the new design temperature for Sarajevo based on the period 2001–2020, aiming to calculate the heating needs of one residential unit. It is done in order to demonstrate the influence of different design outside temperature on the heating needs in a geographical region. The authors of [6] consider climate change when designing infrastructure facilities. Their research aims to assess the flow rate data based on the daily measured values obtained at the Reljevo hydrological station on the Bosna River. The analyses are conducted for two time periods, to obtain the temporal distribution of relevant flow rate data. The coincidence of hydrological parameters along the perceived changes is observed. Monthly and yearly changes of hydrological parameters along the perceived changes are observed based on the analysis of variance, applied to the flow rate data. As a result, maximum flow rates are found in the summer period, which should be considered in flood management. But these both methods are not based on the integration of geographic spatial coordinates with databases used for the research. In these cases, the creation of such integrated data in a preprocessing stage and its implementation in GIS software platforms, would facilitate the subsequent necessary analysis. Such integration would also create an opportunity for easy reflection in the proposed studies of the dynamic changes in the databases.

The paper in [7] presents a research, that is closest to what we offer in this publication. It concerns solving a logistical task related to a methodology for a Local Centers location analysis based on GIS. The proposed method uses a GIS for data collection, spatial analysis, generating alternatives, and producing maps for further analysis. The authors used the GIS for an analysis of the location attributes, including a number of relevant factors, namely spatial position, intermodal connections (road, water, air and rail transport), the size of the available location, topography, local traffic connections, equipment of communal infrastructure, constraints in the area, etc. But the emphasis in our research falls more on the creation of integrated GIS data and empirical determination of logistic time/transport infrastructure parameters, when creating new health care centers.

The proposed article presents the developed cluster method for empirical determining the location and number of necessary new health centers for a certain populated administrative region in Bulgaria. A basic requirement for access to these centers through the existing transport network and within a certain time interval has been met. Spatial data has been prepared for the integration of the transport infrastructure with the geographical coordinates of the settlements, as well as of the existing health facilities. The number of clusters has been empirically optimized to achieve maximum health care coverage of all settlements in the municipality, within the set time interval for movement. The results for different number of clusters representing the new health centers are presented and analysed. They aim to ensure that the people from the settlements in it will be able to receive emergency medical care no more than 10 or 15 minutes away by car. For this purpose, a spatial analysis is applied to select the best locations. For the purpose of the study, a licensed software product, ArtGisPro was used.

2. Creation of shape files with settlements and existing medical centers

At the stage of preliminary preparation of the spatial data, it is necessary to create shape files that represent the relationship between the geographical coordinates of the settlements in the studied administrative region and the parameters of the existing road transport network. With the help of Google Maps, the addresses and coordinates of the medical facilities in the Pleven administrative region have been located. The website of the Ministry of Health [8] was used as a source of information about medical facilities, where all medical facilities in the country are described. Looking at the capabilities of different geo-browsers, we decided to use Google Earth [9] to create a file with basic information about the treatment centers.

Google Earth Pro have been used in this research, because it offers full functionality, including creating files in KML (Keyhole Markup Language) format, intended for sharing. KML format files can be easily converted to SHP format files that are processed directly by ArcGIS software. The obtained KML file contains Placemarks, Ground Overlays, Paths, Polygons, Styles for Geometry, Screen Overlays, etc.

For the purposes of our development, KML files were created, for three medical centers in Pleven region in Bulgaria and one file for all settlements in the same region: SVETA-MARINA-PLEVEN, SVETI-PANTELEYMON-PLEVEN, GEORGI-STRANSKI-PLEVEN and PLEVEN_REGION_POINTs. The KML files are converted to SHP format with point geometry, using Python functions in Jupiter environment. Subsequently each of the obtained shape files, contains the necessary source spatial data, including the spatial geographic coordinates of the existing three medical centers and of all settlements in the region.

3. Development of clustering model

The developed model for determining the location of appropriate new medical centers consists of the following steps:

- Initially, in the used ArcGis software platform, the established shape file with the point layer PLEVEN_REGION_POINTs is loaded, which shows the centroids of the settlements in the Pleven region, shown in Fig.1.

- Add/overlay all three layers of the three existing medical centers SVETA-MARINA-PLEVEN, SVETI-PANTELEYMON-PLEVEN, GEORGI-STRANSKI-PLEVEN and adding a sketch file. to enter symbols and text on the map, shown in Fig.2.

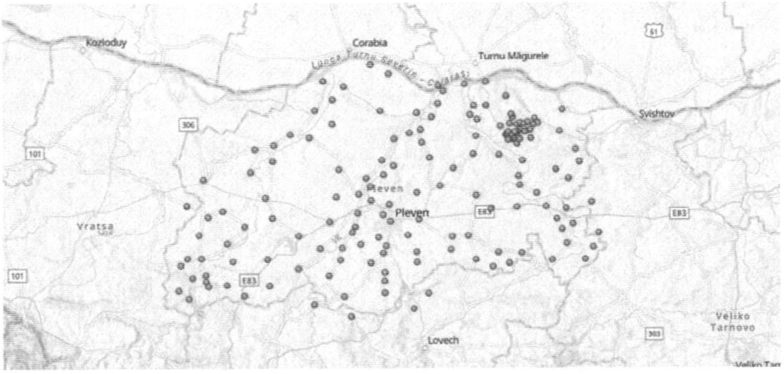

Fig. 1. Point layer with settlements centroids

Fig. 2. Point layer with centroids of the three existing medical centers

- The first part of the task of locating new emergency medical centers is to determine which settlements are within a 10 - minute drive by car, to each of the three existing medical centers. For this purpose, the Create Drive-Time Areas tool was used. Thanks to mathematical calculations, analysis, and overlay method, this tool creates a polygon that includes all the points that can be reached in a fixed time through the existing transport network and depending on whether the distance is covered on foot or by vehicle. The graphical presentation of obtained polygons are shown in Fig.3 for SVETI-PANTELEYMON-PLEVEN & GEORGI-STRANSKI-PLEVEN and in Fig.4 adding SVETA-MARINA-PLEVEN.

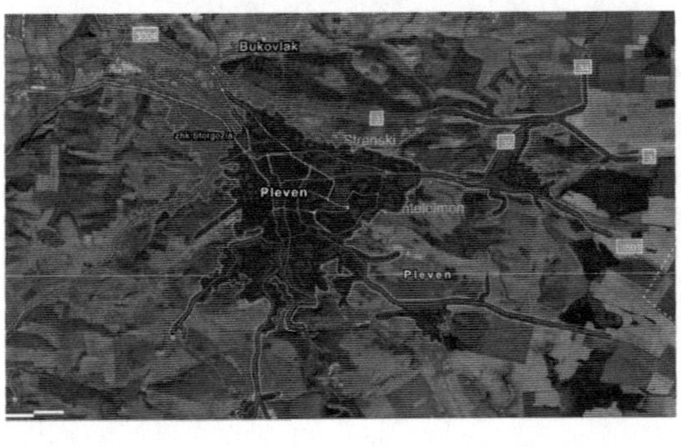

Fig. 3. Settlements within a 10 - minute drive by car to SVETI-PANTELEYMON & GEORGI-STRANSKI

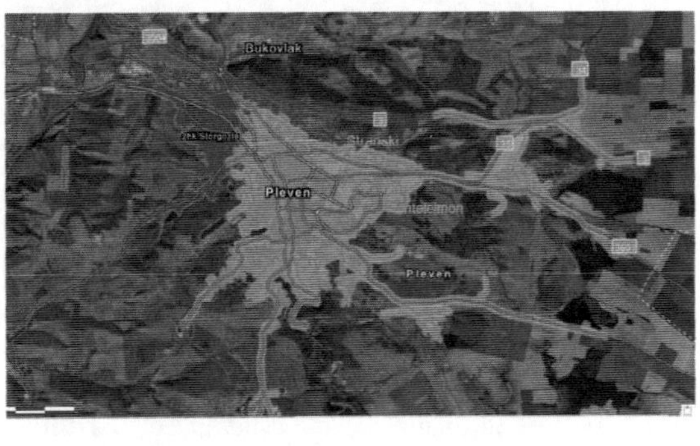

Fig. 4. Settlements within a 10 - minute drive by car to SVETI-PANTELEYMON & GEORGI-STRANSKI & SVETA-MARINA

- The next step is to determine the number of settlements that are outside these 10-minute access zones.

For SVETI-PANTELEYMON these are 158 settlements in 32.35 km^2 area; for SVETA-MARINA these are 156 settlements in 31.76 km^2 area and for GEORGI-STRANSKI again 158 settlements in 34.12 km^2 area. All 158 settlements outside the three 10-minute access zones are represented in Fig.5.

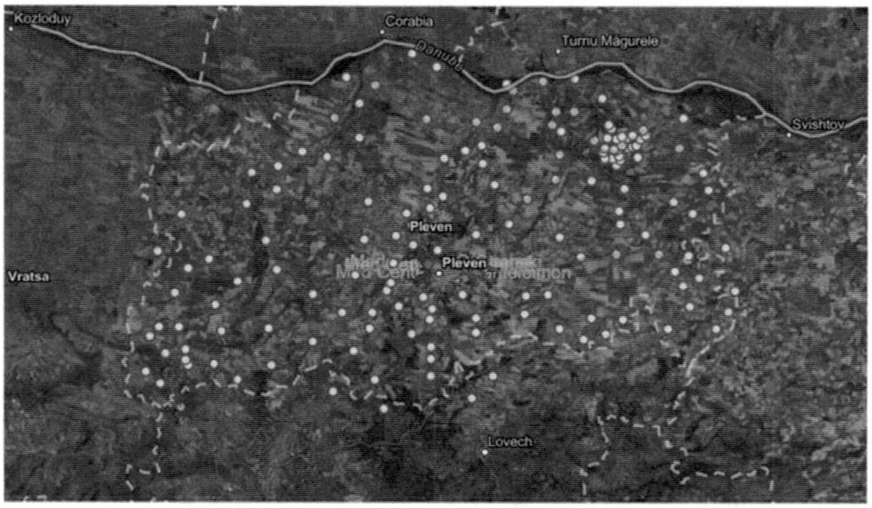

Fig. 5. All settlements outside the three 10-minute access zones

- Different number of new locations are successively assigned, to create clusters of settlements with a distance to the location of 10 minutes by car.

Initially, it was launched with the selection of a small number of new locations/clusters, with the aim of maximum coverage of the remaining settlements without 10-minute access to the three main medical centers.
For choosing three new centers/clusters (given in green colour), the obtained map is shown in Fig. 6. In this case the total number of settlements outside the 10-minute range to the established new three cluster centers is 115.

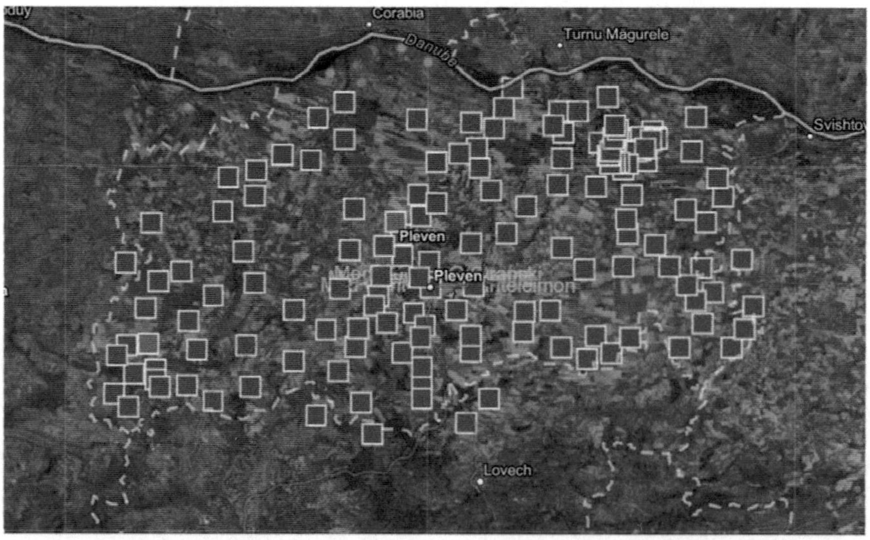

Fig. 6. Distribution of the settlements connected to the three new centers.

For choosing three new centers/clusters, the obtained map is shown in Fig. 6. In this case the total number of settlements outside the 10-minute range to the established new three cluster centers is 115. For choosing ten new centers/clusters (given in green colour), the obtained map is shown in Fig. 7. In this case the total number of settlements outside the 10-minute range to the established new three cluster centers is 95.

Fig. 7. Distribution of the settlements connected to the ten new centers.

For choosing twelve new centers/clusters (given in green colour), the obtained map is shown in Fig. 8. In this case the total number of settlements outside the 10-minute range to the established new three cluster centers is 68.

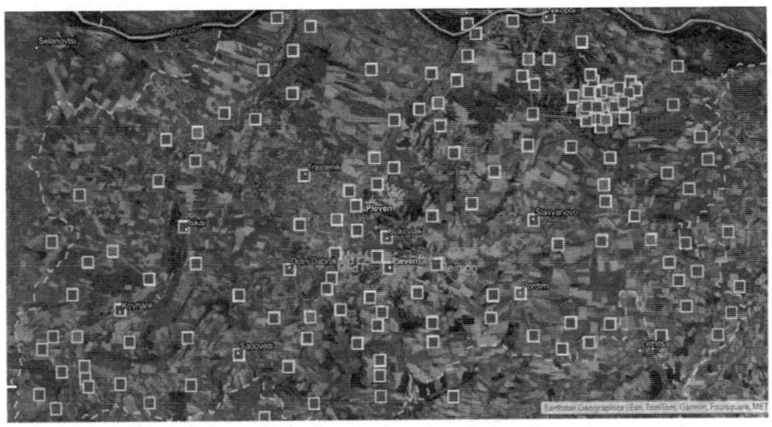

Fig. 8. Distribution of the settlements connected to the twelve new centers

- In the next step, the selected roads from the real road infrastructure to the nearest newly created center are calculated and displayed on the map. Some roads between settlement A and cluster centers B, C and D with corresponding travel time of 15 min, 53 min and 1h & 20 min, are shown in Fig. 9.

Fig. 9. Calculated roads between settlement A and cluster centers B, C and D

4. Summary of results

For the purposes of the study, attempts were made to optimize the ratio between the minimum number of new centers needed, with the maximum number of covered settlements in a certain time interval of car travel and the minimum number of remaining uncovered (in an area outside the clusters) settlements. The best ratio is obtained with 12 clusters and 15 min time range of coverage. Table 1 show the corresponding ratios.

Number of clusters	at 3 clusters	at 10 clusters	at 12 clusters
Number of settlements, 10 minutes away from the respective cluster center	6/ 6/ 31	5/ 3/ 3/ 5/ 4/ 3/ 5/ 5/ 5/ 25	6/ 3/ 7/ 3/ 5/ 5/ 4/ 5/ 6/ 3/ 5/ 38
Total number of settlements outside the 10-minute range to the established cluster centers	115	95	68
Total number of settlements outside the 15-minute range to the established cluster centers	98	72	16

Table 1. Ratios obtained with different clusters and different time range of coverage

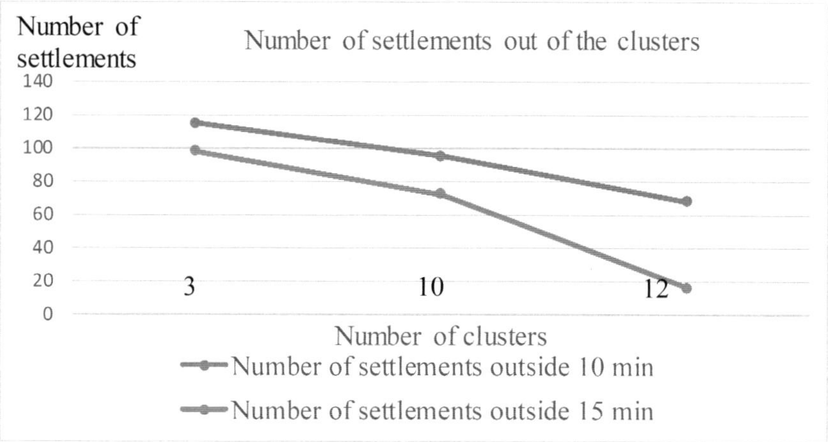

Fig.10. Representation of ratio between number of clusters and number of settlements out of 10 and 15 min access time

5. Conclusion

When there is a need to build new health centers, the *main problem* is how to optimally combine the number of new centers with their territorial location, so that for a certain travel time (the health coverage area) along the available transport network maximum, number of settlements will be served. This article presents the cluster method for determining the location of new health centers for a populated Pleven region in Bulgaria, according to a basic requirement for access to these centers within a certain driving time interval. For this purpose, the necessary shape file was created and a study was conducted for different distances of the settlements.

The proposed method allows conducting different experiments with changing the parameters in the individual steps of the algorithm, such as varying the number of clusters, in order to obtain an optimal distribution for the settlements. The results show that by increasing the number of clusters to 12, an optimal minimum number of settlements remaining outside the range of 15 min for access was obtained. The capabilities of the GIS software also offer the export of the developed method as a web application. This is a prerequisite for easy orientation of the population in the region for quick access to the nearest center. The method can be multiplied for other administrative regions, after preparation of the corresponding shape file.

The *limitations* of this study are given to cases where spatial databases (locations of settlements, data on an up-to-date transport network or information on existing hospital facilities) are missing for certain population regions in a given country. In case such spatial data exists and is available, the proposed method provides a way to create the shape file needed to implement the proposed method. As a continuation of the development, other parameters from the spatial data such as population density, age distribution, etc. can be included. Adding a layer with the age distribution of the population in the settlements and linking the type/specificity of health services to the age distribution is also of interest and can be included in the further development of the method.

6. References

[1] George W. (2009), "Gesundheitsregionen" und "Regionale Gesundheitsversorgung" Gesundheitsnetzwerke initiieren: Kooperationen erfolgreich planen. - Berlin: Med. Wiss. Verl.- Ges., ISBN 978-3-939069-72-0, p. 293-307, Berlin

[2] Iltchev V.& Steno, Al. (2018), "Informational aspects of modelling the Health Regions in Bulgaria", Tagungsband der 8. FDIBA Konferenz, ISSN: 2535-132X (printed), 2535-1338 (online), pp. 89 - 92, Sofia

[3] Dengler, R.; Lakasz, A.; Tsenov, A.; Kollany, Z. & Cassens, M. (2018), "Soziale Lage und Gesundheit: Arteria Danubia ("ArDa")", Public Health Forum: Forschung - Lehre - Praxis, De Gruyter - Springer Verlag, vol:101, iss:4, pp. 1 - 8

[4] Cassenas, M.; Polanyi, Z.& Steno, A. (2021), Book: "Transdisciplinary Perspectives on Public Health in Europe (FOM-Edition, International Series)", Chapter III, Springer Nature Gabler Verlag, DOI 10.1007/978-3-658-33740-7, Softcover ISBN 978-3-658-33739-1

[5] Blažević, R.; Teskeredžić, A. & Zečević, M. (2023). Analysis of Existing Design Temperature for Heating in Sarajevo in Light of Climate Changes, Proceedings of the 34th DAAAM International Symposium, Vienna, Austria, ISSN 1726-9679, ISBN 978-3-902734-41-9, Katalinic, B. (Ed.), pp.0209-0218, Published by DAAAM International, Vienna, DOI: 10.2507/34th.daaam.proceedings.027

[6] Ahmovic, I.; Hadziahmetovic, H.; Blazevic, R. & Alispahic, M.(2023) Statistical Analysis of Hydrological Parameters in the Bosna River Basin, Proceedings of the 34th DAAAM International Symposium, Vienna, Austria, ISSN 1726-9679, ISBN 978-3-902734-41-9, Katalinic, B. (Ed.), pp.0229-0236, Published by DAAAM International, Vienna, DOI: 10.2507/34th.daaam.proceedings.029

[7] Rikalovića, A.; Soares, G. & Ignjatićc, J. (2018), Spatial analysis of logistics center location: a comprehensive approach, Decision Making: Applications in Management and Engineering Vol. 1, Number 1, 2018, pp. 38-50 ISSN: 2560-6018, DOI: 10.31181/dmame180138r

[8] Ministry of Health, https://www.mh.government.bg/bg/administrativni-uslugi/registri/lechebni-zavedeniya/oblast-pleven/

[9] GoogleEarth,https://serc.carleton.edu/introgeo/google_earth/what.html

DOI: 10.2507/35th.daaam.proceedings.006

ENHANCING THUNDERSTORM PREDICTION AND USER PROTECTION: THE EVOLUTION OF DIGITAL PROCESSING SYSTEM ZEUS 2.0

Josip Krzelj, Kresimir Rakic & Zeljko Seremet

This Publication has to be referred as: Krzelj, J[osip]; Rakic, K[resimir] & Seremet, Z[eljko] (2024). Enhancing Thunderstorm Prediction and User Protection: The Evolution of Digital Processing System Zeus 2.0, Proceedings of the 35th DAAAM International Symposium, pp.0040-0047, B. Katalinic (Ed.), Published by DAAAM International, ISBN 978-3-902734-44-0, ISSN 1726-9679, Vienna, Austria
DOI: 10.2507/35th.daaam.proceedings.006

Abstract

Zeus 2.0, an advanced iteration of Hrvatski Telekom's storm forecasting system, represents a significant leap in proactive weather management. This system, integral to the Proactive Care process, integrates a sophisticated algorithm with real-time meteorological data to enhance thunderstorm predictions and mitigate potential damage to user equipment. Building upon the earlier Zeus version, Zeus 2.0 features a refined algorithm that leverages the DBSCAN clustering method for accurate storm forecasting and precise lightning strike detection. It incorporates real-time data from UBIMET, a leading meteorological service known for its high precision, to deliver timely alerts through SMS and the Moj Telekom application. The system's technological foundation includes a robust stack with React for frontend development, Java for backend processes, and PostgreSQL for data management. Key tools such as Apache Kafka and Red Hat OpenShift ensure efficient data processing and system scalability. Zeus 2.0 not only improves user experience with its intuitive interface but also automates critical processes like equipment replacement and troubleshooting through seamless integration with Camunda BPM for business process management. This evolution marks a substantial advance in optimizing user safety and system performance, paving the way for future enhancements in weather prediction and user protection.

Keywords: digital processing system; thunderstorm prediction; DBSCAN algorithm; user experience.

1. Introduction

Unstable weather with high temperatures and storm clouds is ideal for the formation of thunderstorms. This fascinating natural phenomenon can be very dangerous for the safety of people and their property. The development of the Zeus 2.0 system, as a component of the Proactive Care process offered to its users by Hrvatski Telekom in the form of the Moj Telekom application, stems from the need to improve the system for forecasting storms, preventing potential damage to equipment due to lightning strikes, and the process of replacing equipment. Digital Processing System (DPS) Zeus 2.0 represents the evolution of DPS, based on the previous version of Zeus.

This new version was developed to improve the existing algorithm for predicting thunderstorms and allow users timely information and preventive measures to protect their property. The previous version, although useful, proved to be less effective due to limited analysis and lack of user experience. The goal of the project is the integration of meteorological data with the user experience, which enables proactive management and reduces the need for user support interventions. For now, the linear algorithm has been put into production, and by cooperating with meteorological services, Zeus 2.0 lays the foundations for further innovations, with the ultimate goal of optimizing the user experience and reducing the need for user support. This system evolution represents a technological step towards better connectivity and user security.

Zeus 2.0 is designed to provide a comprehensive solution for thunderstorm prediction. Its advanced algorithm enables precise forecasting of weather conditions, while integration with process flows ensures automatic and timely reactions to upcoming storms. In this way, users can turn off their equipment in time, which minimizes the risk of burning or other forms of damage. In addition to technical advantages, Zeus 2.0 also provides an exceptional user experience thanks to a modern and intuitive interface that allows easy access to key information and functionalities. Integration with different services enables faster and more accurate data processing, which results in reliable and accurate information for end users.

In the second chapter of the paper, we will describe the technologies used during the development of the Zeus 2.0 system. The functionalities of the system and its use cases are described in the third chapter. The fourth chapter describes the used algorithm for predicting the thunderstorms, implemented in the Zeus 2.0 system. In the fifth, concluding chapter, an overview of the described system and guidelines for continued research and development are given.

2. Technologies Used in the Application Development

Zeus 2.0 is a network application that uses advanced technologies for thunderstorm prediction, with support for synchronization and actions through Camunda workflows. Integration with different services further improves the user experience, making the application more intuitive and efficient for end users. Globally, Zeus 2.0 implements a wide range of enterprise technologies that ensure high levels of performance and reliability. Technologies used for frontend development include React, a JavaScript library for building user interfaces. Java technology was used for the backend development. It is an object-oriented programming language known for its robustness and portability, which is crucial for large companies. Key elements of the backend support system include Camunda, a business process automation platform that enables process modelling and execution. PostgreSQL is used as a database. Other significant technologies that support the operation of the system are Apache Kafka, a distributed platform for Internet streaming, which enables high throughput and low latency data processing in real-time, and Red Hat OpenShift, a cloud platform for management containers, which enables the orchestration of applications in the Kubernetes environment with additional tools for development and deployment.

2.1. Frontend Development Technologies

Front-end technologies enable the building of user interfaces and interaction with applications, whether on web or mobile platforms. They enable content display, event management and communication with backend services. React is the most powerful JavaScript library used to create user interfaces. It was developed by Facebook and is often used to build single-page applications (SPA) and mobile applications. React was first released in May 2013 and has since become one of the most popular JavaScript libraries for building user interfaces. It is widely adopted by the developer community, and there are many third-party libraries and tools available that integrate with React. React allows developers to create reusable UI components, which can make it easier to build and maintain user interfaces. In addition, React uses a virtual document object model (VDOM), which represents a key innovation in React, improving the performance and efficiency of updating the user interface. VDOM becomes crucial in situations where there are frequent updates to the application state. Speed and efficiency are achieved by optimizing and updating only the parts of the page that have changed, instead of the entire DOM. This reduces the cost of updates, providing a faster user experience. Another important feature of React is JavaScript XML (JSX), which connects Javascript with XML. HTML-like syntax makes development easier for developers familiar with web technologies. JSX allows JavaScript expressions to be embedded, providing dynamism to the code. It is also important to mention React Hooks, which represent a significant step towards a more functional approach to component development [1].

2.2. Backend Development Technologies

Backend technologies are responsible for managing business logic, databases, and communication between servers and client applications. Java is an object-oriented programming language that is known for its portability, robustness and wide application in the development of business applications. Developed at Sun Microsystems, Java was originally conceived as a language for managing household appliances. However, it quickly became apparent that Java has wider possibilities and is adapted for developing secure and platform-independent applications that stand out for their ability that its code written once can run anywhere. This platform-independent nature stems from the use of the Java Virtual Machine (JVM), which allows the execution of Java code on different operating systems without recompilation [2].

In the context of the Zeus 2.0 system implementation, Java proved to be the optimal choice due to its object-oriented nature, high level of abstraction and platform-independent performance. This flexibility enables the development and - implementation on different operating systems without compromising efficiency and ease of maintenance. Java was also chosen for its security, automated memory management, and parallel execution capability, which is extremely important for complex systems such as Zeus 2.0 [3].

2.3. Database Technologies

In today's information age, accessing and handling information is one of the key activities in both professional and private life. Databases are systems for storing, managing and retrieving data in an organized manner. PostgreSQL has its beginnings at Berkeley University in California, in the form of the POSTGRES project (POST inGRES). In its initial phase, POSTGRES was conceived as a database management system that would extend the functionality of the earlier INGRES system, particularly by supporting object-relational concepts and extending SQL [4]. In 2006, PostgreSQL went fully open source, paving the way for a global community of developers, users, and database enthusiasts. The community has become the key force behind the development and improvement of PostgreSQL, thus maintaining a high standard of quality and innovation [5].

PostgreSQL was selected for the Zeus 2.0 project because of its advanced support for SQL standards, robustness, flexibility in handling different types of data, and ability to handle large workloads and complex queries. PostgreSQL is widely used in industry, from small projects and startups to large corporations and organizations that require powerful and scalable databases. Features such as support for JavaScript Object Notation (JSON) binary data types, advanced query management and optimization capabilities, and various enhanced functionalities testify to the constant work on innovation within PostgreSQL.

PostgreSQL is efficient in handling large amounts of data and flexible in data processing. Zeus 2.0 includes a geographic information system, and PostgreSQL provides robust support for spatial data. As an open-source system, PostgreSQL provides transparency and customizability, which is crucial for projects like Zeus 2.0. which require customization according to specific requirements [6].

2.4. UBIMET Data Source

UBIMET is an Austrian company specializing in meteorological services and weather monitoring technologies. UBIMET has built an unparalleled network in Europe, setting a new standard in localization precision and efficient detection. The data is updated every thirty seconds, providing real-time information, unlike traditional radars that update every 10 minutes. The positioning error is less than 100 meters, which ensures exceptional accuracy and reliability. Patented 3D technology is used to analyse strength, polarity, height, and the relationship between clouds and the ground or each other. For the described case, it is important to note that the analysis of the relationship between clouds and the ground is used. This sophisticated approach enables a deeper understanding of atmospheric phenomena [7].

UBIMET stands out for its patented algorithms for predicting lightning behaviour. These internal algorithms enable hazard analysis and timely issuance of severe weather warnings. All these advanced features make UBIMET a reliable partner for monitoring weather conditions with extraordinary precision and speed. The Zeus 2.0 system requires cloud-to-ground lightning strike data, while cloud-to-cloud strike data is not collected. This information is updated once a day and allows for monitoring surveillance and improving the level of preparedness [8].

2.5. Other Technologies

Red Hat OpenShift and similar platforms are used to deploy, manage and scale applications in containers. Entire DevOps processes and all projects can be built on them because they are cloud platforms that enable the development, testing, implementation and management of applications in a highly available and scalable environment. Red Hat Open Shift is the leading hybrid cloud application platform that combines the power of Kubernetes to provide a robust environment for developing, modernizing, deploying, running and managing applications. This innovative platform integrates proven services aimed at overcoming challenges throughout the application lifecycle. The OpenShift Container platform is a platform for developing and running containerized applications. It is intended to enable applications and data centres to scale from a few machines and applications to thousands of machines serving millions of clients. The ELK platform (Elasticsearch, Logstash, Kibana) and Grafana are used for observability, analysis and alarming. We use these powerful tools to monitor and interpret logs, providing an efficient solution for system analysis and immediate response to potential challenges [9].

Apache Kafka is a distributed web streaming platform used to build real-time applications that can process and analyse large amounts of data with low latency. The task of these technologies is to enable reliable and scalable data exchange between systems and applications. Kafka is often used in a big data infrastructure environment. Apache Kafka is an open data stream processing system that was originally developed at LinkedIn to solve the problem of connecting different applications and services. Since its introduction as an open-source project in 2011, it has become a popular tool for processing large amounts of data in real-time. It quickly became a key component of many microservice architectures and data processing platforms.

One of the reasons for using Apache Kafka for Zeus 2.0 is to save log records and history of actions, which enables monitoring and analysis of all system activities. Furthermore, big data integration provides a great advantage because Kafka integrates with this concept, enabling the transfer of large amounts of data to different topics within the Kafka cluster. Connectivity with tools such as ELK, Grafana and Prometheus enables the Zeus 2.0 system to provide detailed insight and monitoring of system performance through visualization of logs and metrics [10].

Camunda is a platform for automating business processes that enables the modelling, execution and monitoring of processes using the Business Process Model and Notation (BPMN), Case Management Model and Notation (CMMN) and Decision Model and Notation (DMN) standards. Camunda is often chosen because of its openness, flexibility, support for standards and ease of integration with existing systems, which enables efficient and scalable automation of business processes, this is one of the key reasons why it was chosen. Camunda engine is a powerful tool for automating business processes. Using the BPMN standard enables visual modelling of business processes, which facilitates collaboration between teams of business analysts and developers. Zeus 2.0 uses these models to define, optimize and implement its business processes [11].

There are several reasons for this:

- Camunda enables the efficient execution of business processes. Each step in the BPMN model represents a specific activity that can be automated or assigned to the user.
- Camunda stands out for its flexibility and enables dynamic adaptation of business processes to changes in real-time.
- Camunda is an open-source program support, which means that it is available for everyone to use, adapt and improve.
- BPMN models provide a clear and visual representation of business processes, facilitating communication between teams, enabling better process understanding among stakeholders, and helping to detect potential problems before implementation.
- Camunda can be integrated with a wide range of technologies enabling smooth coordination between different parts of the system.

The combination of these characteristics makes the Camunda BPMN engine extremely suitable for Zeus 2.0, providing a reliable and efficient environment for managing business processes [12][13].

3. System for Preventive Notification about Thunderstorms Zeus 2.0

One of the DPS systems integrated into the Proactive Care process is Zeus 2.0. The Proactive Care process consists of the following units and systems:

1. Ingestion. To retrieve data from the network, different platforms and systems, advanced systems like Grabber and multiple instances of Flume are used. These systems enable efficient data entry into a big data architecture, ensuring a high level of reliability and scalability.
2. Analysis. In the data analysis phase, sophisticated algorithms and machine learning (ML) models are applied. Java Spark and Spark ML programs on the cloud platform are used, and the big data team uses PySPARK programs with built-in ML models running through AirFlow. Additionally, Proactive Care Analysis hosted on OpenShift provides advanced data analysis to provide deeper insights.
3. Actions. After data analysis, actions are initiated through Proactive Care Action based on the Camunda BPM system. This system enables the implementation of proactive measures based on the results of the analysis, contributing to the continuous optimization of the process. Integration with Camunda BPM provides an agile approach to managing business processes in accordance with the results obtained from big data analysis.

This complete architecture enables an efficient and structured approach from data retrieval, through analysis to the implementation of actions, which ensures process optimization and informed decision-making. Integrations play a key role in the exchange of data between different systems. In the current version of the system, Big Data Kafka is used for efficient and reliable communication between different parts of the system that manipulate large data sets. As for integrations with legacy systems, API integrations are currently being implemented according to PCAC standards, in order to enable smooth and standardized communication with existing systems.

This integration helps maintain compatibility with previous systems and enables efficient data exchange between new and older technology solutions. All these integrations together make the architecture adaptable, resilient and efficient in handling data, both within the system and in interaction with external sources and targets. Part of Proactive Care is a system that performs preventive actions based on data obtained from Big Data Kafka Topic. The system uses a POSTGRES database to store information. System monitoring is done using the ELK stack with a focus on Application Performance Monitoring (APM) with *apmcamprob* sample monitoring. This system was developed in the Java programming language, using the Camunda engine for managing business processes. Camunda provides an environment for modelling, executing and monitoring business processes, while the POSTGRES database serves as a reliable data storage system.

3.1. Zeus 2.0 Use Case 1

A dynamic system was created to monitor meteorological data using the UBIMET API. 1,200 key locations across Croatia were defined, clusters were made, and information about lightning strikes in real time is updated daily through API requests, and comprehensive weather forecasts, daily, weekly, and precise lightning strikes in real time are periodically retrieved. The big data platform regularly captures this data, providing a complete overview of forecasts for the current day and the next 7 days at all 1,200 locations. This amount of information lays the foundation for in-depth analysis and research. The system provides insight into complex patterns of weather changes across the country, providing a valuable resource for making informed decisions and conducting thorough research, with always up-to-date and accurate data.

3.2. Zeus 2.0 Use Case 2

This use case represents a proactive approach to weather monitoring and consists of two parts. Part 1 performs the sending of weather warnings (thunderstorms, etc.) based on prognostic data of the big data platform via SMS or the Moj Telekom application. Rules are also defined for recognizing router burnout.

Sending an automatic notification via the Moj Telekom application or SMS (with a link to install the Moj Telekom application) is done as follows:

- If the user has been logged into the Moj Telekom application for the last 30 days, the notification goes through the application.
- If the user is not logged in, then via SMS with a link to install the Moj Telekom application.

Part 2 starts after the initial thunderstorm warning and covers:

- Detection of burnt equipment via Zeus message.
- If working in group interference, it is set as a priority.
- Implementation of equipment replacement through the Moj Telekom application.
- Alternative channels for equipment replacement: T-Centre or distributors.
- The interference is opened via the chat application.

3.3. Zeus 2.0 Use Case 3

UBIMET continuously monitors data on thunderstorms throughout Croatia. Information is collected about Digital Subscriber Line Access Multiplexer (DSLAM) devices including:

- DSLAM name/slot/port
- Asset data, assetID
- x and y coordinates
- Address (town, city, postal code)
- User segmentation (business or residential)
- The services that users have are also monitored.

Spark is used for continuous analysis of input data. The Spark system checks to see if the user's modem is in the dying gasp status, which signals the modem to turn off power. If the power outage occurred because the user intentionally turned off the modem, those cases are discarded where we get information that the user turned off the power. In situations where the damage was caused by lightning, we do not record any information that the System Problem Analyst (SPA) records. Synchronization is also checked.

A modem is considered synchronized if it is connected to the exchange. Spark sends a message to the Kafka platform for Camunda. Camunda, after receiving a message, starts a certain process according to the defined business logic. This includes notification, automatic replacement or other steps depending on the situation. Camunda can make decisions like waiting a certain amount of time before taking further steps. All these operations are automated to respond quickly and efficiently to problems caused by thunderstorms.

3.4. Camunda Process Flows in Zeus 2.0

Three process flows defined by the BPMN standard are implemented in Camunda:

- SMS alerts, informs the end user about a thunderstorm.
- Burned-out equipment detection, describes the process of confirming the detected burnt-out equipment and notifying the end user.
- Replacement of equipment, opening a complaint in system for reported complaints Donat, and the procedure of replacing the equipment itself.

4. Algorithm for Thunderstorm Prediction

Algorithm Density-Based Spatial Clustering of Applications with Noise (DBSCAN) makes it possible to group data into sets based on the density of the spatial distribution of data. This algorithm identifies subgroups of data based on how densely they are distributed, ignoring areas of low density that are considered noise. The algorithm was first proposed by Martin Ester, Hans-Peter Kriegel, Jörg Sander and Xiaowei Xu in 1996 [14]. DBSCAN is a popular clustering algorithm used in data analysis to identify groups or clusters within a data set. It is especially useful when we have complex data structures with noise and when the number of clusters is not known in advance. This algorithm is used to identify clusters of thunderbolts. Thunderstorms are considered part of the same cluster if their mutual distance is less than 3 km.

DBSCAN is based on two key concepts:

- Eps (ε): Radius of the environment within which adjacent data is searched. If there is a sufficient number of points within this radius, that environment is considered densely populated.
- MinPts: The minimum number of points required within the Eps radius for a point to be considered a centre point of a cluster.

The DBSCAN algorithm groups points based on density, i.e. how close the points are to each other. The work process can be described as follows [15].

- The algorithm randomly selects a point from the data set. For a selected point, the algorithm finds all neighbouring points within the Eps radius.
- Checking the minimum number of points (MinPts): If a point has enough neighbours (≥ MinPts), it is considered the central point of the cluster, and the cluster is expanded to include all neighbours that also meet the density criteria.
- The process continues recursively for all points within the cluster, thus expanding the cluster.
- Points that do not belong to any cluster are considered noise.

Some of the advantages of the DBSCAN algorithm include:

- Does not require a predefined number of clusters.
- Can identify clusters of different shapes and sizes.
- Handles noise and isolated points well.

Disadvantages of the DBSCAN algorithm are:

- Selecting Eps and MinPts can be challenging and may require multiple testing for optimal results.
- Does not work well with data of different densities.

When DBSCAN is applied to lightning clustering, the points (lightning) are analysed based on their spatial proximity. The Eps and MinPts parameters are selected to match the expected lightning density in a given area. The clusters that DBSCAN identifies can help understand how lightning clusters during certain weather conditions or in specific geographic regions, providing useful information for meteorology and related scientific disciplines.

The implementation of the DBSCAN algorithm includes the following steps:

1. Checking the conditions for prediction. For each cluster, it is checked whether the conditions for making a prediction are met. If less than 1 minute has passed since the appearance of lightning or if the cluster contains less than 3 lightning, no prediction is performed.
2. Data processing by minute. Thunderstorm data is sorted by minute. For each minute, a centroid is created that represents the mean value of the lightning positions in that minute.
3. Performing linear regression. Based on the centroid, a linear regression is performed to predict the direction of the storm.
4. Projection and distance calculation. On the line obtained by linear regression, the first and last centroids are mapped by orthogonal projection. The distance between those two points on the line is calculated.
5. Calculation of the predicted distance and angle. Based on the calculated distance and the time difference between the first and last centroid, the predicted distance that the storm will travel in an hour is calculated. The angle and the slope of the direction of the storm movement are calculated.
6. Prediction of the movement point. Based on the last centroid and the calculated distance with the slope, the prediction point of the storm movement is determined.
7. Adjustment of the coefficient. If the storm moves north, the coefficient is 1. In other cases, the coefficient is adjusted according to the geographical background. For lowland areas like Slavonia, if the coefficient is greater than 1.5 and the cluster contains more than 500 lightning strikes, only the data for the last 15 minutes are taken into account and the process is repeated with possible adjustments.

8. Calculation of cluster radius. Finally, the cluster radius is calculated, which is used to determine the area within which alerts will be sent. This approach enables more precise analysis and prediction of storm movements based on lightning data, enabling better planning and response to weather disasters.

The functionality of the DBSCAN algorithm is shown in Figure 1.

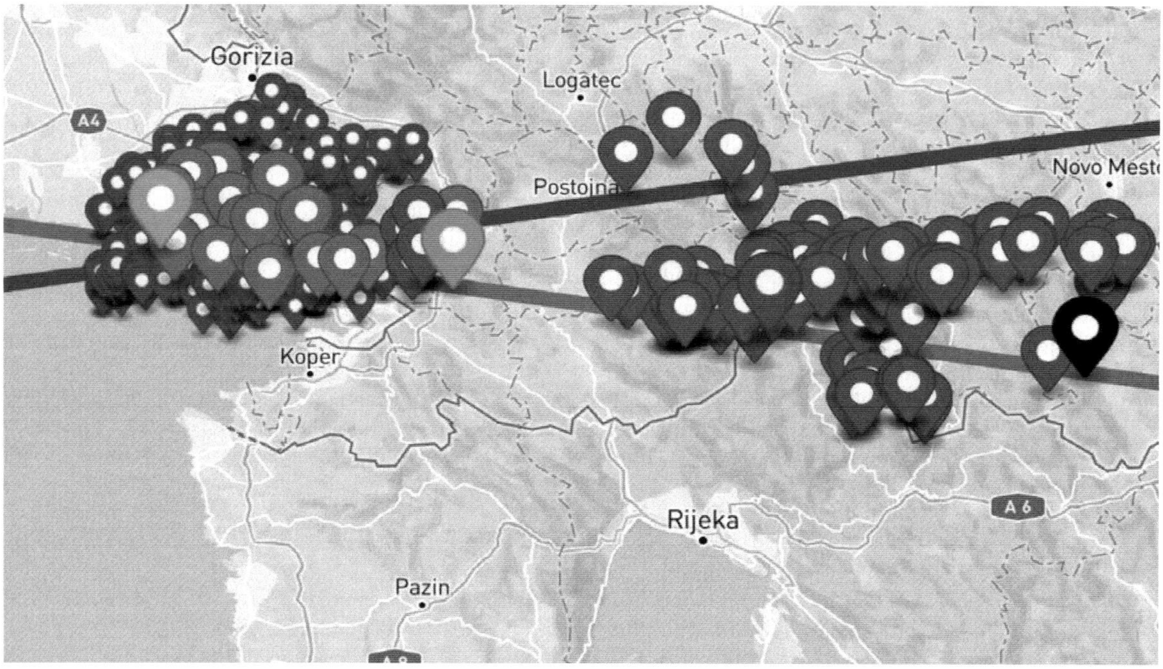

Fig. 1. Showing the direction of lightning on the Google map

Fig. 2. Showing the direction of lightning on the Google map

The red tags are the thunderstorms with which the prediction was calculated, the green are the thunderstorms that occurred after an hour (±15 minutes), the dark blue tag is the prediction for half an hour, and the black tag is the prediction for the hour. The storm is actually moving along the blue line (it's a linear regression of the centroid of the lightning each minute), and the red would be a linear regression in general. The ELK stack is used to monitor and control the very success of the model in such a way that the logs overtake the production environment, and the display map is automatically generated. ELK also allows easy manipulation of time. Logs are generated every 15 minutes [7]. The functionality of the ELK stack is shown in Figure 2.

5. Conclusion

The implementation of the Zeus 2.0 project represents a significant step forward in providing quality service to Hrvatski Telekom users, setting new standards in the telecommunications sector. This initiative enables a quick and efficient reaction to potential difficulties that users may have during thunderstorms. By monitoring and predicting the movement of storms, users are provided with important information and instructions on how to protect themselves from possible service interruptions. It also identifies users whose equipment has burned out, providing them with precise information about the type of disturbance and the steps they need to take. In addition to increasing user satisfaction, the implementation of this project reduces the time required for troubleshooting and enables faster replacement of burned-out equipment. In addition, the project contributes to increasing the safety of users, by informing them in advance about the potential dangers of a lightning strike. There are several directions for further improvement of the Zeus project. For example, using statistics on areas more frequently affected by lightning strikes would allow for the organization of repair technicians as needed, which would further speed up the response to difficulties. Also, consideration of protective equipment for users in such areas or protection of overhead installations could reduce the possibility of equipment damage during storms. In the future, the development of machine learning models that would provide even more accurate data on lightning strikes would be a step further in ensuring optimal safety and user satisfaction. This initiative, open to upgrades and integration with other systems, emphasizes the advantage of continuous improvement and adaptation to new technologies, ensuring that users always get the best possible experience.

6. References

[1] https://react.dev/learn, (2024). React. The Library for Web and Native User Interfaces, Accessed on: 2024-08-21

[2] Kristaly, D. M.; Moraru, S.-A.] & Grigorescu, C.-M. (2011). Multi-Purpose Modular Software Platform for Rapid Development of Web Applications, Annals of DAAAM for 2011 & Proceedings of the 22nd International DAAAM Symposium, Volume 22, No. 1, ISSN 1726-9679 ISBN 978-3-901509-83-4, Editor B. Katalinic, Published by DAAAM International, Vienna, Austria, EU, 2011

[3] https://www.w3schools.com/java, (2024) W3Schools. Java Tutorial, Accessed on: 2024-08-21

[4] Rosu, S. M.; Dragoi, G.; Rosu, L. & Guran, M. (2010). Virtual Enterprise Network Solutions to Support E-learning Sites Development, Chapter 63 in DAAAM International Scientific Book 2010, pp. 725-742, B. Katalinic (Ed.), Published by DAAAM International, ISBN 978-3-901509-74-2, ISSN 1726-9687, Vienna, Austria DOI: 10.2507/daaam.scibook.2010.63

[5] Raguž, A. (2019). Sustav za upravljanje bazom podataka PostgreSQL, Master's thesis, Faculty of Science, University of Zagreb, Zagreb, Croatia

[6] https://www.postgresql.org, PostgreSQL: The World's Most Advanced Open Source Relational Database, Accessed on: 2024-08-21

[7] Eisenbach, S., & Palmer, C. (2017). Digital advances in monitoring and planning for natural hazards along networks and remote track sections. In AusRAIL PLUS 2017, Rail's Digital Revolution, 21-23 November 2017, Brisbane, Qld, Australia.

[8] https://confluence.t.ht.hr/display/OSSAA/Zeus+2.0, (2024). Hrvatski Telekom Intranet. Confluence Zeus, Accessed on: 2024-05-15

[9] https://www.redhat.com/en/technologies/cloud-computing/openshift/container-platform, (2024). Red Hat OpenShift Container Platform, Accessed on: 2024-08-21

[10] https://kafka.apache.org/20/documentation.html, (2024). Kafka 2.0 Documentation, Accessed on: 2024-08-21

[11] https://docs.camunda.io/docs/guides, (2024). Camunda 8 Documentation, Introduction to Camunda, Accessed on: 2024-08-21

[12] https://docs.camunda.io/docs/guides/automating-a-process-using-bpmn, (2024). Camunda 8 Documentation, Design a process using BPMN, Accessed on: 2024-08-21

[13] Rucker, B. (2021). Practical Proces Automation: Orchestration and Integration in Microservices and Cloud Native Architectires, O'Reilly Media, Inc., ISBN: 9781492061458, Newton Massachusetts, USA

[14] Ester, M., Kriegel, H. P., Sander, J., & Xu, X. (1996, August). A density-based algorithm for discovering clusters in large spatial databases with noise. In kdd (Vol. 96, No. 34, pp. 226-231).

[15] Khan, K., Rehman, S. U., Aziz, K., Fong, S., & Sarasvady, S. (2014). DBSCAN: Past, present and future. In The fifth international conference on the applications of digital information and web technologies (ICADIWT 2014) (pp. 232-238). IEEE.

DOI: 10.2507/35th.daaam.proceedings.007

BUSINESS CONTINUITY PLANNING (BCP) AND DISASTER RECOVERY PLANNING (DRP)

Katja Mikulic Premuzic, Vedran Dakic* & Robert Petrunic

This Publication has to be referred as: Dakic, V[edran]; Mikulic Premuzic, K[atja] & Petrunic, R[obert] (2024). Business Continuity Planning (BCP) and Disaster Recovery Planning (DRP), Proceedings of the 35th DAAAM International Symposium, pp.0048-0054, B. Katalinic (Ed.), Published by DAAAM International, ISBN 978-3-902734-44-0, ISSN 1726-9679, Vienna, Austria
DOI: 10.2507/35th.daaam.proceedings.007

Abstract

This paper explores the critical aspects of Business Continuity Planning (BCP) and Disaster Recovery Planning (DRP), highlighting their essential role in ensuring organisational resilience. It defines Business Continuity Planning (BCP) and Disaster Recovery Planning (DRP), emphasising their importance in addressing various operational risks. The paper then systematically explores the processes of risk assessments and business impact analyses, showing how these tools are essential in identifying and avoiding potential disruptions. The paper examines real-world case studies, which provide practical insights into successful and failed implementations of BCP and DRP strategies. These case studies are essential in drawing lessons on proactive measures and typical disaster planning and recovery challenges. The paper also discusses testing methodologies vital for the effectiveness of BCP and DRP, emphasising the need for regular plan updates and exercises. The final part brings together these discoveries, proposing possible paths for further study and progress in BCP and DRP, especially as technology keeps changing and risks evolve.

Keywords: Business Continuity Planning; Disaster Recovery Planning; Risk Assessment; Business Impact Analysis; Organizational Resilience.

1. Introduction

In today's volatile environment, Business Continuity Planning (BCP) and Disaster Recovery Planning (DRP) are essential for maintaining operations and ensuring rapid recovery in the face of disruptions. This paper investigates the strategic planning, risk assessments, impact analyses, and testing methodologies that underpin BCP and DRP. A review of the literature and case studies assesses the effectiveness of these frameworks in actual disaster scenarios, identifies prevalent shortcomings in their implementation, such as outdated systems and insufficient testing, and explores innovative solutions. The insights gained reveal the factors crucial for the success or failure of BCP and DRP initiatives and suggest that ongoing innovation in these strategies is necessary due to the rapid changes in technology and global business conditions. This paper aims to provide a comprehensive understanding of BCP and DRP as vital tools for navigating the complexities of modern business environments, emphasising the need for continuous adaptation and improvement.

2. Business Continuity Planning (BCP) and Disaster Recovery Planning (DRP) Strategies

Business Continuity Planning (BCP) and Disaster Recovery Planning (DRP) are essential for organisational resilience and sustainability in today's business environment. As natural disasters, technological errors, and other disruptions become more frequent and unpredictable, these frameworks have transitioned from optional to critical corporate governance and risk management components. BCP ensures the continuation or quick resumption of crucial business functions, encompassing operational aspects, employee safety, data management, and customer relations. Its goal is to maintain smooth operational continuity during and after crises. Meanwhile, DRP focuses on restoring IT systems and data, which is crucial for operational functionality post-disaster. The importance of BCP and DRP in maintaining organisational resilience and mitigating financial losses during disruptions cannot be overstated. Security Information and Event Management (SIEM) systems are crucial in the realm of Business Continuity Planning (BCP) and Disaster Recovery Planning (DRP), ensuring robust cybersecurity. These systems are central to collecting, normalising, and analysing extensive security data from various network components [6]. The primary function of SIEM systems is to enhance organisational capacity to promptly detect and address potential security threats, thereby bolstering the overall strategy for cybersecurity. Integrating both security information management (SIM) and security event management (SEM), SIEM systems not only manage security data but also facilitate real-time analysis and threat detection [6].

This integration is vital in BCP and DRP, as it ensures that organisations can maintain continuous oversight and quick response capabilities, which is crucial for minimising downtime and protecting critical infrastructure during and after an incident. Critical infrastructure encompasses various technological systems essential for meeting societal needs. These systems are composed of both control systems and systems that are controlled and integral to company processes, social systems (including humans, organisational structures, assets, values, and knowledge), and technological systems (comprising tools, equipment, procedures, and technologies) [7]. As multistage systems, they facilitate bidirectional flows of materials, finances, information, and decisions. Consequently, it is crucial to analyse these systems from the perspective of interactions and interdependencies among their technical, human, social, and organisational components [7]. This analysis includes the consideration of human survival, which can be either active or passive. Passive survival capacity is embedded within the system properties, based on knowledge of environmental defects, depicted through causal chains. On the other hand, active survival is demonstrated by the system's behaviour, considering the uncertainty in forecasting future defects and failures [7].

These strategies help sustain operations and protect the company's reputation by maintaining customer trust during crises. Additionally, they are vital for complying with industry regulations, especially in the finance and healthcare sectors, where data protection and service availability are heavily regulated. An example of a Business Continuity Management plan can be seen in the following figure:

Fig. 1. ISO 22301 business continuity management plan, https://advisera.com/27001academy/knowledgebase/business-continuity-plan-how-to-structure-it-according-to-iso-22301/, accessed on 2024-05-05.

The ongoing complexity of IT infrastructures, including the rise of cloud computing and big data, has made DRP even more significant. It must evolve to address new technological challenges and opportunities. Real-world case studies demonstrate the consequences of inadequate disaster planning and the benefits of robust BCP and DRP. These examples provide crucial lessons in risk assessment and proactive strategy formulation, emphasising the need for continuous adaptation to the dynamic business and technological landscape.

3. Risk Assessments in Business Continuity Planning

Risk assessments are crucial in Business Continuity Planning (BCP) and Disaster Recovery Planning (DRP), serving as the foundation for developing robust strategies to enhance business resilience against disruptions. This process entails qualitative and quantitative methods to identify, evaluate, and prioritise potential risks, thereby shaping effective continuity and recovery strategies.

3.1. Identifying Potential Risks

The process starts by identifying risks that could affect business operations, which include natural disasters, technological failures, cybersecurity threats, and human errors. Risk identification involves analysing potential hazards from various sources, such as historical data, industry reports, and expert opinions. For example, a business in a hurricane-prone area would prioritise this as a significant risk, while a tech-heavy company might focus on cybersecurity threats. Small and medium-sized enterprises (SMEs) must incorporate resilience into their strategic frameworks to address the complexities associated with disaster recovery and business continuity. The implementation of resilience, as explored in [8], entails a strategic management approach to risk factors, enhancing SMEs' capabilities to manage and mitigate risks effectively. This approach includes applying the Delphi technique for iterative feedback and consensus in risk prioritisation, essential for maintaining operational continuity and minimising disruptions in dynamic market conditions [8].

3.2. Qualitative and Quantitative Risk Assessments

Qualitative assessments involve a subjective risk probability and impact analysis, utilising tools like risk matrices and registers to categorise and prioritise risks. This approach is suited for risks that are challenging to measure numerically, such as reputational damage or employee morale [10]. Conversely, quantitative risk assessments use numerical data to estimate risk probability and impact, providing a more objective basis for decision-making. This method is precious for financial risk assessments, where potential losses can be estimated in monetary terms, aiding financial planning and regulatory compliance [10]. Both types of risk assessments are vital for BCP. They allow organisations to understand the risk landscape thoroughly and allocate resources effectively to safeguard critical business functions. Qualitative assessments contribute to a broad understanding of risks and help engage stakeholders and prepare for less quantifiable events [10]. Quantitative assessments provide the precision for developing detailed recovery strategies and financial preparations, such as determining investments in backup systems or insurance. Together, these assessments form the bedrock of a proactive and resilient BCP and DRP strategy [10].

4. Impact Analyses in Business Continuity Planning

The Business Impact Analysis (BIA) identifies, quantifies, and qualitatively assesses the impacts of a loss, interruption, or disruption of business processes on an organisation. It furnishes the necessary data to determine suitable continuity strategies [13]. The most crucial steps in the BIA process are as follows: (1) Identify the management owners and business activities; (2) Select the right personnel to obtain information as efficiently as possible; (3) Identify scenarios that could seriously harm the company's assets, reputation, or financial situation; and (4) Determine the window of time during which interruptions to business operations are unacceptable [13]. Various information-gathering methods, including workshops, questionnaires, and interviews, may be used, depending on the company under analysis. The maximum tolerable period of disruption (MTPD) and the recovery point objectives (RPO) are the two primary deliverables of the BIA for every business activity. This analysis extends beyond simple risk assessment by projecting the potential real-world impacts on organisational functionality. The BIA process is initiated by cataloguing key business processes and assessing their vulnerability to disruptions across various dimensions, such as financial damage, regulatory issues, customer relations, and potential long-term strategic challenges.

4.1. Conducting a Business Impact Analysis

The BIA involves detailed steps, from thoroughly reviewing organisational operations to identifying critical functions and understanding the interdependencies among business units and external entities. It evaluates the potential immediate, intermediate, and long-term impacts of various disruption scenarios like natural disasters, technological failures, or cyberattacks. In cloud computing, assessing and mitigating risks necessitate a structured approach, as illustrated in [12]. Their model incorporates a multi-level grading system that evaluates business continuity risks across several dimensions, including business operations, security levels, system lifecycle, and economic factors. This method ensures that the disaster recovery capabilities are appropriately aligned with the specific needs and vulnerabilities of the cloud environment, thereby enhancing the resilience and responsiveness of the business systems involved [12]. This evaluation guides the prioritisation of recovery efforts to restore essential operations rapidly.

Resource identification for recovery and establishing Recovery Time Objectives (RTO) for each business function is also integral to the BIA. RTOs dictate the maximum tolerable downtime, which is crucial for developing targeted recovery strategies and managing recovery expectations.

4.2. *Significance of Business Impact Analysis*

BIA offers multiple benefits by enabling efficient resource allocation and enhancing the precision of recovery strategies, thereby minimising downtime and accelerating recovery. It provides critical information that improves transparency with stakeholders, employees, customers, investors, and regulators, helping manage expectations and sustain trust during disruptions. Moreover, BIA ensures adherence to regulatory standards and informs decision-making processes in BCP, grounding them in a detailed understanding of potential impacts rather than mere speculation.

In conclusion, BIA is essential for crafting effective Business Continuity Plans. It provides a detailed overview of how disruptions affect business operations, allowing organisations to make informed recovery priorities and resource distribution decisions. This process supports the continuity of operations and facilitates strategic planning and stakeholder management. Regular updates to BIA are necessary to keep pace with changing business landscapes and risk profiles, ensuring that continuity plans remain relevant and robust.

5. Testing Methodologies in Business Continuity and Disaster Recovery Planning

Testing methodologies ensure that Business Continuity Plans (BCP) and Disaster Recovery Plans (DRP) are theoretically sound and practically practical during actual disasters. Regular updates and consistent practice of these plans are critical to adapt to the changing risk environment. Testing in BCP and DRP involves a detailed evaluation of recovery plans to confirm their effectiveness and pinpoint areas needing enhancement. This process often includes various exercises and simulations that check each part of the plan, from communication protocols to technical recovery solutions, aiming to meet Recovery Time Objectives (RTO) and Recovery Point Objectives (RPO). In formulating and testing Business Continuity and Disaster Recovery Plans, it is essential to integrate cybersecurity measures comprehensively. As highlighted in [9], ensuring the security and continuity of business operations, especially in environments with high criticality and availability, is highly important. Integrating cybersecurity throughout the business continuity planning and disaster recovery process helps manage cyber threats and ensures a resilient infrastructure capable of smooth transition and recovery from cyber incidents [9].

Due to the evolving nature of threats and technology, updating and testing BCP and DRP regularly is necessary. Changes in technology, business processes, or regulations could make older plans ineffective. Regular testing helps spot these inefficiencies and allows staff to be trained and familiarised with their roles in disaster scenarios, reducing response times and potential confusion during actual events. In addition to traditional testing methodologies, implementing simulation games has proven effective in enhancing the cybersecurity training aspect of business continuity and disaster recovery plans. As noted in [11], simulation games provide a dynamic platform for personnel to develop critical decision-making and crisis management skills in a controlled yet realistic environment [11]. This method prepares individuals to handle real-world cybersecurity threats and incidents efficiently, ensuring they can maintain operational integrity under adverse conditions [11].

Some key testing processes in BCP and DRP involve:

- Tabletop exercises are discussion-based sessions where team members work through the plan in a simulated disaster scenario. They are critical for assessing the plan's comprehensiveness and training staff on their responsibilities.
- Technical recovery tests: These involve recovering systems and data from backups to check that they can be restored quickly and efficiently, ensuring the integrity and functionality of backup systems.
- Full-scale drills: These extensive exercises mimic a natural disaster as closely as possible, testing the organisation's ability to implement the recovery plan under realistic conditions, often including mobilising resources and personnel.
- Communication tests evaluate the effectiveness of communication channels and protocols, which are crucial during a disaster for maintaining timely and accurate information flow.
- Business process tests assess the ability to sustain or swiftly resume essential business processes during and after a disruption.

Let's now see what these processes will bring to our company and why they are essential to reducing risk.

5.1. *The Impact of Simulations and Drills*

Simulations and drills play a pivotal role in validating the efficacy of BCP and DRP.

They provide numerous benefits:

- Simulations offer a realistic environment to assess the plan's performance, clearly indicating its functionality in a disaster.

- Drills act as practical training sessions for staff, ensuring they understand and can effectively execute their roles under pressure.
- These exercises are essential for identifying weaknesses in the plan related to resources, procedures, or technical capabilities.
- Conducting regular drills can significantly enhance response times, which is crucial for reducing the impact of a disaster.
- Successful simulations and drills boost confidence among all stakeholders—employees, customers, and investors—about the organisation's resilience capabilities.

Testing methodologies form an integral part of the success of BCP and DRP, ensuring that the plans are viable on paper and effective in real scenarios. Regularly updating and practising these plans through tabletop exercises, recovery tests, full-scale drills, communication tests, and business process assessments are vital for maintaining their effectiveness. Such activities are crucial for staff training, spotting plan deficiencies, and enhancing the organisation's resilience in the face of potential disasters.

6. Integration of Case Studies in Business Continuity Planning and Disaster Recovery Planning

The integration of real-life scenarios in Business Continuity Planning (BCP) and Disaster Recovery Planning (DRP) is essential for understanding and enhancing the efficacy of these strategies. Learning from case studies provides invaluable insights into the practical application, challenges, and outcomes of BCP and DRP in real-world situations. These studies serve as a compass, guiding organisations in crafting theoretically sound plans and pragmatically viable plans that are tested against the nuances of real-life disasters and disruptions. The examination of case studies, such as those presented in [1], [2], [3], [4], and [5], highlights the many aspects of disaster recovery and business continuity. These studies show how organisations have succeeded and failed when faced with unexpected problems, such as natural disasters, technology breakdowns, and cyberattacks.

- FAA System Failure [1]: This case study highlights the repercussions of relying on outdated systems, emphasising the need for modernisation and regular updates in BCP and DRP. The incident underscores the importance of having contingency plans for legacy systems and ensuring high availability in critical infrastructure.
- Microsoft Azure/Office Outage [1]: This global disruption showcases the vulnerabilities in cloud-based systems and the need for robust, multi-regional strategies. It stresses the significance of having diversified disaster recovery options, especially for organisations relying heavily on cloud services.
- OVHcloud Data Centre Fire [1]: This event illustrates the catastrophic impact of physical disasters on data centres. It reinforces the 3-2-1 rule of data backup and the importance of off-site backups in safeguarding data against natural calamities.
- NHS Ransomware Attack [1]: This case study demonstrates the devastating effects of cyberattacks on public health infrastructure. It highlights the need for rigorous cybersecurity measures and the importance of quick response mechanisms in BCP and DRP to mitigate the impact of such attacks.
- Houston Wire and Cable's BCDR Success [5]: This example shows the efficacy of a well-implemented BCDR solution, enabling the organisation to maintain operations during extreme weather. The success story emphasises the value of cloud technology and proactive planning in ensuring business continuity amidst natural disasters.

These are just some of the studies we should consider when doing our BCP and DRP, as they're valuable learning resources.

6.1. Successful Recoveries

Examining instances where organisations effectively navigated crises using their Business Continuity and Disaster Recovery Plans (BCP and DRP) provides invaluable insights into the essence of resilient planning. Case studies like those in [2], [4], and [5] exemplify successful recoveries. Houston Wire and Cable's Approach [5]: In the face of an unexpected snowstorm, Houston Wire and Cable demonstrated the robustness of their BCDR solution. They maintained operations despite widespread power outages using Azure cloud technology and proactive planning. The key to their success was the swift transition to a reliable backup system and their readiness for natural disasters, showcasing the importance of cloud-based solutions and the necessity of agility in response strategies. Total Tool's Data Centre Crisis [4]: This case study highlights the importance of having an off-site data centre. When Total Tool faced a robbery that jeopardised their primary data centre, their disaster recovery plan involving an off-site data centre allowed for seamless business operations, illustrating the significance of having a robust, multi-location data storage and recovery strategy.

6.2. *Lessons from Failed Plans*

In contrast, analysing where BCP and DRP have failed offers critical lessons in the pitfalls of disaster recovery planning. Case studies like [1] and [3] reveal common shortcomings.

FAA System Failure [1]: This instance underscores the perils of depending on outdated systems. The failure of the FAA's NOTAM system due to a simple error like a deleted file highlights the need for modern, resilient systems and the importance of regular system updates and backups. NHS Ransomware Attack [3]: This case illustrates the devastating impact of cyberattacks. The NHS's vulnerability to ransomware exposed the lack of adequate cybersecurity measures and swift response mechanisms. This case emphasises the importance of robust security protocols and contingency planning for cyber threats.

6.3. *Synthesis of Learnings*

These case studies offer important lessons. Success in disaster recovery depends on taking proactive steps, like using modern technology and storing data off-site. On the other hand, failures often result from old systems, weak security, and slow response plans. These lessons highlight the need for organisations to stay flexible, regularly update their BCP and DRP with the latest technology, and focus on strong security. Testing and updating these plans periodically are crucial to ensure they remain effective. Analysing these case studies gives us a clear picture of what makes a good BCP and DRP. Preparing for potential issues and adapting to unexpected situations are keys to successful disaster recovery. Failures remind us of the dangers of not planning well. These findings are essential for organisations creating or improving their BCP and DRP. Plans should consider successful and unsuccessful recoveries to build resilience and maintain business continuity in various disruptions.

7. Future Works

Further research could focus on integrating emerging technologies such as artificial intelligence, machine learning, and blockchain in BCP and DRP systems. Investigating their potential to automate risk assessments, predict potential disruptions, and optimise recovery processes could yield significant advancements in this field. Additionally, studies could explore the ethical and practical implications of automating decision-making processes in disaster recovery scenarios, ensuring that these technological solutions do not inadvertently increase organisational risk or create new vulnerabilities.

8. Conclusion

This paper has thoroughly explored Business Continuity Planning (BCP) and Disaster Recovery Planning (DRP), emphasising their indispensable roles in maintaining organisational resilience. The findings reveal that BCP and DRP are dynamic processes requiring ongoing updates and refinements to address evolving risks effectively. Integrating comprehensive risk assessments and detailed business impact analyses is vital for detecting vulnerabilities and formulating effective mitigation strategies. The real-world case studies presented here illustrate both successful implementations and notable failures, underlining the importance of proactive and adaptable strategies for disaster management. These narratives demonstrate the critical need for preparedness and the ability to respond flexibly to unexpected challenges, providing essential lessons for enhancing future strategy. Moreover, this paper underscores that BCP and DRP extend beyond mere disaster survival; they leverage insights from past experiences to thrive in a post-disaster environment. With risks ranging from cyber threats to natural disasters, a forward-thinking approach that anticipates and adapts to change is crucial.

In conclusion, the paper advocates for a proactive, iterative approach to BCP and DRP. Organisations should consider these plans as living documents, continually refined to align with the latest threats and technological changes. By fostering a culture of resilience and staying responsive to the changing world, organisations can prepare themselves to withstand future disruptions and emerge from them stronger and more resilient. This approach is not just about survival; it's about positioning for future success in an unpredictable world.

9. References

[1] https://www.techtarget.com/searchdisasterrecovery/tip/Real-life-business-continuity-failures-Examples-to-study, (2023). Real-life business continuity failures: 4 examples to study, Accessed on: 2024-01-15

[2] https://invenioit.com/continuity/4-real-life-business-continuity-examples/, (2022). 7 Real-Life Business Continuity Examples You'll Want to Read, Accessed on: 2024-01-15

[3] https://square3it.com/real-life-examples-of-business-continuity-plans-that-failed/, (2019). Real-Life Examples Of Business Continuity Plans That Failed, Accessed on: 2024-01-15

[4] https://www.loffler.com/blog/disaster-recovery-case-study-insights, (2022). Insights from Our Business Continuity & Disaster Recovery Case Study, Accessed on: 2024-01-15

[5] https://www.softwareone.com/en/case-studies/global/manufacturing/hwc-business-continuity-and-disaster-recovery-on-azure, (2024). Circuitry with the best shield, Accessed on: 2024-01-15

[6] Suskalo, D.; Moric, Z.; Redzepagic, J. & Regvart, D. (2023). Comparative Analysis of IBM Qradar and Wazuh for Security Information and Event Management, Proceedings of the 34th International DAAAM Symposium 2023

(pp.0096-0102), Published by DAAAM International, ISBN 978-3-902734-41-9, ISSN 1726-9679, Vienna, Austria. DOI: 10.2507/34th.daaam.proceedings.014, in press.

[7] Procházková, D.; Prochazka, J.; Rusko, M.; Mikulova, M. & Ilko, J. (2017). Model for Critical Infrastructure Safety Management, Proceedings of the 28th International DAAAM Symposium 2017 (pp.0602-0610), Published by DAAAM International, ISBN 978-3-902734-11-2, ISSN 1726-9679, Vienna, Austria. DOI: 10.2507/28th.daaam.proceedings.085, in press.

[8] Grujic, J. (2019). Concept of Resilience Implementation in Small and Medium Sized Enterprises (SMEs), Proceedings of the 30th DAAAM International Symposium, Published by DAAAM International, ISBN 978-3-902734-22-8, ISSN 1726-9679, Vienna, Austria. DOI: 10.2507/30th.daaam.proceedings.116, in press.

[9] Altaha, S. & Rahman, M. M. (2023). A Mini Literature Review on Integrating Cybersecurity for Business Continuity, Proceedings of the 2023 International Conference on Artificial Intelligence in Information and Communication (ICAIIC), Published by IEEE, ISBN 978-1-6654-5645-6, Al Hofuf, Al Hassa, Saudi Arabia. DOI: 10.1109/ICAIIC57133.2023.10067127, in press.

[10] Hewitt, J. & Pham, J. (2018). Qualitative Versus Quantitative Methods in Safety Risk Management, Proceedings of the 2018 Annual Reliability and Maintainability Symposium (RAMS), Published by IEEE, ISBN 978-1-5386-2870-6, Reno, USA. DOI: 10.1109/RAM.2018.8463052, in press.

[11] Lang-Muhr, C.; Tjoa, S.; Machherndl, S. & Haslinger, D. (2022). Business Continuity & Disaster Recovery A simulation game for holistic cyber security education, Proceedings of the 2022 IEEE Global Engineering Education Conference (EDUCON), Published by IEEE, ISBN 978-1-6654-4434-7, ISSN 2165-9567, Tunis, Tunisia. DOI: 10.1109/EDUCON52537.2022.9766714, in press.

[12] Zhang, L.; Fang, X.; Chen, Y.; Song, Y. & Qian, S. (2021). Research on business continuity rating model in cloud environment, Proceedings of the 2021 IEEE 5th Advanced Information Technology, Electronic and Automation Control Conference (IAEAC), Published by IEEE, ISBN 978-1-7281-8028-1, ISSN 2689-6621, Chongqing, China. DOI: 10.1109/IAEAC50856.2021.9390871, in press.

[13] Tjoa, S.; Jakoubi, S. & Quirchmayr, G. (2008). Enhancing Business Impact Analysis and Risk Assessment Applying a Risk-Aware Business Process Modeling and Simulation Methodology, Proceedings of the 2008 Third International Conference on Availability, Reliability and Security, Published by IEEE, ISBN 978-0-7695-3102-1, Barcelona, Spain. DOI: 10.1109/ARES.2008.206

DOI: 10.2507/35th.daaam.proceedings.008

APPLICATION OF QUANTUM CRYPTOGRAPHY IN SECURING NETWORK COMMUNICATIONS

Zlatan Moric*, Marijan Milovec & Robert Petrunic

This Publication has to be referred as: Moric, Z[latan]; Milovec, M[arijan] & Petrunic, R[obert] (2024). Application of Quantum Cryptography in Securing Network Communications, Proceedings of the 35th DAAAM International Symposium, pp.0055-0064, B. Katalinic (Ed.), Published by DAAAM International, ISBN 978-3-902734-44-0, ISSN 1726-9679, Vienna, Austria
DOI: 10.2507/35th.daaam.proceedings.008

Abstract

The emergence of quantum computing poses significant risks to conventional network security, prompting researchers to focus on developments in quantum cryptography, particularly Quantum Key Distribution (QKD). According to research, it has been found that traditional security methods, like RSA encryption, could potentially be at risk from quantum algorithms. On the other hand, quantum cryptography is a field of study that leverages the principles of quantum mechanics to improve the security of communication. One of the techniques used in quantum cryptography is Quantum Key Distribution (QKD), which makes use of the unique properties of quantum particles such as photons to securely distribute encryption keys and detect any potential eavesdropping attempts, thanks to the uncertainty principle. Nevertheless, the integration of QKD into current network infrastructures poses significant challenges due to the required technological advancements. This study offers a comprehensive analysis of the fundamental principles underlying quantum cryptography, the obstacles that arise during its implementation, and its potential prospects for protecting against the risks posed by advancements in quantum computing. Furthermore, this study delves into the potential consequences of these technological advancements across different industries and emphasizes the need to create algorithms that can withstand quantum attacks to guarantee strong data security in the age of digitalization.

Keywords: quantum cryptography; Quantum Key Distribution (QKD); secure communications; quantum algorithms.

1. Introduction to Quantum Computing and Network Security

The rapid progress of digital security technologies is propelled by the ongoing development of network threats and the accompanying requirement for strong defence systems. Quantum computing is a major advancement in computational abilities, offering substantial enhancements in processing power but also posing significant problems to current network security frameworks. The fundamental concepts of quantum mechanics, such as superposition and entanglement, present both advantages and risks to the existing state of network security. This study examines the possible impact of quantum computing and quantum cryptography, specifically Quantum Key Distribution (QKD), on network security. These technologies could strengthen security measures and protect against potential dangers posed by quantum technology [1].

1.1. Overview of Traditional Network Security

The field of network security has undergone substantial development since its inception, progressing from basic protective measures to complex systems that are specifically built to combat a wide range of cyber threats in a globally interconnected world. Originally, network security primarily concerned itself with safeguarding physical entry points and subsequently broadened its scope to encompass the diverse range of cyber threats. Traditional network security relies on essential elements such as encryption techniques like AES and RSA, which protect data from unwanted access. Additionally, key exchange protocols are used to guarantee secure communications even when networks may be hacked. Encryption techniques play a crucial role in conventional network security. Symmetric-key methods, such as AES (Advanced Encryption Standard), employ a single key for both encryption and decryption, resulting in efficient operations. However, their effectiveness hinges on the secure transmission of the key. Asymmetric-key algorithms, such as RSA (Rivest–Shamir–Adleman), employ a set of two keys (public and private) to perform encryption and decryption, offering heightened security at the expense of increased computational complexity. The utilization of these cryptographic techniques has played a crucial role in upholding the confidentiality and integrity of data across diverse digital platforms.

Nevertheless, the progression of cyber-attacks has shown fundamental constraints in conventional encryption techniques. Advanced Persistent Threats (APTs), zero-day exploits, and sophisticated malware have posed a threat to the efficiency of traditional security measures. Adversaries consistently devise novel techniques to evade security measures, rendering it progressively more challenging for static, rule-oriented systems to keep up. Traditional network security frameworks encounter continuous difficulties in adjusting to the swiftly changing threat landscape, which requires the creation of security solutions that are more flexible and responsive. An important obstacle is the dependence on the computational complexity of specific mathematical problems, which serves as the foundation for several conventional encryption techniques. For example, RSA encryption is based on the challenge of factoring big prime integers. Nevertheless, the progress in computational capabilities, namely the possible emergence of quantum computers, poses a significant risk to the integrity of fundamental security principles. Quantum computing, due to its exceptional capacity to execute intricate computations at unparalleled velocities, has the potential to undermine these cryptographic systems, necessitating the development of quantum-resistant algorithms and novel methods for safeguarding data [1].

Conventional network security has been vital in safeguarding digital communications and ensuring the integrity of data. Nevertheless, the growing complexity of cyber threats and breakthroughs in technology, such as quantum computing, emphasize the necessity for ongoing development and improvement of security measures to guarantee strong protection in the future.

1.2. Emergence of Quantum Computing and its Impact on Network Security

Quantum computing is a major departure from classical computing methods, as it uses quantum bits (qubits) that take advantage of the principles of quantum superposition and entanglement. Unlike classical bits, which can only be in one of two states (0 or 1), qubits could be in numerous states at the same time. This unique property allows quantum computers to perform many operations simultaneously. Quantum computers possess the ability to solve intricate problems at unparalleled velocities, exceeding the capacities of traditional computers, due to this parallelism. Quantum entanglement significantly improves the efficiency of processing and transmission, allowing for computational power that fundamentally challenges current computational models and cryptographic algorithms [4]. Quantum computing presents a significant risk to conventional cryptography systems, such RSA and ECC, which are the foundation of existing network security frameworks. These cryptographic algorithms are based on the computational complexity of problems such as prime factorization and discrete logarithms. Quantum algorithms, specifically Shor's algorithm, could solve these issues at an exponential speed compared to conventional algorithms. As a result, this poses a threat to the security of systems that rely on them. Hence, the swift progress of quantum computing requires a thorough revaluation and revamp of cryptographic methods to guarantee the ongoing safeguarding of confidential data.

In addition to these risks, quantum computing also stimulates the invention of novel security protocols. Quantum cryptography, specifically Quantum Key Distribution (QKD), utilizes the distinct characteristics of quantum mechanics to create secure channels for communication. Quantum Key Distribution (QKD) employs the concepts of quantum entanglement and the no-cloning theorem to identify any unauthorized interception during the exchange of quantum keys, thus ensuring the integrity of the key distribution process. Protocols such as BB84 demonstrate the application of quantum mechanics in developing encryption techniques that are theoretically impervious to any method of decryption except for a direct quantum attack [8]. Combining quantum cryptography techniques with current network infrastructures poses significant technical and economic obstacles. Quantum cryptography systems necessitate superior single-photon sources and resilient transmission of quantum states across extended distances, which pose significant technical and financial challenges. Moreover, the process of combining these quantum systems with traditional network components requires the development of novel protocols and infrastructures, which presents substantial challenges for achieving general acceptance and use. Quantum computing poses significant difficulties to conventional network security by potentially rendering current encryption techniques ineffective. However, it also introduces new possibilities for developing exceptionally secure communication systems. Developing and applying cryptographic algorithms and approaches that are resistant to quantum computing is crucial to address the risks posed by quantum computing and guarantee the long-term durability of network security protocols.

1.3. The Necessity for Quantum-Safe Security

The progress of quantum computing highlights the urgent requirement for quantum-safe cryptography that can withstand the exceptional powers of quantum machines, potentially making standard encryption approaches outdated. Quantum cryptography, specifically Quantum Key Distribution (QKD), employs quantum features like entanglement and the no-cloning theorem to facilitate secure communication. This method can identify any attempt at eavesdropping by observing its impact on the quantum state of the communication. It is of utmost importance to develop and apply quantum-safe techniques to guarantee the security of communications in the presence of advancing quantum technology. With the progress of quantum computing, it is becoming increasingly important to prioritize quantum cryptography. This requires the development of modern technologies and a reconsideration of network security infrastructure to accommodate quantum communication protocols. Although quantum cryptography, such as QKD, is theoretically strong, it encounters practical obstacles such limited range and scalability that need to be resolved for successful integration into current systems. This research examines methods by which different industries might successfully adjust to these quantum breakthroughs to safeguard future communications. It is crucial to have quantum-safe security techniques to effectively address the increasing dangers brought about by the rapid advancements in quantum computing technology [11].

2. Traditional Cryptography vs. Quantum Cryptography (The Rise of Quantum Cryptography)

The field of cryptography has experienced substantial development, propelled by the necessity to safeguard confidential data in an ever more digitized society. Traditional cryptography, encompassing symmetric-key algorithms such as AES and asymmetric-key algorithms such as RSA, depends on intricate mathematical problems to safeguard data. These technologies have served as the foundation of digital security, protecting communications, financial transactions, and personal data. Nevertheless, the advent of quantum computing presents a major obstacle to these conventional cryptography systems. Quantum computers, due to their capacity to do numerous computations concurrently, have the potential to answer intricate mathematical problems at a significantly accelerated pace compared to conventional computers. Consequently, they pose a threat to the security of systems that depend on these cryptographic techniques. Quantum cryptography utilizes the principles of quantum physics to improve security. Quantum cryptography, in contrast to classical cryptography, utilizes the fundamental laws of quantum mechanics, such as superposition and entanglement, rather than relying on the computational complexity of mathematical problems. Quantum Key Distribution (QKD), the prominent implementation of quantum cryptography, utilizes these concepts to establish secure channels for communication. The BB84 protocol, a fundamental technique in Quantum Key Distribution (QKD), enables two parties to create a mutually agreed secret key, ensuring that any unauthorized interception may be identified, thereby assuring the confidentiality of the communication. The transition from mathematical complexity to physical rules represents a groundbreaking method for ensuring secure connections [11]. Integrating quantum cryptography into current network infrastructures presents numerous technological and economic obstacles. Significant technological developments and investments are necessary for the development of high-quality single-photon sources, efficient transmission of quantum states across large distances, and the integration with classical systems. Furthermore, the incorporation of quantum cryptography systems must tackle concerns regarding scalability and practical implementation to achieve widespread acceptance. To address the risks presented by quantum computing and safeguard the security of digital communications in the future, it is crucial to overcome these obstacles by creating quantum-resistant cryptographic algorithms and integrating quantum cryptographic approaches. Quantum cryptography offers both enhanced security and novel opportunities for secure communication that cannot be achieved using conventional approaches. By harnessing the distinct characteristics of quantum mechanics, it provides a theoretically invulnerable encryption technique, establishing itself as an essential tool in the future of network security protocols. With the continuous advancement of quantum computing, it is becoming increasingly clear that there is a growing need to incorporate quantum-safe security mechanisms. This emphasizes the urgent want for additional research and development in this area. Although classical cryptography has been successful in safeguarding data up to now, the emergence of quantum computing requires a shift towards quantum cryptographic techniques. This progression guarantees that security protocols stay strong against the computing capabilities of future quantum technologies, protecting the integrity and confidentiality of digital communications.

2.1. Key Differences Between Traditional and Quantum Cryptography

Traditional cryptography relies on the computational complexity of mathematical problems to ensure the security of data. Symmetric-key algorithms, such as AES, employ a single key for both encryption and decryption, whereas asymmetric-key methods, like RSA, utilize distinct keys for both operations. These strategies have been fundamental in establishing the core of contemporary digital security, ensuring the protection of delicate information across diverse platforms. Nevertheless, the growing computer capacity and the promise of quantum computing present substantial obstacles to current conventional encryption methods. The security of these approaches relies on the intricate nature of specific mathematical problems, which quantum computers are anticipated to answer with more efficiency, hence compromising the integrity of these systems. Quantum cryptography is a notable change in data security methods that uses the laws of quantum physics to provide a level of security that is theoretically resistant to the computing capabilities of both classical and quantum computers.

The primary distinction between standard and quantum cryptography resides in their core principles and encryption techniques. Quantum cryptography, in contrast to traditional approaches that depend on mathematical intricacy, utilizes the principles of quantum mechanics, offering unique security benefits and difficulties. The number is negative eleven. Quantum Key Distribution (QKD) is a notable implementation of quantum cryptography that utilizes quantum features, such as superposition and entanglement, to ensure the security of key exchange procedures. Any endeavour to covertly listen in on a Quantum Key Distribution (QKD) channel causes changes to the quantum states, rendering it noticeable and so preventing unauthorized data intrusions without detection. This methodology presents a stark contrast to conventional cryptographic methods, which may be susceptible to advancements in processing power. Moreover, the deployment of quantum cryptography systems necessitates surmounting substantial technical and economic obstacles. These requirements encompass the necessity for superior single-photon sources, effective transmission of quantum states across extensive distances, and seamless integration with current classical systems [11]. However, it is crucial to address the issues by creating quantum-resistant cryptographic algorithms and integrating quantum cryptographic approaches. This is necessary to mitigate the risks presented by quantum computing and guarantee the long-term security of digital communications. The emergence of quantum computing requires a shift from conventional to quantum cryptography techniques to ensure strong security measures against the computational capabilities of upcoming quantum technologies, thus protecting the authenticity and secrecy of digital communications.

2.2. The Future of Cryptography

The emergence of quantum computing and the associated progress in quantum cryptography are set to revolutionize the field of information security. The advancement and application of quantum-resistant encryption techniques are crucial for the future of cryptography, as they are specifically designed to withstand the unparalleled computational power of quantum computers. This transition requires a thorough reassessment of existing cryptography procedures and the implementation of innovative approaches that utilize the laws of quantum physics. Quantum Key Distribution (QKD) is a fundamental aspect of the modern era in cryptography. QKD, by harnessing the fundamental principles of quantum mechanics, provides a level of security that is potentially impervious to any classical methods of decryption. The main reason for this is the quantum laws of superposition and entanglement, which guarantee that any attempt to listen in may be identified, thereby maintaining the security of the key exchange procedure. Protocols like BB84 have proven the practical feasibility of QKD, paving the way for the broad implementation of quantum-secure communication channels.

Furthermore, the domain of post-quantum cryptography is undergoing tremendous advancement alongside QKD. This topic of research is dedicated to developing algorithms that maintain their security against both classical and quantum attacks. The research is focused on utilizing structured lattices and hash functions to develop strong encryption techniques that can safeguard data against the computational capability of quantum computers. Adopting these quantum-resistant algorithms is essential to uphold the security of critical communications considering the ongoing advancements in quantum technologies. Integrating quantum cryptography technologies into current network infrastructures poses numerous obstacles. These requirements encompass the necessity for superior single-photon sources, effective long-range transmission of quantum states, and the smooth integration of quantum systems into classical networks. To achieve widespread implementation of quantum cryptography solutions, it is crucial to overcome the technical and economic challenges associated with them. Furthermore, the advancement of hybrid systems that integrate traditional and quantum cryptography methods will be crucial in establishing resilient and flexible security frameworks capable of effectively mitigating existing and emerging threats. The future of cryptography is closely intertwined with the progress made in quantum computing and quantum cryptography. Developing cryptographic algorithms that are resistant to quantum computing and implementing methods for distributing quantum keys are crucial measures for protecting data from the always changing cyber threats. As these technologies develop and advance, they will have a more significant impact on guaranteeing the safety and reliability of digital communications in a world that is influenced by quantum computing.

3. Quantum Key Distribution (QKD): A New Era of Encryption

Quantum Key Distribution (QKD) is considered the forefront of quantum cryptography, signalling a major advancement in the secure distribution of cryptographic keys. This comprehensive investigation will delve into the ways in which quantum mechanics' distinct characteristics are employed by QKD to attain a level of communication security that is theoretically impervious to the vulnerabilities present in conventional cryptographic systems. Robust mechanisms for key exchange that are fundamentally secure against eavesdropping and computational attacks can be achieved by utilizing the principles of quantum superposition and entanglement. The detailed mechanics of Quantum Key Distribution (QKD) will be explored, with a particular emphasis on widely recognized protocols like BB84. The BB84 protocol utilizes the quantum characteristics of photons to securely encode and transmit information. It guarantees that any attempt by an eavesdropper to intercept the information will be detected because it causes disturbance in the quantum states. Quantum superposition is utilized to enable particles to exist in multiple states simultaneously, while quantum entanglement guarantees that particles remain interconnected regardless of the distance between them. This remarkable phenomenon provides a high level of security for key distribution. In addition, this study will analyse potential of Quantum Key Distribution (QKD) to revolutionize secure communication practices is demonstrated, highlighting its capability to offer secure key exchange mechanisms that are immune to both classical and quantum computational attacks. The practical implementation of Quantum Key Distribution (QKD) in various sectors has been found to have the potential to enhance communication security. This technology opens possibilities for more resilient and future-proof cryptographic solutions.

3.1. Exploration of the BB84 Protocol and Security Measures in QKD

The BB84 protocol, formulated by Charles Bennett and Gilles Brassard in 1984, serves as a foundational method for Quantum Key Distribution (QKD). The protocol involves one party generating a random sequence of bits and encoding each bit onto the polarization states of photons using rectilinear or diagonal bases. These photons are then transmitted through a quantum channel to the recipient, who measures the polarization in randomly chosen bases. Subsequently, the two parties publicly compare their bases without disclosing the bits and retain only those where their bases align, thereby creating a sifted key [11]. This process is repeated until a sufficiently long key is generated, ensuring security under the principles of quantum mechanics. Any eavesdropping attempts would alter the quantum states, a change detectable by both parties. To secure the key further, parts of the sifted keys are compared over a classical channel; if a high correlation is observed, it confirms the absence of eavesdropping.

3.2. Practical Implementations and Recent Enhancements of QKD

Practical implementations of Quantum Key Distribution (QKD) have been shown to be effective in enhancing communication security using quantum cryptographic methods. Nevertheless, these implementations encounter challenges such as the presence of noise caused by imperfect detectors and the complexities involved in generating single-photon pulses. To tackle these concerns, researchers have made significant advancements in enhancing the reliability and scalability of QKD systems [4]. Researchers have introduced techniques such as multi-level encoding and the decoy pulse method to mitigate the effects of noise and enhance the overall reliability of QKD protocols. The robustness of the BB84 protocol is enhanced by these modifications, specifically in terms of mitigating imperfections in source preparation and potential side-channel attacks. For example, researchers have proposed the Reference Technique to enhance the security and efficiency of the Quantum Key Distribution (QKD) process. This technique aims to improve the estimation of information leakage and increase the secret-key rate. Real-world applications of Quantum Key Distribution (QKD) encompass a wide range of sectors, highlighting the versatility and effectiveness of this technology. Within the power industry, researchers have developed customized quantum key distribution (QKD) network architectures to specifically tackle the distinct obstacles encountered in power grid environments. These structures are designed to account for communication transmission losses while simultaneously guaranteeing robust security measures [11]. In the field of medical research, advanced protocols such as the EBB84QCP (Enhanced BB84 Quantum Cryptography Protocol) are utilized to ensure the secure transmission of sensitive data across wireless body sensor networks. This protocol provides a reliable and robust solution for maintaining the security of medical data. QKD systems have been adopted by financial institutions to secure critical communication links between front-office and back-office operations, demonstrating the technology's ability to safeguard sensitive financial data. Furthermore, it is worth noting that quantum-secure networks have been implemented in industrial applications to protect manufacturing data. This emphasizes the importance of Quantum Key Distribution (QKD) in improving the security of industrial operations. NVIDIA's exploration of Quantum Key Distribution (QKD) for data centre security highlights the potential of using quantum keys to establish secure communication channels within data centres. This research aims to advance towards a zero-trust security architecture. Overall, the practical applications and recent improvements of QKD demonstrate notable progress in the field of quantum cryptography. The advancements not only tackle the current obstacles, but also broaden the potential use of QKD in different sectors, thereby facilitating the implementation of more secure and robust communication systems in the quantum era.

3.3. Applications of Quantum Cryptography in Various Sectors

Quantum Key Distribution (QKD) has been widely recognized for its value in various sectors. It has been successfully used to secure large-scale sensitive data transfers, including genome sequences. Additionally, QKD has been instrumental in enhancing the security of financial transactions and industrial data transfers [5]. For example, researchers in the power industry have developed a customized Quantum Key Distribution (QKD) network structure to specifically tackle the distinct challenges encountered in power grid environments. This structure effectively deals with communication transmission losses and guarantees strong security [6]. Within the medical field, researchers have utilized the enhanced BB84 protocol (EBB84QCP) to ensure the secure transmission of sensitive data across wireless body sensor networks. The protocol being discussed in this study provides a high level of security that is well-suited for safeguarding sensitive medical information. This, in turn, helps to enhance patient privacy and maintain the integrity of the data [12]. In the financial sector, QKD systems have been implemented to secure critical communication links between front-office and back-office operations. This study highlights the technology's capability to safeguard sensitive financial information against advanced cyber threats, thereby guaranteeing the reliability of financial transactions and communications. QKD has proven to be highly beneficial in industrial applications. One notable example is the use of quantum-secure networks to protect manufacturing data between production facilities. This demonstrates the key role that QKD plays in safeguarding sensitive information in the manufacturing sector. These networks are designed to ensure that sensitive industrial information remains secure during transfer, thus preventing incidents of industrial espionage and data breaches. Furthermore, the investigation conducted by NVIDIA on QKD for data centre security demonstrates the possibility of utilizing quantum keys to establish secure communication channels within data centres. This approach facilitates the advancement of a zero-trust, security-everywhere architecture, which improves the overall security of data centres and guarantees the protection of sensitive data against potential threats posed by quantum computing [5].

4. Comparative Analysis: Quantum vs. Traditional Cryptography

To conduct a comparative analysis between quantum cryptography and traditional cryptography, it is crucial for researchers to have a comprehensive understanding of the fundamental principles underlying each method, as well as the mechanisms they employ to ensure data security. As a researcher, it is important to note that traditional cryptography is based on mathematical algorithms and computational hardness to safeguard information. This is achieved by techniques like symmetric-key encryption, where the same key is used for both encryption and decryption, and asymmetric-key encryption, which involves a pair of public and private keys [5]. These methods have proven to be effective in ensuring security by leveraging the challenge of solving intricate mathematical problems.

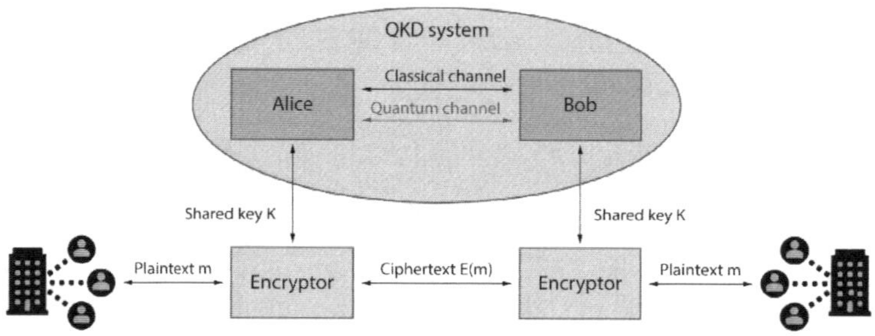

Fig. 1. QKD system

On the other hand, quantum cryptography makes use of the fundamental principles of quantum mechanics, such as superposition and entanglement, to guarantee the security of data. Quantum Key Distribution (QKD), an important application of quantum cryptography, utilizes the properties of quantum particles to identify any potential eavesdropping activities, ensuring the security of the key exchange procedure [11]. Quantum cryptography, in contrast to traditional cryptographic methods, is not susceptible to the increasing computational power of technologies like quantum computing. This is because quantum cryptography relies on the fundamental laws of physics, providing security that is theoretically impossible to break. The transformative potential of quantum cryptography in providing robust and future-proof security solutions is underscored by this distinction.

4.1. Traditional Cryptography: Evolution and Challenges

Traditional cryptography has a long history, dating back to antiquity. One of the earliest known techniques is the Caesar cipher, which was developed around 50 B.C. This field has undergone significant evolution, particularly since the advent of widespread computing in the 1970s. At its core, cryptography is a process that transforms information into ciphertext using intricate algorithms. The researcher explains that symmetric-key algorithms, such as AES (Advanced Encryption Standard), are used in this system. These algorithms use the same key for both encryption and decryption, which helps to maintain data confidentiality if the key is kept secret. Asymmetric-key algorithms, such as RSA (Rivest-Shamir-Adleman), are cryptographic techniques that involve the use of a pair of keys—one public and one private. The public key is used for encryption, while the private key is used for decryption. This allows for secure communication, even when transmitting data over insecure channels [5]. The security of traditional cryptographic methods is heavily dependent on the computational complexity associated with solving specific mathematical problems. For example, RSA encryption relies on the difficulty of factoring large prime numbers, which is currently considered impractical for classical computers due to the time it takes to complete. Nevertheless, the emergence of quantum computing presents a substantial challenge to these conventional approaches. Quantum computers, by utilizing algorithms like Shor's algorithm, possess the capability to efficiently factor large integers. This has the potential to undermine the security of RSA and other cryptographic techniques. The emerging capability of quantum computing requires researchers to reevaluate current cryptographic practices and develop quantum-resistant algorithms. This is crucial to ensure that sensitive information remains protected in the quantum era [5].

4.2. Quantum Cryptography: A New Frontier in Data Security

Quantum cryptography, which was developed as a response to the limitations of conventional cryptography, utilizes principles from quantum mechanics, including the observer effect and the no-cloning theorem, to improve the security of data [1]. The quantum features mentioned in the study offer a high level of security that is theoretically impossible to breach, which is in stark contrast to the vulnerabilities present in traditional methods [7]. Quantum Key Distribution (QKD) is widely recognized as the most prominent application of quantum cryptography. It enables the creation of a shared random secret key between two parties. The method being discussed offers enhanced security in comparison to conventional approaches. This is since any effort to observe or replicate the quantum key would unavoidably modify the system, thereby exposing any attempts at eavesdropping.

QKD protocols, such as BB84, are employed by researchers to leverage the unique quantum properties of photons to achieve secure key exchange. When photons are transmitted, their quantum states are measured by researchers using randomly selected bases. According to research, any attempt by an eavesdropper to intercept the communication would disrupt the states involved, which would then make the intrusion detectable. This helps to ensure the security of the communication process [1]. The fundamental principle of quantum mechanics guarantees that the security of the key is not reliant on computational complexity, as seen in traditional cryptography, but rather on the physical laws that govern quantum particles. In addition, the One-Time Pad (OTP), which is frequently utilized alongside QKD, is widely recognized as the ultimate form of secure cryptography. According to research, the OTP, if implemented correctly, offers encryption that is resistant to attacks, including those from quantum computers. The method described in this study addresses several practical concerns regarding the transmission and management of encryption keys. By using the key only once and then discarding it, the method effectively prevents cryptanalytic attacks and enhances security [7]. By combining Quantum Key Distribution (QKD) with One-Time Pad (OTP), researchers have achieved a level of security that surpasses traditional cryptographic methods. This approach effectively addresses the emerging threats posed by advancements in quantum computing. The advancements in quantum cryptography that have been made highlight the potential for it to revolutionize the field of information security. These advancements offer robust solutions that can be used to protect sensitive data in a world that is becoming more digital and interconnected.

4.3. Distinguishing Factors and the Shift to Quantum Cryptography

The susceptibility of traditional cryptography to the technological advancements of quantum computing is due to its reliance on mathematical complexity. Algorithms like RSA and ECC, which are fundamental to traditional cryptographic techniques, depend on the computational complexity of problems such as integer factorization and elliptic curve discretization. Quantum computing, including algorithms like Shor's, has the potential to solve these problems much faster, which could potentially compromise the security of these systems [5]. On the other hand, quantum cryptography utilizes the principles of quantum mechanics, offering a fundamentally distinct method for ensuring data security. Quantum Key Distribution (QKD) techniques make use of the unique characteristics of quantum particles, such as superposition and entanglement, to provide advanced security capabilities. Quantum Key Distribution (QKD) has the capability to actively detect eavesdropping threats. This is because any attempt to intercept the quantum key will alter its state, thereby revealing the intrusion [11]. The capability to prevent the undetected copying or measurement of quantum states is based on the no-cloning theorem and the observer effect. The transition from traditional to quantum cryptography signifies a profound paradigm shift in the data security landscape. The shift towards quantum computing advancements necessitates significant adaptations in cryptographic practices to effectively mitigate the evolving threats. Quantum cryptography not only mitigates the vulnerabilities inherent in traditional methods but also introduces a level of security based on the immutable laws of physics, offering a theoretically unbreakable encryption solution. This evolution emphasizes the crucial requirement for ongoing research and development in quantum-resistant cryptographic algorithms and the adoption of quantum cryptographic systems to guarantee strong data protection in the quantum era.

5. Challenges in Implementing Quantum Cryptography in Modern Networks

Confronting several significant challenges participates in implementing Quantum Key Distribution (QKD) and other quantum cryptographic techniques within modern networks:

- Infrastructure and Transmission Rates: The management of transmission rates and processing limitations poses a significant challenge in the deployment of quantum cryptography. The requirement for transmitting high-quality single photons over long distances without significant loss presents substantial technical difficulties for researchers. The impact of this necessity on the signal-to-noise ratio and the resulting increase in equipment costs make the implementation of this project both challenging and expensive. The deployment of advanced single-photon detectors and highly stable quantum channels is crucial for maintaining the integrity of the transmitted quantum states.
- Quantum Communication Networks: The establishment of a quantum communication network involves significant expenses, and it is not feasible to modify existing legacy networks to support the required data rates. Developing new networks equipped with advanced single-photon sources is crucial for researchers to facilitate the effective long-distance transmission of quantum states. The integration of quantum repeaters to extend the communication range without significant loss is a critical component in the field of research, although this technology is still in its developmental stages.
- Transferring Quantum States and Protocol Development: The efficient transfer of quantum states, especially at rates higher than 1 Gbit/s, continues to be a significant technical hurdle for researchers. The current technologies, such as Alexandrite lasers, exhibit limitations in terms of intensity and quantum yield. Ionization methods demonstrate potential for facilitating higher data rates; however, additional refinement is necessary for them to become feasible for widespread utilization. Furthermore, it is imperative to advance the development of novel cryptographic protocols that outperform current ones. This will involve introducing additional layers of complexity that are crucial for driving technological advancement. The protocols must ensure robust security while being adaptable to various network conditions and requirements.
- Public Trust and Shared Infrastructure: Establishing public trust in quantum cryptography, particularly in sectors such as government and finance, is crucial for its wider implementation. It is crucial for widespread implementation that

the public perceives quantum cryptographic methods as reliable and secure. Furthermore, it is crucial to address concerns regarding shared infrastructure, specifically the development of network topologies that can facilitate secure and accessible quantum communication for multiple users. The creation of quantum network architectures that can seamlessly integrate with existing classical networks and provide high levels of security and efficiency is a crucial aspect of this research.

Quantum Key Distribution (QKD) and other quantum cryptographic methods pose several challenges when it comes to integrating them into modern networks. These challenges include the expensive nature and technical requirements of setting up quantum communication infrastructures, as well as the need to develop efficient protocols and build public trust in these technologies. Overcoming these challenges is of utmost importance for the successful integration and widespread adoption of quantum cryptography in enhancing network security. The ongoing research and development in quantum technologies is crucial for addressing these issues and ensuring the feasibility and effectiveness of the transition to quantum-secure communications.

6. The Future of Network Security with Quantum Cryptography

Research suggests that the future of network security is closely tied to the rapid progress being made in the field of quantum cryptography. The field of research in security is constantly evolving and has the potential to provide new and creative solutions to both long-standing and emerging security challenges. Quantum cryptography, based on the principles of quantum mechanics, provides advanced security capabilities that exceed those of traditional cryptographic methods. The continuous development in quantum cryptographic protocols, such as Quantum Key Distribution (QKD), and the integration of these protocols into existing network infrastructures, signal a significant shift in how data security will be managed in the digital era.

6.1. Quantum Computing and Network Security

Quantum computing is considered a potential threat to traditional cryptographic algorithms because of its superior computational power, which enables it to solve complex calculations quickly and efficiently, including the mathematical operations that form the basis of these algorithms. Quantum algorithms, like Shor's algorithm, have the capability to factor large integers at an exponentially faster rate compared to classical algorithms. This poses a significant threat to the security of commonly used encryption techniques such as RSA and ECC. Nevertheless, the implementation of quantum computers that can efficiently conduct these tasks on a large scale is hindered by various technical obstacles. At present, quantum computers face limitations due to a restricted number of qubits and encounter substantial challenges concerning error correction and the precise manipulation of multiple qubits within the required time limits. Despite these challenges, the rapid progress in quantum computing highlights the need for researchers to take proactive steps in developing network security systems that can withstand quantum attacks. The potential for quantum computers to break traditional encryption methods highlights the urgent need for cryptographic techniques that can withstand quantum attacks. Quantum Key Distribution (QKD) is a research area that explores the potential of utilizing the principles of quantum mechanics to develop secure key exchange mechanisms. Nevertheless, the improvement of the resilience of QKD systems continues to be a crucial field of study. Fractal geometry is being explored as an innovative approach to enhance the security of Quantum Key Distribution (QKD) systems. Fractal geometry is a promising approach that researchers are exploring to improve the cryptographic strength by leveraging its intricate structural patterns. By utilizing fractals for data encoding, the cryptographic procedure becomes more difficult to decrypt without the precise generative parameters, thereby enhancing the level of security [9]. This method not only extends the cryptographic strength but also introduces new dimensions of complexity that are difficult for adversaries to navigate. Moreover, the burgeoning area of post-quantum cryptography indicates that there is still a pressing requirement to create encryption methods that can withstand quantum attacks, even with the progress made in quantum technologies. Post-quantum cryptography is a field of study that focuses on developing algorithms that can withstand attacks from both classical and quantum computers. The goal is to ensure compatibility and security in a changing technological environment [10]. As a researcher, I would like to mention that there are several techniques that have been developed to provide robust security against quantum advancements. These techniques include lattice-based cryptography, hash-based cryptography, and multivariate polynomial cryptography. Although quantum computing poses considerable challenges to conventional cryptographic methods, it is crucial to develop security systems that are resistant to quantum attacks and can take advantage of quantum enhancements. Research has shown that innovations like fractal-based cryptography and post-quantum cryptographic algorithms have the potential to address the growing concern of data security in the face of advancing quantum computing capabilities.

6.2. Quantum-Resistant Cryptography

As a researcher, it is important to acknowledge the potential risks that quantum computing poses. Therefore, there is a considerable emphasis on the development of quantum-resistant cryptography. The U.S. National Institute of Standards and Technology (NIST) has played a leading role in this initiative by developing a set of cryptographic algorithms that are specifically designed to resist the computational power of future quantum computers. The algorithms, which employ structured lattices and hash functions, have been designed to be resistant to quantum attacks [2].

The structured lattice approach, for instance, is based on the intricate nature of lattice problems, which are widely considered to be challenging for quantum computers to solve. Similarly, hash-based cryptographic techniques are based on the security of hash functions, which are known to be secure under both classical and quantum paradigms. Transitioning to these quantum-resistant algorithms is crucial for researchers to safeguard against the advancing capabilities of quantum computing. The transition discussed in this context encompasses not only the adoption of novel algorithms but also the incorporation of these techniques into pre-existing security infrastructures, guaranteeing a smooth and protected transition. Research suggests that the adoption of post-quantum cryptography is crucial for ensuring data security in the future. Post-quantum cryptography refers to a range of cryptographic techniques that are specifically designed to withstand threats from both traditional and quantum computing. By implementing these approaches, organizations can enhance their data security measures and protect sensitive information from potential attacks. Considering the potential of quantum computing, there is an increasing focus on the development of quantum-resistant cryptography to protect sensitive communications from quantum threats. This research initiative is of utmost importance, considering that traditional encryption techniques might become insufficient soon due to the advancements in quantum computing. The efforts by researchers and institutions like NIST to standardize quantum-resistant algorithms underscore the importance of proactive measures in cryptographic research and development. These efforts are aimed at ensuring that as quantum computing continues to advance, our cryptographic systems can withstand any potential threats, thus safeguarding the integrity and confidentiality of sensitive information. The proactive approach to developing and standardizing quantum-resistant cryptography is crucial for the future of cybersecurity. By preparing for the potential impact of quantum computing now, researchers can ensure that communication systems, financial transactions, and sensitive data remain secure against the sophisticated threats posed by quantum technologies.

6.3. Bridging Quantum and Classical Technologies

The trajectory of network security is expected to adopt a hybrid approach, combining classical and quantum technologies. Researchers are actively exploring how quantum and classical systems can interoperate seamlessly, including the development of hybrid quantum networks [3]. The integration of advanced quantum-resistant algorithms into existing network infrastructures is crucial [1]. Researchers are currently exploring the use of techniques such as photonic integrated circuits and leveraging existing fiber and optical infrastructures to develop scalable quantum solutions. Photonic integrated circuits allow for the integration of optical components onto a single chip, which makes it easier to incorporate quantum technologies into existing network frameworks. Moreover, the utilization of pre-existing fiber optics infrastructure presents a viable and economical approach to deploying quantum networks, eliminating the necessity for creating entirely new physical systems. The objective of this research is to develop quantum networks that could support various applications, including advanced sensing mechanisms that can be used to tackle global issues like climate change. Quantum-enhanced networks have the potential to enhance the precision and efficiency of environmental monitoring systems. This could lead to improved data collection and analysis capabilities, resulting in more accurate information. By combining the strengths of classical and quantum technologies, researchers are exploring the potential of hybrid networks to deliver enhanced security and operational efficiency. These networks have the potential to pave the way for a new era of secure and resilient communications.

7. Future research

Future research in the domain of quantum cryptography should prioritize several key areas to effectively tackle current challenges and take advantage of emerging opportunities. One critical area that needs to be researched is the feasibility of applying quantum cryptography to enhance user-oriented and site-to-site VPN connection security. Considering the limited implementation of these technologies in real-world settings and the scarcity of documented attacks targeting them, this field offers a promising potential for making significant progress in enhancing security in corporate communications. Researching the potential methods for overcoming the current technical limitations of quantum cryptographic systems, such as error correction and qubit control, is essential. Research has shown that the current methods used to attack quantum cryptographic systems primarily focus on exploiting implementation flaws rather than directly breaking the quantum key distribution itself. As quantum cryptography evolves, it is important to consider the potential increase in sophistication and frequency of attacks. This is a critical point that needs to be addressed in future research. Furthermore, it is important for researchers to investigate the potential integration of quantum cryptographic techniques with current classical cryptographic systems to create strong hybrid models. The integration of this technology has the potential to improve the existing security framework by adding an extra layer of protection against both traditional and quantum computational threats. Furthermore, it is imperative to develop new cryptographic protocols that surpass the capabilities of existing ones. This will involve introducing additional layers of complexity that are crucial for advancing technology. Moreover, it is crucial to establish public trust in quantum cryptography, particularly in sectors such as government and finance, to facilitate wider adoption. Researchers should also address shared infrastructure issues, such as developing network topologies that allow secure and accessible quantum communication for multiple users. Our research focuses on developing quantum network architectures that can seamlessly integrate with existing classical networks and provide high levels of security and efficiency. Future research should focus on addressing the technical and practical obstacles related to quantum cryptography, exploring innovative and reliable cryptographic protocols, and promoting public confidence to encourage widespread acceptance. These efforts are crucial for advancing the field and ensuring that network security remains resilient against evolving cyber threats facilitated by quantum computing advancements.

8. Conclusion

The landscape of network security is undergoing a transformative shift with the introduction and integration of quantum cryptography. As discussed in this paper, advancements in quantum computing present significant challenges and opportunities for enhancing security protocols. Traditional cryptographic methods, which have served as the cornerstone of digital security, are increasingly vulnerable to the computational power of quantum computers. In response, the development of quantum-resistant algorithms and the adoption of Quantum Key Distribution (QKD) are critical measures to protect against potential quantum threats. Quantum cryptography, particularly through applications like QKD, leverages the inherent properties of quantum mechanics to provide a new level of security that traditional methods cannot match. This includes utilizing quantum superposition and entanglement to create encryption that is theoretically unbreakable by conventional means. However, integrating this technology into existing systems presents substantial challenges, ranging from technological adaptations to the need for new infrastructure and increased public trust. The seamless fusion of quantum and classical cryptographic technologies is essential. As advancements continue, developing hybrid systems that incorporate robust quantum-resistant protocols will play a crucial role in safeguarding sensitive data against current and future threats. This synergy between new and traditional security methodologies will define the next era of network security, ensuring resilience against an evolving array of cyber threats facilitated by quantum computing advancements. Continuous research and development in quantum cryptography and related fields are vital to maintaining the integrity and security of digital communications in a world increasingly influenced by quantum technologies. As quantum computing capabilities grow, proactive measures and innovative approaches in cryptographic practices will be paramount in protecting sensitive information and securing network infrastructures.

9. References

[1] Tripathi, T.; Awasthi, A.; Singh, S. P. & Chaturvedi, A. (2024). Post Quantum Cryptography & its Comparison with Classical Cryptography, Available from: https://arxiv.org/pdf/2403.19299.pdf Accessed: 2024-05-09, in press.

[2] Morić, Z.; Pinjuh, J. & Kurelic, S. (2021). Securing Communication in Chatter Application, Proceedings of the 32nd DAAAM International Symposium, Published by DAAAM International, ISBN 978-3-902734-33-4, Vienna, Austria. DOI: 10.2507/32nd.daaam.proceedings.013, in press.

[3] Morić, Z.; Branstett, L. & Petrunic, R. (2022). Static-Analysis Techniques of Malware Reverse Engineering, Proceedings of the 33rd DAAAM International Symposium, Published by DAAAM International, ISBN 978-3-902734-36-5, ISSN 1726-9679, Vienna, Austria. DOI: 10.2507/33rd.daaam.proceedings.024, in press.

[4] Pillai, S. V. E. S. & Polimetla, K. (2024). Analyzing the Impact of Quantum Cryptography on Network Security, Proceedings of the 2024 International Conference on Integrated Circuits and Communication Systems (ICICACS), Published by IEEE, ISBN 979-8-3503-1755-8, North Dakota, USA. DOI: 10.1109/ICICACS60521.2024.10498417, in press.

[5] Benaka Santhosha, S.; Sudheesh, K. V.; Kiran & Parameshachari, B. D. (2024). A Comprehensive Study on Various Encryption Techniques for Digital Images in Multimedia Networks, Proceedings of the 2024 International Conference on Integrated Circuits and Communication Systems (ICICACS), Published by IEEE, ISBN 979-8-3503-1755-8, Belagavi, India. DOI: 10.1109/ICICACS60521.2024.10498590, in press.

[6] Mehmood, A.; Shafique, A.; Alawida, M. & Khan, A. N. (2024). Advances and Vulnerabilities in Modern Cryptographic Techniques: A Comprehensive Survey on Cybersecurity in the Domain of Machine, Deep Learning, and Quantum Techniques, Published by IEEE, Abu Dhabi, United Arab Emirates. DOI: 10.1109/ACCESS.2024.3367232, in press.

[7] Jasoliya, H. & Shah, K. (2022). An Exploration to the Quantum Cryptography Technology, Proceedings of the 2022 9th International Conference on Computing for Sustainable Global Development (INDIACom), Published by IEEE, ISBN 978-93-80544-44-1, New Delhi, India. DOI: 10.23919/INDIACom54597.2022.9763109, in press.

[8] Lella, E.; Gatto, A.; Pazienza, A.; Romano, D.; Noviello, P.; Vitulano, F. & Schmid, G. (2022). Cryptography in the Quantum Era, Proceedings of the 2022 IEEE 15th Workshop on Low Temperature Electronics (WOLTE), Published by IEEE, ISBN 978-1-6654-8062-8, Molfetta, Italy. DOI: 0.1109/WOLTE55422.2022.9882585, in press.

[9] Blahová, M.; Mikuličová, M. & Hromada, M. (2020). Utilization of Fractal Geometry Possibilities for Information Systems Security, Proceedings of the 31st DAAAM International Symposium, Published by DAAAM International, ISBN 978-3-902734-29-7, ISSN 1726-9679, Vienna, Austria. DOI: 10.2507/31st.daaam.proceedings.085, in press.

[10] Giroti, I. & Malhotra, M. (2022). Quantum Cryptography: A Pathway to Secure Communication, Proceedings of the 2022 6th International Conference on Computation System and Information Technology for Sustainable Solutions (CSITSS), Published by IEEE, ISBN 978-1-6654-5699-9, Bengaluru, India. DOI: 10.1109/CSITSS57437.2022.10026388, in press.

[11] Shah, N.; Shah, S. & Hargeet, K. (2024). Exploring the Quantum Potential of a 5-Quibit Highly Entangled Quantum State, Proceedings of the 2024 11th International Conference on Computing for Sustainable Global Development (INDIACom), Published by IEEE, ISBN 978-93-80544-51-9, New Delhi, India. DOI: 10.23919/INDIACom61295.2024.10498356, in press.

[12] Anusuya Devi, V. & Kalaivani, V. (2021). Enhanced BB84 quantum cryptography protocol for secure communication in wireless body sensor networks for medical applications, Published by Pers Ubiquitous Comput, Tamilnadu, India. DOI: 10.1007/s00779-021-01546-z, in press.

DOI: 10.2507/35th.daaam.proceedings.009

MACHINE LEARNING FOR SOFTWARE SECURITY TESTING

Zlatan Moric*, Jelena Halic, Damir Regvart & Robert Kopal

This Publication has to be referred as: Moric, Z[latan]; Halic, J[elena]; Regvart, D[amir] & Kopal, R[obert] (2024). Machine Learning for Software Security Testing, Proceedings of the 35th DAAAM International Symposium, pp.0065-0070, B. Katalinic (Ed.), Published by DAAAM International, ISBN 978-3-902734-44-0, ISSN 1726-9679, Vienna, Austria
DOI: 10.2507/35th.daaam.proceedings.009

Abstract

The incorporation of machine learning (ML) into software security testing has become a crucial advancement due to the increasing prevalence of sophisticated cyber-attacks. This integration offers substantial improvements compared to conventional security methods. This article provides a thorough analysis of machine learning methods used to enhance the detection, classification, and management of software vulnerabilities. We explore diverse machine learning techniques and assess their efficacy in automating and improving security operations in various cybersecurity scenarios. The research focuses on analysing the achievements and difficulties faced while training machine learning models for tasks like identifying vulnerabilities and optimizing security controls. This analysis is based on empirical data obtained from current case studies. In addition, we assess crucial areas that require enhancement and suggest strategic approaches for future study with the goal of maximizing the influence of machine learning on software security. This work contributes by doing a thorough investigation of the role of machine learning (ML) in cybersecurity. It presents a convincing argument for the incorporation of ML into security measures to improve their effectiveness.

Keywords: machine learning; software security; software security testing; software vulnerabilities; automated testing.

1. Introduction

In today's digital environment, the widespread presence of advanced cyber threats presents substantial dangers to the security and reliability of software programs and systems. Due to the growing intricacy and frequency of these dangers, it is necessary to employ sophisticated approaches to uncover and reduce software vulnerabilities. Conventional approaches for evaluating software security, however successful, frequently have difficulties in keeping up with the swiftly changing landscape of cyber threats. Therefore, it is increasingly necessary to incorporate innovative technology, such as machine learning (ML), into security testing frameworks to improve their effectiveness and reactivity. Machine learning revolutionizes software security testing by replacing traditional rule-based methods with intelligent, adaptive, and data-driven solutions. Machine learning algorithms could scan large volumes of data to uncover patterns and abnormalities that could signal vulnerabilities. This allows for a more effective and efficient way of detecting and correcting security problems. This study investigates the crucial role of machine learning in enhancing software security testing. It examines different ML techniques and their potential to enhance the identification, classification, and handling of software vulnerabilities. The utilization of machine learning in software security signifies a fundamental change in the way vulnerabilities are identified and controlled. ML algorithms could learn from historical data and adjust to new and emerging threats, unlike older approaches that mainly depend on predetermined rules and signatures.

The ability to adapt is essential in a dynamic environment where cyber threats are always changing and becoming increasingly sophisticated. Machine learning models can improve various areas of software security, such as identifying vulnerabilities, optimizing security controls, and analysing malware. ML algorithms can be trained to identify the attributes of susceptible code, allowing for early identification and prevention of possible security problems before they can be taken advantage of. Furthermore, machine learning has the capability to automate the process of choosing security controls, hence enhancing the efficiency and efficacy of security assessments [3]. To fully harness the potential of machine learning in improving software security, it is necessary to overcome many difficulties. An important obstacle is the interpretability of machine learning models. Several machine learning algorithms function as opaque entities, posing challenges for security analysts to comprehend and have confidence in their determinations. Enhancing the openness of these models is crucial to ensure their seamless integration into security workflows and to simplify the process of identifying and rectifying instances of false positives and negatives [12]. Another significant obstacle is the susceptibility of machine learning models to adversarial attacks, in which malicious actors modify inputs to mislead the model into producing inaccurate predictions. Continuing research is focused on creating strong defences against these threats. Furthermore, the ever-changing nature of cyber threats requires constant assessment and modification of machine learning models to ensure their ongoing effectiveness [13].

The incorporation of machine learning into software security testing provides notable benefits in improving the identification and control of software vulnerabilities. By utilizing the flexible and data-oriented capabilities of machine learning (ML), enterprises can enhance their ability to accurately detect and react to the intricate and ever-changing characteristics of cyber threats. Further investigation and innovation are imperative to surmount the obstacles linked to machine learning (ML) applications in the field of cybersecurity, guaranteeing that these technologies are both efficient and dependable in safeguarding digital systems.

2. Software Vulnerabilities: Threats to Digital Security

Software vulnerabilities, which are typically unforeseen defects or flaws in software code, provide chances for malevolent entities to exploit and compromise the confidentiality, integrity, and availability of sensitive data. Comprehending these weaknesses is crucial for developers, security experts, and companies dedicated to enhancing their digital safeguards. Inadequate input validation is a common source of these vulnerabilities, and therefore is an essential aspect of secure coding methods. Insufficient validation and cleansing of user inputs can lead to SQL injection attacks. These attacks involve the insertion of malicious SQL code into input fields, which can result in unauthorized access to databases, alteration of data, and execution of arbitrary instructions [2]. Authentication and authorization issues present substantial dangers. Insufficiently robust authentication mechanisms and poor authorization checks can lead to unlawful access. Vulnerabilities in these domains might empower malicious actors to obtain unauthorized privileges, retrieve sensitive information, or impersonate genuine users. Security misconfiguration, such as insufficient security configurations, default credentials, or unprotected ports, can provide attackers with unauthorized entry points or methods to increase privileges and exploit additional vulnerabilities in the system [1].

Buffer overflow issues pose significant threats to program security. Buffer overflows happen when a program exceeds the capacity of a buffer and writes more data than it can accommodate, leading to the overwriting of neighbouring memory regions. These vulnerabilities have the potential to result in the execution of arbitrary code, crashes of the program, or manipulation of the application's behaviour. This highlights the importance of careful memory management techniques [2]. Web applications are particularly vulnerable to Cross-Site Request Forgery (CSRF) and Cross-Site Scripting (XSS) vulnerabilities, which are highly worrying. CSRF manipulates a user's browser to do undesired actions on an online application, whereas XSS enables attackers to insert harmful scripts into web sites. Both vulnerabilities can lead to illegal actions, compromise user credentials, and create additional opportunities for exploitation. SQL injection poses a significant threat in this scenario, emphasizing the crucial importance of input validation and sanitation in avoiding the insertion of malicious code into database queries. This is essential for stopping unauthorized access to databases, tampering with data, and compromising sensitive information [1].

These examples demonstrate the range of software security risks, each necessitating comprehensive security evaluations and rigorous validation through testing and monitoring. Software security testing is a complex process that uses a thorough strategy to assess and strengthen the security of software applications. The complexity arises from a range of attack routes, constantly changing software environments, emerging threats, diversified software architectures, regulatory requirements, integration of other components, and the need to balance security with functionality. To address these difficulties, businesses adopt a comprehensive approach that includes static and dynamic analysis, penetration testing, code reviews, security scans, and other measures. Both manual and automated testing methods are employed, with utilization of machine learning becoming increasingly important in improving efficiency and scope of testing [2].

3. Machine Learning Algorithms in Software Security: Use Cases

Machine learning (ML) is a revolutionary method in software security that has made great progress in detecting, classifying, and handling software vulnerabilities. This section examines different scenarios in which machine learning algorithms have been successfully utilized to improve the security of software.

3.1. Optimizing Security Controls

An important use of machine learning in software security is the enhancement of security mechanisms through optimization. A study created an automated decision support system that uses machine learning. It uses historical data from past security evaluations in the banking industry to suggest appropriate security measures for new systems. The validity of this methodology was confirmed through empirical analysis utilizing historical data, resulting in an average recall rate of roughly 94% and a precision rate of around 63%. The results suggest that the system successfully finds the most pertinent security measures while also ensuring that the effort required to validate recommendations remains reasonable. This machine learning methodology greatly assists security analysts by simplifying the process of identifying essential security measures and reducing the likelihood of overlooking crucial measures [3].

The capacity of machine learning to examine vast datasets and identify patterns renders it highly ideal for enhancing security measures. Automating the selection process minimizes the effort on security analysts and improves the overall efficacy of the deployed security controls. This strategy enhances both efficiency and security by ensuring that security measures are current and capable of mitigating the most recent threats. Moreover, the incorporation of machine learning in the optimization of security controls enables ongoing enhancement of the security position. As further data is collected, the machine learning models may be updated to accurately represent the changing threat environment, guaranteeing that the security recommendations stay up-to-date and efficient as time progresses. The capacity to adapt quickly is essential for maintaining strong security measures in response to rapidly evolving cyber threats.

3.2. Software Vulnerability Detection

Machine learning (ML) has been widely used to discover software vulnerabilities, highlighting substantial enhancements in identifying potential security problems. In 2018, a study shown that a machine learning classifier could effectively detect a significant amount of software vulnerabilities by evaluating text attributes taken from C source code functions. This study shown that less intricate characteristics often achieved better results than more intricate ones, attaining an accuracy rate of 74% in comparison to 69% [4].

A 2020 study explored the integration of machine learning, static analysis tools, and software metrics to improve the prediction of vulnerabilities in long-term software projects. This methodology employed many machine learning techniques to forecast the existence and classification of vulnerabilities, elucidating the compromises between precision and recall in vulnerability identification [5]. A distinct study conducted in 2020 concentrated on the timely identification of susceptible code by utilizing machine learning techniques applied to software metrics collected from prominent C/C++ projects like Mozilla Firefox and the Linux Kernel. This study examined different machine learning models for predicting vulnerabilities and emphasized the difficulties in producing a well-balanced and precisely labelled dataset. The challenges of establishing a correlation between software metrics and vulnerabilities were also discussed. The study suggested improving detection models by incorporating supplementary indicators, such as code smells and static code analysis alerts. It also advised refining the models continuously based on developer feedback to boost predictions and minimize false alarms [6]. Additional progress in this domain encompasses the implementation of AIBugHunter in 2022. AIBugHunter is an integrated tool in the Visual Studio Code IDE that employs machine learning to promptly assess, categorize, and rectify software vulnerabilities in C/C++ code. This tool improves the programming process by promptly identifying vulnerabilities, providing explanations, and suggesting remedies [7].

These studies and tools demonstrate the substantial capacity of machine learning to enhance the detection of software vulnerabilities. By harnessing the power of machine learning, the precision and speed of detecting software vulnerabilities are significantly improved, offering strong answers to the constantly changing field of cybersecurity risks.

3.3. Detecting Gaps in Web Application Firewalls

The utilization of machine learning and evolutionary algorithms has played a crucial role in the detection of vulnerabilities in Web Application Firewalls (WAFs), specifically those that could potentially enable SQL injection attacks to evade security safeguards. This study centres on utilizing machine learning to independently produce and develop attack vectors, effectively detecting patterns that can circumvent Web Application Firewalls (WAFs). The results illustrate the effectiveness of this approach in uncovering and taking advantage of WAF weaknesses. By utilizing machine learning, this approach enables ongoing and flexible testing of WAFs against various attack vectors, adjusting as new threats arise. This capacity is essential for preserving strong defence mechanisms in an environment where cyber threats are always advancing. The study utilizes advanced algorithms that imitate the strategies of criminal individuals, therefore revealing possible vulnerabilities in WAFs before they may be exploited in real-life situations [8]. Moreover, the utilization of evolutionary algorithms in combination with machine learning offers a more thorough evaluation of WAF security. It improves traditional static testing methods by incorporating adaptive, data-driven strategies that boost the precision and effectiveness of identifying security vulnerabilities. This novel methodology guarantees that Web Application Firewalls (WAFs) continue to be efficient in countering the most recent and intricate cyber hazards, hence offering an elevated degree of security confidence.

3.4. Malware Detection and Analysis

Malware poses a severe hazard in the digital domain, and over time, strategies to reduce malware risks have greatly advanced to improve digital security. At first, traditional techniques that depended on characteristics like as signatures and heuristics were used for identifying malware. Nevertheless, these conventional methods were insufficient in effectively combating emerging strains of malware that possess advanced obfuscation techniques. Deep Learning (DL) has become widely used in malware detection due to its greater effectiveness in identifying new malware variants compared to traditional methods. Furthermore, techniques based on deep learning expedite the prediction of malware with rapidity and accuracy, while also allowing for the investigation of many forms of malware [9]. Machine learning in malware detection aims to minimize the need for manual effort in classifying malware, focusing on specific and specialized areas. Machine learning algorithms are highly proficient at identifying patterns and similarities in data. However, human interaction and experience are still necessary to create datasets in a format that optimally supports the learning and classification tasks performed by these algorithms. Deep Learning has become a potent tool in identifying malware, surpassing conventional methods by acquiring the ability to detect new types of malwares through the recognition of generalized patterns. Furthermore, recent studies have emphasized the notable progress in automated machine learning (AutoML) for the identification of malware using deep learning techniques. AutoML techniques have been utilized to improve the effectiveness and precision of malware detection systems by automating the process of selecting the most optimal models and hyperparameters. This automation not only simplifies the process of establishing deep learning-based malware detection systems but also enhances their effectiveness and ability to adapt to new variations of malware [10]. The incorporation of machine learning, namely deep learning, in the identification and examination of malicious software has brought about a significant transformation in the domain. These sophisticated techniques provide exceptional performance in detecting and reducing malware threats, therefore greatly improving digital security measures.

3.5. Software Bug Classification

An excellent example of utilizing deep learning in software security can be seen in bug classification. Inaugurated in 2024, this groundbreaking approach entails the establishment of a bug classification system with eight discrete categories, each delineated by precise keywords. Using the developed taxonomy, a significant dataset of 1.36 million software bug resolution reports is annotated using an earth-mover distance technique, following this classification framework. Following that, our methodology utilizes four advanced attention-based deep-learning classification techniques to analyse the annotated bugs. The usefulness of this strategy is demonstrated through empirical testing on a carefully curated dataset. The results demonstrate a mean F1-Score of 84.78%, demonstrating a balanced combination of precision and recall. Additionally, there is a mean macro-average ROC of 98.25%, suggesting strong overall performance in bug classification [11]. This research is leading the way in the growing interest in utilizing deep learning techniques to categorize bugs based on resolution reports. While machine learning methods have been used before in this field, this study highlights the improved performance capabilities of deep learning techniques. The objective is to utilize these sophisticated deep-learning techniques to greatly enhance the effectiveness of bug classification. To summarize, this paper introduces a promising and new method for categorizing bugs, highlighting the substantial potential of deep learning in this domain. The use of advanced deep learning techniques has the potential to improve the effectiveness and reliability of software bug classification, leading to more secure software solutions.

4. Challenges in Machine Learning Models for Cybersecurity

The description of machine learning (ML) models poses a substantial obstacle in the field of cybersecurity, where the clarity of algorithmic decision-making is important. A considerable number of machine learning models function as "black boxes," employing intricate and frequently obscure procedures to handle inputs and generate outputs. The absence of transparency is particularly troublesome in security situations, when comprehending the rationale behind model decisions is essential, particularly when addressing false alarms. Cybersecurity analysts frequently choose interpretable and comprehensible systems, such as rule-based and signature-based systems. These models provide more detailed understanding of their operations, facilitating the tuning and optimization process to decrease the occurrence of false positives and negatives. On the other hand, neural-based approaches, which lack transparency, provide challenges in adjusting and reducing risks because of their unclear characteristics. The opacity of the models' operations presents challenges in monitoring, assessing, and correcting them, as the underlying rationales for actions are not visible. The multifunctionality and complexity of current digital systems greatly increase their susceptibility, highlighting the importance of openness in efficiently managing and reducing potential security risks. The need for interpretability is not solely a matter of preference, but rather crucial for the efficient administration and dependability of security systems. Analysts must possess a comprehensive understanding of the underlying cause that led to the activation of an alarm to effectively mitigate prospective dangers and prevent the possibility of disregarding or misinterpreting any malevolent behaviours [12]. Furthermore, machine learning models are susceptible to adversarial attacks when malicious actors deliberately modify input data to induce the model to malfunction. The intrinsic intricacies of these systems contribute to their vulnerability to sophisticated adversarial attacks, necessitating the development of effective defence mechanisms as a crucial topic of continuing research.

This manipulation has the potential to cause ML models to misclassify or provide inaccurate outputs, which poses a substantial risk in security testing. These flaws might be exploited to bypass detection or deceive security measures. Enhancing the resilience of ML models against such assaults requires the creation of strong algorithmic defences, careful design, and ongoing vigilance to ensure their ability to survive sophisticated manipulation attempts. Another persistent obstacle is the requirement for continuous assessment and verification of ML models to uphold their efficacy against evolving security risks. This necessitates the incorporation of systems that enable consistent monitoring, evaluation, and adjustment of machine learning models over a period. Implementing these procedures guarantees the continued efficacy of the models in the face of emerging dangers and the ever-changing digital environment. It is essential to establish precise and measurable measures for evaluating the efficiency of machine learning models in the context of software security testing. These indicators are essential for objectively evaluating model performance, making informed decisions, and consistently enhancing the security posture of software systems. Measuring the true impact and effectiveness of ML models is difficult without using measurable measures in a thorough and methodical manner.

5. Future Works

Addressing the challenges identified in the application of machine learning (ML) for cybersecurity requires a comprehensive approach focusing on several critical areas of development. Future research should prioritize enhancing the interpretability of ML models. This involves not only improving the performance of these models but also making their decision-making processes transparent. Techniques such as feature importance scoring, which highlights the variables most influential in model predictions, model visualization, which provides graphical representations of model functions, and the development of explainable artificial intelligence (XAI) methods, are crucial. These approaches can help demystify the inner workings of complex models like neural networks, making them more accessible and understandable to cybersecurity analysts who rely on them to make critical security decisions.

Another vital area of future work is the development of robust defences against adversarial attacks. ML models in cybersecurity settings are particularly vulnerable to such attacks, where malicious inputs designed to deceive models can lead to incorrect outputs. Research into developing new defensive algorithms that can detect and neutralize these manipulation attempts is crucial. Incorporating adversarial training, a technique in which models learn from adversarial examples during training, can enhance their resilience by preparing them to recognize and resist these attacks better [12]. Additionally, establishing standardized frameworks for the continuous evaluation and validation of ML models is highly important. These frameworks should facilitate regular updates based on the latest threat intelligence and include adaptive mechanisms that allow models to evolve in response to emerging cyber threats. Continuous evaluation ensures that models remain effective as the landscape of threats changes and as new vulnerabilities are discovered. Moreover, there is a pressing need to develop standardized, quantifiable metrics for assessing the effectiveness of ML models in cybersecurity applications. These metrics should comprehensively measure various aspects of model performance, including accuracy, precision, recall, and robustness against attacks. Such metrics provide a basis for objectively evaluating a model's impact and effectiveness, enabling informed decisions about deploying and refining ML solutions in security settings [13].

Collaboration across fields such as data science, cybersecurity, psychology, and ethics can yield innovative solutions that are technically effective, user-friendly, and ethically sound. This multidisciplinary approach can help tackle the complex challenges of ML and cybersecurity, leading to the development of robust, understandable, and morally responsible ML solutions.

6. Conclusion

This paper explores the significant advancements and applications of machine learning (ML) in software security testing, highlighting a pivotal shift towards more sophisticated, adaptable, and data-centric methods to enhance vulnerability detection and security protocols. A comprehensive review of various ML algorithms has provided insights into their implementation across different cybersecurity tasks, such as optimizing security controls, detecting vulnerabilities, and analysing malware, demonstrating the depth and utility of these technologies in contemporary security practices. A primary takeaway from this discussion is ML's indispensable role in addressing the deficiencies of conventional security testing methods, which often fall short of matching the complexity and evolving nature of cyber threats. By integrating ML, organizations can improve their precision in identifying threats more efficiently and on a larger scale than traditional methods allow. For example, the application of ML in bug detection showcases its ability to effectively pinpoint security flaws with considerable accuracy, thereby hardening the security frameworks of software systems.

However, this paper also highlights several challenges accompanying ML use in cybersecurity, such as the difficulty in interpreting ML models and their susceptibility to adversarial attacks. These issues underscore the critical need for continued research focused on improving the transparency and security of ML models to ensure they are both understandable and resilient against targeted attacks. Moving forward, the paper outlines essential steps for overcoming these challenges, stressing the importance of developing interpretable ML models that provide clear and actionable insights into their operations and enhancing the defence mechanisms of these models against adversarial threats. It also advocates for creating robust frameworks for continuously evaluating and validating ML models to maintain their efficacy as cyber threats evolve and new security practices emerge.

While adopting machine learning in software security testing offers substantial benefits in enhanced threat detection and system security, it also necessitates a thoughtful approach to tackle the associated challenges. The success of ML in enhancing cybersecurity will hinge on both technological innovations and the strategic deployment of these technologies within comprehensive and evolving security infrastructures.

7. References

[1] https://signmycode.com/blog/top-software-vulnerabilities-of-2022-and-how-to-prevent-them, (2022). Top Software Vulnerabilities of 2022-23 and How to Prevent Them?, SignMyCode, Accessed on: 2024-01-15

[2] Jimenez W., Mammar A., Cavalli A. (2010). Software Vulnerabilities, Prevention and Detection Methods: A Review 1., Available from: https://www.researchgate.net/ publication/ 253704494_ Software_ Vulnerabilities_ Prevention_and_ Detection_Methods_A_Review_ 1, Accessed: 2024-01-16, in press.

[3] Bettaieb, S.; Shin, S. Y.; Sabetzadeh, M.; Briand L. C.; Garceau, M. & Meyers, A. (2020). Using machine learning to assist with the selection of security controls during security assessment, Empirical Software Engineering (EMSE), Vol. 25, pp. 2550-2582., ISSN 1573-7616, DOI: 10.1007/s10664-020-09814-x, in press.

[4] Chernis B. & Verma R. (2018). Machine Learning Methods for Software Vulnerability Detection, Proceedings of the Fourth ACM International Workshop on Security and Privacy Analytics, March 2018, pp. 31–39, DOI: 10.1145/3180445.3180453, in press.

[5] Pereira, J. D.; Campos, J. R. & Vieira, M. (2021). Machine Learning to Combine Static Analysis Alerts with Software Metrics to Detect Security Vulnerabilities: An Empirical Study, Proceedings of the 17th European Dependable Computing Conference (EDCC), Published by IEEE, ISBN 978-1-6654-3671-7, ISSN 2641-810X, Munich, Germany. DOI: 10.1109/EDCC53658.2021.00008., in press.

[6] Medeiros, N.; Ivaki, N.; Costa, P. & Vieira M. (2020). Vulnerable Code Detection Using Software Metrics and Machine Learning, Published by IEEE, In: IEEE Access, Vol. 8, pp. 219174-219198, ISSN 2169-3536, DOI: 10.1109/ACCESS.2020.3041181, in press.

[7] Fu, M.; Tantithamthavorn, C.; Le, T.; Kume, Y.; Nguyen, V.; Phung, D. & Grundy, J. (2023). AIBugHunter: A Practical Tool for Predicting, Classifying and Repairing Software Vulnerabilities, Empirical Software Engineering (EMSE), Vol. 29, No. 4, DOI: 10.1007/s10664-023-10346-3, in press.

[8] Appelt, D.; Nguyen, C. D.; Panichella, A. & Briand L. C. (2018). A Machine-Learning-Driven Evolutionary Approach for Testing Web Application Firewalls, In: IEEE Transactions on Reliability, Vol. 67, No. 3, pp. 733-757., ISSN 1558-1721, DOI: 10.1109/TR.2018.2805763, in press.

[9] Brown, A.; Gupta, M. & Abdelsalam, M. (2024). Automated Machine Learning for Deep Learning Based Malware Detection, Computers & Security, Volume 137, ISSN 0167-4048, DOI: 10.1016/j.cose.2023.103582., in press

[10] Chio, C. & Freeman, D. (2018). Machine Learning and Security: Protecting Systems with Data and Algorithms, Published by O'Reilly Media, ISBN 978-1491979907, in press.

[11] Prakash Meher, J.; Biswas, S. & Mall, R. (2024). Deep Learning-Based Software Bug Classification, Information and Software Technology, Vol. 166, ISSN 0950-5849, DOI: 10.1016/j.infsof.2023.107350., in press.

[12] Peraković, D.; Hušnjak, S. & Remenar, V. (2012). Research of Security Threats in the Use of Modern Terminal Devices. Annals of DAAAM for 2012, Proceedings of the 23rd International DAAAM Symposium, Published by DAAAM International, Vol. 23, No. 1, ISBN 978-3-901509-91-9, ISSN 2304-1382, Vienna, Austria, in press.

[13] Dakić V.; Redžepagić J.; Bašić M. (2022). CI/CD Toolset Security, 33rd DAAAM International Symposium on Intelligent Manufacturing and Automation, Published by DAAAM International, ISBN 978-3-902734-36-5, ISSN 1726-9679, Vienna, Austria, DOI: 10.2507/33rd.daaam.proceedings.022, in press.

DOI: 10.2507/35th.daaam.proceedings.010

BIOMETRICS AVAILABLE ON MOBILE PHONES AND POSSIBLE DANGERS IT PRESENTS

Zlatan Moric*, Ivan Trojnar & Robert Petrunic

This Publication has to be referred as: Moric, Z[latan]; Trojnar, I[van] & Petrunic, R[obert] (2024). Biometrics Available on Mobile Phones and Possible Dangers it Presents, Proceedings of the 35th DAAAM International Symposium, pp.0071-0080, B. Katalinic (Ed.), Published by DAAAM International, ISBN 978-3-902734-44-0, ISSN 1726-9679, Vienna, Austria
DOI: 10.2507/35th.daaam.proceedings.010

Abstract

Biometric authentication technologies are becoming more common on mobile devices, providing a user-friendly and safe alternative to traditional password-based systems. The paper thoroughly examines different mobile biometric authentication approaches, assessing their advantages and weaknesses. The text conducts a thorough assessment of the security vulnerabilities linked to these technologies, including the possibility of data breaches, surveillance, and discrimination. It also suggests methods to reduce these risks. The significance of strong security frameworks and user education is emphasized to improve the safety and effectiveness of biometric systems. The conversation explores the duality of biometric authentication on mobile platforms, emphasizing the need to strike a balance between technological progress and the protection of user privacy and data security. This study emphasizes the necessity of continuous research and development to improve these technologies, guaranteeing that their advantages remain greater than the inherent security threats.

Keywords: mobile biometrics; biometric security risks; biometric authentication; privacy concerns.

1. Introduction

Biometric authentication has become a crucial part of mobile device security, revolutionizing the methods used to access devices and protect critical information. The appeal of this technology stems from its ability to combine convenience with heightened security measures by utilizing distinct physical or behavioural characteristics, such as fingerprints, facial features, and iris patterns, as means of user identification. This research explores the rapid spread of biometric technologies on mobile platforms and analyses their increasing popularity. Amidst a time characterized by frequent breaches of conventional passwords [13], biometric authentication arises as a rapid and seemingly more reliable substitute. However, it raises significant issues surrounding the safeguarding of biometric data. In contrast to passwords, which can be changed if they are compromised, biometric data is unchangeable and hence poses enduring security hazards. As the use of these technologies becomes more widespread, it is crucial to carefully examine their weaknesses, especially in the areas of fingerprint, facial, and iris identification [1]. This study evaluates the resilience and constraints of each biometric technique employed on mobile devices, with a focus on potential security concerns such as data breaches, unwanted spying, and inherent biases.

2. Biometric Authentication Techniques

Biometric authentication techniques are a major advancement in mobile security, using distinct physiological and behavioural traits to confirm a user's identification [3]. This section examines the primary techniques now utilized: fingerprint recognition, facial recognition, voice recognition, and iris scanning, each providing unique benefits and facing various obstacles.

2.1. Fingerprint Recognition

Fingerprint recognition is utilized to verify the identity of users by analysing the unique ridge patterns and minutiae spots on their fingerprints. This approach is widely recognized for its exceptional precision, which significantly decreases the chances of false positives. As a result, it improves both security and user convenience [3]. Technological advancements like Apple's Touch ID demonstrate the capacity of this method to offer fast device access, combining user pleasure and improved security.

Nevertheless, fingerprint detection is not without its obstacles. The primary issue is its vulnerability to counterfeiting; sensors may be deceived by high-quality reproductions or latent prints. To avoid unwanted access and usage, it is necessary to employ strong security measures such as advanced encryption and secure data storage techniques [2]. Furthermore, continuous research and development play a vital role in consistently enhancing sensor technology, with the objective of mitigating growing security risks and enhancing the accuracy and dependability of the system as time progresses [2]. Leading smartphone manufacturers have extensively embraced fingerprint identification, easily integrating this technology into numerous products. Advancements in sensor technology persistently improve accuracy and speed, rendering fingerprint recognition a dependable element of mobile security.

2.2. Facial Recognition

Facial recognition technology utilizes the functionalities of front-facing cameras on mobile devices to verify the identity of users by analysing their facial characteristics. The contactless nature of this approach has led to a significant increase in its utilization, especially during the COVID-19 epidemic [6]. Facial recognition improves user experience by facilitating a smooth interaction, removing the requirement for physical contact or manual input. Although facial recognition technology has numerous benefits, it also encounters substantial obstacles. The system's accuracy and dependability can be significantly impacted by changes in illumination, facial angles, and expressions, making it highly sensitive to these fluctuations [8]. Furthermore, this technology is prone to biases in the algorithms, leading to unequal performance among various demographic groups. The susceptibility to spoofing attempts, such as the utilization of pictures or 3D masks to imitate a genuine user's face, poses a significant security hazard [9].

In order to tackle these weaknesses, continuous progress in sensor technology and algorithm enhancement is crucial. The purpose of these enhancements is to strengthen the resilience of facial recognition systems against external variations and advanced spoofing techniques, thereby guaranteeing secure and reliable authentication of users in various settings and scenarios.

2.3. Voice Recognition

Voice recognition technology verifies the identity of persons by examining distinct vocal attributes such as pitch, tone, and speaking rhythm [8]. The approach is becoming more highly regarded for its usefulness in remote authentication situations, making it easier to operate without using hands and improving accessibility, especially for people with physical disabilities. Voice recognition's adaptability allows it to be used on multiple secure mobile platforms, providing added user convenience while maintaining security. Nevertheless, voice recognition is prone to numerous security flaws. It can be vulnerable to spoofing techniques like replay attacks, in which recorded speech samples of the authorized user are played back to gain unauthorized access [8]. Furthermore, environmental variables such as ambient noise and the user's physical well-being, such as having a cold or sore throat, can have a negative impact on the precision of voice recognition systems [8].

Continuous improvements in artificial intelligence and machine learning are essential for reducing these hazards. These developments are designed to enhance the systems' capacity to adjust to tiny variations in a user's voice, thereby ensuring optimal levels of security and dependability. Furthermore, it is crucial to employ strong encryption and secure storage methods for voiceprints to prevent illegal entry and guarantee user privacy [8].

2.4. Iris Scans

Iris scanning technology capitalises on the unique patterns found in the coloured ring of the human eye to authenticate identity, offering an exceptionally high degree of accuracy. This method is quick and non-invasive, and advanced iris scanners can operate effectively under a variety of lighting conditions, and even accurately identify individuals who are wearing glasses or contact lenses [6][11].

Despite its numerous advantages, iris scanning faces several challenges. The requirement for high-resolution imaging equipment can lead to increased costs, potentially limiting broader application [4]. Privacy concerns are substantial with iris scanning, as the unauthorised use of iris data could result in severe security breaches such as identity theft. Furthermore, certain medical conditions affecting the eyes or suboptimal alignment during the scanning process can reduce the efficacy of iris recognition systems [4]. Continuous improvements in technology are essential to address these issues. Enhancements in camera technology and scanning algorithms are being developed to expand the usability and reduce the limitations associated with distance and angle of scan [11]. Ensuring robust encryption and secure storage of iris data, alongside stringent access controls, are pivotal to safeguarding the privacy and security of the information [4].

3. In-depth analysis of Security Risks

Biometric authentication systems offer improved user security and convenience, but they also have inherent vulnerabilities that need to be addressed to maintain system integrity and reliability. This section explores the precise attacks and security obstacles encountered by each biometric authentication technique, such as facial recognition, fingerprint recognition, iris scanning, and voice recognition.

3.1. Facial Recognition Vulnerabilities

Despite its advanced nature, facial recognition technology is susceptible to severe weaknesses that might undermine its security. The main threats include:

1. **Photo Attacks**: In the initial stages, facial recognition systems were susceptible to photo attacks, in which a still image of the user's face might be exploited to deceive the system. Technological advancements have brought forward techniques to combat these attacks, such as utilizing the analysis of depth, texture, and authentic motions such as blinking or subtle facial expressions to ascertain liveness. These systems utilize intricate algorithms to differentiate between two-dimensional photographs and three-dimensional face structure of a live individual [9].
2. **Video Replay Attacks**: These attacks entail the utilization of a video recording of the authorized user to trick the facial recognition system. Video attacks are more advanced than photo attacks because videos can imitate realistic motions, which can possibly deceive systems without advanced liveness detection capabilities. Contemporary methods mitigate this danger by examining intricate biometric indicators that are distinctive and challenging to duplicate, such as thermal imaging for detecting skin temperature and micro-expression analysis for observing involuntary face movements [9].
3. **3D Mask Attacks**: Sophisticated 3D masks may be created to accurately mimic an individual's face, including lifelike skin texture and intricate facial characteristics. These masks present a substantial obstacle for facial recognition systems that do not employ multimodal biometric verification. To reduce this risk, several systems utilize multi-factor authentication procedures that necessitate supplementary verification techniques beyond facial recognition, such as iris scans or voice verification. In addition, systems may utilize depth-sensing technologies and thermal sensors to identify the lack of normal human skin characteristics and the distinct thermal pattern linked to living human skin [4].
4. **Deepfake Technology**: The emergence of deepfake technology, employing advanced machine learning algorithms to fabricate counterfeit photos and videos, presents a heightened level of danger to facial recognition systems. These technologies have the capability to create facial images that exhibit genuine blinking, smiling, or movement, which enables them to bypass security systems that do not regularly update their detection algorithms. It is necessary for systems to constantly develop and integrate AI-powered anomaly detection approaches that can differentiate between authentic human behaviours and those artificially created by AI [9].

The vulnerabilities linked to facial recognition technology emphasize the need for ongoing improvements in security protocols. Developers and security professionals must improve their countermeasures as attackers continue to create more advanced methods to hack these systems. This involves the utilization of multi-modal biometric analysis, using artificial intelligence to identify and adjust to emerging risks, and including supplementary levels of authentication to enhance security. Ensuring the resilience of facial recognition technology not only safeguards individual privacy but also protects the integrity of systems that depend on biometric authentication for secure access. By taking proactive measures to address these weaknesses, we may improve the dependability and credibility of facial recognition technology in an ever more digitalized environment.

3.2. Fingerprint Recognition Vulnerabilities

Fingerprint identification, although generally dependable in different situations, has inherent weaknesses that could potentially be manipulated:

1. **Latent Prints**: A prevalent weakness is the reliance on latent prints, which are fingerprints left on surfaces and collected via methods like dusting or adhesive lifting. These prints can thereafter be used to fabricate models or

duplicates with the intention of misleading fingerprint sensors. To address this issue, contemporary fingerprint scanners integrate liveness detection technology that may evaluate indicators of vitality, such as blood circulation or perspiration, which are not present in a copied fingerprint [2].

2. **Fingerprint Replication**: A notable danger arises from the ability to duplicate fingerprints using high-resolution photos. Assailants could fabricate intricate duplicates using substances such as silicone or gelatine that accurately imitate the tactile sensation of human skin. Advanced scanners employ ultrasonic imaging or capacitive sensors to identify the distinctive electrical conductivity and sub-dermal characteristics found in genuine human skin, which are challenging to precisely reproduce [2].
3. **Sensor Bypassing**: Advanced adversaries may try to completely circumvent the fingerprint scanner by exploiting hardware or software vulnerabilities. Manufacturers improve the security characteristics of scanners by adding encryption to the sensor and using secure communication protocols between the scanner and the device's main processor [5].
4. **Database Breaches**: Storing fingerprint data in a centralized database makes it susceptible to cyberattacks. Security breaches have the potential to result in the permanent compromise of fingerprint data. To protect this information, it is crucial to use data encryption, utilize decentralized storage options, and enforce strict access controls [5].

The weaknesses inherent in fingerprint identification systems emphasize the critical necessity for ongoing developments in biometric security technologies. Given the high sensitivity of biometric data, it is crucial to prioritize protection against its misuse. To successfully address these weaknesses, it is crucial to advance the development of more sophisticated liveness detection, improved encryption technologies, and strong security standards. By fortifying these specific areas, we may greatly improve the security and dependability of fingerprint-based identification systems, guaranteeing their continued effectiveness and safety in a constantly changing landscape of potential dangers.

3.3. Iris Scanning Vulnerabilities

Iris scanning is lauded for its exceptional precision and non-intrusive characteristics, although it is not impervious to security vulnerabilities:

1. **High-Resolution Image Capture and Replication**: One of the major vulnerabilities is the unauthorized acquisition and duplication of high-resolution iris pictures. Adversaries can duplicate an iris by utilizing precise pictures, especially those captured with high-resolution cameras in the infrared range. Contemporary iris scanning systems now include liveness detection elements that examine the authentic movement of the iris and its reactive reaction to light. This helps distinguish between a genuine eye and a replicated image [5].
2. **Contact lens spoofing**: A notable danger in which attackers utilize contact lenses that are printed with the iris pattern of someone else. Iris scanners employ intricate textural analysis to identify the existence of a contact lens and scrutinize reflections in the eye that may suggest the use of a contact lens [10].
3. **Distance and Angle Limitations**: Iris scanners may encounter constraints regarding the distance and angle of usage. Insufficient scanning angles or distances might result in scans that are not precise or comprehensive, which in turn raises the likelihood of false rejections or, even worse, false acceptances [11]. Progress is being made in camera technology and scanning algorithms to address these restrictions, enabling more versatile and precise scanning in different ambient situations.
4. **Privacy Concerns and Data Security**: Like other biometric technologies, iris scanning gives rise to substantial privacy concerns. Illegitimate access to stored iris data might result in identity theft and other security breaches. To reduce these risks, it is essential to utilize strong encryption and safe storage for biometric data, enforce stringent access rules, and regularly perform security audits [4].

Continuous improvements in biometric security procedures are required due to the security vulnerabilities connected with iris scanning. To mitigate against these weaknesses, it is crucial to implement enhanced liveness detection, enhance camera technologies, and enforce strict data protection rules. To maintain the advantages of iris scanning as a secure and dependable biometric authentication method, it is important to tackle these problems and prevent potential security threats from overshadowing the benefits of this technology.

3.4. Voice Recognition Vulnerabilities

Voice recognition technology, although it enables convenient and accessible authentication, is prone to many specific vulnerabilities:

1. **Recording and Playback Attacks**: The simplest approach to deceive speech recognition systems entails capturing the voice of an authorized user and replaying it to obtain unauthorized entry. These assaults can go around systems that do not use advanced liveness recognition technology that can differentiate between live speech and recordings [12].

2. **Voice Synthesis and Deepfake Audio**: Advancements in artificial intelligence have resulted in the creation of speech synthesis technologies capable of producing audio snippets that closely imitate the vocal patterns of individuals. The emergence of AI-generated voices presents a notable obstacle for speech recognition systems, particularly if they are capable of accurately imitating distinct vocal intonations and patterns that are specific to the authentic user [12].
3. **Background Noise and Clarity Issues**: Disturbances may have a substantial impact on the accuracy of voice recognition, leading to decreased performance. Systems need to have the ability to separate the speech from background noise and ensure that voice commands are captured clearly enough to accurately authenticate.
4. **Health-Related Voice Changes**: Whether temporary or permanent, might result in authentication failures for users. These changes may occur because of health conditions such as a cold or sore throat. For voice recognition systems to remain accurate over time, they must be able to adapt and recognize variations in voice timbre.

The dangers inherent in voice recognition systems underscore the imperative for ongoing improvements in technology and security standards. Integrating sophisticated liveness recognition capabilities and developing robust noise-cancelling algorithms are essential for enhancing system resilience against external and internal influences that impact speech quality. In addition, using multi-factor authentication solutions can offer an extra level of protection, guaranteeing that speech recognition technology remains a dependable and secure approach for user authentication. To effectively mitigate these vulnerabilities, we can take proactive measures to prevent future exploits and strengthen user confidence in voice-based biometric systems.

4. Comprehensive Mitigation Strategies

To successfully reduce the security concerns linked to biometric authentication systems, it is crucial to adopt a comprehensive strategy that combines innovative technology, strong policy frameworks, and ongoing user education. The strategies provided below detail comprehensive measures to bolster the security and integrity of biometric systems:

1. Utilizing innovative encryption algorithms is essential for ensuring the security of data both when it is stored and when it is being transferred. Utilizing tamper-resistant hardware modules is an effective method to safeguard biometric data from illegal access and tampering.
2. Liveness detection technologies are necessary in biometric systems to prevent spoofing attacks. These technologies can analyse physiological indicators of life or subtle behavioural patterns that are challenging to reproduce using artificial reproductions or digital alterations [9].
3. Multi-factor Authentication (MFA) is the process of combining biometric technologies with other authentication factors, such as passwords or security tokens, to greatly decrease the likelihood of unwanted access. MFA guarantees that the system cannot be breached by compromising a single element alone [13].
4. Ensuring the regular updating of biometric systems with the most recent security patches and software updates is essential to protect against known vulnerabilities. Regular audits and updates efficiently tackle recently developing dangers in a timely manner.
5. The implementation of a privacy-by-design strategy in the development and deployment of biometric systems guarantees that privacy concerns are incorporated at every step, along with minimal data retention. To reduce the risk of data breaches, it is important to limit the amount of personal data stored and collected to only what is necessary for the intended purpose [12].
6. User Education and Awareness: The process of instructing users on the proper and secure utilization of biometric systems, as well as informing them about the potential dangers associated with it, is of utmost importance in protecting biometric data. Awareness campaigns can educate users on essential protocols, such as the significance of frequent device security upgrades and the prevention of prevalent phishing attacks.
7. Legal and Regulatory Compliance: Adhering to international data protection requirements, such as GDPR in Europe or CCPA in California, is crucial for upholding stringent privacy and security measures. Organizations need to stay updated on legislative developments to verify that their biometric systems are complying, and that user data is protected according to the most recent legal requirements.

Organizations can improve the security, dependability, and user trust in biometric systems by applying these thorough mitigation techniques. To fully utilize the capabilities of biometric technologies and minimize the dangers involved, it is crucial to constantly enhance and adapt to the ever-changing cyber threat landscape.

5. Case Studies: Real-World Incidents of Biometric Compromise

The susceptibility of biometric systems to security breaches can have significant ramifications. Examining actual occurrences in the real world yields useful knowledge about possible vulnerabilities and guides the creation of stronger security protocols. Here are 5 noteworthy occurrences [17]:

1. In 2015, the United States Office of Personnel Management (OPM) suffered a significant data breach, resulting in the exposure of the fingerprints of 5.6 million persons. This incident emphasized the enduring hazards linked to the storage of biometric data, as compromised biometric information cannot be reset in the same manner as a password.
2. India's Aadhaar system, which is one of the largest biometric databases in the world, has seen repeated breaches that have impacted millions of inhabitants. The occurrence of these breaches was frequently enabled by software vulnerabilities or inadequate security standards, underscoring the crucial requirement for rigorous security measures in extensive biometric databases.
3. The Suprema Biostar 2 Breach in 2019 compromised the biometric access control systems of Suprema, a prominent company specializing in biometric security. Insufficient security measures, namely regarding database encryption and access control, led to the exposure of more than one million fingerprint records and other personal data items.
4. In 2020, a notable occurrence of voiceprint phishing fraud took place, where attackers successfully infiltrated a bank's voiceprint system by executing a phishing assault. The compromise enabled attackers to circumvent the biometric security of voice authentication and conduct deceitful transactions, highlighting the necessity for multi-factor authentication and ongoing employee training to counteract social engineering strategies.
5. In 2021, a hospital's iris scanning system was hacked because of a software vulnerability that had not been fixed. This led to the unauthorized access to patients' iris data and personal health information. This incident highlights the criticality of regularly updating software and managing vulnerabilities, particularly in systems that deal with sensitive biometric and personal data.

These case studies illustrate the diverse methods by which biometric systems can be exploited and the extensive ramifications of such breaches. The importance of implementing a thorough security strategy that encompasses strong encryption, frequent software upgrades, multi-factor authentication, and rigorous training and awareness initiatives is emphasized. Implementing such steps is crucial for safeguarding sensitive biometric data and upholding the integrity of biometric authentication systems.

6. Ethical and Privacy Concerns in Mobile Biometrics

The incorporation of biometric technologies into mobile devices gives rise to significant ethical and privacy concerns that necessitate comprehensive scrutiny. The fundamental issues at hand are cantered on the ownership of data, obtaining consent, and the possibility of its exploitation.

1. Data Ownership and Control: Biometric data possesses an intrinsic quality of being personal and sensitive. Users commonly produce this data, but it is frequently held and controlled by device manufacturers or application developers, raising questions regarding the true ownership of this data. It is crucial to consider the ethical consequences of data ownership, with a focus on empowering users to maintain control over their biometric information and comprehend its usage.
2. The idea of informed consent is essential when considering biometric data. Users must possess complete knowledge regarding the specific biometric data being gathered, the intended objective of its collection, and the individuals or entities authorized to access it. Transparency is crucial not just for ethical purposes, but also for adhering to worldwide privacy rules such as GDPR, which prioritize individuals' rights to comprehend and manage their personal data.
3. Potential for Misuse: The utilization of biometric data goes beyond verifying the identity of users; it possesses the capacity for creating profiles and conducting surveillance, which can result in discriminatory practices and other forms of mistreatment. Law enforcement or government entities have the potential to utilize biometric data in manners that encroach upon personal privacy or violate civil liberties. Strong protections and well-defined legal frameworks are essential to prevent the improper use of biometric technologies.
4. The utmost importance is in securely storing and managing biometric data to mitigate the risks of breaches, which might result in irreparable consequences. Unlike passwords, biometric data is immutable once it is compromised. It is crucial to encrypt biometric data and retain it for the minimum required duration to reduce privacy threats.
5. Effect on Vulnerable Populations: The use of biometric systems may have a greater impact on certain populations compared to others. Refugees or those who are poor may lack a comprehensive understanding of the consequences associated with the gathering of biometric data, or they may lack the resources to choose not to participate in such programs. It is necessary to implement specific measures and safeguards to prevent the exploitation or unfair targeting of certain groups.

It is crucial to consider and resolve the ethical and privacy issues related to mobile biometrics to build confidence and get acceptance from users. A comprehensive approach is necessary, which incorporates strict data protection measures, clear regulations, and ethical norms that prioritize user consent and data security. Through proactive management of these challenges, developers and policymakers may guarantee the responsible and helpful use of mobile biometric technology for all users.

7. User Perception and Acceptance Studies of Biometric Authentication

Understanding user perception and acceptance is pivotal for the effective implementation of biometric authentication technologies. This section examines empirical research and theoretical models to highlight factors influencing user attitudes towards biometric systems and the broader implications for consumer adoption [16].

7.1. Factors Influencing User Perception and Acceptance

User perception and acceptability play a significant role in the effective adoption of biometric authentication systems. The subsequent elements play a crucial role in influencing users' perception and acceptance of biometric technologies:

1. **Privacy Concerns**: Users frequently express anxiety about the confidentiality of their biometric data. The primary focus is on the concerns about the collection, storage, usage, and sharing of biometric information. Users are more inclined to embrace biometric solutions when they are provided with the guarantee that their data is being safely and confidentially managed.
2. **Trust and Reliability**: The acceptance of biometric technologies is greatly influenced by the trust and reliability placed in them. Users require assurance that the technology will consistently and accurately detect them without any errors, such as false rejections or misidentifications. Trust is established by being transparent about the system's abilities and limitations.
3. **Usability**: The level of simplicity and intuitive nature of biometric technologies significantly influences their acceptance and implementation. Users choose systems that are user-friendly and can be easily incorporated into their daily routines. The ease provided by biometrics, such as expedited authentication procedures, amplifies consumer contentment and approval.
4. **Awareness and Knowledge**: User adoption is influenced by the understanding of biometric technologies. Disseminating awareness campaigns and educational initiatives that elucidate the advantages and functioning of biometric systems help clarify the technology and assuage apprehensions or misunderstandings. Users that possess a high level of knowledge are more inclined to recognize and value the benefits and increased safety measures offered by biometrics.

Gaining a comprehensive understanding of these issues and responding appropriately to deal with them is crucial for improving the acceptance and utilization of biometric technologies. Stakeholders can promote a good perception and wider acceptance of biometric systems by implementing strong privacy measures, enhancing transparency, and improving the user experience. Moreover, it is imperative to provide education to the public and strictly comply with rigorous regulations in order to establish trust and confidence among users.

7.2. Societal Attitudes and Perceptions

The perception of biometric authentication in society is shaped by a multifaceted environment, which is influenced by elements such as cultural norms, privacy concerns, and technological familiarity. This section examines the intricate viewpoints that influence the level of societal approval or opposition towards biometric technologies [16].

1. **General Acceptance with Reservations**: Biometric authentication is generally accepted, albeit with misgivings. It is widely recognized for its convenience and improved security. However, concerns remain, especially surrounding privacy and the possibility for biometric data to be misused. Research suggests that users value the effectiveness and safety of biometrics, but they are still concerned about the management and protection of their personal data.
2. **Cultural Variations**: The attitudes towards biometrics might range greatly in various cultural situations. Certain cultures with a strong belief in government and a heightened focus on security may have a more positive perception of biometric technologies. On the other hand, civilizations that have had worries about privacy in the past or where people have poor faith in authority may show more doubt and opposition.
3. **Impact of Previous Technological Experiences**: The impact of prior technological experiences: Societies that have had positive experiences and achieved favourable outcomes from earlier technology adoptions are more likely to be receptive to the adoption of biometric technologies. In contrast, areas that have had security breaches or improper use of technology tend to adopt a more careful approach towards biometrics. They prioritize the implementation of strong security measures and the establishment of clear and enforceable rules.
4. **Demographic Influences**: Demographic factors can exert an influence on people's attitudes towards biometrics, with different demographic groupings often exhibiting varied perspectives. Typically, younger folks who are more knowledgeable about technology are more likely to embrace biometric authentication, considering it as a forward-thinking measure towards modernity. Individuals belonging to older age groups or those with limited technological experience may exhibit more reluctance due to privacy concerns and the intricacy of novel systems.

Gaining insight into cultural views towards biometric authentication is essential for developing targeted methods that address specific concerns and improve public acceptance. Understanding and valuing the cultural, demographic, and regulatory differences can help improve the application and integration of biometric technologies in society. By actively involving communities in discussions and education on the advantages and protective measures of biometrics, we may effectively reduce opposition and foster a favourable atmosphere for the use of these technologies.

7.3. Implications for Consumer Devices

The use of biometric technologies into consumer electronics has extensive ramifications that impact design, user experience, and adherence to regulations. This section thoroughly examines these aspects [16]:

1. **Design Considerations**: Manufacturers should give priority to user-centric design principles when incorporating biometric technologies into devices. This entails not only guaranteeing the biometric systems possess a high degree of dependability and speed, but also ensuring they are user-friendly and unobtrusive. The design must be inclusive of many user populations, including individuals with physical limitations or limited familiarity with complex technology, to ensure accessibility and user-friendliness.
2. **User Education and Engagement**: To optimize the adoption and correct utilization of biometric systems, manufacturers must allocate resources towards extensive user education. This entails furnishing explicit guidelines on the proper and secure use of biometric features, together with full disclosure of the implemented data security protocols. Providing users with information regarding the advantages and restrictions of biometric technology can help reduce worries and misunderstandings, resulting in a user base that is more educated.
3. **Regulatory Compliance**: Adhering to international data protection rules, such as GDPR in the EU or CCPA in California, is essential due to the collection and processing of extremely sensitive personal data by biometric technologies. Manufacturers are obligated to guarantee that biometric data is gathered, stored, and handled in accordance with these regulations, which encompass requirements for user permission, data reduction, and the right to data erasure.
4. **Balancing Security with Convenience**: Striking a balance between security and convenience is crucial when considering the use of biometrics for authentication. While biometrics provide a more comfortable and secure approach compared to previous techniques, manufacturers must exercise caution in managing these two factors. Implementing measures such as multi-factor authentication to improve security should not excessively burden the user experience. On the other hand, it is important to ensure that attempts to improve usability do not weaken the effectiveness of security measures.
5. **Privacy Protection**: Due to the sensitive nature of biometric data, it is imperative for manufacturers to incorporate robust privacy protection mechanisms. This encompasses employing sophisticated encryption methods to safeguard data both when it is stored and while it is being transmitted, guaranteeing that biometric data is not retained on centralized systems that are susceptible to unauthorized access, and furnishing users with explicit privacy options that empower them to manage their own data.

The incorporation of biometric technologies into consumer electronics has profound and diverse effects. Manufacturers may improve the security and functionality of their devices and gain trust and happiness from consumers by thoroughly addressing design, education, regulatory, and privacy concerns. Effectively overcoming these issues will be crucial for the general acceptance and sustained prosperity of biometric technology in the consumer market.

8. Challenges, Opportunities and Limitations

The incorporation of biometric authentication technology into consumer electronics presents a multifaceted set of technical obstacles, security considerations, and prospects for advancement, coupled with substantial regulatory and ethical ramifications. Ongoing difficulties arise from technical challenges such as sensor accuracy, environmental adaptability, and the computational demands of processing secure biometric data [14]. To maintain consistent performance in diverse settings, it is necessary to have strong hardware that can handle complex encryption and process data in real-time, considering the technical limits. Additionally, the extremely sensitive nature of biometric data gives rise to significant security concerns, underscoring the importance of implementing strict data protection measures, regular system updates, and diligent security policies to minimize the danger of unauthorized access and misuse [13]. Although there are difficulties, biometric technologies present favourable prospects for advancing security enhancements and user interface design. This drives progress in areas such as liveness detection and anti-spoofing techniques, ultimately promoting greater consumer trust and market acceptance [15]. Nevertheless, the acceptance of users continues to be a significant constraint, driven by apprehensions about privacy, cultural perspectives, and confidence in technology. This issue can be resolved by providing extensive user training and ensuring clear and open communication about the use and safeguarding of biometric data [16]. In addition, the swift implementation of these technologies brings up significant regulatory and ethical concerns, requiring adherence to global privacy regulations and thoughtful examination of the ethical consequences of surveillance and monitoring methods [17].

The global market's willingness to embrace these technologies differs, necessitating customized solutions that align with local technological infrastructures, regulatory frameworks, and consumer tastes to promote acceptance and integration [16]. To effectively address the challenges and opportunities of biometric authentication technologies, it is necessary to adopt a comprehensive strategy that encompasses ongoing technological advancements, strict adherence to security standards, ethical practices, and proactive regulatory compliance. These factors are crucial for maximizing the benefits of biometric solutions in enhancing authentication security. This study offers extensive knowledge on mobile biometrics, but its findings are limited due to its dependence on current technology, which is undergoing significant advancements. Furthermore, the results mostly pertain to situations that were simulated in controlled settings, which may not completely depict intricate relationships that occur in the actual world.

9. Future Works

The advancement of biometric authentication technologies offers a rich vein for future research, particularly in enhancing security and user interface adaptability. Key areas include developing sophisticated liveness detection techniques that effectively differentiate real users from synthetic replicas, integrating these systems seamlessly with multi-factor authentication to bolster security without sacrificing convenience, and employing artificial intelligence and machine learning to improve the accuracy and adaptability of biometric systems [13, 15]. Additionally, the exploration of privacy-preserving biometric systems is crucial; this involves investigating anonymization techniques, homomorphic encryption, and secure multi-party computation that can process biometric data while safeguarding user privacy [16]. The increasing ubiquity of biometric technologies across various devices also calls for research into cross-device and cross-platform compatibility to ensure a uniform user experience and consistent security measures across different technologies and platforms [14].

Moreover, the evolving landscape of biometric technologies necessitates continuous updates to regulatory and ethical frameworks to address the emerging challenges of biometric surveillance, consent, and privacy, ensuring that technological advancements align with global standards and societal values [17]. Addressing these areas in future studies will significantly contribute to refining biometric authentication technologies, making them more secure, efficient, and ethically grounded while enhancing their acceptance and integration into daily use.

10. Conclusion

This research has thoroughly analysed the various features of biometric authentication technologies, including their benefits, weaknesses, and wider societal consequences. By conducting a comprehensive examination of different biometric techniques, such as fingerprint, facial, iris, and voice recognition systems, we have pinpointed the primary technological and security obstacles linked to each method [2, 4, 5, 8, 9, 10]. Our study of actual instances of biometric breaches offers concrete proof of the possible dangers and emphasizes the need for strong security measures to safeguard sensitive biometric information [17].

Moreover, this work has made a valuable contribution to the academic discussion by methodically examining the ethical and privacy issues that arise from the use of biometric technologies in consumer gadgets. The text highlights the significance of dealing with these concerns by obtaining informed consent, minimizing data collection, and adhering to regulatory requirements to increase user confidence and acceptance [16]. Analysis of user perception and acceptability has shown that the adoption of biometric systems is significantly influenced by privacy concerns, faith in technology, and cultural factors [16]. This understanding is essential for customizing biometric technologies that are responsive to user requirements and social standards. In addition, the study has described detailed measures to reduce the impact of potential risks by implementing sophisticated encryption techniques, employing multi-factor authentication, and regularly updating the software to improve the security and dependability of biometric systems [13]. Implementing these measures is crucial for reducing the risks connected with biometric data breaches and guaranteeing the reliability of biometric authentication systems.

Ultimately, this study has made a comprehensive scientific contribution to the comprehension of biometric technologies, emphasizing their capabilities as well as their constraints. The findings emphasize the necessity for continuous research and development to tackle growing difficulties and guarantee that biometric authentication technologies progress in a way that is both safe and user-friendly, while also being socially acceptable. As we progress in our technological skills, it is crucial that we enhance our ethical and security structures to protect personal privacy and uphold public confidence in these systems.

11. References

[1] Zhao, K.; Jin, H.; Zou, D.; Chen, G. & Dai, W. (2013). Feasibility of Deploying Biometric Encryption in Mobile Cloud Computing, Proceedings of the 8th Annual ChinaGrid Conference, 2013, Los Alamitos, CA, USA, ISBN 978-0-7695-5058-9, ISSN 1949-1328, Published by IEEE, DOI: 10.1109/ChinaGrid.2013.10, in press.

[2] Deshmukh, P. & Mohod, S. (2020). Biometric Jammer: Prevention of Fake Acquisition of Fingerprint for Security Enhancement, Proceedings of the 2020 International Conference on Emerging Trends in Information Technology

and Engineering (ic-ETITE), ISBN 978-1-7281-4142-8, Published by IEEE, Vellore, India. DOI: 10.1109/ic-ETITE47903.2020.18, in press.

[3] Yildrim, N. & Varol, A. (2018). Android-Based Mobile Application Development for Web Login Authentication Using Fingerprint Recognition Feature, Proceedings of the 2015 23rd Signal Processing and Communications Applications Conference (SIU), ISSN 2165-0608, ISBN, 978-1-4673-7386-9, Published by IEEE, Malatya, Turkey. DOI: 10.1109/SIU.2015.7130436, in press.

[4] Wang, Q. & Wang, D. (2020). Understanding Failures in Security Proofs of Multi-Factor Authentication for Mobile Devices, Journal of IEEE Transactions on Information Forensics and Security, Vol. 18, December 2020, pp. 597-612, ISSN 1556-6021, Published by IEEE, DOI: 10.1109/TIFS.2022.3227753, in press.

[5] Belkhede, M.; Gulhane, V. & Dr. Bajaj, Preeti (2012). Biometric Mechanism for Enhanced Security of Online Transaction on Android System: A Design Approach, Proceedings of the 2012 14th International Conference on Advanced Communication Technology (ICACT), ISSN 1738-9445, ISBN 978-89-5519-163-9, pp. 1193-1197, Published by IEEE, PyeongChang, Korea, in press.

[6] Keykhaie, S. & Pierre, S. (2023). Lightweight and Secure Face-Based Active Authentication for Mobile Users, Journal of IEEE Transactions on Mobile Computing, Vol. 22, No. 3, March 2023, pp. 1551-1565., ISSN 1558-0660, Published by IEEE, DOI: 10.1109/TMC.2021.3106256, in press.

[7] Ahamad, S.S.; Sastry, V.N. & Nair, M. (2013). A Biometric based Secure Mobile Payment Framework, Proceedings of the 2013 4th International Conference on Computer and Communication Technology (ICCCT), ISBN 978-1-4799-1572-9, Published by IEEE, Allahabad, India. DOI: 10.1109/ICCCT.2013.6749634, in press.

[8] Islam, F. M. & Islam, N. M. (2012). A Biometrics-Based Secure Architecture for Mobile Computing, Proceedings of the 2012 IEEE Long Island Systems, Applications and Technology Conference (LISAT), ISBN 978-1-4577-1343-9, Published by IEEE, Farmingdale, NY, USA. DOI: 10.1109/LISAT.2012.6223108, in press.

[9] Yu, Z.; Cai, R.; Li, Z.; Yang, W., Shi, J. & Kot, A.C. (2024). Benchmarking Joint Face Spoofing and Forgery Detection with Visual and Physiological Cues, Journal of IEEE Transactions on Dependable and Secure Computing, pp. 1-15, ISSN 1941-0018, Published by IEEE, DOI: 10.1109/TDSC.2024.3352049, in press.

[10] Bowyer, W. K. & Doyle, S. J. (2014). Cosmetic Contact Lenses and Iris Recognition Spoofing, Journal of IEEE Biometrics Compendium, IEEE RFIC Virtual Journal, IEEE RFID Virtual Journal, Vol. 47, No. 5, May 2014, pp. 96-98., ISSN 1558-0814, Published by IEEE, DOI: 10.1109/MC.2014.118, in press.

[11] Abidin, Z. Z.; Manaf, M.; Shibghatullah, A. S. & Anawar, S. (2013). Iris Recognition Failure in Biometrics: A Review, Proceedings of the Kolokium Siswazah Sains Komputer Dan Matematik 2013 Peringkat Kebangsaan (SISKOM2013), in press.

[12] Gong, Y. & Poellabauer, C. (2018). An Overview of Vulnerabilities of Voice Controlled Systems, Proceedings of the 1st International Workshop on Security and Privacy for the Internet-of-Things (IoTSec), Published by Cornell University, DOI: 10.48550/arXiv.1803.09156, in press.

[13] Redzepagic, J.; Dakic, V.; Stanesic, J. & Basic, M. (2023). Analysis of Password Security Policies and Their Implications on Real-Life Security, Proceedings of the 34th DAAAM International Symposium, pp.0077-0081, Published by DAAAM International, ISBN 978-3-902734-41-9, ISSN 1726-9679, Vienna, Austria. DOI: 10.2507/34th.daaam.proceedings.011, in press.

[14] Lovric, E.; Moric, Z.; Redzepagic, J. & Regvart, D. (2023). Detecting Security Vulnerabilities on Internet-connected Devices, Proceedings of the 34th DAAAM International Symposium, pp.0103-0110, Published by DAAAM International, ISBN 978-3-902734-41-9, ISSN 1726-9679, Vienna, Austria. DOI: 10.2507/34th.daaam.proceedings.015, in press.

[15] Luo, J.; Hu, F. & Wang, R. (2019). 3D Face Recognition Based on Deep Learning, Proceedings of the 2019 IEEE International Conference on Mechatronics and Automation (ICMA), ISBN 978-1-7281-1699-0, ISSN 2152-744X, Published by IEEE, Tianjin, China. DOI: 10.1109/ICMA.2019.8816269, in press.

[16] Rukhiran, M.; Wong-In, S.; Netinant, P. (2023). User Acceptance Factors Related to Biometric Recognition Technologies of Examination Attendance in Higher Education: TAM Model, Journal of Sustainability 2023, Vol. 15, pp. 3092, published by MDPI, DOI: /10.3390/su15043092, in press.

[17] Jaiswal, S.; Dr. Bhadauria, S. S. & Dr. Jadon, R. S. (2011). Biometric: Case Study, Journal of Global Research in Computer Science, Vol. 2, No. 10, ISSN 229-371X, in press.

DOI: 10.2507/35th.daaam.proceedings.011

SPREAD2INNO ONE-STOP SHOP WEB PLATFORM: FRAMEWORK, FUNCTIONALITIES AND USER ENGAGEMENT

Damir Haskovic, Violeta Heinze, Celine Rabe & Paulina Villalobos Quesada

This Publication has to be referred as: Haskovic, D[amir]; Heinze, V[ioleta]; Rabe, C[eline] & Villalobos Quesada, P[aulina] (2024). SPREAD2INNO One-Stop Shop Web Platform: Framework, Functionalities and User Engagement, Proceedings of the 35th DAAAM International Symposium, pp.0081-0090, B. Katalinic (Ed.), Published by DAAAM International, ISBN 978-3-902734-44-0, ISSN 1726-9679, Vienna, Austria
DOI: 10.2507/35th.daaam.proceedings.011

Abstract

This paper describes the development framework and the architecture of the SPREAD2INNO One-stop shop web platform. The implementation took place in the frame of the Horizon Europe funded SPREAD2INNO project, focused on making 'modest' and 'moderate' innovation ecosystems more attractive and competitive. Therefore, the main objective of the web platform is to facilitate information exchange between different innovation ecosystems, targeting business acceleration and innovation service providers, SMEs and start-ups from all over Europe. To achieve this objective, the platform's main concept is based on the online community of practice paradigm. To this end, the platform was designed with a modular structure keeping it flexible and extendable for various contexts. This allows the platform to be more easily recognised and found by all interested stakeholders. Based on the analysed user engagement metrics and behavioural information, the platform has achieved, within a relatively short project timeframe, a steady visitor traffic flow along with satisfactory user engagement and networking activities.

Keywords: Web platform; Modules; Online community of practice; Innovation ecosystems; Startups; EU

1. Introduction

In today's competitive landscape, innovative companies achieve success through the support of a network that provides additional resources, information, capabilities, data, customers and knowledge. In this context, an innovation ecosystem refers to a range of value creating interactions and relationships between sets of interconnected organisations [1] and other entities that coevolve capabilities around a shared set of technologies, knowledge, or skills, and work cooperatively and competitively to develop new products and services [2]. Connectivity within and between innovation ecosystems greatly contributes to sustainable business growth with high societal value. To this end, the European Union aims to create more connected and efficient innovation ecosystems to support the scaling up of companies, encourage innovation and stimulate cooperation among national, regional and local innovation actors [3].

However, European innovation ecosystems are facing increasing global competition while the support mechanism for start-ups and SMEs is lacking capacity and cooperation, which can cause financial challenges, regional disparity and scaling-up risks. As a result, start-ups from better-connected innovation ecosystems benefit from more local accelerators with greater funding opportunities and business support services compared to their counterparts elsewhere [4].

According to the European Union's 2024 European Innovation Scoreboard (EIS), a substantial geographical gap still persists in innovation activities among European countries and regions, especially between southern and eastern Europe, which underperform relative to western and northern Europe [5]. This geographical gap often requires start-ups from less connected ecosystems to move to more established hubs elsewhere in Europe. Consequently, there are fewer scale-ups in less connected innovation ecosystems, creating unbalanced business activity and employment opportunities. Based on a systematic comparison of research and innovation achievements across EU member states and neighbouring regions, the EIS has defined four distinct groups: Innovation Leaders, Strong Innovators, Moderate Innovators, and Emerging Innovators. Therefore, there is a growing need to make 'modest' and 'moderate' innovation ecosystems more attractive to international companies interested in setting up their businesses outside the established centres and tapping into the potential of local ecosystems [6].

As an answer to these challenges, six projects (AccelerAction | 101072073, BEYOND | 101071410, S3E | 101072135, Food-scalEUp | 101072248, ACCELERO | 101072092 and SPREAD2INNO | 101071977) were funded by the European Commission as part of the Horizon Europe framework programme and the call „Expanding Acceleration Ecosystems (HORIZON-EIE-2021-SCALEUP-01-01)" [6], [7]. The SPREAD2INNO project was carried out within a period of 24 months by a consortium of seven partners from six EU countries (Germany, Italy, Belgium, Greece, Bulgaria, Austria) [7]. The SPREAD2INNO consortium consists of actors from business acceleration, innovation hubs, research institutions and competence centres, science parks, business network and consultancy. The project was coordinated by Fondazione E. Amaldi (Italy) and the partners include Fraunhofer IPK (Germany), European Business Angels Network (Belgium), Teseas (Italy), Cleantech Bulgaria (Bulgaria), MINDS & SPARKS GmbH (Austria) and Patras Science Park (Greece.).

To elevate the scalability potential of European businesses, SPREAD2INNO brought together different innovation stakeholders from crucial European innovation regions. The goal was to implement a scalable, replicable and holistic programme for the provision of high-quality business services, and collaboration between different entities of well-connected and less connected regional innovation ecosystems, with immediate impacts on a local and EU level. In order to create a sustainable and replicable model easily adopted in the rest of EU, the project implemented a set of activities for enhancing capacity and competitiveness of both business acceleration entities and SMEs and startups. The unique and interdisciplinary SPREAD2INNO methodology has helped to enhance business activities in less developed innovation regions, reinforced their acceleration capacity and enabled entrepreneurs to scale their concept to a fully concrete solution with strong market potential, subsequently enhancing the development and competitiveness of these regions.

To achieve the specific objectives of SPREAD2INNO, the project also implemented the SPREAD2INNO One-stop shop web platform to facilitate information exchange between different innovation ecosystems, targeting business acceleration and innovation service providers, SMEs and start-ups from all over Europe. On the platform, all relevant stakeholders are able to come together to connect through networking tools and find business growth initiatives and opportunities in their own and other regions across Europe. The matchmaking and partnering functionalities can benefit both businesses and innovation stakeholders, even investors, for commercial reasons and knowledge exchange. Furthermore, as the content is crucial for bringing the platform to life, a structured approach for producing content in the form of news articles, promotion kits, photo and video materials as well as startup event information has been carried out. The platform was designed with a modular structure to keep it flexible and extendable for various contexts. All modules and content-rich awareness channels are optimised for search engines (Google, Bing etc.) to ensure strong visibility and to facilitate long-term cooperation. Here, the platform provides access to various content, resources, information and users. This allows the platform to be more easily recognised and found by all interested stakeholders.

The purpose of this paper is to present the development framework and the architecture of the SPREAD2INNO One-stop shop web platform. Furthermore, the paper highlights the underlying modular approach of the platform which allows the integration of new content, services as well as the extension of capabilities with new functionalities within short iterations. The paper is organized as follows: Section 2 describes the online community of practice model. Section 3 presents the main concept, architecture and technical environment. Section 4 introduces the main modules (Knowledge, Directory, Networking and Administration) as well as their functionalities. Section 5 analyses the user engagement metrics and behavioural information regarding the platforms' users. The paper concludes with the conclusion statement and outlines next steps and future work.

2. Community of Practice concept

The term "*Community of Practice*" was proposed in 1991 by Jean Lave and Étienne Wenger when they debated the aspect of learning in the context of social relations [8]. In doing so, they showed that - in addition to structures or models - participation in a community is crucial for knowledge acquisition. The model of communities of practice is an attempt to show the "anatomy" of the interlocking of individual learning processes with those of the further development of the embedding social community. Therefore, the term describes a practice-oriented community of actors who are informally connected, face similar tasks and want to learn from each other through interaction.

2.1. Structure

In the community of practice, a certain structure is established, where everyone gets a specific role [9]. This role is determined on the basis of their activity and the acceptance or rejection by other members.

As a result, the "*Identity*" is being formed by the distribution of tasks, which are negotiated among the members. Thus, active and less active members as well as moderators and experts develop on the basis of communication processes. These moderators should not be imagined as rigidly assigned positions within the network of communities of practice. They also represent situational-negotiated functions and tasks within the community. Sub-groups can also be formed or external persons can be integrated as guests.

2.2. Criteria

A list of criteria exists, which limit the idea of communities of practice and make them identifiable. The members participate in a joint undertaking (*"Enterprise"*), they build together in a tool pool (*"Shared Repertoire of Tools"*), they negotiate standards and they participate in a common practice. Here, tasks are performed for the further development of the group, but also of each individual who has joined this community for a specific reason. The criteria include but are not limited to [10]:

- Having related enterprises
- Facing similar conditions
- Sharing historical roots or artefacts
- Very quick setup of a problem to be discussed
- Local lore, shared stories and inside jokes

2.3. Benefits

Communities of practice offer a variety of benefits: Relationships within teams or between organizations are strengthened, the flow of communication is improved, implicit knowledge is made visible, and learning is taking place faster and more efficient than in conventional training. Via the connections among the group within their community of practice, participants can acquire social capital in the process of sharing expertise that can provide value to the group and the individual [11].

2.4. Online Community

The online community of practice (also known as virtual community of practice) is a specific variation which uses the World Wide Web. It meets the predefined criteria of Lave and Wenger and therefore includes active members in a specific area of interest. To assist in the creation, sharing and negotiation of knowledge, social structures are created online within the community [12].

3. Technical framework and architecture

This section presents the technical foundations of the SPREAD2INNO One-stop shop web platform. Furthermore, the selected Content Management System (CMS), Plugin Framework Herbert as well as the overall architecture, plugins, services, tools and development approach will be described.

3.1. Concept

WordPress was used as the content management system as it provides a large number of important modules and functions, such as user and content administration. Since the standard version of the CMS system was insufficient for the use cases, the CMS was expanded with the Herbert plug-in framework as well as with an open-source framework called BuddyPress, which allows the implementation of community-like features if needed. For the front-end, depending on the needs of the project, the SPREAD2INNO platform can have public or restricted modules:

- **Public**: All visitors of the platform can access and view the provided content without the need to register
- **Restricted**: If users want to browse or filter through the provided content, they must first become a registered member of the platform

3.2. Architecture overview

Fig. 1. summarizes the architecture of the platform. Despite being separated into different sub-systems, the platform uses a single database where all content and user data are stored safely and securely using state-of-the-art database technologies and encryption. All modules and sub-systems can be accessed by the end user via a single-entry point, providing usability and simplicity. Technically, the web platform is built on established PHP Framework and CMS System WordPress including open-source libraries such as Bootstrap. It is a CSS, HTML and JS framework, which reduces the development effort of modern web application. MySQL is the preferred database system to be used to store and manage information. The Backend consist of a user and role management system including administration panels to handle the system efficiently.

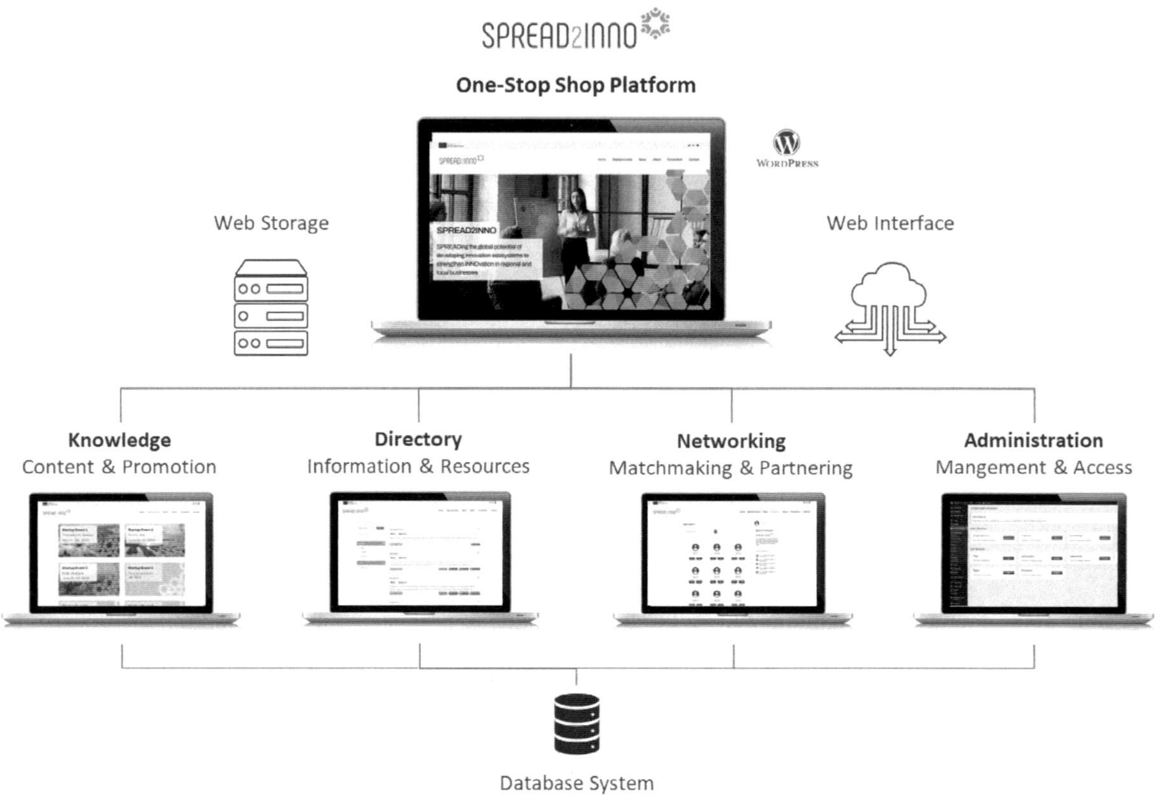

Fig. 1. SPREAD2INNO One-stop shop web platform architecture

3.3. Technical environment

In order to be most efficient, the One-stop shop is implemented with a flexible and scalable architecture and built on the basis of use cases, which ensures high intuitiveness and usability of the provided interfaces and functionalities. For the technical implementation, several programming platforms are used – including the WordPress Plugin Framework Herbert, Bootstrap, MySQL and WordPress. Table 1. provides an overview of the development environment based on software components and libraries used for the back-end and the front-end of the platform. The following sub-sections describe the key components of the development framework: the WordPress Plugin Framework, the JavaScript Library, the Google ReCAPTCHA framework and the WordPress Plugin that allow the implementation of a multilingual platform.

Resource Type	Name	Version
HTTP server	Apache	2.4.6
Programming Language	PHP	7.2.34
Database	MYSQL	15.1
PHP CMS	WordPress	6.2.2
Theme	QI	1.1
HTML, CSS & JS Framework	Bootstrap	4.7.0
JavaScript Library	jQuery	3.5.1
JavaScript Library	Google ReCAPTCHA	2.0
Operating System	Linux CentOS	7.4
WordPress Plugin Framework	Herbert	0.9
WordPress Plugin	WPML (WordPress Multilingual Plugin)	3.0
PHP Template engine	Twig	1.3
JavaScript Library	Leaflet.js	1.7.1

Table 1. Back-end and front-end technologies

Herbert: WordPress plugin framework which strongly relies on state-of-the-art stack technologies like symphony, Illuminate, twig and a modern, object-oriented programming approach [13]. Herbert makes the plugin development process faster and more organised and uses aspects and patterns of the Laravel programming framework. It was used mainly for the "Directory" module.

Google ReCAPTCHA: Free service that protects websites from spam and abuse. reCAPTCHA uses an advanced risk analysis engine and adaptive CAPTCHAs to keep automated software from engaging in abusive activities [14]. It does this while letting valid users pass through in a simplified manner. Every time CAPTCHAs are solved, that human effort helps digitize text, annotate images, and build machine learning datasets.

Twig: Template engine for the PHP programming language [15]. Its syntax originates from Jinja and Django templates. It's an open-source product licensed under a BSD License. Symfony2 PHP framework comes with a bundled support for Twig as its default template engine. It allows the creation of frontend pages and elements in a more elegant and efficient way without the need to mix several Web Development elements.

WPML (WordPress Multilingual Plugin): WPML is a plugin for WordPress [16]. WPML lets authors write content in different languages and translate content. It also includes advanced features for translation management and an interface for professional content translation. Although it is at this point not foreseen that this functionality will be used, it is implemented and ready if needed.

User and content management: As for the user and content management, an architecture was set up to ensure that all parameters of the users are taken into account. This architecture is based on the standard WordPress User Management System but was extensively extended by the "Networking" Modules which are provided by the free and open source BuddyPress plugin for WordPress.

WordPress: Free and open-source blogging tool, which can also be used as a fully functional content management system [17]. WordPress features a very intuitive and user-friendly administration interface. The simplicity in the usage of the CMS has a huge impact on the popularity of the software. In addition, the CMS features search engine optimized (SEO) URLs without the need to install any third-party extensions.

BuddyPress: Free and open-source plugin for WordPress which offers a networking and social media themed extension to WordPress [18]. It allows the creation of a user profile, user groups, forums and discussions among registered users of the platform as well as an activity stream, where user's recent activities such as the creation of new groups, new registered members and new discussions will be shown.

MVC Patterns: Model–View–Controller (usually known as MVC) is an architectural pattern used for developing user interfaces that divides an application into three interconnected parts, the Model, the View and Controller [19]. The MVC design pattern decouples these major components allowing for code reuse and parallel development. The Model consists of several model classes. Each represents a basic unit within the data structure used. They also provide basic data operations that are not part of the basic program logic.

4. Modules and functionalities

The SPREAD2INNO One-stop shop platform homepage, is accessible via the following link: https://www.spread2inno.eu/. It serves as a project website and web platform hybrid, meaning that it features awareness raising information about the SPREAD2INNO project (about, background, project facts, structure, deliverables, objectives, impacts, consortium, contact form), as well as promotional sections dedicated to the startup events, innovation academies and any other project-organised event or activity.

Generally, the platform development concept is differentiating between "*Information*" modules, which provide content in the form of articles, images, videos or similar content (such as short snippets), and "*Services*" modules, which are more interactive and allow users to engage with the platform, use various filters, search, visualisation or registration functions. The web platform is structured into three main areas, with an additional area for administration purposes, as shown in Fig. 2.

On a technical framework level, the comparison is based on:

- **Information**: Information modules are managed via a backend panel of the CMS (WordPress). The information-based content can be provided on the web platform with visual elements.
- **Services**: Services modules are managed via a backend panel (Admin Centre) and can be provided by the web platform host or by third parties. The services content is based on programmed lists and detailed views of entities connected to a database.

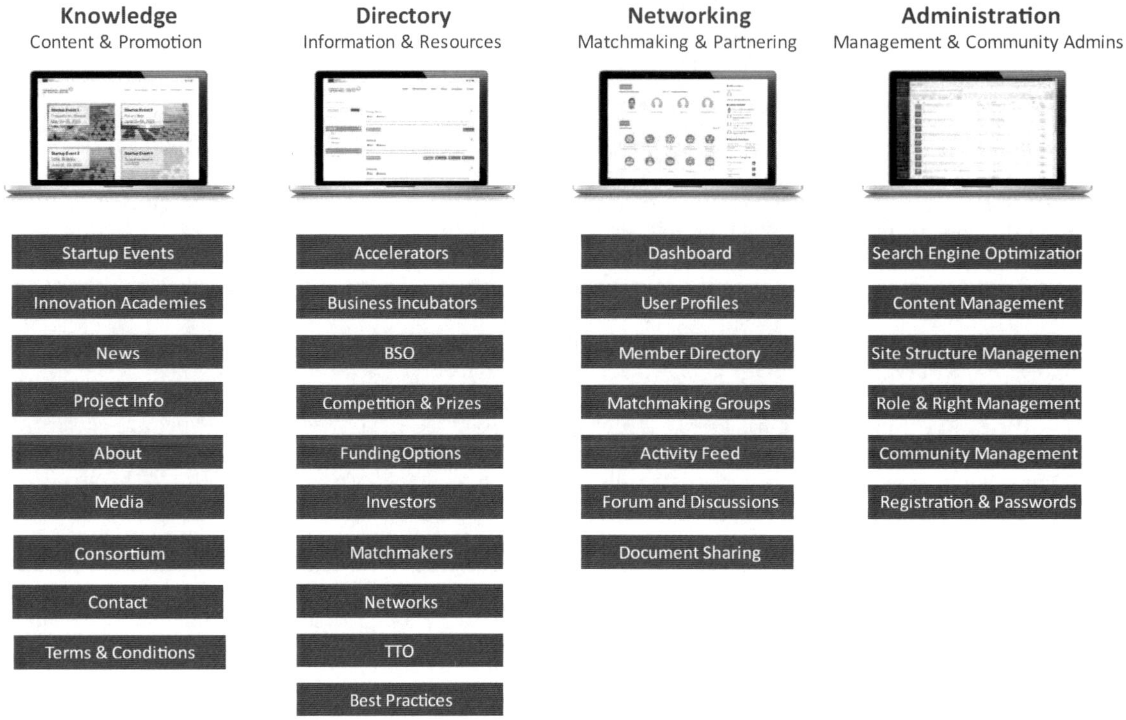

Fig. 2. SPREAD2INNO One-stop shop web platform modules

4.1. Knowledge

A steady creation of interesting content is crucial in order to be found online, to draw attention and to keep the targeted stakeholder groups coming back to the SPREAD2INNO web platform. As experience from other (EU) projects shows, online platforms need to show continuous activity and provide not only a set of tools but also content – especially if they are intended to be more than just a repository of tools.

Homepage: The top header displays the "Funded by the European Union" disclaimer logo as well as links to the project's social media channels (Twitter/X, LinkedIn and YouTube). Below, users can find the menu, followed by a dynamic header image. If the user has not been logged in, the "Register Here" banner is displayed. Otherwise, if the user has already completed the registration procedure and is logged in, "View my Profile" banner is displayed. The rest of the homepage consists of dashboard widgets which provide a dynamic overview of project related information.

Startup Events: The startup event page functions as a promotional site of the web platform. All six startup events were announced and published on the web platform, along with pre-event and post-event information, as well as interlinking to external services and platforms. Each event has its own dedicated page and follows a pre-established template, which present the header image, event information, photos, promotional videos and information about the winning startups.

Innovation Academies: Similar to the startup events, the innovation academies page functions as a promotional site of the web platform. Both innovation academies were announced and published on the SPREAD2INNO web platform, along with pre-event and post-event information, photos, promotional video and winning startups.

News: News articles contain both text and images to showcase quick and digestible information on specific upcoming events, noteworthy project milestones, new publications or recent developments. Furthermore, keywords are added for search engine optimization. The articles are tagged and can be searched as well as filtered for ensuring a broader usage. For writing the articles, a structured template was used, allowing for a quick implementation on the platform.

Project Info: The Project Info section introduces the SPREAD2INNO project, its main aims, an overview of activities and the main project objectives.

About: The About section presents the SPREAD2INNO project concept, the expected project journey, three main phases, impacts, background, project facts and structure, as well as a list of public project deliverables and publications, which are available for download.

Media: The Media section enables visitors to download SPREAD2INNO logo package, business cards, leaflets and stickers to be used for dissemination and awareness raising purposes.

Consortium: The consortium page lists each consortium member along with a description and their corresponding organisation website and social media channel links.

Contact: Details and the visualised location of the project coordinator as well as a web form for getting in touch with the project team connected to the office@spread2inno.eu email address. The contact form requires a name, email, subject and main body to be filled out.

Terms and conditions: In accordance with the General Data Protection Regulation (GDPR), the website also contains a detailed privacy policy including information about the (tracking) cookies enabled on the website. The "Terms & Conditions" page is linked from the footer section of the webpage, providing contact details to the role of data processing and controller entity.

4.2. Directory

The "Directory" module represents dynamic, interactive content that allows users to engage with the platform and to use various filter, search and visualisation functions. The database for the "Directory" module has been created as a result of mapping out the best practices and stakeholders in the regional innovation ecosystems of Bulgaria, Greece, and Italy as well as on EU-level. As a result, the "Directory" module offers networking and growth opportunities to business acceleration and innovation service providers, SMEs, and start-ups throughout the project's targeted low and medium innovation countries in Europe. All relevant stakeholders are able to connect through networking and find business growth initiatives and opportunities in their own and other regions across Europe. Furthermore, the "Directory" module also features an interactive map view, facilitating a fast, aggregated overview of all collected entities. Finally, an important functionality has been implemented which enables external stakeholders to add new entries in the directory and as such help the project to expand the database continuously beyond the end of the SPREAD2INNO project.

Currently, the directory includes the following entities:

- **Accelerators**: A total of 24 accelerators were identified, seven of which operate at EU level, six in Bulgaria, four in Greece, six in Italy and one which is accessible to companies from Bulgaria, Greece and Italy likewise.
- **Business incubators**: In the stakeholder analysis, 16 business incubators have been identified - three of them on EU level, eight active in Greece, two that operate in Italy and Bulgaria respectively, and one business incubator that operates in all three of the countries
- **Business Support Organisations (BSO)**: In the stakeholder analysis 67, Business Support Organisations have been identified, with 16 of them on EU level, eleven that operate in Greece, 34 in Italy and four in Bulgaria.
- **Competition and prizes**: In the frame of the stakeholder analysis, two prizes and seven competitions on the EU level as well as one Greek competition have been identified.
- **Funding options**: In the frame of the stakeholder analysis, six funding options have been identified, with five on the EU level and one active in Greece, Italy and Bulgaria
- **Investors for venture capital**: In the stakeholder analysis, 21 investors, one of which operates at the EU level, seven in Greece, one in Bulgaria, nine in Italy, two in Bulgaria and Greece and one that operates in all three regions were identified.
- **Matchmakers**: In the frame of the stakeholder analysis, six matchmakers have been identified, five of which are active on the EU level and one operating in Italy, Greece and Bulgaria
- **Networks**: In the stakeholder analysis, 30 networks in total with eight of them on the EU level, four that operate in Greece, 17 in Italy and one in Bulgaria were identified.
- **Technology Transfer Offices (TTO)**: In the stakeholder analysis, eight Technology Transfer Offices in total, with three of them being operated in Greece, four active in Italy and one operating in Bulgaria, were identified.
- **Best practices**: In the analysis, 18 best practices with six from Greece, five from Italy and seven from Bulgaria were identified.

4.3. Networking

The "Networking" module is based on the Community of Practice approach. It represents a virtual space and a meeting point for all innovation ecosystem stakeholders, business acceleration and innovation service providers, as well as SMEs and start-ups from all over Europe, as they represent one of the most important aspects for sustainability of the project. Here, the SPREAD2INNO One-stop shop networking module represents a single point of reference for information exchange between different innovation ecosystems in regards to opportunities, events and funding accessible in specific regions.

User profiles: Once registered, each user has a unique username and password, enabling them to personalise their profile. As such, under the "Profile" section of the side menu, users are able to:

- View existing data: first name, surname, user, organisation
- Edit and change aforementioned data, as well as defining organisation and role fields as public or private
- Upload or take a photo in addition to deleting the existing one
- Add a cover photo or delete the existing one

Member directory: All registered members on the SPREAD2INNO web platform can connect and create various public as well private groups and initiate discussions on related topics or issues. Furthermore, they can invite members to the group, share material and also actively connect with other users. Each group owner has the option to set privacy standards for the group, so that only group members can see the posts within a group.

Matchmaking groups: Groups can represent a virtual gathering space for beneficiaries, helping them to find a funding/investment opportunity as well as potential investor located at any level (local, regional, national and EU); accessing innovation agencies, business incubators, mentors and programs; interacting with business providers, SMEs and start-ups matched according to needs and available opportunities and initiatives. Furthermore, registered users can create a group on a specific topic to which they can invite other registered or non-registered users, and/or join existing groups. Within a group, users can share documents and start discussions in the form of public chats or as private conversations. The groups can also be open or closed, in order to let each team individually decide which information they want to make public and what they want to keep confidential.

Activity feed: To allow a quick overview of what has recently happened, the networking module includes a dedicated "Activities" feed, where the latest activities are shown (new members, profile updates, new question, likes, shared files etc.).

Forum and discussions: The cooperation capacity can be extended through the implemented forum, where project participants as well as other actors are able to debate, discuss and share both challenges and best practices, but also have an overview of accelerator programs, financial resources, mentors, events in partner ecosystems and share feedback. To use the forum, it is necessary for users to register on the platform and be logged into their account. The forum can be structured into general topics of relevance and allow users to add new threads to start discussions.

4.4. Administration

The Administration provides the backend system that consists of a user and role management system, Search Engine Optimisation including administration panels to handle the system efficiently. Here, the user and content management are based on the standard WordPress User Management System which enables creation of user profiles, user groups, forums and discussions among registered users of the platform.

5. User engagement

This section outlines analytical and behavioural information regarding the platform's users (registered members and visitors). Trends and insights are pointed out and displayed along with corresponding screenshots. Throughout the duration of the project, (starting from the web platform launch in July, 2023 up to early December, 2024), the web platform was visited numerous times by users from all over Europe as well as internationally (most notably from the US, India, China), as shown in Table 2. Fig. 3. below displays an excerpt from the Google Analytics services, which were used to analyse web platform activity over time. The graph indicates that the platform has had a steady traffic of visitors. Furthermore, there are numerous spikes overlapping with the six organised local events (2 per Italy, Bulgaria and Greece), two innovation academies as well as the training week. The last spike in November/December 2024 period overlaps with the final event as well as with the additional bonus contest round for startups registered on the platform.

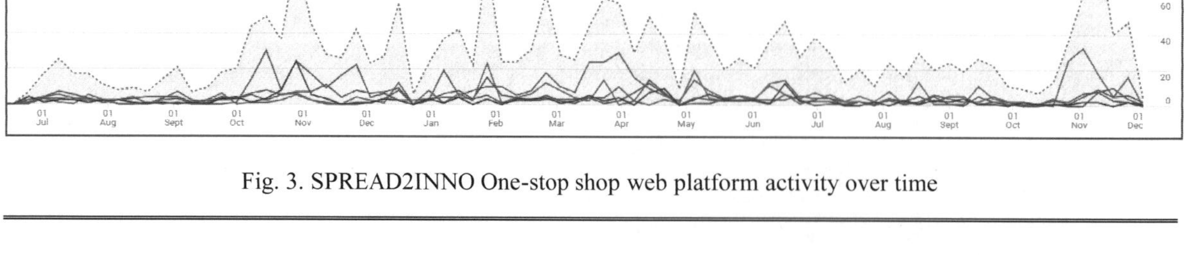

Fig. 3. SPREAD2INNO One-stop shop web platform activity over time

As shown in Table 2., total number of distinct active users who visited the web platform is 1700+, while the number of new users who interacted with or launched the platform for the first time is almost identical (98.7%). The number of engaged sessions that lasted 10 seconds or longer, or had 1 or more key events or 2 or more page or screen views is 2000+. The percentage of sessions that were engaged sessions equates to an engagement rate of over 54% while the average number of engaged sessions per active user is 1.18. The average engagement time that the web platform was in focus in an active user's browser equates to 1 minute and 25 seconds. The total number of times users triggered an event is 23000+. Overall, the highest engagement statistics were achieved in Italy, Greece, Bulgaria and Austria, coinciding with the innovation ecosystems targeted by the SPREAD2INNO project as well as locations of consortium partners.

Country	Active users	New users	Engaged sessions	Engagement rate	Engaged sessions per active user	Average engagement time per active user	Event count
US	360	359	64	17.78%	0.18	00:00:04	1236
Italy	299	293	464	59.49%	1.55	00:01:37	4616
Greece	228	223	382	67.73%	1.68	00:01:51	3845
Austria	152	152	515	65.11%	3.39	00:04:07	6604
Bulgaria	127	124	187	56.16%	1.47	00:02:01	2143
Netherlands	78	73	23	29.49%	0.29	00:00:13	292
Germany	73	68	90	56.96%	1.23	00:01:09	960
Finland	51	51	9	17.65%	0.18	00:00:03	164
France	50	46	17	31.48%	0.34	00:00:17	241
India	29	29	21	56.76%	0.72	00:00:58	282
...	...	...	...	...	...	...	...
TOTAL	**1718**	**1695**	**2027**	**54.11%**	**1.18**	**1m 25s**	**23173**

Table 2. SPREAD2INNO One-stop shop web platform engagement per country (Top 10 + all other)

A prominent feature of the SPREAD2INNO web platform is the activity feed. It not only shows the past activities that have taken place on the platform, but also allows members to search through them, create posts, see favourites, mention other members as well as to engage in public discussions. Table 3 summarises the most important activities to provide a quick overview of registered users' activities on the SPREAD2INNO web platform. Although the interface design was focused on providing a modern layout that is based on a responsive web design approach suitable for both desktop and mobile, the majority of users (83.30%) have interacted with the platform using their desktop web browsers, as shown in Table. 3. and the corresponding pie chart.

Description	Amount
Member registrations	70
Members joined a group	28
Member changed profile image	23
Member connections established	23
Profile updates	11
New posts published	10
Status update post	7
Discussions started (Q&A)	7
Groups created	4
Total	**183**

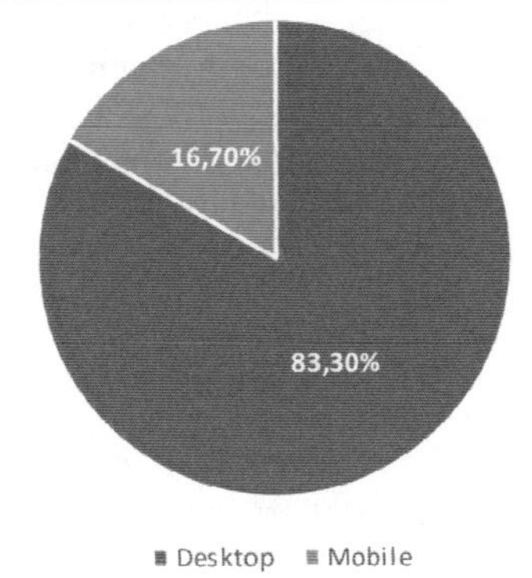

Table 3. Types of networking activities (Left); Active users by device category (Right) | Jul, 2023 – Dec, 2024

5. Conclusion

This paper presented the main development steps of the SPREAD2INNO One-stop shop web platform, showcasing the main concept, technical framework and architecture, individual modules and their functionalities. In this context, the main objective of the developed web platform was to facilitate the information exchange between different innovation ecosystems, targeted to business acceleration and innovation service providers, SMEs and start-ups from all over Europe, enabling them to connect and find business growth initiatives and opportunities.

To achieve this objective, the platform's main concept was based on the practice-oriented community of actors who are informally connected, face similar tasks and want to learn from each other through interaction. To this end, the platform was designed in a modular structure to keep it flexible and extendable for various contexts. All modules and content-rich awareness channels are optimised for search engines (Google, Bing, etc.) to ensure strong visibility and to facilitate long-term cooperation. Here, the platform provides access to various content, resources, information and users. This makes it possible to be more easily recognised and found by all interested stakeholders.

Based on the analysed user engagement metrics and behavioural information, the platform has achieved a steady visitor traffic flow as well as satisfactory user engagement and networking activities within a relatively short project timeframe. In this context, the Directory module provides a filterable database (also visualised as a map) of best practices and stakeholders in the regional innovation ecosystems. Since the directory content is public, all visitors can access and view the provided content without the need to register. This has resulted in the highest number of sessions and active users. Therefore, such repositories of relevant innovation ecosystem knowledge in combination with local events, programs and matchmaking opportunities can help to ensure the longevity and sustainability.

Future work will focus on iterative updates and refinements of the web platform, based on continuous testing, project requirements, consortium decisions and end-user engagement and feedback. Furthermore, ensured community engagement and maintenance of the web platform will play an essential role in scaling up the innovation ecosystems, sustainability and exploitation. As a result, the SPREAD2INNO web platform will be hosted for at least two more years after the end of the project in January 2025.

6. Acknowledgments

The SPREAD2INNO project [project number: 101071977] is funded by the European Union under the call HORIZON-EIE-2021-SCALEUP-01-01. Views and opinions expressed are however those of the author(s) only and do not necessarily reflect those of the European Union or the European Innovation Council and SMEs Executive Agency. Neither the European Union nor the granting authority can be held responsible for them.

7. References

[1] Autio, E., & Thomas, L. (2014). Innovation ecosystems (pp. 204-288). The Oxford handbook of innovation management.

[2] Moore, J. F. (1993). Predators and prey: a new ecology of competition. Harvard business review, 71(3), 75-86.

[3] European Innovation Ecosystems—European Commission. (2024, April 15). https://research-and-innovation.ec.europa.eu/funding/funding-opportunities/funding-programmes-and-open-calls/horizon-europe/european-innovation-ecosystems_en

[4] Pinheiro, F. L., Balland, P. A., Boschma, R., & Hartmann, D. (2022). The dark side of the geography of innovation: relatedness, complexity and regional inequality in Europe. Regional Studies, 1-16.

[5] European innovation scoreboard—European Commission. (n.d.). Retrieved December 2, 2024, from https://research-and-innovation.ec.europa.eu/statistics/performance-indicators/european-innovation-scoreboard_en

[6] EU Funding & Tenders Portal. (n.d.). Retrieved December 2, 2024, from https://ec.europa.eu/info/funding-tenders/opportunities/portal/screen/opportunities/topic-details/horizon-eie-2021-scaleup-01-01

[7] SPREADing the global potential of developing innovation ecosystems to strengthen INNOvation in regional and local businesses | SPREAD2INNO Project | Fact Sheet | HORIZON. (n.d.). CORDIS | European Commission. Retrieved December 2, 2024, from https://cordis.europa.eu/project/id/101071977

[8] Lave, J., & Wenger, E. (1991). Communities of practice: Creating learning environments for educators.

[9] Li, L. C., Grimshaw, J. M., Nielsen, C., Judd, M., Coyte, P. C., & Graham, I. D. (2009). Evolution of Wenger's concept of community of practice. Implementation science, 4, 1-8.

[10] Barab, S., Scott, J., Del Valle Martin, R., & Fang, F. (2012). Coming to terms with communities of practice: A definition and operational criteria.

[11] Fontaine, M. A., & Millen, D. R. (2004). Understanding the benefits and impact of communities of practice. In Knowledge networks: Innovation through communities of practice (pp. 1-13). IGI Global.

[12] Zhang, W., & Watts, S. (2008). Online communities as communities of practice: a case study. Journal of knowledge management, 12(4), 55-71.

[13] Getherbert/herbert. (2024). [PHP]. Herbert. https://github.com/getherbert/herbert (Original work published 2014)

[14] reCAPTCHA. (n.d.). reCAPTCHA. Retrieved December 2, 2024, from https://www.google.com/recaptcha/about/

[15] Twig—The flexible, fast, and secure PHP template engine. (n.d.). Retrieved December 2, 2024, from https://twig.symfony.com/

[16] WPML - The WordPress Multilingual Plugin. (n.d.). WPML. Retrieved December 2, 2024, from https://wpml.org/

[17] Blog Tool, Publishing Platform, and CMS. (2024, November 21). WordPress.Org. https://wordpress.org/

[18] BuddyPress.org. (n.d.). Retrieved December 2, 2024, from https://buddypress.org/

[19] Model View Controller Pattern Definition & Erklärung | Datenbank, DWH & BI Lexikon. (n.d.). Datenbanken - für Anfänger und Profis. Retrieved December 2, 2024, from https://www.datenbanken-verstehen.de/lexikon/model-view-controller-pattern/

DOI: 10.2507/35th.daaam.proceedings.012

IMPROVING DESIGN PARAMETERS OF DRIVING SYSTEMS

Vyarka Ronkova & Antoaneta Dobreva

This Publication has to be referred as: Dobreva, A[ntoaneta] & Ronkova, V[yarka] (2024). Improving Design Parameters of Driving Systems, Proceedings of the 35th DAAAM International Symposium, pp.0091-0095, B. Katalinic (Ed.), Published by DAAAM International, ISBN 978-3-902734-44-0, ISSN 1726-9679, Vienna, Austria
DOI: 10.2507/35th.daaam.proceedings.012

Abstract

Based upon the literature survey and upon the professional experience of the authors, the main objective of the research has been determined. Through analysing the contact and bending stresses in gear set meshing, the dimensions and the volume of the elements of driving systems are to be significantly reduced. An advanced methodology in six stages has been created and validated. The authors introduce two main design variables through which the improvement of the ratio between volume and load capacity is to be implemented. The possible boundaries of changing the macro-geometry indicators are determined by the permissible stresses according to the criteria: contact and bending strength in the gear set meshing. A significant improvement of the design parameters and of ratio between the volume and load capacity of the mechanism investigated has been achieved. Practical recommendations and conclusions have been deduced.

Keywords: Design parameters; Driving systems; Volume and load capacity; Advanced methodology.

1. Introduction

Gear trains are important machine components in driving systems. In the area of engineering design, the optimization of driving systems is critical, concerning the development of energy efficient mechanical systems. Gear optimization and the investigation of weight, volume and load capacity of these components are important areas of scientific studies.

An interesting investigation of an objective function has been done in order to determine some limits on the volume/weight ratio of a gearbox in [1]. The advantages of Matlab have been used in this study. Significant improvements in the optimization process of a pair of cylindrical gears have been introduced by Marcelin [2] suggesting to find solutions using a genetic algorithm. The module, face width and number of gear teeth have been used as design variables. The weight of the gear pair was used to evaluate each unit (i.e., objective function). The process also includes consideration of additional design variables: helix angle of the meshing, addendum coefficients.

Gologlu and Zeyveli [3] suggest a solution of the problem of optimal weight of a two-stage reducer. Power, input frequency, helix angle and gear ratio are included in the input databases, while gear modules, face width, and teeth number of each gear pair are used as design variables. However, the specific way of introducing boundary parameters leads to contradiction in the results obtained. Different types of wear and cooling conditions and the influence of these phenomena on the quality of design components have been thoroughly investigated in [4]. An interesting point of view is presented in [5] concerning the surface roughness of mechanical components. Six different kinds of materials have been considered.

Later on, the authors' team of [6] has continued the investigation in this area, applying precision and conventional cooling systems. Based upon the literature survey made, it can be concluded that the problem of the optimal ratio between weight and volume of the driving systems has been thoroughly investigated. However, most of the scientific papers have been focused on the results and efficiency of the algorithms applied [7], [8] and at the same time, no specific suggestions, offering definite guidelines and recommendations have been made.

Taking into account the literature analysis and based upon the professional experience of the authors' team, the research problem can be defined. It is possible to introduce practical approaches to improve the optimal usage of material and to reduce the volume of the driving systems by applying advanced methodology. The objective of the research is to achieve a maximum reduction of the dimensions and volume of the driving systems elements by analysing the contact and bending stresses in involute gear trains.

2. Prerequisite for solving the problem defined

The experience of the authors in the field of design theory, systems approach and new product development includes some contribution in the area of developing optimization procedures for improving the different parameters of transmissions. Original theoretical models supporting the improvement of transmission design have been presented in [9], [10]. The main objective of this research in the field of power transmissions is to improve the operating parameters of driving systems.

A study of planetary gear trains and internal meshing has been presented in [11]. Different methods for the complex analysis of internal meshing and for determining the permissible and optimal values of its geometry parameters according to different sets of criteria have been examined, applied and discussed. Improved and updated procedures for evaluation and comparative analysis of different design solutions to a certain problem, involving several main stages, are described in details in [12], [13] and [14].

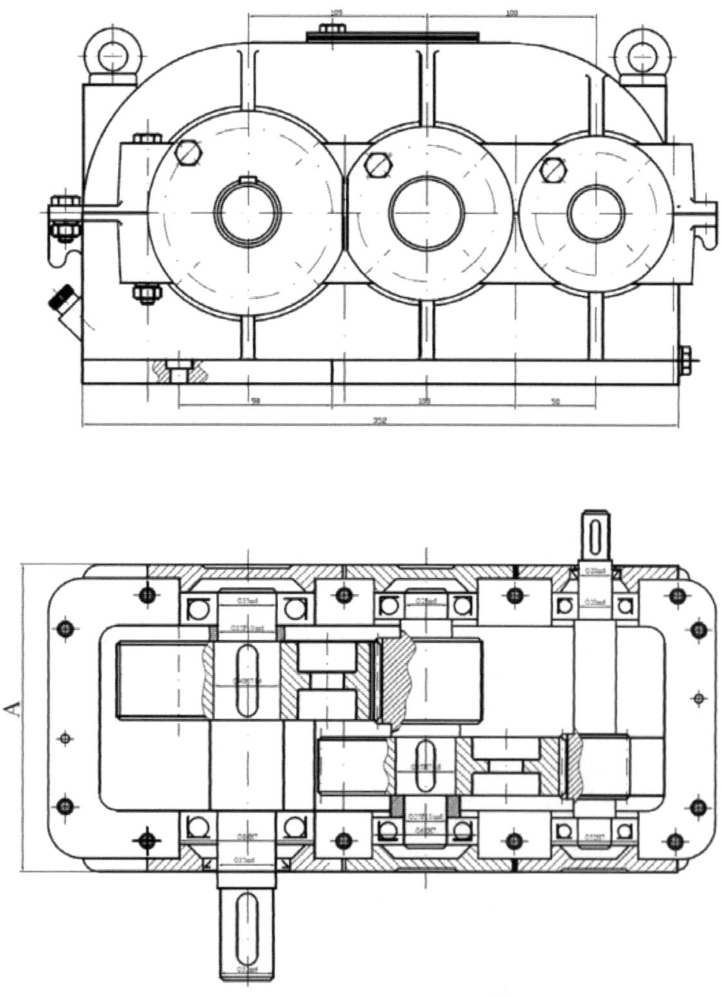

Fig. 1. Model of driving mechanism, [15]

Important components of optimization of involute gears have been considered in [15] and [16]. Models and drawings of various products have been presented as investigation results, Fig. 1. The contributions described are significant prerequisites for solving the problem defined and for creating an advanced methodology for improving the macro-geometry parameters of driving systems.

3. Advanced methodology

The methodology created goes through the following six stages:

- During the first stage the driving system input parameters are set as a database.
- The second stage includes verification calculations of the load capacity according to the criteria: contact strength and bending strength in the involute meshing.
- Based upon the results of the analytical calculations, the permissible boundaries of the changes in the macro-geometry indicators are to be defined in the third stage.
- During the fourth stage, main indicators are introduced as design variables, aiming to improve the ratio between volume and load capacity of the driving system components.
- In stage five, sequences of calculations are to be carried out in order to determine the influence of the main design variables defined in the previous stage.
- During the last stage, the results of the minimization of the volume and the geometry parameters of the driving system, obtained through defining permissible boundaries are to be presented and analyzed.

The authors' team considers that the research methodology presented is appropriate and adequate to the scientific problem which is to be solved. The selection of the determinant geometry parameter has a significant influence. A comparative analysis with the investigations presented in [15] and [16] indicates the application of similar optimization procedures. These procedures have been performed based upon the following geometry parameters: variation with the value of module; varying the total number of teeth; changes in the addendum coefficients of the involute profile: x_1 and x_2. The research presented in this publication considers the variation of meshing width as a function of the module to be a determining parameter in the design improvement of driving systems.

4. Application case study and results obtained

The methodology created has been applied for improving the design parameters of an involute two-stage reducer shown in Fig. 1. The input parameters are set as a database and include: rotational frequency and torque of the driven shaft, gear ratios, number of gear teeth and modules for both stages. The possible boundaries of the changes of the macro-geometry indicators are determined by the permissible stresses according to the criteria: contact strength and bending strength. The authors introduce two main indicators as design variables through which the improvement of ratio between the volume and the load capacity is to be carried out: face width (for the first b_{w1} and for the second b_{w2} stage of the reducer) and the geometry parameter A (indicated in Fig.1). This geometry parameter represents the width of the two-stage gearbox housing.

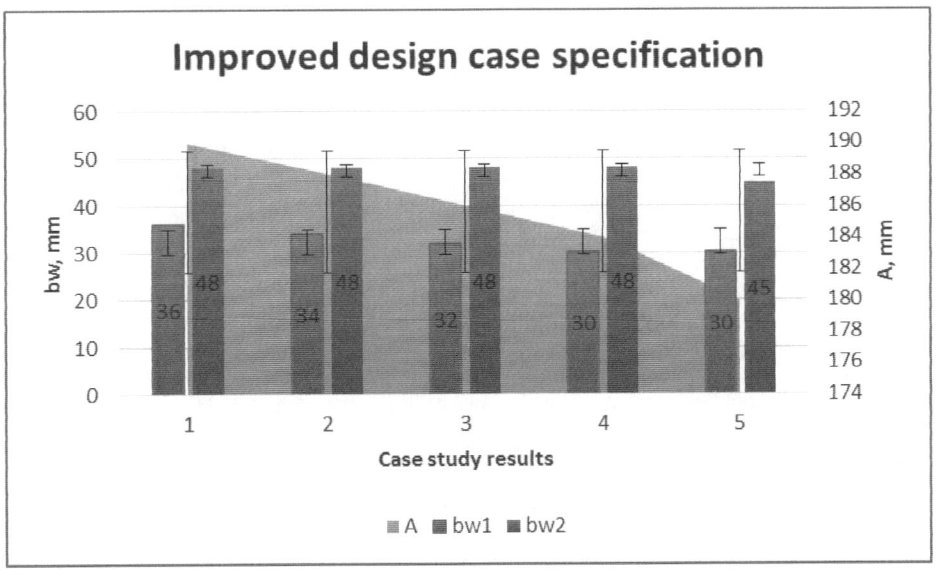

Fig. 2. Case study results

The results obtained are shown in a graphical way in Fig. 2. The reduction of the parameter "face width of the teeth meshing (for the first and for the second stage of the reducer)" is significant. It is decreased based upon its linear dependence on the module of the relevant stage. The recommendation is that the face width is to be reduced to the possible minimum value, defined by the strength calculation. Besides, this value is to be equal to or greater than 15.m (m – module of the relevant stage).

Applying the methodology described, the macro-geometry indicator "A" has been reduced significantly: with a value greater than 9%. The ratio between volume and load capacity in this application case study has been improved significantly, Fig. 2. Therefore, the objective of the research presented has been accomplished: a maximum reduction of the size and volume of the elements of the driving systems has been achieved. The results obtained are clear, explicit and significant. Besides, they are presented in a comprehensible and unambiguous way, which makes them appropriate for further applications in the engineering practice and engineering design education.

The limitations of the research implemented are connected to the selected main macro-geometry indicator "A". The intentions of the authors' team are to distinguish the influence of this parameter in future studies in this area, and to establish the comparative impact factor of other parameters with influence on the load capacity, the material and energy efficiency in different design versions of reducers and driving systems.

5. Conclusions

The research problem is clearly and precisely defined: by analysing the contact and bending stresses in involute gear trains, a maximum reduction concerning the dimensions and the volume of the driving systems elements is to be achieved. Based upon the results obtained, the following conclusions can be deduced:

An advanced methodology including 6 stages has been created and applied for a specific application case study: involute two-stage reducer. A significant improvement of the design parameters of this driving mechanism has been achieved (over 9%), which also leads to an improvement in the ratio between the volume and the load capacity of the mechanism investigated. It is recommended that the face width be reduced to the possible minimum value, defined by the strength calculation. Besides, this value is to be equal to or greater than 15.m (m – module of the relevant stage).

Applying the methodology presented, the expected results and their application will lead to improving the design versions of driving systems concerning their reduced overall dimensions and volume. These results will affect the total weight of the design versions and will lead to lightweight design. Theses outcomes achieved will contribute to the improvement of economic indicators, such as prime cost and price of the final product.

The future research work of the authors' team envisages further application of a methodology created through defining other main indicators and design variables. Investigations of two-stage involute helical gear trains and driving systems operating with different torque values are planned. Housing details will be presented and juxtaposed using CAD systems. Calculations of the volumes of the relevant housing are to be impelemented in order to realize the most accurate assessment of the efficiency, achieved as a result of the methodology applied.

6. Acknowledgments

The authors express gratitude and acknowledgements for the support of the scientific fund at the University of Ruse, Bulgaria.

7. References

[1] Golabi, S., Fesharaki, J. & Yazdipoor, M. (2014). Gear train optimization based on minimum volume/weight design, Mech. Mach. Theory 73, pp. 197–217, doi: 10.1016/j.mechmachtheory.2013.11.002

[2] Marcelin, S. (2001). Genetic optimisation of gears, Int. J. Adv. Manuf. Technol. 17, pp. 910–915, doi: 10.10 07/s0 0170 0170101

[3] Gologlu, C. & Zeyveli, M. (2009). A genetic approach to automate preliminary design of gear drives, Comput. Ind. Eng. 57, pp. 1043–1051, doi: 10.1016/j.cie. 20 09.04.0 06

[4] Bleicher, F.; Finkeldei, D. & Siller, A. (2016). Machining of Difficult-To-Cut Materials. Proceedings of the 27th International DAAAM Symposium 2016. DAAAM International Vienna, p.p. 0473–0479. ISBN 978-3-902734-08-2, doi:10.2507/27th.daaam.proceedings.071

[5] Fulemova, J. & Janda, Z. (2012). Selection of cutting material for milling of ferritic-martensitic stainless steel tool wear vs. roughness of machined surface. Proceedings of the 23rd International DAAAM Symposium 2012. DAAAM International Vienna, p.p. 0361–0364. ISBN 978-3-901509-91-9,

[6] Lukas, A. & Sykora, J. (2022). Comparison of Optimized and Conventional Cooling When Turning Stainless Steel, Proceedings of the 33rd DAAAM International Symposium, pp.0511-0518, B. Katalinic (Ed.), Published by DAAAM International, ISBN 978-3-902734-36-5, ISSN 1726-9679, Vienna, Austria, DOI: 10.2507/33rd.daaam.proceedings.071

[7] Mendi, M., Baskal, T., Boran, K. & Boran, F. (2010). Optimization of module, shaft diameter and rolling bearing for spur gear through genetic algorithm, Expert Syst. Appl. 37, pp. 8058–8064, doi: 10.1016/j.eswa.2010.05.082

[8] Tudose, L., Buiga, C.,Tefanache, S. & Sóbester, A. (2010). Automated optimal design of a two-stage helical gear reducer, Struct. Multidiscip. Optim. 42, pp. 429–435, doi: 10.10 07/s0 0158-010- 0504- z

[9] Stoyanov, S., V. Dobrev, V. & Dobreva, A. (2017). Investigating Dynamic Behavior of Planetary Gear Trains through the Systematic Approach, VDI Berichte, 2, pp. 127 - 132

[10] Stoyanov, S., V. Dobrev, V. & Dobreva, A. (2017). Finite Element Contact Modelling of Planetary Gear Trains, Mat. Science and Eng., 252, pp. 012034 – 38

[11] Dobreva, A. & Stoyanov, S. (2012). Optimization Research of Gear Trains with Internal Meshing, University Publishing Centre, Ruse, 137 pp

[12] Stoyanov, S. & A. Dobreva, A. (2010). Development, Design and Optimization of Planetary Gear Trains for Vehicles – Computer Aided Frequency Analysis of Planetary Gears, VDI – Berichte, 2108.2, pp. 1423 – 1426

[13] Dobreva, A. (2013). Theoretical Investigation of the Energy Efficiency of Planetary Gear Trains, Springer Verlag Dordreeht, pp. 289-298

[14] Dobreva, A. & Dobrev, V. (2018). Innovative Methodology for Decreasing Mechanical Losses in Vehicles, Proc. of the 4th International Cong. of Aut. and Trans. Eng., Springer Verlag, pp 234 – 242

[15] Angelova, E., Ronkova, V. & Nenov, P. (2010). Increasing Load Capacity of Cylindrical Gear by Optimizing Their Geometric Parameters, INMATEH – Agricultural Engineering, vol 31, №2, pp. 40-46, ISBN 2068-2239

[16] Nenov P., Angelova, E., Kamenov, K. & Ronkova, V. (2013). Visualization Packages for Quick Accumulation of Experience in High Quality Design Of Reducers, 3 rd International Symposium on Computing in Science & Engineering (ISCSE 2013), pp. 193-198, ISBN 2147-9097

DOI: 10.2507/35th.daaam.proceedings.013

NEW INSIGHT INTO BLACK CARBON ANALYSIS OF STRONG AIR POLLUTION EPISODES

Adnan Masic, Emina Dzaferovic-Masic, Mirnes Durakovic & Halim Prcanovic

This Publication has to be referred as: Masic, A[dnan]; Dzaferovic-Masic, E[mina]; Durakovic, M[irnes] & Prcanovic, H[alim] (2024). New Insight into Black Carbon Analysis of Strong Air Pollution Episodes, Proceedings of the 35th DAAAM International Symposium, pp.0096-0101, B. Katalinic (Ed.), Published by DAAAM International, ISBN 978-3-902734-44-0, ISSN 1726-9679, Vienna, Austria
DOI: 10.2507/35th.daaam.proceedings.013

Abstract

Continuous measurement and analysis of black carbon concentrations in ambient air during a two-year period in Sarajevo and Zenica valleys was performed. The central apparatus for this analysis consists of two AE33 aethalometers, which work on optical principles of the attenuation of light by black carbon deposits on filter tape at seven different wavelengths. Measuring solely black carbon concentration would be performed at one wavelength, but expanding the measurements at seven different wavelengths provides an extensive analysis of dependence of attenuation on the wavelength of light. This enables the derivation of the Angstrom exponent, whose values differentiate the origin of black carbon from biomass burning (wood and coal) from liquid fossil fuel (such as diesel oil). This is the first analysis of this type performed simultaneously in Sarajevo and Zenica valley under strong air pollution episodes.

Keywords: Black carbon; Air pollution; Angstrom exponent; Temperature inversion; Source apportionment.

1. Introduction

Black carbon (BC) is a component of air pollution that originates from incomplete combustion of fossil fuels (FF) and biomass burning (BB). It is argued that BC is the second largest anthropogenic climate change factor [1], [2] (carbon dioxide being first). The negative health effects of black carbon are well documented [3], [4]. Previous research articles include Evaluation with a photoacoustic instrument [5], method comparisons [6], vertical profiles over Milan [7], diesel exhaust [8], wood combustion particles [9], emissions from fresh biomass burning [10], and source apportionment [11]. Western Balkans countries are highly affected by air pollution and BC pollution. Those countries that are not members of the European Union may have missing regulations on BC and air pollution and sparse and incomplete monitoring. Bosnia and Herzegovina is a good example: it has strong air pollution due to several factors, some of which are natural, while others are anthropogenic. Sarajevo, the capital of Bosnia and Herzegovina, is well known for air pollution caused by the burning of FF and BM and unfavourable microclimate (temperature inversions in winter time), while the city of Zenica has a steel foundry in the city centre and has been a synonym for air pollution for decades.

This research is advanced measurement and analysis of long-term campaigns in Sarajevo and Zenica simultaneously, which is a very valuable asset in an attempt to understand and solve the air pollution problem in Bosnia and Herzegovina.

2. Experimental setup

Black carbon has the ability to absorb solar radiation (almost the entire spectrum). This property enables the construction of optical-based devices, called aethalometers, for the measurement of BC concentrations. We have used AE33 aethalometers from Aerosol Magee Scientific (Berkeley, California, United States of America). At the beginning of the air sampling line, the inlet cyclone separates particles greater than 2.5 μm in diameter. BC particles are collected on filter tape, and the light absorption method is used to calculate BC concentration at 880 nm wavelength, as described in our previous work [12]. The black carbon facility is one component of a larger project of our research group, where different aspects of air pollution are observed and compared, including toxic gases [13], [14], and particulate matter [15]-[21]. We have used this technique previously [22], but this time, a new location is included: the city of Zenica. Figure 1 shows two measurement locations: one in Sarajevo (Faculty of Mechanical Engineering, N 43° 51' 15", E 18° 23' 46") and one in Zenica (Radakovo, N 44° 11' 43", E 17° 55' 55").

Although the BC is measured at the wavelength 880 nm, our measurements were performed at 7 different wavelengths: 370, 470, 520, 590, 660, 880, and 950 nm. The reason for this is the following: if we have measurements at multiple wavelengths, we can deduce the Angstrom exponent, which is the coefficient of (logarithmic) dependence of optical attenuation on the wavelength of light. It is possible then to deduce biomass burning (BB) as a percentage of total BC. This technique can be used for source apportionment since it distinguishes solid from liquid fuels as the source of BC.

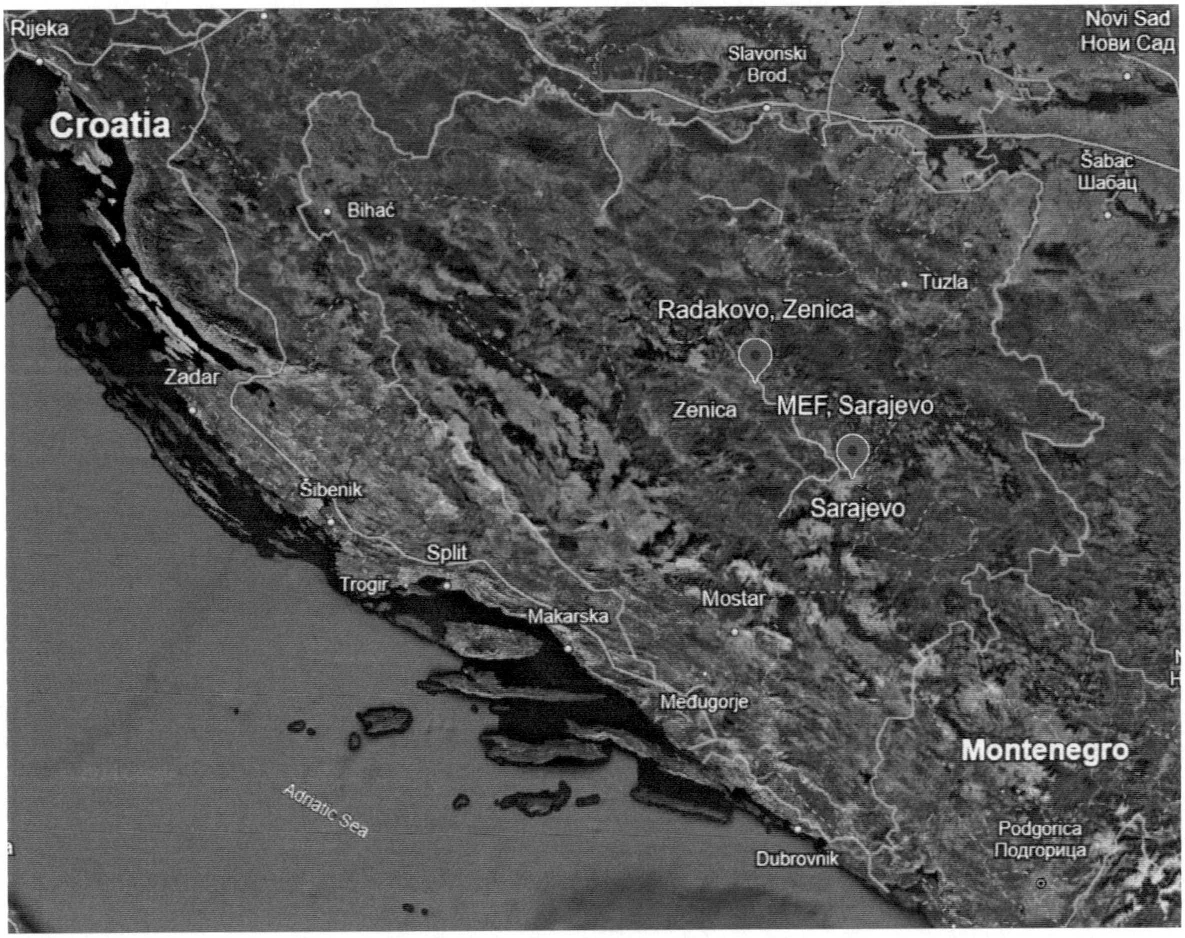

Fig. 1. Locations of measurements

3. Results

The measurements were performed in Sarajevo and Zenica simultaneously in 2023 and 2024 (up to 07/OCT/2024). A rigorous quality assurance (QA) procedure was implemented: all measurements are checked against multiple criteria (flow rate ratio, optical performance, etc.), and only those with near-perfect status are kept.

That is the reason why we have some gaps in data, but the quality of such data is the highest possible. Figure 2 shows measurements in Sarajevo. Hourly and daily average values are plotted. The average value of BC concentration was 2.943 μg/m³. It is observable from Figure 2 that the highest peaks are in winter months, namely in December, where we have measured the highest value of daily average BC concentration of 35.212 μg/m³.

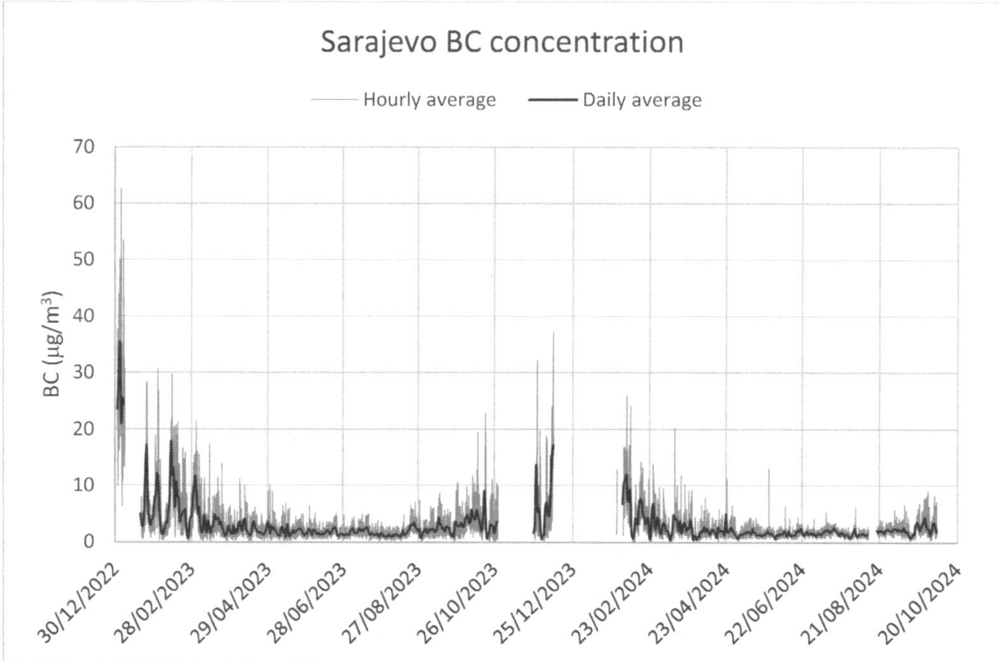

Fig. 2. BC concentration in Sarajevo

Figure 3 shows the BB percentage during the same period. While BB is above 50% during the winter, it drops to less than 20% during the summertime. This clearly indicates the primary source of air pollution in Sarajevo: burning solid fuels such as wood and coal for domestic heating! The average BB during the entire period was 27.48%.

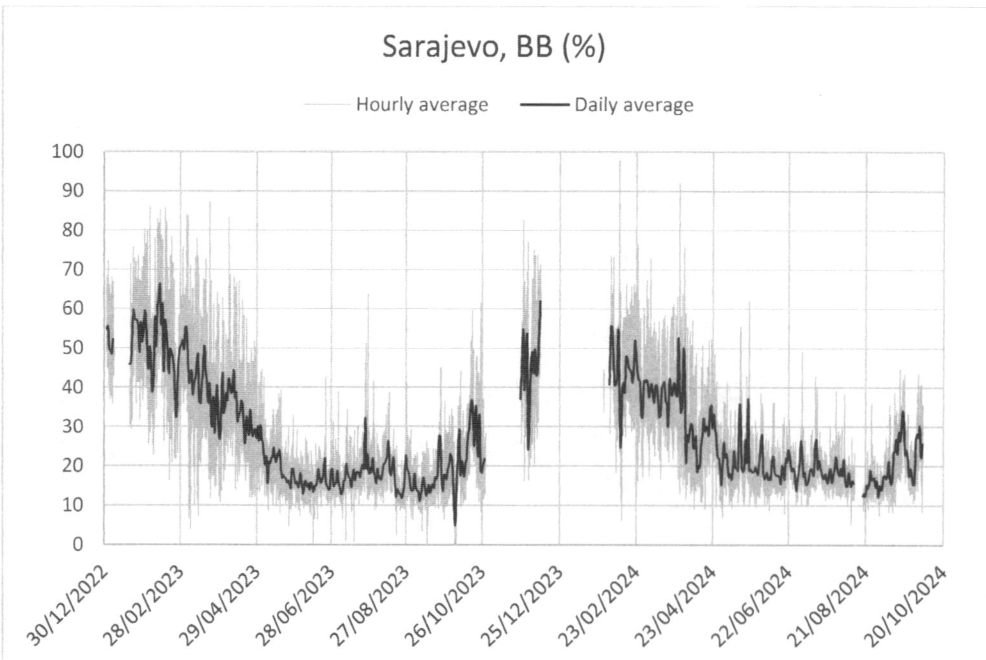

Fig. 3. BB percentage in Sarajevo

Measurements in Zenica are shown in Figure 4. The average value of BC concentration was 2.064 μg/m^3, while the highest measured daily average value was 11.971 μg/m^3. BB percentage was around 20% during the summer and around 60% during the winter. A similar conclusion about sources remains: the primary source of BC during the winter is the burning of solid fuels. The average BB during the entire period was 36.15%. Interesting comparisons could be drawn. While there was a greater BC concentration in Sarajevo than in Zenica, the BB percentage was lower in Sarajevo than in Zenica. Since BB is one of the main sources of BC [23], it follows that BC concentration captured in Sarajevo comes from other sources, not just BB. Zenica valley is known for its industrial sites, which would explain the difference in BB percentage in Sarajevo and Zenica. This needs further analysis and clarification.

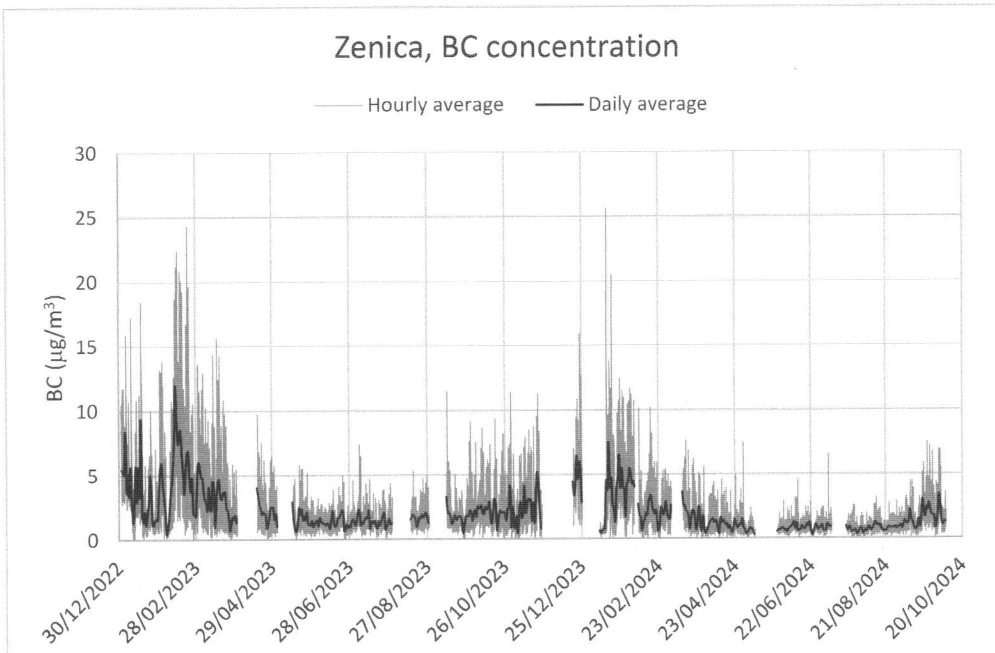

Fig. 4. BB percentage in Sarajevo

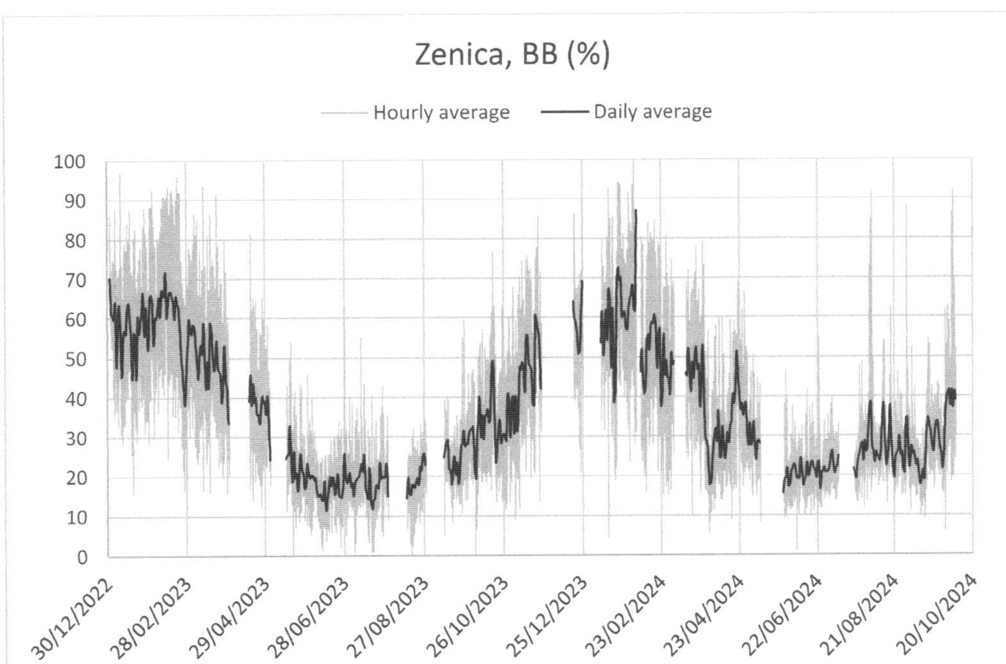

Fig. 5. BB percentage in Sarajevo

4. Conclusions

The long-term BC measurement and analysis campaign in Sarajevo and Zenica was performed using two AE33 aethalometers. The results are auspicious and very promising: it is possible to distinguish now the primary source of black carbon (and consequently the air pollution) using the Angstrom exponent, which is calculated from measurements of 7 different wavelengths of light. A very similar pattern is observed in Sarajevo and Zenica: BB drops to about 20% during the summer and rises above 50% during the winter. Average concentrations of BC were high, particularly during the winter. This research should continue in the future. Every new winter season is of particular interest. Comparison with EU member states would be indicative as well.

5. References

[1] Ramanathan, V. & Carmichael, Gregory. (2008). Global and Regional Climate Changes Due to Black Carbon. Nat Geosci. 1. DOI: 10.1038/ngeo156.

[2] Bond, T. C. et al. (2013), Bounding the role of black carbon in the climate system: A scientific assessment, J. Geophys. Res. Atmos., 118, 5380– 5552, DOI: 10.1002/jgrd.50171.

[3] Bové, H.; Bongaerts, E.; Slenders, E. et al. (2019). Ambient black carbon particles reach the fetal side of human placenta. Nat Commun 10, 3866 (2019). DOI: 10.1038/s41467-019-11654-3

[4] Janssen, N. A.; Hoek, G.; Simic-Lawson. M.; Fischer, P.; van Bree, L.; ten Brink, H.; Keuken, M.; Atkinson, R. W.; Anderson, H. R.; Brunekreef, B. & Cassee, F. R. (2011). Black carbon as an additional indicator of the adverse health effects of airborne particles compared with PM10 and PM2.5. Environ Health Perspect. 119(12):1691-9. DOI: 10.1289/ehp.1003369

[5] Arnott, W. P.; Hamasha, K.; Moosmüller, H.; Sheridan, P. J. & Ogren, J. A. (2005). Towards Aerosol Light-Absorption Measurements with a 7-Wavelength Aethalometer: Evaluation with a Photoacoustic Instrument and 3-Wavelength Nephelometer, Aerosol Science and Technology, 39:1, 17-29, DOI: 10.1080/027868290901972

[6] Babich, P.; Davey, M.; Allen, G. & Koutrakis, P. (2000). Method Comparisons for Particulate Nitrate, Elemental Carbon, and PM2.5 Mass in Seven U.S. Cities, Journal of the Air & Waste Management Association, 50:7, 1095-1105, DOI: 10.1080/10473289.2000.10464152

[7] Ferrero, L.; Mocnik, G.; Ferrini, B.S.; Perrone, M.G.; Sangiorgi, G. & Bolzacchini, E. (2011). Vertical profiles of aerosol absorption coefficient from micro-Aethalometer data and Mie calculation over Milan, Science of The Total Environment, Volume 409, Issue 14, 2011, Pages 2824-2837, ISSN 0048-9697, DOI: 10.1016/j.scitotenv.2011.04.022.

[8] Ning, Z.; Chan, K.L.; Wong, K.C.;.Westerdahl, D.; Močnik, G.; Zhou, J..H. & Cheung, C. S. (2013). Black carbon mass size distributions of diesel exhaust and urban aerosols measured using differential mobility analyzer in tandem with Aethalometer, Atmospheric Environment, Volume 80, 2013, Pages 31-40, ISSN 1352-2310, DOI: 10.1016/j.atmosenv.2013.07.037.

[9] Wang, Y.; Hopke, P. K.; Rattigan, O. V.; Xia, X.; Chalupa, D. C. & Utell, M. J. (2011). Characterization of Residential Wood Combustion Particles Using the Two-Wavelength Aethalometer Environmental Science & Technology 2011 45 (17), 7387-7393, DOI: 10.1021/es2013984

[10] Pan, X.; Kanaya, Y.; Taketani, F.; Miyakawa, T.; Inomata, S.; Komazaki, Y.; Tanimoto, H.; Wang, Z.; Uno, I. & Wang, Z. (2017). Emission characteristics of refractory black carbon aerosols from fresh biomass burning: a perspective from laboratory experiments, Atmos. Chem. Phys., 17, 13001–13016, DOI: 10.5194/acp-17-13001-2017

[11] Zotter, P.; Herich, H.; Gysel, M.; El-Haddad, I.; Zhang, Y.; Močnik, G.; Hüglin, C.; Baltensperger, U.; Szidat, S. & Prévôt, A. S. H. (2017). Evaluation of the absorption Ångström exponents for traffic and wood burning in the Aethalometer-based source apportionment using radiocarbon measurements of ambient aerosol, Atmos. Chem. Phys., 17, 4229–4249, DOI: 10.5194/acp-17-4229-2017

[12] Masic, A.; Bibic, D.; Kamenica, I.; Bektesevic, J.; Haziabdic, V. & Mehuljic, M. (2020). Mobile Measurements of Black Carbon for Source Apportionment Studies, Proceedings of the 31st DAAAM International Symposium, pp.0227-0231, B. Katalinic (Ed.), Published by DAAAM International, ISBN 978-3-902734-29-7, ISSN 1726-9679, Vienna, Austria DOI: 10.2507/31st.daaam.proceedings.031

[13] Masic, A.; Bibic, D.; Pikula, B. & Razic, F. (2018). New approach of measuring toxic gases concentrations: principle of operation, Proceedings of the 29th DAAAM International Symposium, pp.0882-0887, B. Katalinic (Ed.), Published by DAAAM International, ISBN 978-3-902734-20-4, ISSN 1726-9679, Vienna, Austria, DOI: 10.2507/29th.daaam.proceedings.127

[14] Bibic, D.; Pikula, B.; Masic, A. & Razic, F. (2018). New approach of measuring toxic gases concentrations: application examples, Proceedings of the 29th DAAAM International Symposium, pp.0876-0881, B. Katalinic (Ed.), Published by DAAAM International, ISBN 978-3-902734-20-4, ISSN 1726-9679, Vienna, Austria, DOI: 10.2507/29th.daaam.proceedings.126

[15] Masic, A.; Pikula, B.; Bibic, D.; Musemic, R. & Halac, A. (2018). Calibration and Assessment of Low-cost Dust sensors, Proceedings of the 29th DAAAM International Symposium, pp. 523-528, B. Katalinic (Ed.), Published by

DAAAM International, ISBN 978-3-902734-xx-x, ISSN 1726-9679, Vienna, Austria, DOI: 10.2507/29th.daaam.proceedings.075

[16] Masic, A.; Bibic, D. & Pikula, B. (2019). On the applicability of low-cost sensors for measurements of aerosol concentrations, Proceedings of the 30th DAAAM International Symposium, pp.0452-0456, B. Katalinic (Ed.), Published by DAAAM International, ISBN 978-3-902734-22-8, ISSN 1726-9679, Vienna, Austria DOI: 10.2507/30th.daaam.proceedings.060

[17] Masic, A.; Pikula, B. & Bibic, D. (2017). Mobile Measurements of Particulate Matter Concentrations in Urban Area, Proceedings of the 28th DAAAM International Symposium, pp.0452-0456, B. Katalinic (Ed.), Published by DAAAM International, ISBN 978-3-902734-11-2, ISSN 1726-9679, Vienna, Austria DOI: 10.2507/28th.daaam.proceedings.063

[18] Masic, A.; Bibic, D.; Pikula, B.; Dzaferovic-Masic, E. & Razic, F. (2018). Visualization of Data from Network of Sensors: Appropriate Spatial Interpolation Method, Proceedings of the 29th DAAAM International Symposium, pp.0529-0533, B. Katalinic (Ed.), Published by DAAAM International, ISBN 978-3-902734-20-4, ISSN 1726-9679, Vienna, Austria DOI: 10.2507/29th.daaam.proceedings.076

[19] Masic. A.; Bibic, D.; Pikula, B.; Blazevic, A.; Huremovic, J. & Zero, S. (2020). Evaluation of optical particulate matter sensors under realistic conditions of strong and mild urban pollution. Atmospheric Measurement Techniques, 13, 6427–6443, DOI: 10.5194/amt-13-6427-2020

[20] Masic, A.; Pikula, B.; Haziabdic, V.; Mehuljic, M. & Bektesevic, J. (2020). Evaluation of Face Mask Filtering Efficiency Using High-End Scanning Mobility Particle Sizer, Proceedings of the 31st DAAAM International Symposium, pp.0224-0226, B. Katalinic (Ed.), Published by DAAAM International, ISBN 978-3-902734-29-7, ISSN 1726-9679, Vienna, Austria DOI: 10.2507/31st.daaam.proceedings.030

[21] Masic, A.; Kepnik, G.; Bektesevic, J.; Mehuljic, M.; Saric, I. & Haziabdic, V. (2020). The Network of Smart Sensors for Indoor Air Quality Monitoring, Proceedings of the 31st DAAAM International Symposium, pp.0232-0235, B. Katalinic (Ed.), Published by DAAAM International, ISBN 978-3-902734-29-7, ISSN 1726-9679, Vienna, Austria DOI: 10.2507/31st.daaam.proceedings.032

[22] Masic, A.; Kepnik, G.; Bibic, D.; Mesic E. & Pikula, B. (2021). Long-term measurement of black carbon concentration and source apportionment, Proceedings of the 32nd DAAAM International Symposium, pp.0131-0135, B. Katalinic (Ed.), Published by DAAAM International, ISBN 978-3-902734-33-4, ISSN 1726-9679, Vienna, Austria DOI: 10.2507/32nd.daaam.proceedings.020

[23] Katz, S. D.; Kelly, R. P. ; Robinson, R. S. et al. (2024). Biomass burning is a source of modern black carbon to equatorial Atlantic Ocean sediments. Communications Earth Environment 5, 536 DOI: 10.1038/s43247-024-01642-x

DOI: 10.2507/35th.daaam.proceedings.014

POSSIBILITIES OF UTILIZING ARTIFICIAL INTELLIGENCE FOR CONTROLLING TECHNICAL PARAMETERS OF PRIMARY AND SECONDARY COOLING CIRCUITS OF VVER-1000(1200) NUCLEAR REACTORS

Vyacheslav V. Potekhin, Amirhossein Bahrami & Anshuman Mishra

Peter the Great St. Petersburg Polytechnic University

This Publication has to be referred as: Potekhin, V[yacheslav]; Bahrami, A[mirhossein] & Mishra, A[nshuman] (2024). Possibilities of Utilizing Artificial Intelligence for Controlling Technical Parameters of Primary and Secondary Cooling Circuits of Vver-1000(1200) Nuclear Reactors, Proceedings of the 35th DAAAM International Symposium, pp.0102-0111, B. Katalinic (Ed.), Published by DAAAM International, ISBN 978-3-902734-44-0, ISSN 1726-9679, Vienna, Austria
DOI: 10.2507/35th.daaam.proceedings.014

Abstract

In the following article, the possibility of using intelligent control systems (more specifically fuzzy logic controllers) to regulate the technological parameters of the 1st and 2nd loops of VVER-1000 (1200) type reactors has been analyzed. To get a comparative analysis of the regulation of fuzzy-logic-based controllers, as compared to usual PI-controllers, the comparison of the pressure regulation in the purge-recharge system (TK) of the 1st loop of the VVER-1000 reactor, by the control systems mentioned earlier was simulated using the "MatLab-Simulink" software. The ability of both types of controllers to support and maintain the required pressure level of 180 kgf/cm^2 (about 17.6 MPa) was analyzed and conclusions were drawn regarding the advantages and disadvantages of a fuzzy logic-based control system controlling the mentioned reactor parameter as compared to the usual PI-controller.

Keywords: Intelligent control systems; Reactor; regulator; Machine learning; fuzzy logic

1. Introduction

Instrumentation and control systems in a nuclear power station (NPP) can be considered as the nervous system of the plant. Through its various constituent elements (e.g. equipment, modules, subsystems, redundancies, systems, etc.), the plant I&C system senses basic parameters, monitors performance, integrates information, and makes automatic adjustments to plant operations as necessary [1]. However, at present, the majority of components utilized in the I&C systems in NPPs around the world are based on old technologies of the 1960s-70s. Since those years, there have been significant developments in instrumentation and control technologies due to the rapid advancements in digital electronics and software engineering.

These developments have led to more modern instrumentation and more automated instrumentation and control systems being adopted across different industries. One such modern control system is the "Intelligent control systems" which control the parameters of the object of control, relying on the principles of artificial intelligence (AI) [2].

In this article, the possibilities of adopting such intelligent control systems for the I&C systems of NPPs will be investigated. The architecture of the I&C system, its instruments, and personnel together serve as the central nervous system of the NPP. Utilizing its various components, the I&C system determines all the important parameters, and the performance of the station and performs corrective action on the parameters, when necessary, automatically. The system also reacts to abnormalities during the operation of the plant and takes action to ensure safety. The primary goal of the I&C system of the NPP is to provide energy production at the highest possible efficiency while ensuring the safety of the plant personnel as well as the surroundings.

To serve the purpose of the I&C system mentioned above, hundreds of technical, physical, and chemical parameters such as pressure, temperature, neutron flux, etc. of the NPP must be continuously kept within their design limits. For this purpose, in different parts of the station, various instruments of the I&C system such as valves, motors, etc. The control systems of the nuclear power plants have three primary functions.

- Regulate parameters of the station and enable the plant personnel, to take appropriate actions in time.
- Provide automatic regulation of various systems and subsystems of the NPP.
- Protect the station in the event of parameters reaching levels beyond the margins of safety due to errors made by either man or machine.

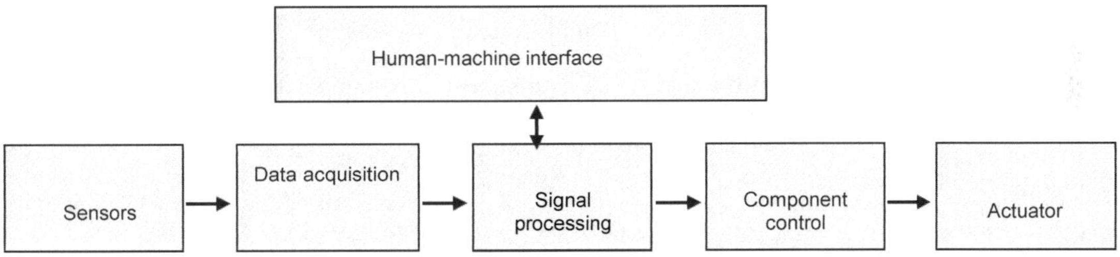

Fig. 1. Block diagram of functionality of a typical control system

In the diagram above, the overview of the functionality of the I&C system of the NPP is shown. The I&C systems rely on several sensors installed in various parts of the station to monitor different plant parameters such as temperature, pressure, neutron flux, etc. in real-time to help keep them within design limits. The type of sensor depends on the parameter it monitors. The measured value is then compared with the set-point value and depending on the difference, the system undertakes corrective action.

Depending upon their functionalities, the various components of the I&C system of NPP can be classified into the following groups [1].

- Sensors;
- Operative control, regulation, and monitoring systems;
- Safety systems;
- Human-machine interfaces;
- Actuators.

The I&C system utilizes different types of sensors such as pressure, temperature, and flow rate sensors, sensors of neutron flux, radiation, and chemical parameters such as concentration of boric acid, etc. Overall, the sensors make up the first layer of interfaces of the I&C systems. The next layer consists of transmitters, electronic devices, and signal processing systems. The third part is made up of actuation devices that can manipulate the various parameters of the NPP and are used to carry out the corrective actions commanded by the automated control system.

The next important aspect of the I&C system is field communication. This layer connects the surface interface devices such as sensors and actuators with the centralized control systems. Field communication in NPPs is carried out in accordance with different communication protocols, depending upon the modernity of the station as well as the type of data collected and the environment in which the sensor is placed. The following protocols are used to different degrees for field communication in NPPs.

- Analog communication (4-20mA DC): The majority of field communication in NPPs around the world is still carried out through analogue communication channels, using 4-20mA DC signals.
- HART (Highway Addressable Remote Transducer protocol): This is a protocol that enables digital signals to be transmitted over already existing 4-20mA DC analogue communication lines by using frequency shift keying (FSK) to superimpose digital signals on top of an analogue DC signal. This protocol is extremely popular in today's process industries.

- Digital communication (wired): Several digital communication protocols that enable the transmission of information over wires, such as SPI, RS-232, I2C, etc. Using such communication protocols information can be transmitted between transmitters and electronic signal processing devices.
- Digital communication (wireless): Such protocols and corresponding devices are rarely ever used for field communication inside a nuclear power plant, except for in some cases, transmission of information about diagnosis of station equipment. However, wireless digital communication is more suitable for devices used in radiation monitoring outside the premises of the NPP.

Cables, penetration, and junction boxes are used for the exchange of data as well as commands between the field communication devices and data transmission devices. Another important aspect of the I&C system in the NPP is the "process monitoring and control system", it consists of the data acquisition systems and the control systems that regulate the various processes.

- Data Acquisition Systems (DAQs): These systems are utilized for the acquisition, archiving, and processing of signals in control, monitoring, and safety systems.
- Control systems: Typical single-loop regulators with a set-point value, such as the proportional-integral-differential (PID) controller are almost universally used as closed-loop control systems. Such regulators are very widely used in NPPs across the world. The PID controller tries to rectify the error value between the measured output and the desired set-point value by calculating the difference between the two values and then commanding the appropriate corrective action which would rapidly minimize the error.
- Relay-based centralized control systems: These are control systems of older standards that use relay circuits to build logic.
- Centralized electronic control systems: Electronic systems are usually a set of discrete functional blocks that are put together in an architecture to satisfy the requirements of the control systems. These blocks could be analogue input and output cards, binary input/output cards, cards used to carry out specific logical functions or special purpose cards.
- Distributed control systems: Distributed control systems are digital systems that perform the same functionality as centralized systems but, in this case, they use decentralized elements or subsystems to control distributed processes or plant systems. Hence the processing capabilities are shared between distributed processors, and the system database is also distributed amongst the various processing units.

Overall, the I&C systems of NPPs across the world rely on traditional PID controllers to carry out their functions and maintain the different parameters of the station within their design limits.

2. Overview of intelligent control systems

Advancements in the field of automation and developments of control systems have always been aimed at creating more intelligent and automated systems so that the human participation aspect in their functionality would move into more specialized positions.

With the arrival of artificial intelligence (AI), steps were taken to utilize its principles to develop more advanced and automated control systems.

Traditionally control systems are developed using the linear modelling theory, however, there exist systems that could be non-linear in their characteristics. Modelling traditional control systems for such inherently non-linear systems is a difficult task to carry out. Even though there exist methods to construct appropriate models for inherently non-linear systems using the linear modelling theory, they come with certain limitations such as the construction of regulators varying from each system to another. A universal model can be created for linear systems of any order 'n', however doing so for non-linear systems is impossible. Again, due to the dynamics of the construction, the stability of such systems can't be guaranteed.

The term "intellectual control system" refers to any combination of hardware and software, which is joined by general information process, operating autonomously or in man-machine mode, and capable to synthesize the control goal and to find rational ways to achieve the control goal [3]. Intelligent control systems can be classified into the following groups based on their principle of regulation;

- Neural network-based control systems;
- Machine learning-based control systems;
- Fuzzy control;
- Bayesian control;
- Reinforcement learning-based control systems;
- Neuro-fuzzy control;
- Expert systems;
- Genetic control.

3. Researching the possibilities of utilizing intelligent control systems (fuzzy control) for components of VVER-1000 (1200) type reactors

Fuzzy control is one of the types of intelligent control that regulates and controls the control object using the principles of fuzzy logic. In recent years, this regulation method has received particular attention among various methods of parameter regulation [4]. In this case this paper, the possibility of using fuzzy control for pressure regulation in the 1st cooling circuit of the VVER-1000 nuclear reactor was investigated.

The first cooling circuit contains the purge-recharge system (TK), which is used to provide boric acid, which acts as a neutron absorber in the moderator as well as pump the water of the first cooling circuit into the water purifying systems. This system must operate at the working pressure of the first cooling circuit, which in the case of a VVER-1000, is 16.5 MPa. The automatic regulator TKC-02 is responsible for maintaining the working pressure of the first cooling circuit in the purge-recharge system. The regulator carries out its task using the PI (proportional-integral) principle of regulation. The regulatory valve "TK81S02" acts as the actuator for the pressure control system [5].

To understand and investigate the possibilities of using fuzzy control as the working principle of regulator TKC-02 instead of PI, we first need to put forward mathematical modelling of the process of pressure regulation using a regulatory valve. In the picture above (fig2.) a simplified diagram of the actuator and the water supply tube in which the pressure is to be regulated is given. As shown in the diagram, the valve, TK81S02 acts as the actuator for the pressure regulator TKC02. In this article, a 15m segment of the water supply tube that goes through the purge-discharge system of the VVER-1000 nuclear reactor is considered for simulating pressure control within [6].

In this case, the pressure in the tube shown in fig2 is the control object and the actuator is the valve, TK81S02. The process of pressure control using a valve can be mathematically represented using the following differential equation [7];

$$\left(\frac{a_{nn}}{a_l + a_{rv}}\right)\Delta P' + \Delta P = b_{rv}\left(\frac{a_l}{a_l + a_{rv}}\right)\Delta S_{rv} \qquad (1)$$

Here, constants $a_{nn}, a_l, a_{rv}, b_{rv}$ are obtained from the constructional parameters of the regulatory valve TK81S02 and the feedwater tube through which the water flows into the purge-recharge system.

Parameter	Value
Working pressure	17,65 MPa
Working temperature	350^{0}C
Internal diameter of the feedwater tube	0.120м
Sum of local hydraulic resistance coefficients ($\Sigma\xi$	93,69
Hydraulic resistance coefficient of the valve ($\xi_{рк}$)	15,7
Water flow rate ($\overline{M}$)	6,94 кг/с = 42,71 м3/час
Water density (ρ)	585,26 кг/м3
Length of the feedwater tube (L)	15м
Internal diameter of the valve	109мм
S_t(Area of the cross-section of the tube)	0,011 м2
S_{rv}(Cross-section of the regulatory valve)	9,331x10^{-3} м2

Table 1. Parameters of the water (moderator), regulatory valve and the feedwater tube

The constants in (1) can be obtained in the following way;

$$a_{nn} = \frac{L}{S_t} = 363.636 \qquad (2)$$

$$a_l = 2K_l M = 2\frac{\Sigma\xi}{2\rho S_{tp}^2} M \approx 9181,1 \qquad (3)$$

$$K_{rv} = \frac{\xi_{rv}}{2\rho} = 0.0134 \qquad (4)$$

$$a_{rv} = \frac{2K_{rv}\overline{M}}{S_{rv}{}^2} = 2136.18 \qquad (5)$$

$$\mathrm{B}_{rv} = \frac{2\mathrm{K}_{rv}\overline{\mathrm{M}}^2}{S_{rv}{}^3} = \mathbf{1.58x10^6} \tag{6}$$

Using Laplace transform on (1) and considering the change in pressure as output and change in cross-section of the valve as input, the following transfer function can represent the pressure control process using a regulatory valve;

$$W(S) = \mathrm{K}\frac{1}{1+TS} \tag{7}$$

In this case, the time constant, T will be

$$T = \left(\frac{a_{nn}}{a_l + a_{rv}}\right) \approx 0.12 \tag{8}$$

$$K = b_{rv}\left(\frac{a_l}{a_l + a_{rv}}\right) \approx 1.288x10^6 \tag{9}$$

Now the transfer function can be written as;

$$W(S) = \mathrm{K}\frac{1}{1+TS} = \frac{1.288x10^6}{1+0.12S} \tag{10}$$

4. Investigation using MATLAB

To simulate the pressure regulation process using a PI-controller as well as that using a fuzzy controller, a model of the pressure controller TKC-02 was created in the "MatLab-Simulink" environment.

In that environment, the responses of the pressure control system "TKC-02" with the usual PI controller and with a fuzzy logic controller were obtained and compared. In the given case, the actuator is considered non-inertial, i.e. as a proportional component.

$$W_{ac}(S) = K_{ac} \tag{11}$$

Here, $K_{ac} = 0.4271\ m^3/hr$ (converting % opening of the valve into the water flow rate through it.)
At first pressure control using a PI-regulator was analysed and the response was obtained. Here, the proportional and integral amplification factors were taken as $K_p = 0.18$ and $K_I = 4.7$.

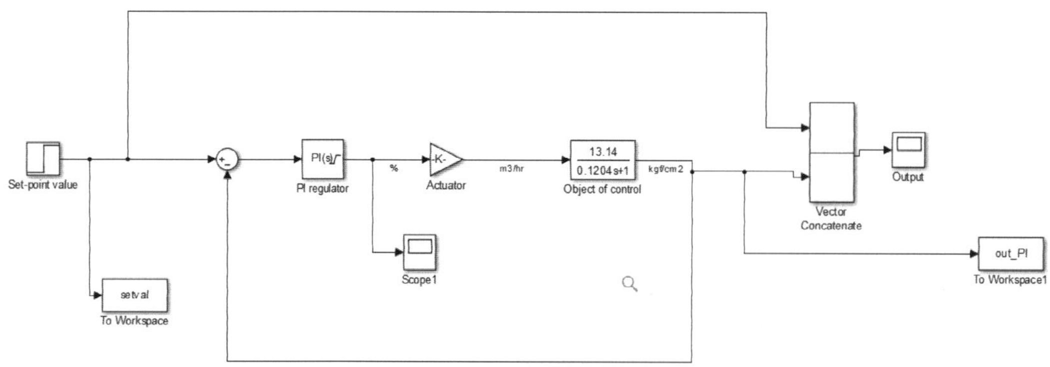

Fig. 2. Block diagram of pressure regulator TKC-02 with PI-controller

In the figure above, the schematic diagram of pressure regulation by the pressure control system TKC-02, utilizing a PI-controller is shown. For this simulation, the set-point value is taken as 180 kgf/cm^2 (≈17.65 MPa, a value above the normal working pressure of the first cooling circuit of VVER-1000). The diagram above contains the control object and actuator blocks, created using the tools available in the MatLab-Simulink library, in accordance with the transfer functions obtained from (11) and (12) respectively.

The input signal in this case is provided as a step signal of magnitude of 180kgf/cm^2. The water flow in the tube and thereby the pressure on the water can be regulated by the degree of opening of the valve (which can be expressed in %). The PI-regulator controls the % of opening of the valve and thereby the rate of flow of water through it.

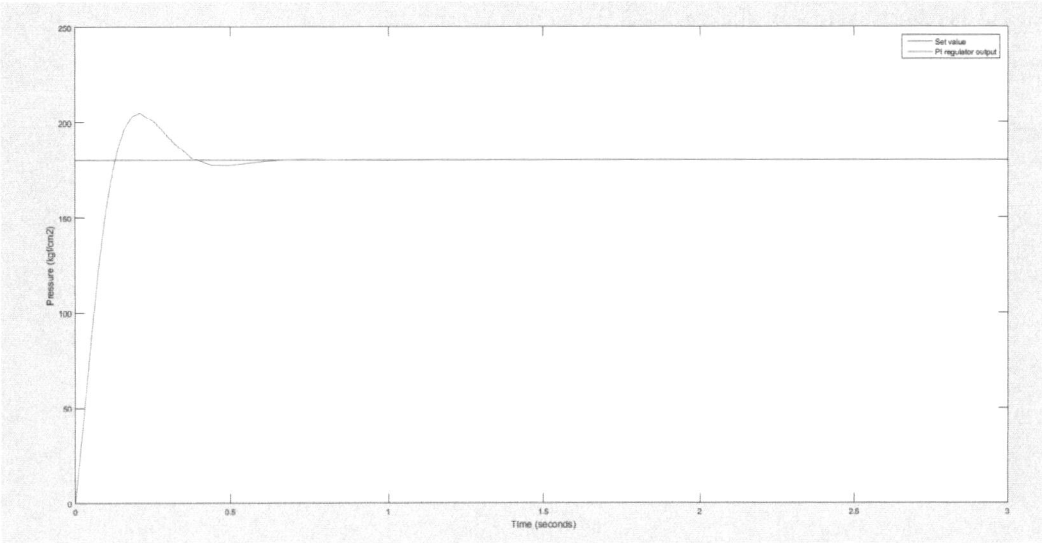

Fig. 3. System response with PI-controller

The above figure shows the response of the control system incorporating the PI-controller to the input of the set-point value of 180 kgf/cm^2. From the MatLab simulated output, it is clear that the control system can maintain the set-point value without any significant steady-state error. The output rises to the set-point value within *1.2 seconds.* It should also be noted that the output from the PI controller also included 'overshoot' (i.e. output exceeding the set-point value).

Next, we consider the case of a fuzzy controller to carry out the same task of maintaining the set-point value of pressure. To simulate the fuzzy controller, a set of fuzzy rules must be established for the fuzzy controller to carry out its task.

Deviation (e) (kgf/cm2)	**Level of valve opening (IN_AC) (%)**
Large deviation (225<e <325)	80% <IN_AC <100%
Medium deviation (150 <e <250)	60% <IN_AC <85%
Low deviation (75 <e <175)	40% < IN_AC <65%
Very low or no deviation (0 <e <100)	40% < IN_AC <65%
Overshoot (-137.5 <e <37.5)	0 < IN_AC<25%

Table 2. Fuzzy rules for pressure regulation

With the rules in place, the model of the fuzzy-control-based pressure regulator was constructed.

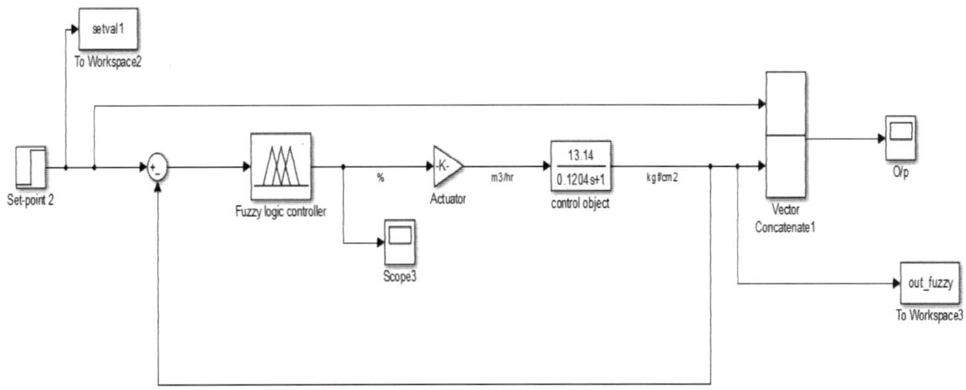

Fig. 4. Pressure controller TKC-02 with fuzzy controller

In the figure above, the block diagram of the pressure regulator based on a fuzzy controller is shown. The actuator and control object blocks are the same as those in fig4 and the set-point value of 180 kgf/cm^2 is also the same. The fuzzy controller, in this case, regulates the % of opening of the regulatory valve, which can manipulate the rate of water flow through the tube as well as pressure. With the fuzzy rules of table 2, the % of opening of the regulatory valve was controlled and thereby pressure was regulated, with the following response.

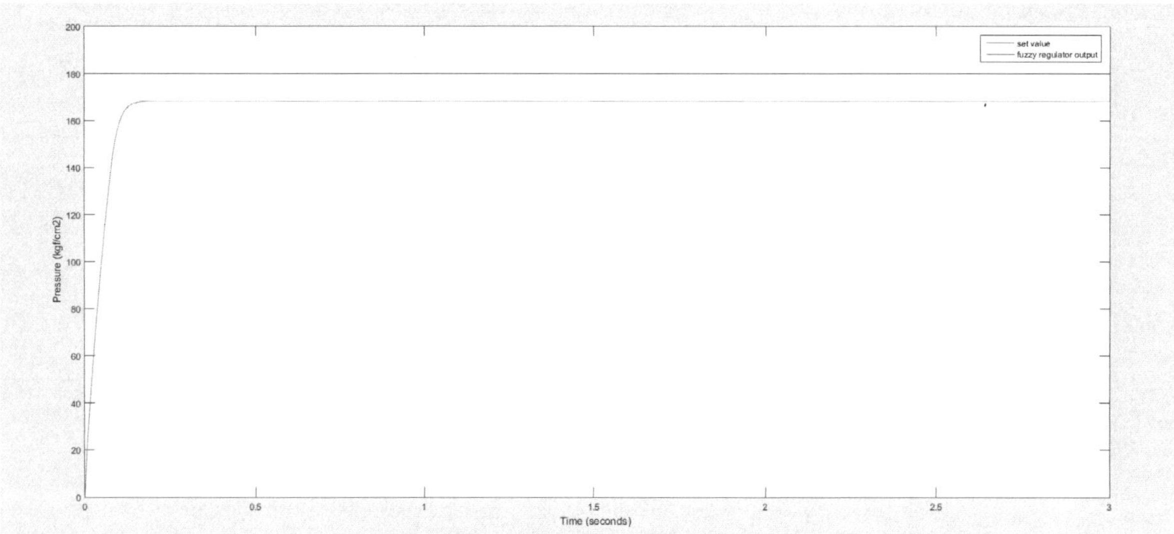

Fig. 5. System response with fuzzy controller

In fig6, the response of the fuzzy-logic-based pressure control system to the set point value is shown. In this case, it can be seen that there exists a certain value of steady-state error. For a set-point value of 180 kgf/cm^2, the fuzzy-control-based system maintains the pressure level at around 170 kgf/cm^2, and therefore a steady-state error of about *5%* exists in the case of the fuzzy-control-based system. In this case, the output value rises to the steady state within *0.4 seconds.*

5. Possibilities of using deep learning

Deep learning (DL) is an advanced subset of artificial intelligence that may be used to regulate and optimize the technical parameters of both main and secondary cooling circuits in VVER-1000/1200 nuclear reactors. Deep neural networks (DNNs) can represent very complicated, nonlinear interactions between input and output variables, which is critical for dealing with the intricate dynamics of nuclear cooling systems.
Deep learning algorithms must be extremely resilient, trustworthy, and interpretable to maintain safety in nuclear reactor cooling circuits. The architectures listed below are recommended for the task:

5.1. Recurrent Neural Networks (RNNs) and Long Short-Term Memory (LSTM):

Cooling systems work in dynamic, time-dependent settings in which past temperature, flow rate, and pressure values have a direct influence on present and future states. RNNs and LSTMs are ideal for this since they are built to handle time series data and sequential dependencies.
RNNs are intended for time-series data, where each input depends on the prior ones. They do this by including loops into their design, which allows information to persist through a concealed state vector throughout time steps. Mathematically, the hidden state $h(t)$ at time step t is computed as follows:

$$\boldsymbol{h(t) = \sigma(W_{hh}h_{t-1} + W_{xh}x_t + b_h)} \tag{12}$$

Where $x(t)$ is the input at the time step t, h_{t-1} is the hidden state from the previous time step, W_{xh} and W_{hh} are weight matrices for input-to-hidden and hidden-to-hidden connections, b_h is a bias term, and σ is an activation function (usually a tanh or ReLU).

The output y_t at time step t is then calculated by:

$$\boldsymbol{y_t = W_{hy}h_t + b_y} \tag{13}$$

Although RNNs are capable of representing sequential data, they often suffer from the vanishing gradient problem, which makes it challenging to comprehend relationships over large distances. This is where LSTMs are useful. [8]

5.2. Long Short-Term Memory (LSTM)

LSTMs perform better than regular RNNs because they have gates that control the information flow, which enables them to retain data for longer periods. The forget gate, input gate, cell state, and output gate are the four main parts of an LSTM. [9] The LSTM cell is updated as follows:

$$\boldsymbol{f_t = \sigma\left(W_f \cdot [h_{t-1}, x_t] + b_f\right)} \tag{14}$$

$$\boldsymbol{i_t = \sigma(W_i \cdot [h_{t-1}, x_t] + b_i)} \tag{15}$$

$$\boldsymbol{\tilde{C}_t = tanh(W_C \cdot [h_{t-1}, x_t] + b_C)} \tag{16}$$

$$\boldsymbol{C_t = f_t * C_{t-1} + i_t * \tilde{C}_t} \tag{17}$$

$$\boldsymbol{o_t = \sigma(W_o \cdot [h_{t-1}, x_t] + b_o)} \tag{18}$$

$$\boldsymbol{h_t = o_t * tanh(C_t)} \tag{19}$$

Where f_t, i_t, and o_t are the forget, input, and output gates, respectively, $\tilde{C}_t$ is the candidate cell state, C_t is the updated cell state, h_t is the hidden state. LSTMs well capture long-range temporal dependencies, and their ability to retain memory across lengthy sequences makes them useful for managing nuclear cooling systems, where the system's behaviour is strongly influenced by its past states. [10]

Time-varying elements of a nuclear reactor's cooling system, such as temperature, pressure, and coolant flow rate, show temporal relationships. To support preventative maintenance or optimize control operations, LSTMs may learn these relationships and forecast future states based on previous data. We should structure the deep learning algorithms using the same input-output mapping: Deviation (e) (kgf/cm^3) as the input and Valve Opening Level (IN_AC) (%) as the output to give a meaningful comparison between fuzzy logic and deep learning techniques. In this context, let's go over the algorithms in depth, highlighting how each may regulate the valve depending on variations in pressure.

RNNs and LSTMs are particularly useful for time-dependent data. Here, the deviation in pressure $e(t)$ at time t is used as input to predict the necessary valve opening level (IN_AC). Given the deviation data $e(t)$ over time, the RNN/LSTM can predict the next valve opening adjustment $_{AC}(t+1)$, ensuring the system compensates for deviation efficiently. [12]

$$\boldsymbol{h_t = \sigma(W_{hh}h_{t-1} + W_{xh}e_t + b_h)} \tag{20}$$

$$\boldsymbol{{}_{AC}(t+1) = W_{hy}h_t + b_y} \tag{21}$$

Where $e(t)$ represents the deviation in pressure at the time t, the hidden state h_t stores the cumulative effect of past deviations, the output $_{AC}(t+1)$ represents the predicted valve opening level based on both current and past deviations. LSTMs can handle scenarios where the pressure deviation may fluctuate or vary slowly since they take into account the history of deviations (e.g., how rapidly or slowly deviations build). However, fuzzy logic offers more transparent and understandable principles for in-the-moment decision-making, and they are less computationally intensive and sophisticated than fuzzy logic.

5.3. Summary of Comparison

Method	Input	Output	Key Strengths	Key Limitations
Fuzzy Logic	Deviation (e) (kgf/cm²)	Valve opening (IN_AC) (%)	Transparent, easy to understand and implement	Limited ability to handle complex temporal relationships
RNN/LSTM	Deviation (e) over time	Valve opening prediction	Captures time dependencies effectively	Computationally complex, harder to interpret

Table 3. Comparison of key advantages and disadvantages of fuzzy logic and deep learning-based control systems

6. Conclusion

- Here it must be noted that the response of the system in the case of fuzzy-control-based regulation was significantly faster.
- The fuzzy logic-based control system's output reached its steady state 3 times faster than that in the case of the PI-controller-based control system.
- However, in the case of the PI-controller-based system, it was possible to obtain an output without any steady-state error.
- In the case of fuzzy-logic-based regulation, a steady state error of ~5% was present.
- There are possibilities and certain advantages of using recurring neural networks and LSTM-based controllers which provide advantages such as the ability to predict the position of the valve.

From the above conclusions, we can see that the fuzzy-logic-based controller can give faster responses as compared to a traditional PI-controller-based system. As far as control systems are concerned, a faster output with less steady-state error is always a desired outcome. In this case, however, there was indeed a steady-state error involved in the case of the fuzzy logic-based controller. However, fuzzy control is just one aspect of intelligent control, there are other methods such as neural-network-based control systems, that can also be applied for such applications.

Also, the fuzzy rules used as the basis of the fuzzy controller could be made more accurate, which can reduce the steady-state error. The preliminary investigation regarding the usage of recurring neural networks and LSTM-based controllers also highlights certain possible advantages as well as complexities that require further research and investigation.

As mentioned before, the nuclear industry suffers from the usage of older automation standards and therefore further research on possible applications of modern automation technologies like intelligent control in nuclear power plants must be continued. Faster and more effective control systems can reduce the chances of irregularities and hazards occurring in nuclear infrastructures. Also, NPPs and their related infrastructure require effective automation much more than any other industry due to the nuclear radiation and reactions occurring in its core parts making them inaccessible to humans, even with protective equipment. However, the newer automation technologies must also satisfy various safety requirements and standards set by nuclear regulators such as the IAEA (International Atomic Energy Agency) to be approved for installation in NPPs. And since rudimentary research in this direction, as in the case of the given article provide encouraging results, research must continue in this direction to find more concrete data, using real-time simulation techniques and programs.

7. References

[1] Core Knowledge on Instrumentation and Control Systems in Nuclear Power Plants, IAEA Nuclear Energy Series, No. NP-T-3.12, ISBN: 978-92-0-113710-4,

[2] Lu, C. (2020). Nuclear Power Plants With Artificial Intelligence in Industry 4.0 Era: Top-Level Design and Current Applications—A Systemic Review, IEEE Access, Vol. 8, pp. 194315-194332, doi: 10.1109/ACCESS.2020.3032529.

[3] Vassilyev, S.N.; Kelina, A.Yu., Kudinov, Y.I. & Pashchenko, F.F. (2016). Intelligent control systems, XIIth International Symposium «Intelligent Systems», INTELS'16, 5-7 October Moscow, Russia.

[4] Pawelczyk, M. (1995). Integrating different control modes in a fuzzy controller, Proceedings of 1995 IEEE International Conference on Fuzzy Systems, Vol.3, pp. 1435-1440, doi: 10.1109/FUZZY.1995.409868.

[5] Ministerstvo Rossiiskoy Federatsiy. Tekhnologicheskiye systemyi reaktornovo otdeleniya – Ocnovnoye Oborudovaniye reaktornovo otdeleniya, Rosenergoatom, Moscow. (In Russian)

[6] Ministerstvo Rossiiskoy Federatsiy. Tekhnologicheskiye systemyi reaktornovo otdeleniya - Vspomogatelnye systemyi, Rosenergoatom, Moscow. (In Russian)

[7] Demchenko, V.A. (2001). Avtomatizatsiya i modelirovaniye tekhnologicheskikh protsessov AEC i TEC, ISBN: 966-549-541-0, Odessa. (In Russian)

[8] Hochreiter S. & Schmidhuber J. (1997). Long Short-Term Memory, Neural Comput, Vol. 9, No. 8, pp. 1735–1780, doi: 10.1162/neco.1997.9.8.1735.

[9] Graves, A. & Schmidhuber, J. (2005). Framewise phoneme classification with bidirectional LSTM and other neural network architectures, Neural Networks, Vol. 18, No. 5–6, pp. 602–610, doi: 10.1016/j.neunet.2005.06.042.

[10] Sutskever, I.; Vinyals, O. & Le, Q. V. (2014). Sequence to Sequence Learning with Neural Networks, [Online]. Available: https://arxiv.org/abs/1409.3215

[11] Greff, K.; Srivastava, R. K., Koutnik, J., Steunebrink, B. R. & Schmidhuber, J. (2017). "LSTM: A Search Space Odyssey," IEEE Trans Neural Netw Learn Syst, Vol. 28, No. 10, pp. 2222–2232, doi: 10.1109/TNNLS.2016.2582924.

[12] Tiainen, T.; Miettinen, J.O., Viitala, R., Hiekkanen, K. (2019). Digital Twwin and Virtual Sensor for a Rotor System, Proceedings of the 30th Internal DAAM Symposium, pp.1115-1121, doi:10.2507/30th.daaam.proceedings.156.

DOI: 10.2507/35th.daaam.proceedings.015

IMPLEMENTATION OF AN AI-POWERED LANGUAGE ASSISTANT FOR ENHANCING ENGINEERING STUDENTS' COMMUNICATIVE PROFICIENCY THROUGH MOBILE TECHNOLOGIES

Oleg Ipatov, Darina Barinova, Antonina Andreeva,
Maria Odinokaya, & Elena Krylova

Peter the Great St.Petersburg Polytechnic University

This Publication has to be referred as: Ipatov, O[leg]; Barinova, D[arina]; Andreeva, A[ntonina]; Odinokaya, M[aria] & Krylova, E[lena] (2024). Implementation of an AI-powered Language Assistant for Enhancing Engineering Students' Communicative Proficiency through Mobile Technologies, Proceedings of the 35th DAAAM International Symposium, pp.0112-0120, B. Katalinic (Ed.), Published by DAAAM International, ISBN 978-3-902734-44-0, ISSN 1726-9679, Vienna, Austria
DOI: 10.2507/35th.daaam.proceedings.015

Abstract

This study investigates the impact of an AI-powered language assistant on improving the communicative proficiency of engineering students through mobile technologies. Recognising the critical role of communicative skills in engineering education, this research addresses the gap between technical expertise and effective communication. By integrating an AI-driven chatbot designed for language learning into the curriculum, the study explores its efficacy in enhancing students' abilities in technical writing, oral communication, and interdisciplinary collaboration. A quasi-experimental design involving 100 undergraduate engineering students from the Institute of Energy at Peter the Great St. Petersburg Polytechnic University was employed, with data collected through pre- and post-intervention assessments, surveys, and interviews. The findings demonstrate significant improvements in the experimental group's language proficiency, particularly in speaking and listening skills. High levels of student engagement and positive feedback underscore the potential of AI-powered tools to transform traditional language education, making it more interactive and accessible. The study highlights the importance of integrating advanced technological solutions in educational practices to prepare engineering students for the global professional environment.

Keywords: AI-powered language assistant; engineering students; mobile technologies; chatbot; glossary function; innovative educational tool.

1. Introduction

Communicative proficiency is a critical competency for engineering students, encompassing the ability to effectively convey technical information, collaborate in multidisciplinary teams, and engage with diverse stakeholders. In the context of engineering education, the significance of communicative proficiency extends beyond mere linguistic capabilities.

It is instrumental in bridging the gap between complex technical concepts and practical applications, thereby enhancing both individual and collective problem-solving skills [1]. Effective communication is paramount for the dissemination and exchange of technical knowledge. Engineering students often engage in activities that require the articulation of intricate ideas through reports, presentations, and documentation. Mastery in technical writing and oral communication ensures that students can convey their findings and ideas clearly and concisely, which is crucial for academic success and professional practice (Sampath & Sethuraman, 2017). Poor communicative skills can lead to misunderstandings, errors, and inefficiencies, thereby impeding the progress of projects and research endeavors.

The collaborative nature of engineering projects necessitates proficient interpersonal communication. Engineers frequently work in teams comprising individuals from various disciplines and cultural backgrounds. Communicative proficiency enables students to collaborate effectively, fostering a productive team environment. It facilitates the negotiation of ideas, conflict resolution, and the synthesis of diverse perspectives, ultimately leading to innovative solutions [2], [3]. Inadequate communication can hinder teamwork, resulting in disjointed efforts and suboptimal outcomes.

The globalised nature of the engineering profession requires students to interact with international colleagues, clients, and regulatory bodies. Proficiency in communication, particularly in English as the lingua franca of the engineering domain, is essential for engaging in cross-border collaborations and staying abreast of global advancements [4]. This competency also encompasses the ability to understand and be understood by non-technical stakeholders, which is vital for project management, client relations, and public communication.

In addition to enhancing professional practice, communicative proficiency contributes to personal development and lifelong learning. It empowers students to participate in academic and professional discourse communities, access a broader range of resources, and continually update their knowledge and skills. This aligns with the dynamic and evolving nature of the engineering field, where ongoing education and adaptability are key to career longevity and advancement [5].

Given the critical importance of communicative proficiency, engineering education programs must prioritise the development of these skills, integrating communicative training into the curriculum to prepare students for the multifaceted demands of the engineering profession. In recent years, significant advancements in artificial intelligence (AI), particularly the development of AI-powered language assistants, have introduced novel approaches to enhancing communicative proficiency in educational settings. These sophisticated tools utilise AI, including machine learning and natural language processing, to provide interactive, personalised, and context-aware language learning experiences. Their integration into educational settings can transform traditional language education methodologies, addressing several key areas of improvement [6], [7].

This study focuses on engineering students and aims to investigate the potential role of AI-powered language assistants in enhancing their communicative proficiency. Given the advancements in AI and the specific educational context, this research explores the impact and effectiveness of using AI-powered language assistants for engineering students, with a particular emphasis on improving their communicative skills in English as a foreign language. By leveraging AI technology, this study seeks to provide insights into how these tools can address the communication challenges faced by engineering students and enhance their overall educational experience.

2. Literature Review

Communicative proficiency is increasingly recognised as a critical competency for engineering students. Traditionally, engineering education has focused on developing technical skills and knowledge; however, the ability to communicate effectively is essential for professional success [1]. Effective communication skills are necessary for various activities, including technical report writing, presentations, team collaborations, and interactions with non-technical stakeholders. According to Shuman, Besterfield-Sacre, and McGourty, engineering graduates must possess strong communication skills to effectively convey complex technical information and work collaboratively in multidisciplinary teams. The Accreditation Board for Engineering and Technology (ABET) has also emphasized the importance of communication skills, integrating them into their criteria for accrediting engineering [8], [9]. Despite this recognition, many engineering curricula still lack sufficient focus on developing these skills, resulting in a gap that needs to be addressed to better prepare students for the professional environment.

The application of artificial intelligence (AI) in language learning is grounded in several theoretical foundations, including constructivist and behaviourist learning theories. AI technologies, particularly machine learning and natural language processing, enable the creation of adaptive learning environments that can provide personalised instruction and feedback. The constructivist theory, which posits that learners construct knowledge through active engagement and interaction with their environment [10], is well-supported by AI-driven tools that offer interactive and immersive learning experiences. Behaviourist principles, emphasizing reinforcement and feedback [11], are also applied in AI systems that provide immediate corrective feedback to learners.

AI applications in language learning encompass various tools and platforms designed to enhance different aspects of language proficiency. These include automated speech recognition systems, intelligent tutoring systems, and conversational agents, such as chatbots, which can simulate human-like interactions and provide learners with opportunities to practice language skills in a controlled environment. Research has shown that these tools can significantly improve learners' language abilities by providing them with personalised, context-specific feedback [12].

Mobile learning (m-learning) has emerged as a significant trend in education, driven by the widespread availability of smartphones and other mobile devices. M-learning facilitates learning anytime and anywhere, offering flexibility that traditional classroom settings cannot match. According to Smith, Traxler, and Scott, mobile technologies support informal, spontaneous, and contextual learning, making them particularly effective for language acquisition. Studies have demonstrated that mobile apps and platforms can enhance language learning by providing interactive and engaging content, fostering collaboration, and enabling access to authentic language resources [13], [14].

The integration of AI and mobile technologies in language learning presents unique opportunities for enhancing the educational experiences of engineering students. AI-powered language learning tools can address the specific needs of these students by providing tailored exercises that focus on technical vocabulary and communication skills relevant to their field. Mobile technologies, with their ability to deliver content in a flexible and accessible manner, complement these tools by facilitating continuous learning beyond the classroom [15].

The current research builds on these theoretical foundations and empirical studies to explore the effectiveness of integrating AI-powered language learning tools with mobile technologies in improving the oral foreign language communicative skills of engineering students. By leveraging the strengths of both AI and mobile learning, this study aims to provide insights into how these technologies can be used to address the communication challenges faced by engineering students and enhance their overall educational experience.

3. Methodology

The theoretical foundation of this study is anchored in the methodological principles of foreign language instruction in both mixed and online learning environments, principles of professionally oriented language teaching, and core principles of the communicative approach to foreign language teaching. General scientific methods of theoretical research were employed, including an analysis of domestic and foreign literature on the topic. The primary research methods included questionnaire surveys and analysis of scientific literature.

This study utilised a quasi-experimental research design to evaluate the effectiveness of an AI-powered language assistant in enhancing the communicative proficiency of engineering students. The design incorporated both quantitative and qualitative methods to provide a comprehensive assessment of the intervention. The study was structured into three phases: pre-intervention, intervention, and post-intervention, allowing for the measurement of changes in language proficiency and the collection of detailed feedback from participants.

The study adopted a mixed-methods approach, combining quantitative data from standardised tests and usage analytics with qualitative data from surveys and interviews to triangulate findings and provide a comprehensive understanding of the AI-powered language assistant's impact. Quantitative data were collected through standardised language proficiency tests and usage analytics, while qualitative data were obtained from surveys and interviews with participants. This approach ensured that both measurable outcomes and subjective experiences of the students were captured.

Participants in this study were undergraduate engineering students at the Institute of Energy at Peter the Great Saint-Petersburg Polytechnic University. A total of 100 students were recruited, with 50 students assigned to the experimental group (using the AI-powered language assistant) and 50 students to the control group (utilising traditional language learning methods). Stratified random sampling was employed to ensure a representative sample in terms of gender, academic performance, and language proficiency levels. The AI-powered language assistant was implemented as a supplementary tool in the language learning curriculum for the experimental group. This tool was accessible via mobile devices, allowing students to engage with the assistant outside regular classroom hours. The control group continued to use traditional language learning resources, such as textbooks and in-class activities.

The AI-powered language assistant used in this study was a ChatGPT-based chatbot designed to support language learning through personalised and interactive experiences. It incorporated natural language processing and machine learning algorithms to provide real-time feedback, adaptive learning pathways, and conversational practice.

The "Eng-Rus IEEE Energetics Glossary" (@Energy_Glossary_bot via Telegram messenger) chatbot (see Fig.1) within the domain of language education and professional training in higher education institutions represents a promising practice. This approach has the potential to substantially enhance the educational experience for both students and educators. The chatbot provides immediate access to definitions and explanations of terms, thereby facilitating a deeper comprehension of course material among students. Moreover, it enables students to swiftly find answers to their questions, enriching their knowledge base. Educators can also leverage the glossary chatbot to develop interactive assignments, allowing students to assess their knowledge and skills. This contributes to more effective assimilation and retention of information.

The chatbot, integrated with the "Eng-Rus IEEE Energetics Glossary", serves as a valuable tool that grants users access to the interpretation of energy-related terms from English to Russian and vice versa. This chatbot can be utilised in the context of professional language training for future energy engineers in the following ways:

1. Rapid translation of English terms into Russian: Users can input any term into the chatbot, which will not only provide the Russian version of the term but also suggest various interpretations.
2. Retrieval of relevant usage examples: The chatbot displays examples of term usage in specific contexts, aiding in a deeper understanding of the term and its applications in discourse.
3. Search for synonyms and derivatives: The chatbot offers all related derivatives and synonymous terms and phrases in both Russian and English.

4. Responses to professional thematic inquiries: For instance, in response to the question, "Can you explain the role of turbines in converting geothermal heat into electrical energy?" the chatbot replies, "Certainly. A turbine represents a crucial component in the process of electricity generation, not merely in geothermal systems but also in various other energy generation systems. In the context of generating geothermal energy or 'geotermal'naya energiya' in Russian, a turbine (or 'turbina' in Russian) plays the role of a converter that changes thermal energy into mechanical energy."

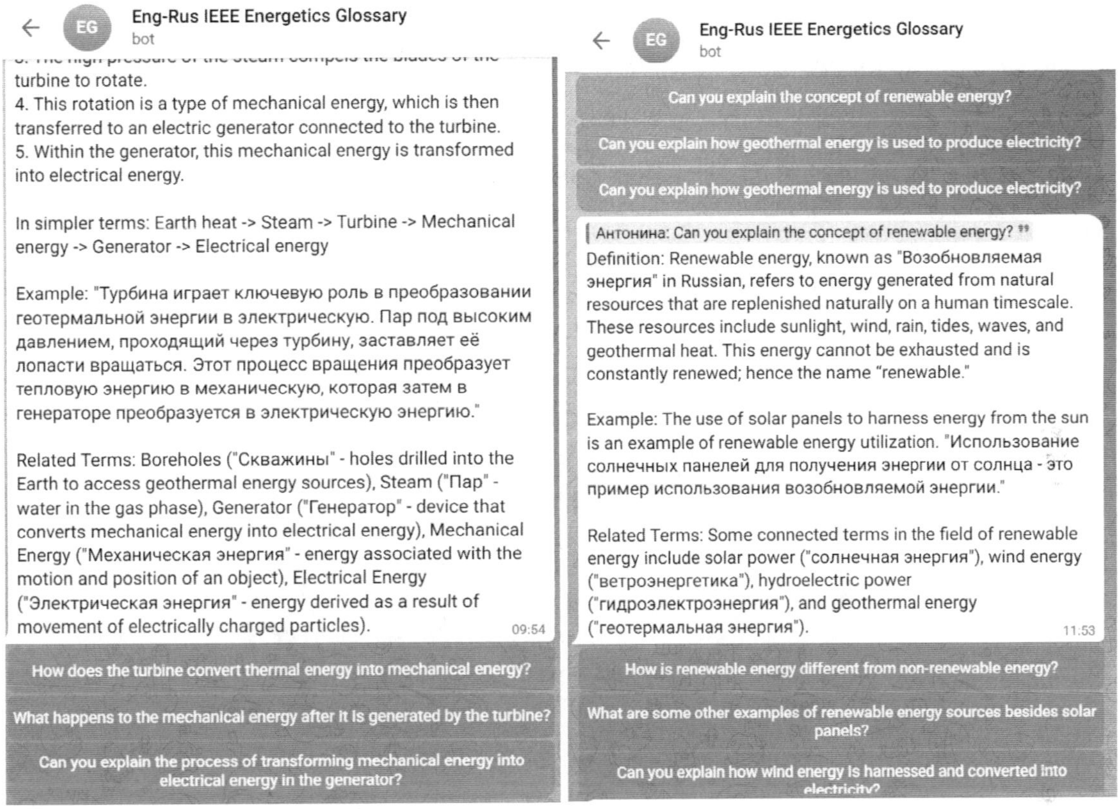

Fig. 1. The "Eng-Rus IEEE Energetics Glossary" chatbot

To maximise accessibility and engagement, the AI tool is integrated with mobile technologies, allowing students to use it on smartphones and tablets. The mobile integration ensures that learners can practice language skills anytime and anywhere, aligning with modern educational trends towards flexible and mobile learning environments.

Prior to the intervention, students in the experimental group received comprehensive training on how to use the AI-powered language assistant. This training covered the functionalities of the chatbot, best practices for effective usage, and troubleshooting common issues. Orientation materials were also provided in the form of user manuals and video tutorials to support independent learning.

Data collection involved multiple methods to gather comprehensive information on the impact of the AI-powered language assistant. These methods included pre- and post-intervention language proficiency tests, surveys, interviews, and usage analytics from the AI tool. The combination of these methods provided both quantitative measures of language proficiency and qualitative insights into student experiences.

Language proficiency assessments were conducted before and after the intervention to measure changes in students' language skills. The assessments included standardised tests covering speaking, listening, reading, and writing. The pre-intervention assessment established a baseline for each student, while the post-intervention assessment evaluated the impact of the AI tool on their language proficiency.

Surveys were administered to both the experimental and control groups to gather data on student perceptions, engagement, and satisfaction with their respective learning methods. Additionally, semi-structured interviews were conducted with a subset of participants from the experimental group to gain deeper insights into their experiences with the AI-powered language assistant. These qualitative data helped to understand the subjective impact of the AI tool and identify areas for improvement.

Usage analytics from the AI-powered language assistant were collected to monitor how students interacted with the tool. Metrics such as frequency of use, types of activities completed, and time spent on different exercises were analysed to identify patterns and correlations with learning outcomes. This data provided valuable insights into student engagement and the effectiveness of different features of the AI tool.

This study's methodology was designed to rigorously assess the effectiveness of an AI-powered language assistant in enhancing the communicative proficiency of engineering students. By combining quantitative and qualitative methods, the research aimed to provide a comprehensive evaluation of the intervention, offering insights that can inform future implementations of AI technologies in language education.

4. Results

The quantitative findings from this study provide a detailed analysis of the impact of the AI-powered language assistant on the communicative proficiency of engineering students. Data were collected from pre- and post-intervention assessments as well as from usage analytics of the AI tool. The pre- and post-intervention assessments revealed significant improvements in the communicative proficiency scores of the students in the experimental group (see Fig.2). The average overall language proficiency score increased by 15% from pre-intervention to post-intervention. Specifically, the experimental group showed notable gains in speaking (+20%), listening (+18%), reading (+12%), and writing (+10%) skills. In contrast, the control group exhibited only marginal improvements with an average overall increase of 5%. These results suggest that the AI-powered language assistant substantially enhanced the language skills of the students compared to traditional learning methods. The chart clearly shows that the experimental group, which used the AI-powered language assistant, had substantial gains in language proficiency across all skill categories compared to the control group.

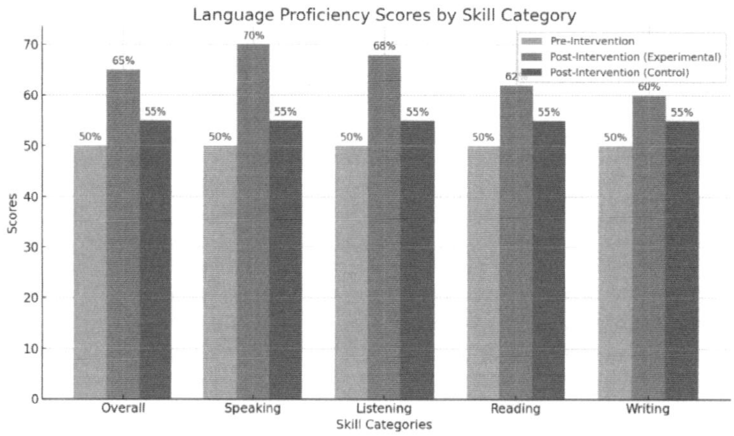

Fig. 2. The pre- and post-intervention assessments

Usage analytics from the AI tool provided insights into student engagement and usage patterns. On average, students in the experimental group used the AI assistant for 45 minutes per day. The most frequently used features were grammar and vocabulary (40% of total usage), spelling correction (30%), and conversational practice (30%). Engagement metrics indicated high levels of sustained interaction, with 85% of students using the tool consistently throughout the intervention period (see Fig.3). Higher usage frequency correlated positively with greater improvements in language proficiency, highlighting the importance of regular practice.

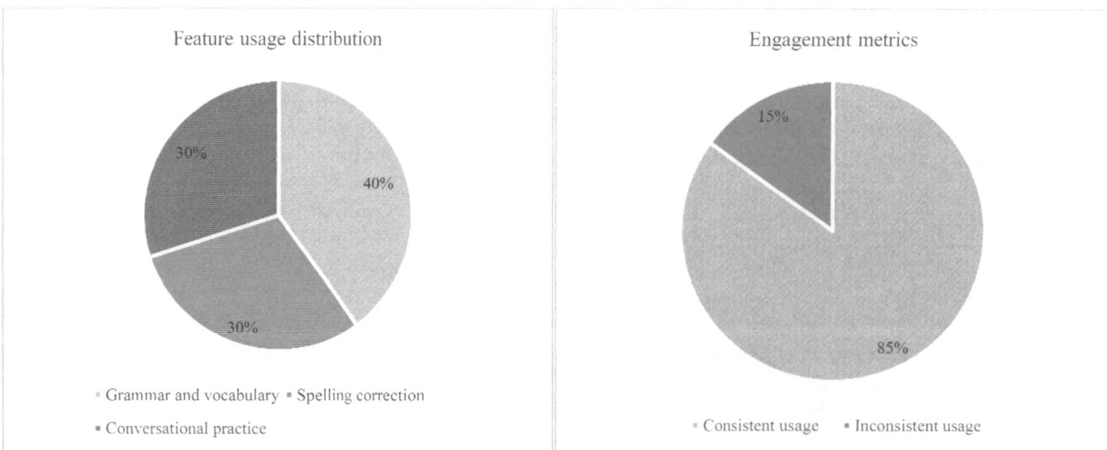

Fig. 3. The visualisations of the usage analytics and engagement metrics

Qualitative findings were derived from surveys and semi-structured interviews conducted with students in the experimental group. These findings provided valuable insights into student feedback and perceptions regarding the AI-powered language assistant. Student feedback was overwhelmingly positive, with 90% of participants expressing satisfaction with the AI tool. Key themes that emerged from the surveys and interviews included appreciation for the personalised feedback, the interactive nature of the learning activities, and the flexibility offered by mobile access. Students reported that the AI assistant made language learning more engaging and less intimidating, as they could practice at their own pace and receive immediate corrections. Additionally, many students highlighted the convenience of using the tool on their mobile devices, allowing them to integrate language practice into their daily routines seamlessly.
The qualitative data indicated that the AI-powered language assistant had a positive impact on the overall learning experience and student motivation. Students reported feeling more motivated to engage with language learning due to the gamified elements and interactive exercises provided by the AI tool. The ability to track progress and set personal learning goals within the app was also cited as a motivating factor. Moreover, students felt that their confidence in using English had increased, particularly in speaking and listening skills, as the AI assistant provided a safe environment for practice without the fear of judgment.

The results of this study demonstrate that the AI-powered language assistant effectively enhanced the communicative proficiency of engineering students. The quantitative improvements in language proficiency scores combined with high levels of engagement and positive qualitative feedback suggest that AI-powered tools can play a significant role in improving language learning outcomes. These findings underscore the potential of integrating AI technologies into educational practices to create more personalised, engaging, and effective learning experiences.

5. Discussion

The findings of this study provide compelling evidence for the effectiveness of AI-powered language assistants in enhancing the communicative proficiency of engineering students. The significant improvements observed in the experimental group's language proficiency scores highlight the potential of AI tools to effectively supplement traditional language learning methods. This discussion explores the implications of these results, addresses potential limitations, and suggests directions for future research.

The substantial gains in speaking, listening, reading, and writing skills observed in the experimental group underscore the value of integrating AI-powered language assistants into engineering curricula. The 15% overall increase in language proficiency, particularly the 20% improvement in speaking skills, suggests that these tools can address specific challenges faced by engineering students in mastering technical and professional communication.
The positive impact on speaking and listening skills is particularly noteworthy, given that these areas are often the most challenging for non-native speakers. The AI assistant's ability to provide real-time, personalised feedback likely contributed to these gains by allowing students to practice conversational English in a low-pressure environment. This aligns with constructivist learning theories, which emphasize the importance of active engagement and interaction in the learning process [10], [16].

Moreover, the integration of mobile technologies facilitated continuous learning, allowing students to practice language skills anytime and anywhere (Dawadi et al., 2024). This flexibility is crucial in the context of engineering education, where students often juggle demanding coursework and project deadlines. The ability to integrate language practice seamlessly into daily routines can enhance learning outcomes and foster a more positive attitude toward language acquisition.

The high levels of student engagement and satisfaction with the AI-powered language assistant, as indicated by the qualitative data, suggest that these tools can significantly enhance the overall learning experience. The interactive and gamified elements of the AI assistant, along with the convenience of mobile access, contributed to sustained student interaction and motivation. These findings are consistent with previous research on mobile learning, which highlights the benefits of flexibility and accessibility in promoting language acquisition [13].

The positive feedback regarding personalised feedback and progress tracking further supports the use of AI tools in language education. Students' ability to set personal learning goals and monitor their progress can foster a sense of ownership and responsibility for their learning, which is crucial for sustained motivation and success [17].
The study's results underscore the critical role of communicative proficiency in engineering education and professional practice. Effective communication is essential for collaborating in multidisciplinary teams, engaging with diverse stakeholders, and disseminating technical information [3]. The AI-powered language assistant's focus on technical vocabulary and professional communication skills directly addresses these needs, preparing students for real-world engineering challenges.

The ability to practice and receive feedback on technical terms and professional discourse is particularly valuable in the engineering context. The "Eng-Rus IEEE Energetics Glossary" chatbot, for example, provided students with immediate access to technical definitions and usage examples, enhancing their understanding and retention of course material. This practical application of AI technology aligns with behaviourist principles, which emphasize reinforcement and feedback as key components of the learning process [11], [18]. While the study's findings are promising, several limitations should be acknowledged. The quasi-experimental design, while robust, cannot completely rule out potential confounding variables.

Additionally, the study was conducted with a specific sample of engineering students at a single institution, which may limit the generalisability of the results. Future research should aim to replicate these findings across diverse educational contexts and with larger sample sizes. Another limitation is the reliance on self-reported data for qualitative insights. While the feedback from surveys and interviews provided valuable information, self-reported data can be subject to biases. Future studies could incorporate more objective measures of engagement and motivation, such as observational data or third-party assessments.

Further research is also needed to explore the long-term effects of using AI-powered language assistants. While the study demonstrated significant short-term gains in language proficiency, it is important to assess whether these improvements are sustained over time and translate into enhanced professional performance.

Concerns have been raised regarding the effectiveness of banning the use of ChatGPT, which could lead to stringent measures against students, such as expulsion, physical closed-book examinations, and written tests (Crawford et al., 2023). Restricting access to this tool may inadvertently increase its allure. Attempting to circumvent artificial intelligence by designing tasks beyond its current capabilities may prove futile, given the advancements in newer model versions like ChatGPT, which can analyse images and provide extensive descriptions. While such an approach demands additional effort from higher education faculty, they can assess students' knowledge through oral examinations based on unique assignments.

We concur with researchers who advocate for educating students on the responsible use of chatbots rather than imposing bans [12], [17], [18]. Students should be instructed on the importance of academic integrity, ethics, and personal accountability, as they are responsible for the quality of their work. Allowing students to choose topics that genuinely interest them, where their voices are heard and their opinions respected, is crucial [19]. Implementing authentic assessments that provide students with a creative, substantive, and motivating learning experience, and evaluating their skills and knowledge in real-world situations is recommended.

Incorporating artificial intelligence tools into discussions and assignments is vital, teaching students the prudent use and limitations of text processing and demonstrating poor-quality text examples to highlight the value of human writing. Emphasizing the importance of motivation and the writing process within the context of intellectual growth is necessary [12]. Updating assessment methods and transitioning to modern formats such as oral presentations, group projects, self-assessments, peer reviews, results-based evaluations, and the creation of web pages, videos, and animations by students is essential.

Educating students on the effective use of chatbots in higher education necessitates clear guidelines and expectations. It is crucial to instruct students on the responsible use of chatbots, including authorship and ethical considerations [19]. Generative artificial intelligence can assist students in developing critical thinking skills through iterative interaction and the application of their abilities in new contexts. Students should be encouraged to use ChatGPT within reasonable limits and to critically evaluate its outputs [16].

Encouraging broad, multi-faceted dialogue among various stakeholders, including students, learning and teaching experts, faculty from all disciplines, IT specialists, career center staff, industry and community representatives, legal and external experts, and government officials, and subsequently implementing the outcomes of these discussions in policies, guidelines, and instructional materials underscores the importance of digital literacy in education. This includes the use of AI tools such as chatbots and Grammarly, which employs AI to review texts and suggest improvements.

Fostering extensive and diverse exchanges of opinions among stakeholders, including students, faculty across disciplines, learning, teaching, and IT experts, career center staff, industry, community representatives, legal and external experts, and government officials [18], and implementing the results in policies, guidelines, materials, and instructional aids underscores the significance of integrating digital literacy. This encompasses the use of AI tools like chatbots. Updating academic honesty policies and developing clear guidelines for the use of language models, including proper usage and consequences of violations, is imperative. The University of Tasmania provides an example of AI utilisation, but adherence to academic integrity rules when using generative AI is necessary [17]. The use of chatbots and other AI tools represents only the initial steps in developing more powerful technologies in higher education, potentially impacting various aspects of academic life.

6. Conclusion

The findings of this study underscore the significant impact of AI-powered language assistants on enhancing the communicative proficiency of engineering students. Through the integration of advanced AI technologies and mobile learning, the experimental group demonstrated substantial improvements in language skills compared to traditional learning methods. This research highlights the transformative potential of AI in addressing the specific communication challenges faced by engineering students, preparing them for the demands of professional practice in a globalised and interdisciplinary environment. The quantitative results reveal that the AI-powered language assistant facilitated significant gains in speaking, listening, reading, and writing skills. The observed 15% overall increase in language proficiency, particularly the 20% improvement in speaking skills, suggests that personalised, real-time feedback provided by the AI tool is highly effective in fostering language development.

These findings align with the principles of constructivist and behaviourist learning theories, emphasizing the importance of interactive, immersive learning experiences and immediate corrective feedback. Qualitative data from surveys and interviews further reinforce the positive impact of the AI-powered language assistant on student engagement and motivation. The interactive, gamified elements of the AI tool, combined with the convenience of mobile access, contributed to high levels of sustained interaction and a positive learning experience. Students appreciated the personalised feedback, the ability to track progress, and the flexibility to practice language skills at their own pace, indicating that these features enhance motivation and confidence in language learning.

The integration of AI tools into engineering education not only improves language proficiency but also prepares students for real-world challenges by focusing on technical vocabulary and professional communication skills. The "Eng-Rus IEEE Energetics Glossary" chatbot exemplifies the practical application of AI in providing immediate access to technical definitions and usage examples, thereby enhancing students' understanding and retention of course material.

Despite the promising results, this study acknowledges certain limitations, including the quasi-experimental design and the specific sample of engineering students, which may affect the generalisability of the findings. Future research should aim to replicate these results across diverse educational contexts and with larger sample sizes to validate the effectiveness of AI-powered language assistants. Additionally, exploring the long-term effects of these tools on language proficiency and professional performance will provide deeper insights into their sustained impact.

The study also addresses concerns about the ethical use of AI tools in education. It emphasizes the importance of educating students on responsible AI use, academic integrity, and personal accountability. The study advocates for authentic assessments and innovative teaching methods that incorporate AI technologies, enabling students to develop critical thinking skills and apply their knowledge in real-world situations.

In conclusion, the integration of AI-powered language assistants into engineering education represents a promising advancement in enhancing communicative proficiency. The substantial improvements in language skills, coupled with high levels of student engagement and satisfaction, demonstrate the potential of AI to revolutionise language learning. By leveraging AI technologies, educational institutions can provide more personalised, engaging, and effective learning experiences, ultimately better preparing students for the multifaceted demands of the engineering profession.

7. References

[1] Hua, C. Innovative Approaches in English Language Teaching: Integrating Communicative Methods and Technology for Enhanced Proficiency. CHR (2024) Vol. 32: 214-220. DOI: 10.54254/2753-7064/32/20240075.

[2] Sampath, S., Sethuraman, M. (2017). The Role of Oral Communicative Tasks (OCT) in Developing the Spoken Proficiency of Engineering Students. Advances in Language and Literary Studies. 8. 161. 10.7575/aiac.alls.v.8n.2p.161.

[3] Mercier, E., Goldstein, M., Baligar, P., Rajarathinam, R. (2023). Collaborative Learning in Engineering Education. 10.4324/9781003287483-23.

[4] Odinokaya, M., Andreeva, A., Mikhailova, O., Petrov, M., Pyatnitsky, N. (2020). Modern aspects of the implementation of interactive technologies in a multidisciplinary university. E3S Web of Conferences. 164. 12011. 10.1051/e3sconf/202016412011.

[5] Litynska, V., Romanovska, L., Kravchyna, T., Shandruk, S., Sopotnytska, O., Kondratska, L. (2024). Personnel Career Advancement Factors: Evidence from Educational Institutions. European Journal of Sustainable Development. 13. 228. 10.14207/ejsd.2024.v13n2p228.

[6] Vancova, H. (2024). AI and AI-powered tools for pronunciation training. Journal of Language and Cultural Education. 11. 12-24. 10.2478/jolace-2023-0022.

[7] Ipatov, O., Barinova, D., Ni, O., Saramaha, S., Odinokaya, M. (2022). New Teaching Approaches in Conditions of Globalisation in Digitized Higher Engineering Education, Proceedings of the 33rd DAAAM International Symposium, pp.0215-0220, B. Katalinic (Ed.), Published by DAAAM International, ISBN 978-3-902734-36-5, ISSN 1726-9679, Vienna, Austria DOI: 10.2507/33rd.daaam.proceedings.030.

[8] Shuman, L., Besterfield-Sacre, M., McGourty, J. (2005). The ABET 'professional skills'—Can they be taught? Can they be assessed?. Journal of Engineering Education. 94. 10.1002/j.2168-9830.2005.tb00828.x.

[9] Criteria for accrediting engineering programs, 2019 - 2020 (2018) ABET. Available at: https://www.abet.org/accreditation/accreditation-criteria/criteria-for-accrediting-engineering-programs-2019-2020/ (Accessed: 14 July 2024).

[10] Mayer, S. J. (2008). Dewey's dynamic integration of Vygotsky and Piaget. Education and Culture, 24(2), 6-24.

[11] Neto, M., Mayer, P. (2011). Skinner and the asymmetry between reinforcement and punishment. Acta Comportamentalia. 19. 21-32.

[12] Gimpel, H., Hall, K., Decker, S., Eymann, T., Lämmermann, L., Mädche, A., Röglinger, R., Ruiner, C., Schoch, M., Schoop, M., Urbach, N., Vandirk, S. (2023) Unlocking the power of generative AI models and systems such as GPT-4 and ChatGPT for higher education: A guide for students and lecturers. University of Hohenheim.

[13] Smith, M., Traxler, J., Scott, H. (2022). When Classrooms Aren't an Option: Researching Mobile Language Learning through Disruption.

[14] Dawadi, S., Kukulska-Hulme, A., Seargeant, P. (2024). English for the EDI generation: predicting and tracking the role of English and digital/mobile technologies in higher education across East and South Asia. 3091-3091. 10.21125/inted.2024.0831.

[15] Perkins, M. (2023). Academic Integrity considerations of AI Large Language Models in the post-pandemic era: ChatGPT and beyond. Journal of University Teaching & Learning Practice. – № 20(2).

[16] Ali, J. K. M., Shamsan, M. A. A., Hezam, T. A., & Mohammed,A. A. (2023). Impact of ChatGPT on learning motivation: Teachers and students' voices. Journal of English Studies in Arabia Felix, 2023. – № 2(1). – P. 41-49.

[17] Crawford, J., Cowling, M., Allen, K. A. (2023). Leadership is needed for ethical ChatGPT: Character, assessment, and learning using artificial intelligence (AI). Journal of University Teaching & Learning Practice, 2023. – № 20(3).

[18] Malinka, K., Perešíni, M., Firc, A., Hujňák, O., Januš, F. (2023). On the educational impact of ChatGPT: Is Artificial Intelligence ready to obtain a university degree? ArXiv, 2023. – URL: https://arxiv.org/pdf/2303.11146.pdf.

[19] Karpovich I. A., Krepkaia T. N., Voronova L. S., Combarros-Fuertes P. Novice university educators: Professional, psychological and motivational spheres of adaptation. Proceedings of the IOP Conference Series: Materials Science and Engineering. 2020. 012136. DOI: 10.1088/1757-899X/940/1/012136

DOI: 10.2507/35th.daaam.proceedings.016

A REVIEW OF FFF 3D PRINTED POLYMER MATERIALS: FILAMENTS, PHYSICAL ASPECTS IN THE FORMATION, MECHANICAL PROPERTIES AND STANDARD TESTING

Adi Pandžić, Damir Hodžić, Ismar Hajro & Petar Tasić

This Publication has to be referred as: Pandzic, A[di]; Hodzic, D[amir]; Hajro, I[smar] & Tasic, P[etar] (2024). A Review of FFF 3D Printed Polymer Materials: Filaments, Physical Aspects in the Formation, Mechanical Properties and Standard Testing, Proceedings of the 35th DAAAM International Symposium, pp.0121-0134, B. Katalinic (Ed.), Published by DAAAM International, ISBN 978-3-902734-44-0, ISSN 1726-9679, Vienna, Austria
DOI: 10.2507/35th.daaam.proceedings.016

Abstract

The objective of this paper is to provide a comprehensive description and definition of Fused Filament Fabrication (FFF) 3D printed polymer materials, based on an extensive review of the relevant literature and previous research. The review covers the fundamental materials used in FFF technology, known as filaments, and provides a detailed analysis of the physical aspects involved in the formation of FFF printed polymer materials. Furthermore, the paper presents the key factors that influence the mechanical properties of FFF printed polymer materials and describes the testing methods used to evaluate these properties, including the relevant testing standards. The research findings aim to contribute to the development of a deeper understanding of FFF printed polymer materials and their properties, as well as to provide guidance for future research in this field.

Keywords: FFF; Polymer; 3D Print; Printed Materials; FDM Materials

1. Introduction

The advancement of modern industry hinges on the continual pursuit of innovative and cutting-edge research activities, particularly as they relate to production processes, materials, and product design. In today's manufacturing landscape, market competition demands not only low cost and high quality but also complex products with shorter life cycles, shorter delivery times, and greater customization, all while requiring less skilled labor. As a result, the demand for sophisticated and intricate products has never been higher, posing significant challenges to both design and manufacturing. To address these challenges, there is a strong imperative to invest in the design, development, and implementation of new technologies that can meet these ever-evolving demands [1].

One technology that is particularly well-suited to meet these requirements is additive manufacturing (AM). AM, also known as 3D printing, is a disruptive technology that is currently driving innovation across a range of areas, from design and engineering to materials and manufacturing. By reducing costs and waste and increasing efficiency, AM is poised to reshape the manufacturing industry and has the potential to generate significant industrial, economic, and social impact. Its versatility and flexibility make it an ideal solution for producing complex, customized products with short lead times, which are becoming increasingly important in today's fast-paced business environment [2].

AM technologies are typically based on the layer-by-layer addition of materials to create finished products. There are now several different AM technologies available, each with its own unique features and advantages. One of the most popular AM technologies today is Fused Deposition Modeling (FDM), also known as Fused Filament Fabrication (FFF) or material extrusion technology. This method involves the extrusion of thermoplastic materials through a heated nozzle to create each material layer of the product (Figure 1). FDM technology is valued for its versatility, accuracy, and ability to produce functional prototypes and end-use parts quickly and cost-effectively. Other popular AM technologies include Selective Laser Sintering (SLS), Stereolithography (SLA), and Digital Light Processing (DLP), each of which has its own strengths and applications [3], [4], [5]..

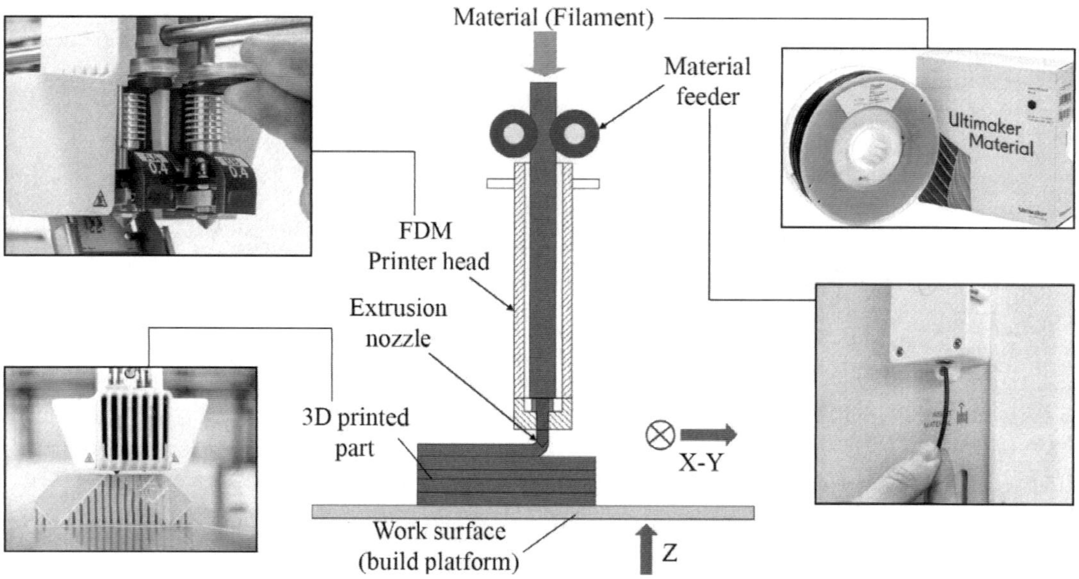

Fig. 1. Simplified presentation of FDM printing process [6]

AM is currently the subject of extensive research, with a particular focus on FDM printed polymer materials. According to the findings of a study [7], it is not surprising that polymers account for over half (51%) of the parts produced using additive manufacturing (AM) technologies (Figure 2). One reason for this is the prevalence of FDM 3D printers, which tend to be more affordable than other AM technologies [19]. As a result, FDM technology has become increasingly popular for producing prototypes, tooling, and end-use parts in a range of applications, from aerospace to medical devices. Ongoing research is focused on further improving the accuracy, speed, and reliability of FDM printing, as well as exploring new materials and applications for this versatile technology.

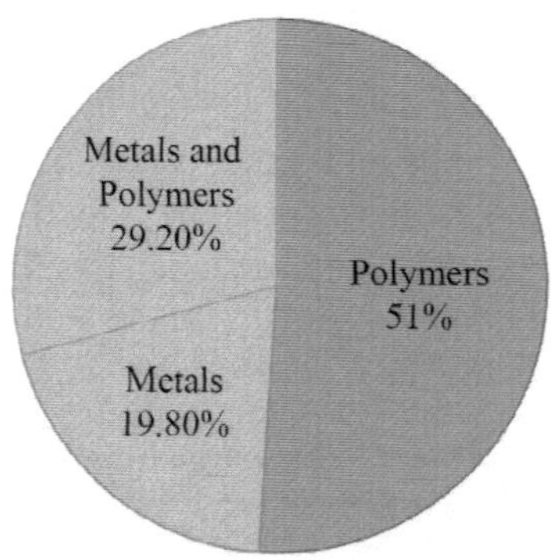

Fig. 2. Materials used for products currently manufactured in the AM industry [7]

Additive Manufacturing (AM) heavily relies on the choice of materials. The way materials are used in AM differs significantly from conventional manufacturing methods, making material selection and preparation a critical aspect of the AM process. Specific AM machines and technologies are optimized for particular materials, shapes, and conditions. A wide variety of materials suitable for Fused Deposition Modeling (FDM) are currently available on the market, with the selection continually expanding [8], [9].

In general, FDM materials can be categorized into three groups: thermoplastic polymers, composite materials with a polymer base, and metal materials. Thermoplastic polymers such as PLA (Polylactic Acid), ABS (Acrylonitrile Butadiene Styrene), PC (Polycarbonate), PETG (Polyethylene terephthalate glycol), Nylon (PA6, Polyamide), Tough PLA, and PEEK (Poly-Ether-Ether-Ketone) are among the most widely researched and tested FDM printed materials [2], [7], [10], [11], [12]. Obtaining the "best" mechanical properties of FDM printed polymer materials is a significant research challenge. The goal is to identify and adjust the manufacturing parameters to achieve the desired mechanical properties, high production speed, high quality, required surface roughness, and low cost.

The mechanical properties of FDM printed polymer materials are significantly affected by the type of material and manufacturing parameters, making their prediction a complex task that is mainly based on experimental research. A thorough review of the literature and prior studies indicates that these factors have a notable impact on the mechanical properties. Therefore, a comprehensive understanding of the interplay between these factors is necessary for achieving the desired mechanical properties of FDM printed polymer materials [13]. Fused Deposition Modeling (FDM) is one of the most promising additive manufacturing (AM) technologies for 3D printing. However, FDM is a complex process that depends on many parameters, leading to variations in the properties of FDM printed materials [14], [15], [16]. A critical literature review suggests that product quality is a function of various parameters that affect product quality and material properties. Each of these parameters is interrelated, making it challenging to understand the impact of the combination of parameters [17]. Therefore, to achieve the desired material properties of the printed part, it is crucial to investigate the influence of various parameters in the FDM process on the mechanical properties of the printed material. By choosing optimal parameters, the desired quality of the part can be achieved [12], [18].

When analyzing the parameters influencing the mechanical properties of FDM printed materials, the focus is on factors such as the orientation of the piece during production, layer thickness, shape and density of the infill, angle and width of the raster, distance between the rasters (air gap), temperature, and printing speed [11, 15, 16, 19–21]. In addition, Ultimaker, in partnership with TU Delft University, conducted a study called "Tensile strength of 3D printed PLA: a systematic investigation", which identified four parameters that significantly affect the strength of FDM printed materials, namely, layer height, 3D printing speed, temperature, and infill density [22].

Therefore, understanding the impact of these parameters on the mechanical properties of FDM printed materials is essential for optimizing the process and achieving high-quality printed parts. Ongoing research efforts aim to develop more sophisticated modeling techniques that can predict material properties accurately, enabling faster and more cost-effective optimization of the FDM process.

2. FDM Polymer Materials – Filaments

Currently, the market offers a diverse range of FDM filaments, including pure polymers, polymer blends, and composites. Moreover, new types of filaments are continuously being developed. Typically, FDM filaments consist of thermoplastic polymers with an amorphous structure, such as ABS and PC. However, semi-crystalline polymers, such as PET, PLA, and PEEK, are also utilized. Figure 3 provides a comprehensive overview of commercially available FDM polymer filaments, as well as those still under research and development [23], [24], [25].

ABS is a highly regarded FDM material due to its toughness, impact resistance, and widespread availability. It is an amorphous thermoplastic composed of polybutadiene and styrene-acrylonitrile copolymer phases, which contribute to its toughness and dimensional accuracy, respectively. However, when working with larger or complex geometries, ABS material tends to shrink, making it necessary to use a 3D printer with a heated working surface and closed chamber to control the temperature during printing [24], [26], [27]. ASA (Acrylonitrile Styrene Acrylate) is a viable alternative to ABS, utilizing acrylate rubber instead of polybutadiene. ASA material is highly resistant to weather conditions, making it a suitable choice for outdoor products [28], [29].

PLA material is among the most popular, widely used, and extensively researched FDM materials, along with ABS. It is a semi-crystalline, biodegradable thermoplastic derived from renewable sources. PLA is better suited for FDM technology than ABS due to its lower coefficient of thermal expansion and lower required printing temperature. FDM printed PLA samples exhibit higher tensile strength and a better finish than ABS samples. Moreover, PLA can be printed using inexpensive desktop FDM printers without the need for substrate heating or a closed printing chamber with controlled temperature. It is a reliable and environmentally friendly material that does not release harmful particles during printing, making it ideal for use in homes [24], [30], [31].

Despite the many advantages of PLA, one of its biggest limitations is low heat resistance and susceptibility to deformation at relatively low temperatures (around 50°C), that can limit its applicability. Additionally, PLA is relatively brittle compared to other FDM materials. To address these limitations, improved PLA materials, such as HTPA (high temperature PLA) from Proto-paste and Tough PLA from Ultimaker, have been developed. HTPA offers properties similar to PLA but can withstand higher temperatures, while Tough PLA is a technical PLA material with improved toughness and plasticity similar to ABS. These materials provide a means of overcoming the limitations of traditional PLA and expanding its potential applications [12], [24], [32].

The materials used for support structures play a critical role in determining the visual quality of the final FDM product. In single-head FDM printers, the basic material of the product and the support structure must be the same, and the support structures are removed mechanically after production, which can create defects such as increased roughness where supports were attached.

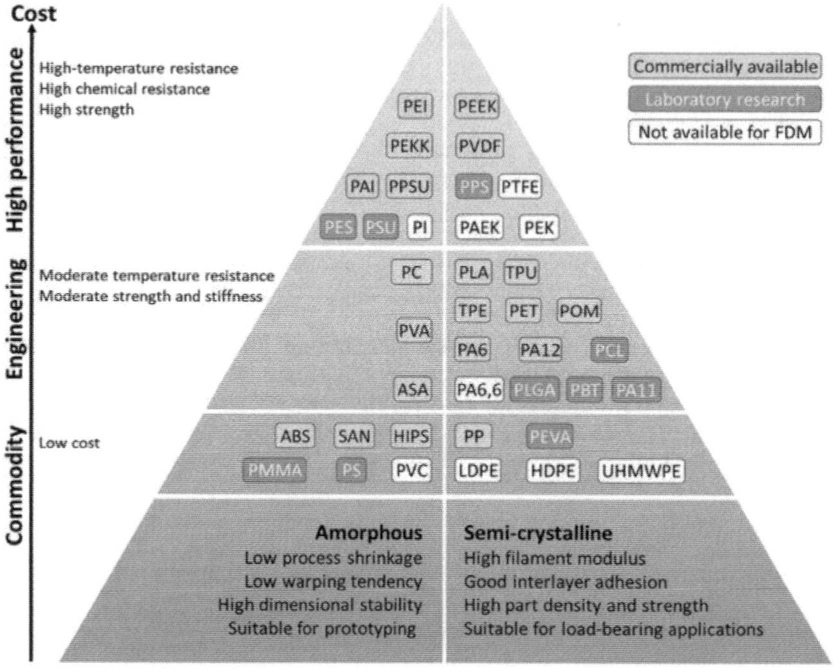

Fig. 3. Overview of FDM polymer materials – filaments [24]

In contrast, dual-head FDM printers allow for the use of a different material for support structures, which can be dissolved in water or another medium to reduce the likelihood of defects (Figure 4). The most common soluble FDM materials for support structures are PVA (Polyvinyl Alcohol) and HIPS (High Impact Polystyrene), the latter of which is often used as a support material for ABS and dissolves in d-limonene. Choosing the right support material is therefore important for producing high-quality FDM products [24], [33]. Amorphous polymers with low coefficients of thermal expansion and minimal shrinkage during fabrication have already been widely adopted in FDM technology. In contrast, semi-crystalline polymers such as PA12 (Polyamide 12), PET, and PP (polypropylene), which offer better mechanical properties and higher temperature resistance, remain less explored and commercialized due to their challenging deformation and shrinkage during 3D printing. However, the integration of semi-crystalline polymers into FDM technology is crucial for its evolution from prototyping to direct production of functional components under varying loads. Thus, further research and development efforts are needed to overcome the obstacles and unleash the full potential of semi-crystalline polymers in FDM technology [24], [35]. High-performance polymers such as PEI (polyetherimide) and PEEK (polyether ether ketone) are widely recognized for their exceptional strength, making them excellent choices for FDM printing. In fact, FDM printed PEI and PEEK materials can achieve impressive tensile strengths of 90 – 110 MPa, respectively [11], [36]. By comparison, the tensile strength of ABS, PLA and PC FDM printed materials ranges from 30 to 70 MPa [24], [37], [38].

Fig. 4. An example of an FDM printed product with a support structure made of PVA material after manufacturing (left) and after dissolving the PVA material in water (right) [34]

PEI and PEEK materials have high melting temperatures, requiring nozzle temperatures exceeding 360°C. As nozzle temperature increases, the porosity of printed parts decreases (360-420°C) and crystallinity increases (380-480°C), leading to significant improvements in mechanical properties [24], [36], [39]. However, it is crucial to avoid excessive nozzle temperature and overheating, as this can cause polymer degradation, thereby worsening the mechanical properties of FDM printed materials [40]. The high-temperature requirements of extruding high-performance polymers make these materials more demanding to fabricate with FDM technologies due to the large temperature differences between the newly deposited and pre-deposited material. To minimize these temperature differences, FDM printing of these materials often requires a heated work surface and a closed FDM device, where the working temperature can be controlled in the working space. A study [39] demonstrated that an increase in the working space temperature of the FDM printer from 25°C to 200°C improved the crystallinity and mechanical properties of the FDM printed PEEK material.

Moreover, the mechanical properties of high-performance FDM printed polymer materials can be significantly enhanced by subsequent thermal treatments [24], [39]. Therefore, high-performance polymers such as PEI and PEEK offer promising potential for the future of FDM printing technology, particularly to produce components that require exceptional strength and durability. FDM elastomers, or rubber-like elastic materials suitable for FDM technology, are available on the market in the form of TPE (thermoplastic elastomer) and TPU (thermoplastic polyurethane) materials. TPU is known for its high strength, high elasticity, and high elongation at break, with a Shore A hardness of 95 [41], [42]. On the other hand, TPE is a hybrid polymer FDM material consisting of hard and soft phases, exhibiting rubber-like elasticity. The hard phase is a semi-crystalline thermoplastic segment that makes it easier to process with FDM technology. Meanwhile, the soft phase is an amorphous elastomeric segment that provides the elastic properties. TPE has a lower Shore A hardness of around 70 compared to TPU [43].

However, using TPE or TPU filaments for FDM poses challenges due to their high elasticity and deformability, making them susceptible to deformation when caught by pressure rollers during the filament's transfer to the printer head. Axial stress can also cause bending when it arrives at the printer head for melting and further extrusion under axial force [24]. In order to improve the properties of pure FDM polymer materials, various blends with amorphous and semi-crystalline polymers are being developed to achieve improved strength, hardness, toughness, plasticity, printability, and higher temperature resistance, as well as functional properties such as electrical conductivity, antibacterial properties, and shape memory [44], [45], [46], [47]. The most popular blends for FDM are PC/ABS and PC/PEI (Ultem 9085), which were originally developed for injection molding applications. The PC/ABS blend is a widely used thermoplastic in the traditional polymer industry, and its FDM printed parts exhibit high strength, toughness, and heat resistance [24], [45], [46], [47], [48]. On the other hand, PC/PEI is a high-performance and fire-resistant polymer certified for use in the automotive, aerospace, and medical industries. PC/PEI is currently the standard PEI material used in FDM technology due to its better viscosity than pure PEI (Ultem 1010) [24], [49], [50].

Material	Tensile strength [MPa]	Elastic modulus [GPa]	Strain at brake [%]	Manufacturer
PLA	48	3,04	2,5	Stratasys
ABS	28	2,4	8	Stratasys
PC	57	1,94	4,8	Stratasys
PETG	50	1,87	6,4	Ultimaker
Nylon	27,8	0,57	210	Ultimaker
ASA	33	2,14	5,9	Stratasys
CPE	41.1	1,5	5.1	Ultimaker
PVA	-	-	-	Ultimaker
PP	8,7	0,22	>300	Ultimaker
TPU	16,8	0,015	552	Stratasys
TPE	-	-	600	MatterHackers
PEEK	98	4	45	Victrex
ULTEM	80	2,86	3,8	Sabic
PEKK	61	3,8	20	Arkema
PA12	50	1,51	30	Stratasys
PA6	67,6	2,32	38	Stratasys
HTPLA	67,4	5,6	2	Proto-Pasta
ABS-X	23,8	0,45	13,3	3D Republika
ABS/PC	41	1,9	6	Stratasys
PC/PEI (Ultem 9085)	69	2,51	5,4	Stratasys
PLA-X	53,2	0,43	15	3D Republika
PEI (Ultem 1010)	80	3	4	Stratasys

Table 1. Overview of FDM printed polymer materials and their mechanical properties

The development of polymer mixtures in the form of FDM filaments is constrained by the composition of the polymers involved and the need to maintain appropriate thermal and rheological properties. Achieving a high-quality 3D print requires careful adjustment of the mixture composition and FDM processing parameters. Furthermore, the blends must be processed into filament, wound on a roll, before they can be used for FDM 3D printing. However, FDM systems based on screw extrusion are being developed that allow for the use of a wider spectrum of polymer mixtures, facilitating the production of products for various applications [24]. Table 1. presents a summary of the mechanical properties of selected FDM pure polymer materials, as well as blends that have been developed to enhance the performance of FDM printed parts. The table provides data on key mechanical properties such as tensile strength, elastic modulus and strain at break.

3. Physical aspects in the formation of FDM printed polymer materials

During the FDM printing process, the filament is first heated to a semi-molten state and then extruded through a nozzle that follows a predefined path based on G-code instructions. After extrusion, thermal energy is transferred through conduction (within and between the printed material), convection, and radiation (between the printed material and the external environment). The heating and subsequent cooling processes involved in semi-molten filament resolidification are characterized by high thermal gradients that induce residual stresses in the printed material. These stresses can lead to distortions and deformations of printed products, both during and after the printing process. Physical phenomena of FDM printed material that occur during the FDM process are presented on Figure 5.

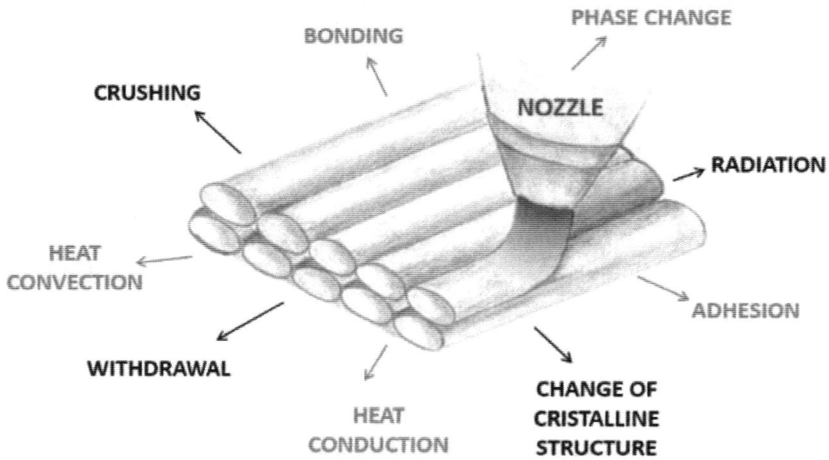

Fig. 5. Physical phenomena of FDM printed material that occur during the FDM process (blue color represents primary and black color secondary physical effects) [3]

The quality of FDM printed products is influenced by a variety of factors, including thermal gradients. Experimental evidence has demonstrated that these gradients can have significant impacts on both the shape and dimensions of the final product, as well as its material mechanical properties such as yield strength, tensile strength, strain, and others [3, 24]. The FDM printing process involves extruding polymer material onto a work surface, which may be heated depending on the type of material being used. As the extruded material cools, it transitions from a liquid to a solid state. During this phase, heat transfer occurs through three main mechanisms, as described in [3]:

- Conduction, resulting from the temperature gradient that develops within the FDM material;
- Convection, which occurs between the FDM material and its surroundings;
- Radiation, involving the emission of heat energy from the FDM material.

The process of phase change in polymeric materials is highly dependent on their molecular structure. As previously discussed, polymers can be either crystalline (semi-crystalline) or amorphous, each with its own unique characteristics. Two types of phase changes can occur in polymers, depending on their molecular structure:

- Melting is a typical process of crystalline polymers, where ordered molecular chains transition from a solid to a liquid state at the melting temperature (Tm). To achieve this, a certain amount of energy is required to overcome the intermolecular forces, typically supplied through thermal energy.
- Glass transition is a typical process of amorphous polymers, which includes most FDM thermoplastic materials. With increasing temperature, molecular disorder increases in amorphous polymers. The glass transition temperature (Tg) is the point at which amorphous polymers exhibit a "rubbery" behavior and become viscous enough to be extruded through nozzle in FDM technology.

Figure 6 illustrates the different phase change processes for crystalline and amorphous polymers. Notably, the glass transition process does not require latent heat to change phases, in contrast to the melting process [51].

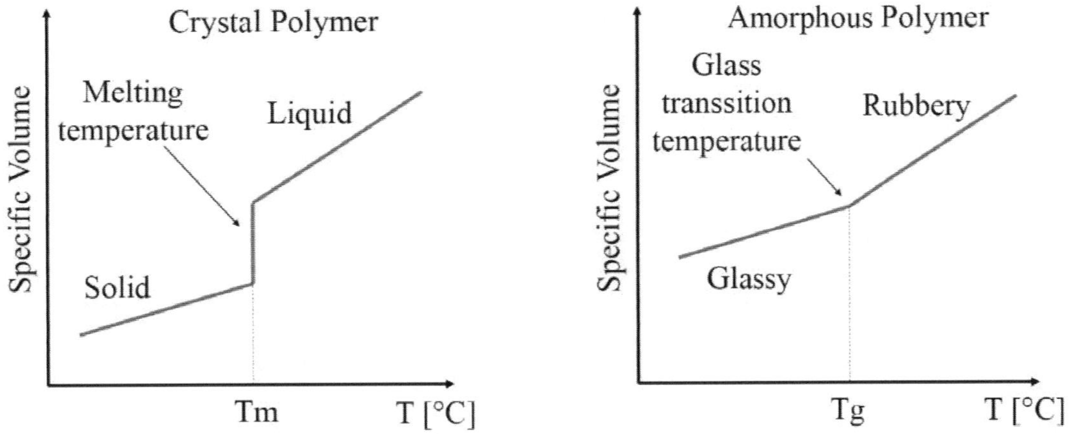

Fig. 6. Phase changes in crystalline and amorphous polymers

After the extrusion process, the FDM filament material is deposited onto the work surface in the form of rasters (threads). To solidify the material, it is cooled to a temperature below its glass transition temperature (Tg) or melting temperature (Tm). When depositing the next filament raster in the same plane or a new layer, the boundary between the previously deposited raster and the new raster heats up to Tg due to the heat transfer from the new raster. This causes the previously deposited raster to soften, allowing the newly deposited raster to bond with it. This process creates a cohesive force between the connected rasters (Figure 7).

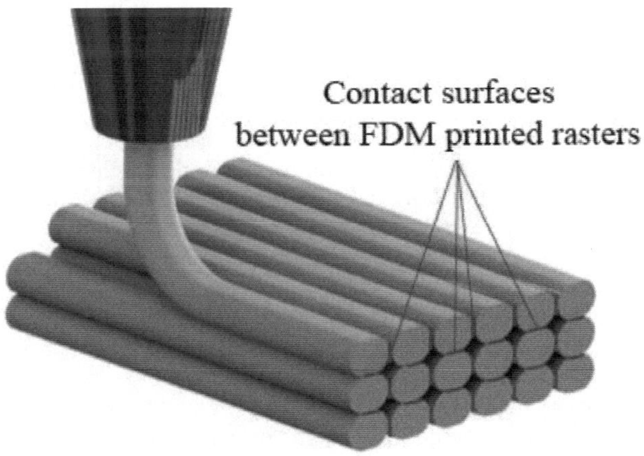

Fig. 7. Extrusion and deposition of FDM printed filament rasters

In the process of creating FDM printed polymer material, raster bonding plays a crucial role in determining the final mechanical properties of the FDM printed material. Specifically, the mutual bonding of printed rasters is essential for creating a strong and durable printed material. During the FDM printing process, the temperature of the deposited material must be precisely controlled to ensure proper solidification and bonding between adjacent rasters. If the bonding between raster is weak or inconsistent, the printed material may exhibit inferior mechanical properties such as reduced strength or durability. Therefore, it is essential to understand and optimize the bonding behavior of the FDM printing process to produce high-quality printed materials of products with desired mechanical properties [25], [52], [53].

Figure 8 depicts a schematic representation of the bonding process between adjacent extruded polymer rasters during the FDM process. Initially, when the rasters come into contact, they undergo a process called "necking," during which polymer chains diffuse between the rasters. The success of this bonding process depends on several factors, including the rheological properties of the polymer, viscosity, thermal conductivity, and cooling rate. In particular, the temperature at which the FDM process is conducted plays a critical role in the interlayer bonding process. Higher temperatures can help facilitate the diffusion of polymer chains between the rasters, leading to a stronger bond between adjacent rasters [24], [25], [54].

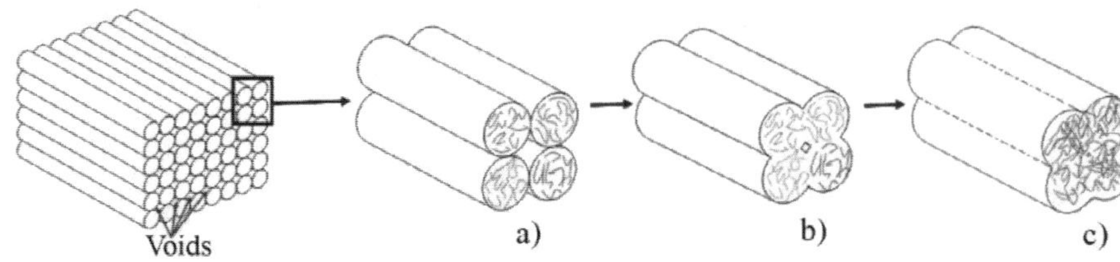

Fig. 8. Schematic representation of bonding of FDM printed filaments: (a) surface contact, (b) neck formation and growth, (c) diffusion and randomization of molecular chains [52]

FDM technology involves two types of bonding for printed rasters, as illustrated in Figure 9 [3]:

- Intra-layer bonding occurs between adjacent FDM printed rasters within the same layer. This type of bonding creates a typical neck between two adjacent printed rasters, which results in the final dimensions (L2) being smaller than the nominal ones (L1).
- Inter-layer bonding occurs between layers that are connected to each other along the z-axis. This type of bonding also forms a neck between adjacent printed rasters, causing the final height (h2) to be slightly less than the nominal height (h1).

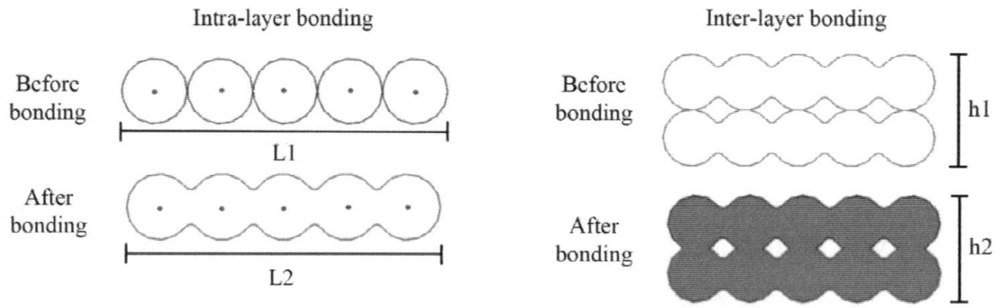

Fig. 9. Bonding of FDM printed rasters within a layer and between layers during the material formation process [3]

The time interval between the deposition of two printed rasters is a critical factor affecting bonding strength. Longer time intervals lead to weaker diffusion of the polymer chain, resulting in weaker bonding between mutually touching rasters. This time-dependent behavior is evident from the anisotropic properties of most FDM printed materials, where interlayer bonding is weaker than intralayer bonding. Bonding strength is crucial for the thermal and electrical properties of FDM printed materials, as air gaps between printed filaments can significantly affect these properties. During the FDM process, the printed rasters undergo cooling shrinkage, which is influenced by the thermal properties of the base material, including its thermal conductivity, heat capacity, coefficient of thermal expansion, and crystallinity (for semi-crystalline materials). Uneven shrinkage can cause warping or warpage of the first layers of the printed product in contact with the working surface. This can lead to delamination or even breakage of the printed material, as shown in Figure 10. To prevent these problems, the working surface on which printing is done is heated, and the workspace of the 3D printer is enclosed in a chamber with a controlled working temperature. Other methods may also be used [24].

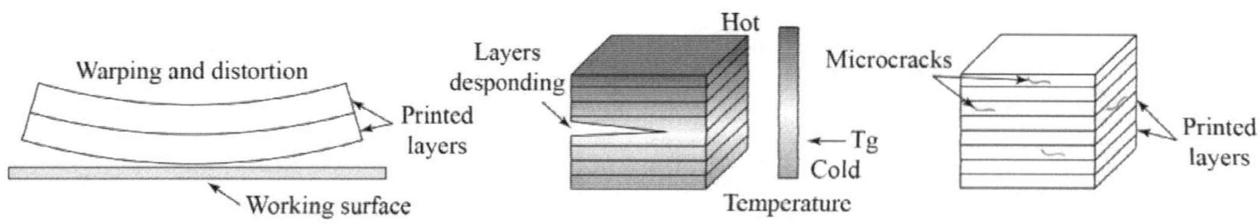

Fig. 10. Defects in FDM printed polymer materials due to uneven shrinkage [55]

During the FDM process, the base material, which is the FDM polymer filament, undergoes various thermal cycles and phase changes. The filament is then applied in layers composed of printed rasters, where their mutual bonding plays a crucial role in the formation of the final FDM printed material. Along with many other factors, this bonding determines the properties of the printed material. Figure 11 highlights the key properties that an FDM polymer material should possess to meet the requirements of the FDM printing process and the final product. These properties include filament diameter, roundness, and tolerance, as well as mechanical, thermal, and electrical properties. Ensuring that the base material satisfies these properties can result in a printed product that meets the desired specifications.

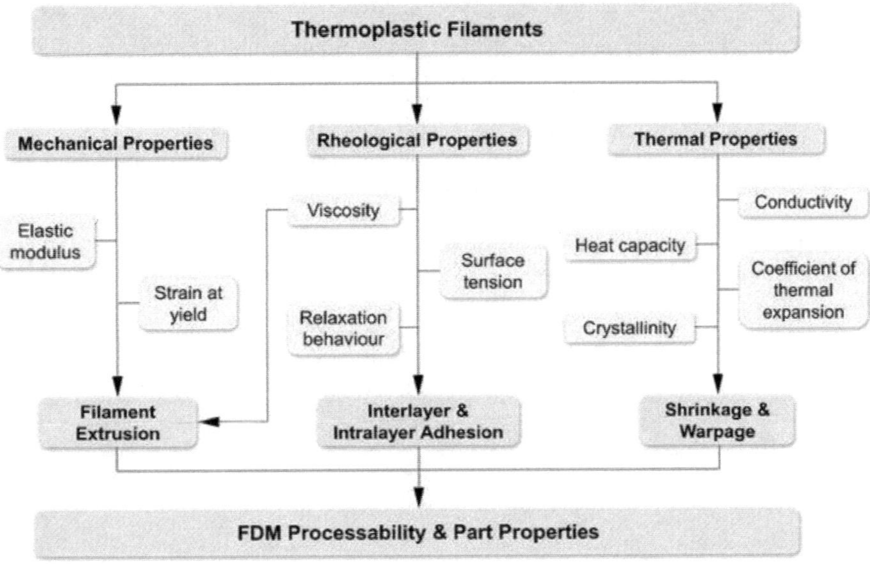

Fig. 11. Key properties of FDM filament and their influence on the properties of FDM printed material as well as the final product [24]

4. Mechanical properties of FDM printed polymer materials

To establish the minimum dimensions of a product made from specific materials, while simultaneously ensuring its mechanical durability under load, a comprehensive understanding of product material mechanical properties is crucial. This knowledge allows for the prevention of undesired deformations and potential fractures when exposed to external loads. The mechanical properties of a material reveal the correlation between applied external loads and resultant deformations. Essential mechanical properties to consider include strength (such as tensile, compressive, and flexural strength), hardness, toughness, fatigue, and creep. Accurate determination of a material's mechanical properties requires laboratory testing under actual operating conditions, as closely as possible, to ensure the validity of the results [56].

When discussing FDM printed polymer materials, it is essential to emphasize the fundamental principle of this technology: the product is fabricated directly by adding materials, layer by layer, based on a digital CAD model of the product. This stands in stark contrast to traditional manufacturing methods such as turning, milling, and cutting, where material is removed from a pre-existing piece with known mechanical properties, which is then shaped into the final product [57], [58]. This fundamental difference underscores the importance of understanding the unique printed material properties and limitations of FDM printing when designing and manufacturing products.

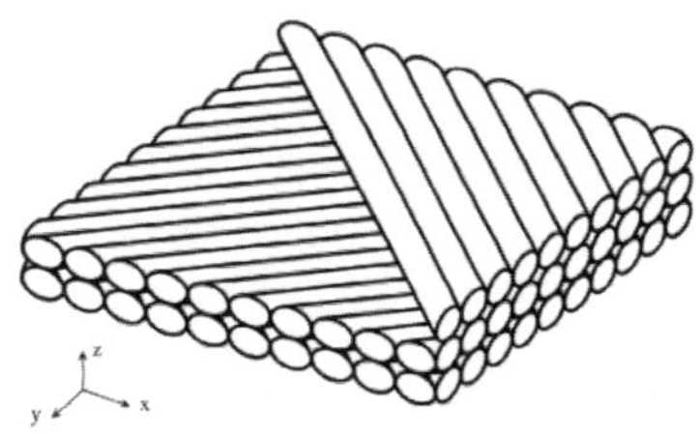

Fig. 12. Structure of FDM printed material [59]

When analyzing the mechanical properties of FDM printed polymer materials, it is important to differentiate between the filament (base material) and the FDM printed polymer material that makes up the final product. The filament, which is wound on rolls, serves as the fundamental building block of the FDM process. In contrast, the FDM printed polymer material is created by heating and extruding the filament and applying it layer by layer, with each layer composed of extruded rasters of the base material. The filament undergoes the entire FDM process, including being heated to a viscous state and extruded onto the working surface during the initial layer formation, and then further extruded onto the already printed material (as depicted in Figure 12). Various factors, such as the properties of the thermoplastic filament, extrusion parameters, process parameters, structural parameters, and external conditions (humidity, ambient temperature, etc.), can directly or indirectly affect the mechanical properties of the FDM printed polymer material. As a result, there is a considerable focus on researching the influence of these parameters on the mechanical properties of FDM printed materials, as well as their underlying mechanisms [57], [58], [60], [61], [62].

Understanding these factors is crucial to optimize the mechanical properties of FDM printed polymer materials and to ensure their suitability for specific applications. In their papers, Dawoud et al. [63], Popescu et al. [30], Kristiawan et al. [17], Sheoran et al. [64], Rahmatabadi et al. [13], Shanmugam et al. [65], Waseem et al. [66], Zisopol et al. [67], and Mohamed et al. [68] have conducted a comprehensive literature review to identify factors that influence the mechanical properties of FDM printed polymer materials. Based on their studies, it can be concluded that infill design impact the mechanical properties of FDM printed polymer materials, including tensile, flexural, and compressive strength, hardness, toughness, fatigue, and creep (Figure 13). It is important to note that while the figure below illustrates the influence of infill design on different mechanical properties, also, there are numerous other factors (layer height, printing speed, raster orientation, etc.) that can also directly or indirectly affect the mechanical properties of FDM printed polymer materials. Therefore, researchers and engineers should consider a wide range of factors when designing and optimizing FDM printed polymer materials for specific applications.

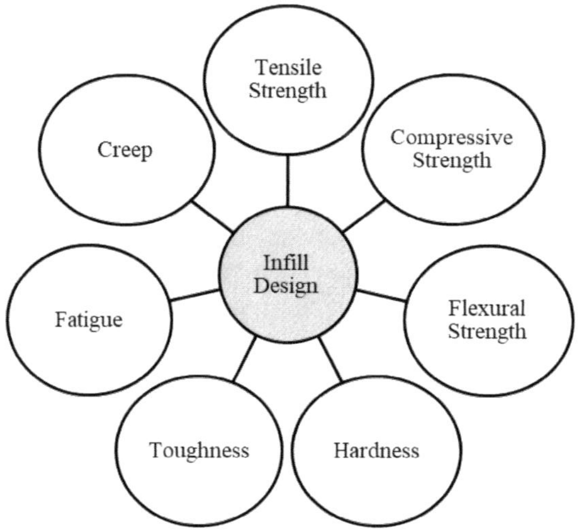

Fig. 13. Influence of infill design on the mechanical properties of FDM printed polymer materials

It is evident that understanding the factors that influence the mechanical properties of FDM printed polymer materials and promoting standardization are critical. Currently, manufacturers include mechanical properties of the material in their material declarations, but few disclose details about the FDM printing and testing process. As a result, the impact of various factors on mechanical and other properties of FDM printed materials is not always apparent. It is crucial to conduct research in this area to better comprehend and analyze the mechanical properties of FDM printed materials. Such research will significantly contribute to standardization and the overall advancement of this technology.

5. Mechanical testing of FDM printed polymer materials

The increased prevalence and application of additive manufacturing (AM) technologies in various industries and everyday life have led to the development of AM materials. As a result, there is a growing need for standardization in this field. Many manufacturers of AM materials are available in the market, each seeking to distinguish their products by quality, strength, and other properties. However, to ensure product quality and distinguish materials and their properties authoritatively, there must be universal definitions of high quality, strength, and other properties, which can only be achieved through standardization.

In their research titled " Analysis of General and Specific Standardization Developments in Additive Manufacturing from a Materials and Technological Approach " [69], Dominguez A.G. et al. provided an overview of the current standards in the field of AM, including ISO 17296-3:2014 and ISO/ASTM 52921:2013. ISO 17296-3, " Additive manufacturing — General principles — Part 3: Main characteristics and corresponding test methods", outlines the main characteristics of AM and the corresponding test methods. On the other hand, ISO/ASTM 52921:2013, " Standard terminology for additive manufacturing — Coordinate systems and test methodologies", focuses on the standardization of terminology for test results reports. While these standards do not specifically address the mechanical testing of FDM printed polymer materials, they are still important for establishing common terminology and test methods in the AM field.

Research and testing of mechanical and other properties of AM materials are conducted intensively, with a focus on FDM printed polymer materials. However, despite the research efforts, specific standards for characterizing and testing the mechanical properties of FDM printed polymer materials have not yet been developed. Generally, research is based on existing ISO and ASTM standards for polymer materials [7], [69], [70]. Therefore, there is a pressing need for the development of specific standards for FDM printed polymer materials to ensure the quality and reliability of products made using this technology.

Through a detailed review of previous research, Dizon J.R.C. et al. [7], Popescu D. et al. [30], and Forster A.M. [71] studied the topic of standards used for mechanical testing of FDM printed polymer materials. Based on their findings and the literature review in this paper, the figure below (Fig 14) presents the most commonly used ISO and ASTM standards for testing the characteristic mechanical properties of FDM printed polymer materials.

Tensile properties	•ISO 527 •ASTM D638
Bending properties	•ISO 178 •ASTM D790
Compression properties	•ISO 604 •ASTM D695
Creep properties	•ISO 899 •ASTM D2990
Fatigue properties	•ISO 13003 •ASTM D7774
Impact testing	•ISO 179, ISO 180 •ASTM D6110, ASTM D256

Fig. 14. An overview of the most commonly used standards for testing the mechanical properties of FDM printed polymer materials through previous research

It is important to note that the current standards for testing the mechanical properties of polymer materials are primarily intended for those created through traditional methods such as injection molding. However, with a lack of standards available for testing the properties of FDM printed polymer materials, these existing standards are often used to evaluate printed materials with a 100% infill density or materials with a infill structure. In a paper by Juan Claver A.G.D. and others [72], the applicability of these standards for characterizing the mechanical properties of FDM printed materials was analyzed. The authors cautioned that when evaluating the mechanical properties of FDM printed polymer materials according to existing standards, one should consider their limitations due to the unique characteristics of these materials. FDM printed materials can be produced as solid materials with cavities, materials with a infill structure, or lattice and cellular structures, which require specific testing methods.

Despite these limitations, in most studies, these existing standards are still utilized to evaluate the properties of FDM printed materials. Additionally, Forster A.M. and the NIST Institute conducted a comprehensive study [71] entitled "Materials testing standards for additive manufacturing of polymer materials: State of the art and standards applicability ". The study concluded that while most existing standards for polymeric materials can be applied to test AM materials, specific guidelines are needed for measuring properties in engineering cases enabled by AM technologies.

It is crucial to note, and it has already been mentioned before, that the mechanical properties of FDM printed polymer materials are highly dependent on various factors, such as the characteristics of the base material (filament), 3D printing parameters, and external conditions. The research results [73] showed that the color of the material can affect its tensile mechanical properties, even if the material is from the same manufacturer. In research [74], significant differences were observed in the tensile mechanical properties of PLA materials from different manufacturers, even though it was the same material. Moreover, the research [75] demonstrated that FDM printed polymer PLA materials made on different models of 3D printers exhibit different tensile properties. The paper [8] analyzed the influence of different deformation rates during testing on tension and showed that tensile properties are dependent on the rate of deformation during testing. Thus, when specifying the mechanical properties of FDM printed polymer materials, it is essential to specify several factors, including the type, manufacturer, and color of the material, 3D printing parameters and 3D printer model, the standard by which the test was performed, the geometry of the sample, and the rate of deformation during the test, as well as external factors during which the examination was carried out. Simply providing numerical values of mechanical properties for FDM printed polymer materials is insufficient. Today, more and more material manufacturers (such as Ultimaker, Stratasys, 3D Republika, etc.) provide information about the filament's characteristics, in addition to the mechanical properties, 3D printing parameters, external factors, and the standard according to which the properties were tested.

6. Conclusion

When considering the materials used in products created by FDM technology, it is important to understand that the properties of the filament (input material) and the resulting FDM printed polymer material can differ due to various influencing factors. The physical aspects of the formation process are crucial to understanding the final properties of the FDM printed materials in finished products.

Regarding the mechanical properties of FDM printed materials, it can be concluded that various factors including the thermoplastic filament, extrusion and process parameters, structural parameters, and external factors such as humidity and ambient temperature have a direct or indirect influence on these properties. It is noteworthy that current mechanical property testing standards for FDM printed materials are primarily intended for polymer materials produced by conventional methods such as injection molding. Considering the unique physical aspects and influencing factors involved in the formation of FDM printed polymer materials, it is recommended that additional guidelines are needed for FDM printed materials, especially for cases involving materials with a infill structure, as well as lattice and cellular materials which are becoming more common in modern manufacturing. Understanding the interplay between these factors is important for achieving consistent and predictable material properties in FDM printing technology.

Also, Material manufacturers should provide complete information when defining the mechanical and other properties of FDM printed polymer materials on their declaration. The basic influencing factors such as the color of the material, the model of the 3D printer used, key printing parameters, and the standard and method for examination of characteristics must be listed. Mere numerical values of mechanical and other properties are insufficient to define FDM printed polymer materials properly. This information can be easily achieved through standardization of FDM technologies. Standardization is necessary to ensure reliable and predictable mechanical and other properties of FDM printed polymer materials. Therefore, material manufacturers should prioritize providing complete information when defining the mechanical and other properties of FDM printed polymer materials on their declaration, to aid in the creation of quality products.

Recommendation for future research entails extending the detailed analysis beyond polymers to encompass a broader range of materials, including composites, metals, and others. Furthermore, exploring alternative additive manufacturing (AM) technologies in conjunction with different materials is advised.

7. Acknowledgments

Acknowledgments to UltiMaker company for their invaluable support not only for this research but also for advancing research in the broader field of FDM 3D printed materials.

8. References

[1] Abdulhameed O, Al-Ahmari A, Ameen W and Mian S H 2019 Additive manufacturing: Challenges, trends, and applications Adv. Mech. Eng. 11 1–27

[2] Alghamdi S S, John S, Choudhury N R and Dutta N K 2021 Additive manufacturing of polymer materials: Progress, promise and challenges Polymers (Basel). 13 1–39

[3] Cattenone A 2018 Analysis and Simulation of Additive manufacturing processes (Faculty of Engineering - University of Pavia)

[4] Banjanin B 2018 Characterisation of manufacturing parameters of embossing dies produced by 3D printing technique (Ph.D. thesis, University of Novi Sad, Faculty of Technical Sciences)

[5] Pilipović A 2012 Influence of processing parameters on the properties of polymer prototype (Doctoral thesis, Faculty of Mechanical Engineering and Naval Architecture, University of Zagreb)

[6] Adi Pandžić 2022 Contribution to the evaluation of mechanical properties of FDM printed polymeric materials with different infill design (Doctoral thesis, Faculty of Mechanical Engineering, University of Sarajevo)

[7] Dizon J R C, Espera A H, Chen Q and Advincula R C 2018 Mechanical characterization of 3D-printed polymers Addit. Manuf. 20 44–67

[8] Hodzic D, Pandzic A, Hajro I and Tasic P 2020 Strain rate influence on mechanical characteristics of FDM 3D printed materials 31st DAAAM International Symposium vol 31 pp 168–75

[9] Tofail S A M, Koumoulos E P, Bandyopadhyay A, Bose S, O'Donoghue L and Charitidis C 2018 Additive manufacturing: scientific and technological challenges, market uptake and opportunities Mater. Today 21 22–37

[10] Dizon J R C, Gache C C L, Cascolan H M S, Cancino L T and Advincula R C 2021 Post- Processing of 3D-Printed Polymers Technologies 9 61

[11] Algarni M and Ghazali S 2021 Comparative study of the sensitivity of PLA, ABS, PEEK, and PETG's mechanical properties to FDM printing process parameters Crystals 11

[12] Naveed N 2021 Investigating the material properties and microstructural changes of fused filament fabricated PLA and tough-PLA parts Polymers (Basel). 13

[13] Rahmatabadi D, Aminzadeh A, Aberoumand M and Moradi M 2021 Mechanical Characterization of Fused Deposition Modeling (FDM) 3D Printed Parts Fused Deposition Modeling Based 3D Printing ed H K Dave and J P Davim (Springer) p 517

[14] Baich L 2016 Impact of Infill Design on Mechanical Strength and Production Cost in Material Extrusion Based Additive Manufacturing (Master of Science Thesis, Youngstown State University)

[15] Chaudhry M S and Czekanski A 2020 Evaluating FDM process parameter sensitive mechanical performance of elastomers at various strain rates of loading Materials (Basel). 13 1–10

[16] Camposeco-Negrete C 2020 Optimization of printing parameters in fused deposition modeling for improving part quality and process sustainability Int. J. Adv. Manuf. Technol. 108 2131–47

[17] Kristiawan R B, Imaduddin F, Ariawan D, Ubaidillah and Arifin Z 2021 A review on the fused deposition modeling (FDM) 3D printing: Filament processing, materials, and printing parameters Open Eng. 11 639–49
[18] Alafaghani A and Qattawi A 2018 Investigating the effect of fused deposition modeling processing parameters using Taguchi design of experiment method J. Manuf. Process. 36 164–74
[19] Harpool T D 2014 Observing the Effects of Infill Shapes on the Tensile Characteristics of 3D Printed Plastic Parts (Master of Science Thesis, Wichita State University)
[20] Dev S and Srivastava R 2021 Optimization of fused deposition modeling (FDM) process parameters for flexural strength Mater. Today Proc. 44 3012–6
[21] Park K, Kim G, No H, Jeon H W and Kremer G E O 2020 Identification of optimal process parameter settings based on manufacturing performance for fused filament fabrication of CFR-PEEK Appl. Sci. 10 1–23
[22] Floor J 2015 Getting a Grip on Ultimaker 2, Tensile Strength of 3D Printed PLA: A Systematic Investigation (TU Delft, Ultimaker)
[23] Gibson I, Rosen D, Stucker B and Khorasani M 2021 Chapter 14: Materials for additive manufacturing Additive Manufacturing Technologies (Springer Nature Switzerland AG 2021) pp 379–428
[24] Tan L J, Zhu W and Zhou K 2020 Recent Progress on Polymer Materials for Additive Manufacturing Adv. Funct. Mater. 30 1–54
[25] Wu H, Fahy W P, Kim S, Kim H, Zhao N, Pilato L, Kafi A, Bateman S and Koo J H 2020 Recent developments in polymers/polymer nanocomposites for additive manufacturing Prog. Mater. Sci. 111
[26] Dilberoglu U M, Simsek S and Yaman U 2019 Shrinkage compensation approach proposed for ABS material in FDM process Mater. Manuf. Process. 34
[27] Zhu J, Hu Y, Tang Y and Wang B 2017 Effects of styrene–acrylonitrile contents on the properties of ABS/SAN blends for fused deposition modeling J. Appl. Polym. Sci. 134
[28] Raam Kumar S, Sridhar S, Venkatraman R and Venkatesan M 2020 Polymer additive manufacturing of ASA structure: Influence of printing parameters on mechanical properties Mater. Today Proc. 39 1316–9
[29] Guessasma S, Belhabib S and Nouri H 2019 Printability and tensile performance of 3D printed polyethylene terephthalate glycol using fused deposition modelling Polymers (Basel). 11
[30] Popescu D, Zapciu A, Amza C, Baciu F and Marinescu R 2018 FDM process parameters influence over the mechanical properties of polymer specimens: A review Polym. Test. 69 157–66
[31] Pandzic A and Hodzic D 2021 Mechanical properties comparison of PLA, tough PLA and PC 3D printed materials with infill structure – Influence of infill pattern on tensile mechanical properties IOP Conference Series: Materials Science and Engineering vol 1208 p 012019
[32] Akhoundi B, Nabipour M, Hajami F and Shakoori D 2020 An Experimental Study of Nozzle Temperature and Heat Treatment (Annealing) Effects on Mechanical Properties of High-Temperature Polylactic Acid in Fused Deposition Modeling Polym. Eng. Sci. 60
[33] Goh W H and Hashimoto M 2018 Dual sacrificial molding: Fabricating 3D microchannels with overhang and helical features Micromachines 9
[34] Ultimaker B.V. 2022 Ultimaker Materials Tehnical Data Sheet
[35] Spoerk M, Holzer C and Gonzalez-Gutierrez J 2020 Material extrusion-based additive manufacturing of polypropylene: A review on how to improve dimensional inaccuracy and warpage J. Appl. Polym. Sci. 137
[36] Ding S, Zou B, Wang P and Ding H 2019 Effects of nozzle temperature and building orientation on mechanical properties and microstructure of PEEK and PEI printed by 3D-FDM Polym. Test. 78 105948
[37] Kuznetsov V E, Tavitov A G, Urzhumtsev O D, Mikhalin M V. and Moiseev A I 2019 Hardware factors influencing strength of parts obtained by fused filament fabrication Polymers (Basel). 11
[38] Cress A K, Huynh J, Anderson E H, O'neill R, Schneider Y and Keleş Ö 2021 Effect of recycling on the mechanical behavior and structure of additively manufactured acrylonitrile butadiene styrene (ABS) J. Clean. Prod. 279
[39] Yang C, Tian X, Li D, Cao Y, Zhao F and Shi C 2017 Influence of thermal processing conditions in 3D printing on the crystallinity and mechanical properties of PEEK material J. Mater. Process. Technol. 248 1–7
[40] Zhao Y, Zhao K, Li Y and Chen F 2020 Mechanical characterization of biocompatible PEEK by FDM J. Manuf. Process. 56 28–42
[41] Ferretti P, Leon-Cardenas C, Sali M, Santi G M, Frizziero L, Donnici G and Liverani A 2021 Application of TPU-sourced 3d printed FDM organs for improving the realism in surgical planning and training Proceedings of the International Conference on Industrial Engineering and Operations Management
[42] Wang J, Yang B, Lin X, Gao L, Liu T, Lu Y and Wang R 2020 Research of TPU materials for 3D printing aiming at non-pneumatic tires by FDM method Polymers (Basel). 12
[43] Awasthi P and Banerjee S S 2021 Fused deposition modeling of thermoplastic elastomeric materials: Challenges and opportunities Addit. Manuf. 46 102177
[44] Vinyas M, Athul S J, Harursampath D and Nguyen Thoi T 2019 Experimental evaluation of the mechanical and thermal properties of 3D printed PLA and its composites Mater. Res. Express 6
[45] Zhou Y G, Zou J R, Wu H H and Xu B P 2020 Balance between bonding and deposition during fused deposition modeling of polycarbonate and acrylonitrile-butadiene-styrene composites Polym. Compos. 41
[46] Salem Bala A, bin Wahab S and binti Ahmad M 2016 Elements and Materials Improve the FDM Products: A Review Adv. Eng. Forum 16

[47] Alhijjaj M, Belton P and Qi S 2016 An investigation into the use of polymer blends to improve the printability of and regulate drug release from pharmaceutical solid dispersions prepared via fused deposition modeling (FDM) 3D printing Eur. J. Pharm. Biopharm. 108 111–25

[48] Mohamed O A, Masood S H and Bhowmik J L 2017 Experimental investigation of time- dependent mechanical properties of PC-ABS prototypes processed by FDM additive manufacturing process Mater. Lett. 193 58–62

[49] Blanco I, Cicala G, Ognibene G, Rapisarda M and Recca A 2018 Thermal properties of polyetherimide/polycarbonate blends for advanced applications Polym. Degrad. Stab. 154

[50] Cicala G, Ognibene G, Portuesi S, Blanco I, Rapisarda M, Pergolizzi E and Recca G 2018 Comparison of Ultem 9085 used in fused deposition modelling (FDM) with polytherimide blends Materials (Basel). 11

[51] Turner B N, Strong R and Gold S A 2014 A review of melt extrusion additive manufacturing processes: I. Process design and modeling Rapid Prototyp. J. 20

[52] Kumar N, Jain P K, Tandon P and Pandey P M 2018 The effect of process parameters on tensile behavior of 3D printed flexible parts of ethylene vinyl acetate (EVA) J. Manuf. Process. 35 317–26

[53] Goh G D, Yap Y L, Tan H K J, Sing S L, Goh G L and Yeong W Y 2019 Process–Structure– Properties in Polymer Additive Manufacturing via Material Extrusion: A Review Crit. Rev. Solid State Mater. Sci. 45 113–33

[54] Blok L G, Longana M L, Yu H and Woods B K S 2018 An investigation into 3D printing of fibre reinforced thermoplastic composites Addit. Manuf. 22

[55] Hassen A A and Kirka M M 2018 Additive manufacturings the rise of a technology and the need for quality control and inspection techniques Mater. Eval. 76 438–53

[56] Pašić O 2008 Materials (Unpublished book, Faculty of Mechanical Engineering, University of Sarajevo)

[57] Syrlybayev D, Zharylkassyn B, Seisekulova A, Akhmetov M, Perveen A and Talamona D 2021 Optimisation of strength properties of FDM printed parts—A critical review Polymers (Basel). 13

[58] Solomon I J, Sevvel P and Gunasekaran J 2021 A review on the various processing parameters in FDM Mater. Today Proc. 37 509–14

[59] Jager A, Johannesmann S, Claes L, Webersen M, Henning B and Kupnik M 2017 Evaluating the influence of 3D-printing parameters on acoustic material properties IEEE International Ultrasonics Symposium, IUS

[60] Narang R and Chhabra D 2017 Analysis of Process Parameters of Fused Deposition Modeling (FDM) Technique Int. J. Futur. Revolut. Comput. Sci. Commun. Eng. IJFRCSCE

[61] Dey A and Yodo N 2019 A systematic survey of FDM process parameter optimization and their influence on part characteristics J. Manuf. Mater. Process. 3

[62] Enemuoh E U, Duginski S, Feyen C and Menta V G 2021 Effect of process parameters on energy consumption, physical, and mechanical properties of fused deposition modeling Polymers (Basel). 13 1–16

[63] Dawoud M, Taha I and Ebeid S J 2016 Mechanical behaviour of ABS: An experimental study using FDM and injection moulding techniques J. Manuf. Process. 21 39–45

[64] Jaisingh Sheoran A and Kumar H 2020 Fused Deposition modeling process parameters optimization and effect on mechanical properties and part quality: Review and reflection on present research Mater. Today Proc. 21 1659–72

[65] Shanmugam V, Das O, Babu K, Marimuthu U, Veerasimman A, Johnson D J, Neisiany R E, Hedenqvist M S, Ramakrishna S and Berto F 2021 Fatigue behaviour of FDM-3D printed polymers, polymeric composites and architected cellular materials Int. J. Fatigue 143 106007

[66] Waseem M, Salah B, Habib T, Saleem W, Abas M, Khan R, Ghani U and Siddiqi M U R 2020 Multi-response optimization of tensile creep behavior of PLA 3D printed parts using categorical response surface methodology Polymers (Basel). 12 1–16

[67] Zisopol D G, Nae I, Portoaca A I and Ramadan I 2021 A Theoretical and Experimental Research on the Influence of FDM Parameters on Tensile Strength and Hardness of Parts Made of Polylactic Acid Eng. Technol. Appl. Sci. Res. 11

[68] Mohamed O A, Masood S H and Bhowmik J L 2017 Influence of processing parameters on creep and recovery behavior of FDM manufactured part using definitive screening design and ANN Rapid Prototyp. J. 23 998–1010

[69] Garcia-Dominguez A, Claver J, Camacho A M and Sebastian M A 2020 Analysis of General and Specific Standardization Developments in Additive Manufacturing from a Materials and Technological Approach IEEE Access 8

[70] Allen M, Chen W and Wang C 2016 3D Printing Standards and Verification Services Appl. Innov. Rev.

[71] Forster A M 2015 Materials testing standards for additive manufacturing of polymer materials: State of the art and standards applicability Additive Manufacturing Materials: Standards, Testing and Applicability

[72] García-Domínguez A, Claver J, Camacho A M and Sebastián M A 2020 Considerations on the applicability of test methods for mechanical characterization of materials manufactured by FDM Materials (Basel). 13

[73] Pandzic A, Hodzic D and Milovanovic A 2019 Influence of material colour on mechanical properties of PLA material in FDM technology 30th DAAAM International Symposium vol 30 pp 555–61

[74] Hodžić D, Pandžić A, Hajro I and Tasić P 2020 Strength comparison of FDM 3D printed PLA made by different manufacturers TEM J. 9 966–70

[75] Pandzic A, Hodzic D, Hajro I and Tasic P 2020 Strength properties of PLA material obtained by different models of FDM 3D printer 31st DAAAM International Symposium vol 31 pp 315–22

DOI: 10.2507/35th.daaam.proceedings.017

INFLUENCE OF NOZZLE DIAMETER ON TENSILE AND FLEXURAL MECHANICAL PROPERTIES OF FDM 3D PRINTED PET-CF MATERIAL

Adi Pandžić, Damir Hodžić, Petar Tasić & Ismar Hajro

This Publication has to be referred as: Pandzic, A[di]; Hodzic, D[amir]; Tasic, P[etar] & Hajro, I[smar] (2024). Influence of Nozzle Diameter on Tensile and Flexural Mechanical Properties of FDM 3D Printed PET-CF Material, Proceedings of the 35th DAAAM International Symposium, pp.0135-0142, B. Katalinic (Ed.), Published by DAAAM International, ISBN 978-3-902734-44-0, ISSN 1726-9679, Vienna, Austria
DOI: 10.2507/35th.daaam.proceedings.017

Abstract

Fused Deposition Modeling (FDM) is a widely researched and applied technology in the additive manufacturing industry. Extensive research is conducted on FDM due to the numerous factors influencing the material properties of 3D printed products. This study focuses on one such critical parameter: nozzle diameter. Nozzle diameter impacts not only the mechanical properties of FDM printed materials but also surface quality and dimensional stability. Specifically, this research investigates the effect of nozzle diameter on the tensile and flexural mechanical properties of PET-CF material (PET reinforced with chopped carbon fibers), along with its influence on printing time, which directly affects the cost of the final product. The study results demonstrated that nozzle diameter significantly affects the tensile and flexural mechanical properties of FDM printed PET-CF material. Additionally, nozzle diameter was found to influence the total printing time.

Keywords: FDM; 3D Print; PET-CF; Composite; AM Material.

1. Introduction

Fused Deposition Modeling additive manufacturing (AM) technology is one of the most used today due to its versatility, accessibility, cost, ease of use and ability to fabricate complex geometries. In general, additive manufacturing is opposed to traditional subtractive manufacturing methods, it is characterized as a process of creating product from 3D-model data (CAD 3D model), typically layer by layer. This process is also referred to by various synonyms including additive fabrication, additive processes, additive techniques, additive layer manufacturing, layer manufacturing, and freeform fabrication [1], [2]. FDM 3D printing operates by feeding a thermoplastic filament from a spool through a feeder to an extrusion head, where it is melted and deposited layer by layer on the print bed to form the final product (Figure 1). FDM materials (filaments) include a variety of thermoplastics and composites, with common options being PLA (Polylactic Acid), PC (Polycarbonate), PET (Polyethylene Terephthalate), PP (Polypropylene), ABS (Acrylonitrile Butadiene Styrene), TPU (Thermoplastic Polyurethane), and Polyamide (Nylon) [3].

FDM technology is widely used in education, research, and various engineering and industrial sectors, including automotive, aerospace, medical and dental applications. It is also popular in maker spaces and start-up hubs, revolutionizing the traditional methods of developing prototypes and products [4].

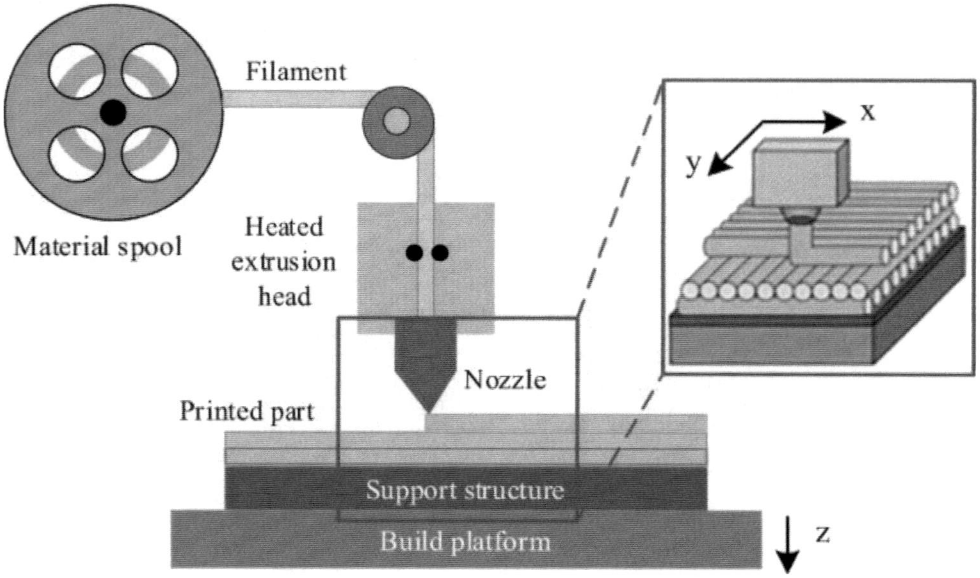

Fig. 1. FDM 3D printing technology process [5]

The advancement of FDM and AM technologies is closely tied to the research and development of the materials used in these processes. Unlike traditional manufacturing methods, AM technologies create materials during the product formation, making the final properties highly dependent on manufacturing parameters and other factors. Key parameters that significantly influence the mechanical properties of FDM 3D printed materials include printing temperature, nozzle diameter, printing speed, layer height, infill density, and printing orientation. These factors are frequently analyzed due to their substantial impact on the final product's performance [6], [7], [8].

When considering the diameter of the nozzle as an influential factor in FDM 3D printing technology, previous research indicates that it significantly affects the mechanical and physical properties of the FDM printed material. This includes impacts on surface roughness, geometric accuracy of the printed product, and printing time [9], [10]. In the article [11], authors examine the impact of printing nozzle diameter on the tensile strength of samples produced using FDM technology. Tests were conducted with different nozzle diameters and infill types, revealing that tensile strength generally increases with nozzle diameter. Specifically, samples with 50% infill density showed higher strength with larger nozzles, while those with 100% infill density exhibited decreased strength as nozzle diameter increased. Larger nozzles also reduced printing time but compromised visual quality.

In study [12], authors tested nozzle sizes of 0.4 mm, 0.6 mm, and 0.8 mm, and found that larger nozzle sizes generally improved the fatigue life of the ABS specimens. Specifically, specimens printed with a 0.8 mm nozzle diameter demonstrated the longest vibration time before fracture. Larger nozzle sizes contribute to fewer micro voids and defects within the material, enhancing fatigue resistance. However, the printing orientation and environmental temperature had a more significant impact on fatigue performance than nozzle size.

The research [13] aimed to determine the effect of nozzle diameter on the spatial structure (raster width) and selected mechanical properties (tensile strength, Charpy impact strength and impact tensile strength) of PLA samples produced by Fused Filament Fabrication (FFF). Different nozzles with diameters of 0.2 mm, 0.4 mm, 0.8 mm, and 1.2 mm were tested, with a constant layer height of 0.2 mm and 100% infill. The study found that nozzle diameters greater than 0.4 mm resulted in increased air voids between print paths, which influenced impact strength by absorbing energy during tests. Samples produced with 0.4 mm and 0.8 mm nozzles showed higher tensile strength, while those with 0.2 mm nozzles showed the lowest mechanical properties. This suggests that larger nozzle diameters enhance layer interconnection and mechanical strength, particularly when the layer height is appropriately matched to the nozzle diameter.

The authors in [14] analyzed the influence of nozzle diameter on the flexural strength of sintered alumina (Al2O3) disks created using FDM printing of a ceramic–polymer filament. Statistical analysis revealed that a nozzle diameter of 0.6 mm provided the smallest variation in mass and geometric dimensions of the ceramic samples. The best mechanical properties were achieved with a layer height of 0.4 mm, as this minimized voids between horizontal layers, enhancing strength. The authors found a non-linear relationship between the weight of sintered samples and nozzle diameter. Using a 0.6 mm nozzle diameter offers a balance between maintaining accurate geometric dimensions and maximizing flexural strength, making it a preferred choice for producing high-strength sintered alumina components.

The authors in [15] conducted a thorough examination of the impact of nozzle diameter on the tensile strength of 3D-printed polylactic acid (PLA) parts, focusing on six nozzle sizes ranging from 0.3 to 0.8 mm. The study, performed using an open-source FFF 3D printer, revealed a significant correlation between nozzle size and tensile strength. Smaller nozzles resulted in parts with higher tensile strength due to finer layers and improved interlayer adhesion.

However, the study also highlighted the trade-off between tensile strength and printing time associated with smaller nozzle sizes, underscoring the importance of balancing these factors when optimizing 3D printing processes for specific applications. The authors in study [16] examined the influence of nozzle diameter on the tensile and flexural properties of 3D printed PLA material specimens using a range of diameters (0.3 mm, 0.4 mm, 0.5 mm, 0.6 mm, and 0.8 mm). Tensile and flexural tests were conducted, revealing that both strengths generally increase with nozzle diameter, up to a certain point. The maximum tensile strength of 33.32 MPa was achieved with a 0.6 mm nozzle, while the maximum flexural strength of 76.76 MPa occurred with a 0.5 mm nozzle.

In research [17] authors investigated the impact of different printing parameters, including nozzle diameter, on the fatigue behavior of polylactic acid (PLA) in FDM 3D printing. Standard samples were produced with varying print speeds, temperatures, and nozzle diameters, and subjected to rotary bending fatigue tests. The study revealed that nozzle diameter significantly influenced the fatigue resistance of the specimens, with larger diameters decreasing resistance due to inappropriate layer thickness ratios. However, lower print temperatures increased fatigue lifetime. A cubic model was developed to predict fatigue lifetime based on stress level, print speed, temperature, and nozzle diameter, with significant interactions observed among parameters.

The objective of this study was to assess the impact of two different nozzle diameters (0.4 mm and 0.6 mm) on the tensile and flexural mechanical properties of FDM 3D printed PET-CF material. The analysis focused on key mechanical characteristics including maximum strength, modulus of elasticity, and total deformation prior to specimen failure in both tensile and flexural tests. Furthermore, the study investigated the influence of nozzle diameter on printing time, with detailed findings presented in subsequent sections.

2. Materials and Methods

This research focuses on the experimental analysis of the influence of nozzle diameter on the tensile and flexural mechanical properties of FDM 3D-printed PET-CF material. The experimental procedure is illustrated in the flow diagram shown in the figure below.

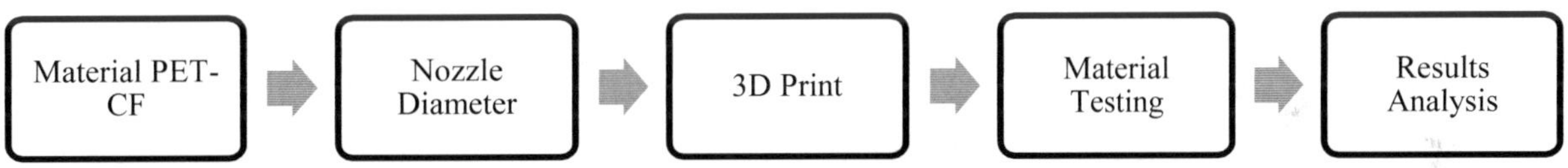

Fig. 2. Experimental procedure flow diagram

The material analyzed in this study was PET-CF from Ultimaker, a carbon fiber-reinforced variant of Polyethylene Terephthalate containing chopped carbon fibers. PET-CF is a composite material that retains the excellent printability of PETG while offering enhanced stiffness, as well as improved temperature and chemical resistance, which can be further improved through annealing. This material is suitable for applications such as tooling, functional prototyping, manufacturing aids, and more. The mechanical properties of PET-CF material by the manufacturer are presented in the table below [18].

Mechanical properties	Test method	Typical value (XY, flat)
Tensile modulus	ASTM D3039 (1 mm / min)	4342 ± 89 MPa
Tensile strength	ASTM D3039 (5 mm / min)	50,6 ± 0,6 MPa
Elongation at break	ASTM D3039 (5 mm / min)	5,5 ± 0.6%
Flexural modulus	ISO 178 (1 mm / min)	5743 ± 150 MPa
Flexural strength	ISO 178 (5 mm / min)	102,8 ± 2,6 MPa
Flexural strain at brake	ISO 178 (5 mm / min)	No break (>10%)
All properties are for specimens printed in flat position and with 0,6 mm CC Core nozzle diameter.		

Table 1. Ultimaker PET-CF FDM 3D Printed material properties by manufacturer

The 3D CAD models of the testing specimens were designed in Solidworks 2023 in accordance with ISO 527-2 for tensile testing and ISO 178 for flexural testing, as illustrated in Fig. 3. These CAD models were then converted into STL format using Solidworks, enabling their use in slicer software for 3D printing preparation.

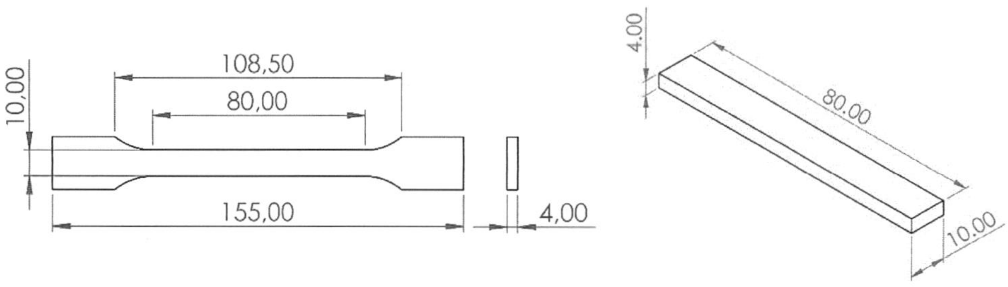

Fig. 3. Tensile (left) and flexural testing specimen (right)

Specimens and printing parameters were prepared and defined using CURA 5.6.0 slicer software. All specimens were oriented flat for printing. Due to the composite nature of the material, which includes carbon fibers, a CC Core nozzle was used for printing, with nozzle sizes of 0.4 mm and 0.6 mm. All specimens are 3D printed on Ultimaker S5 Pro with printing parameters presented in Table 2.

Printing parameters	
Layer height	0,15 mm
Infill density	100 %
Printing temperature	270 °C
Build plate temperature	80 °C
Print speed	30 mm/s
Fan speed	10 %
CC Core nozzle diameter	0,4 mm and 0,6 mm
Printing parameters are defined using CURA "Normal profile" 0,15 mm	

Table 2. Printing parameters used for 3D printing testing specimens with 0,4 mm and 0,6 mm nozzle diameter

For each type of testing, ten replicas of each test sample were printed and subsequently tested for tensile and flexural properties using a Shimadzu AGS-X universal testing machine with a capacity of 10 kN (Fig. 4).

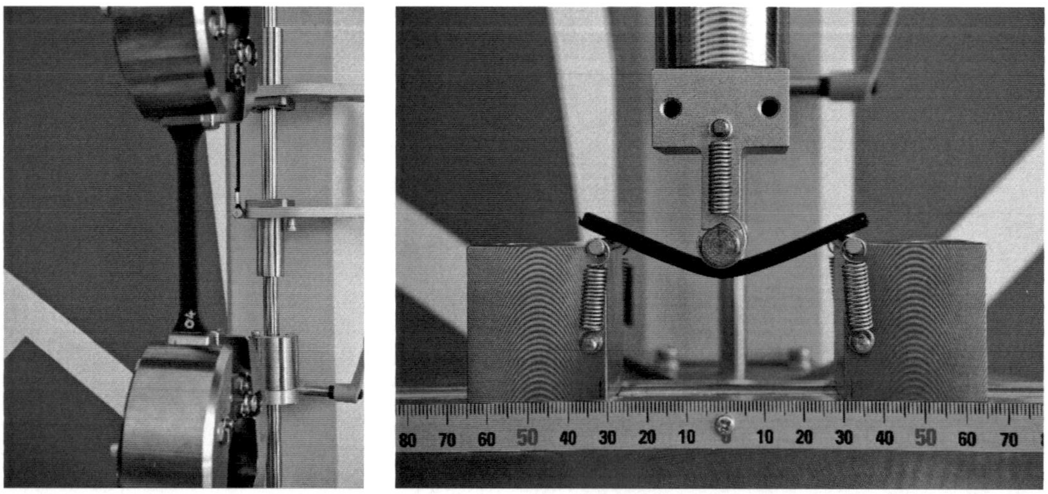

Fig. 4. Tensile (left) and flexural (right) testing on AGS-X universal testing machine

Data collection and monitoring were conducted using Shimadzu's Trapezium-X software, which facilitated the plotting of stress-strain diagrams. The printing time for each sample was determined based on calculations provided by the CURA software. Subsequently, all data were statistically analyzed using Excel.

3. Results and Analysis

Tensile and flexural tests were performed with a constant testing speed of 5 mm/min, according to ISO 527-2 and ISO 178 standards. Stress-strain diagrams were plotted and presented on Fig. 5.

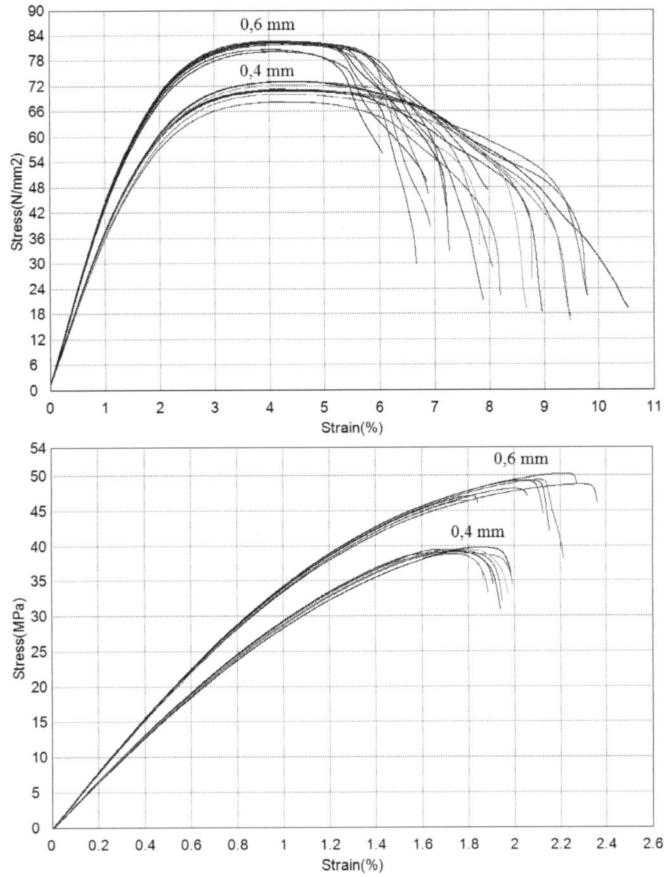

Fig. 5. Stress-strain diagrams for tensile (upper) and flexural (lower) mechanical properties for different nozzle diameter

The analysis of the study results focused on evaluating the impact of nozzle diameter on tensile and flexural strength, the modulus of elasticity in both tension and flexural, the maximum deformation before brake during tensile and flexural testing, and the total printing time.

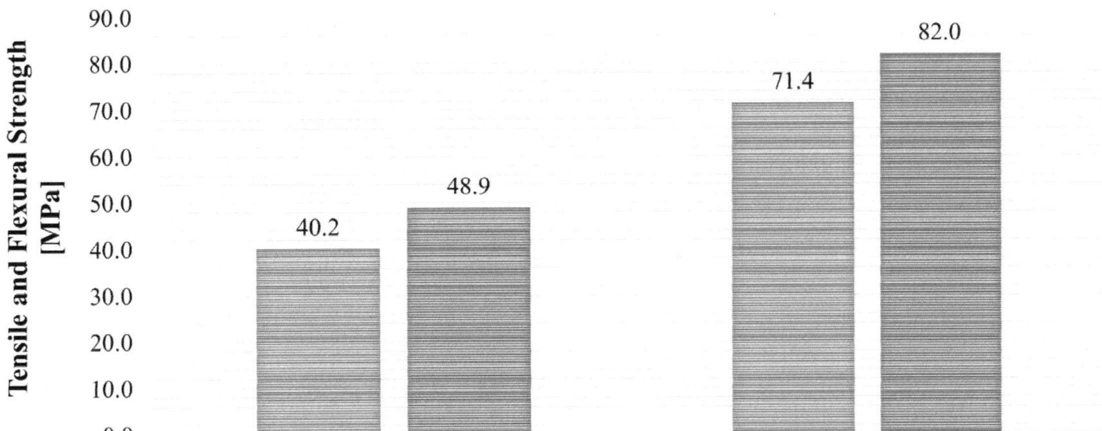

	Tensile Rm (MPa)	Flexural Rm (MPa)
Nozzle 04 mm	40.2	71.4
Nozzle 06 mm	48.9	82.0

Fig. 6. Tensile and flexural strength of PET-CF FDM 3D printed material with different nozzle diameter

From the Fig. 6., it can be concluded that nozzle diameter affects the tensile and flexural strength of FDM 3D-printed PET-CF material. The diagram indicates that using a 0.6 mm diameter nozzle increases tensile strength up to 21.6% and flexural strength up to 14.8%.

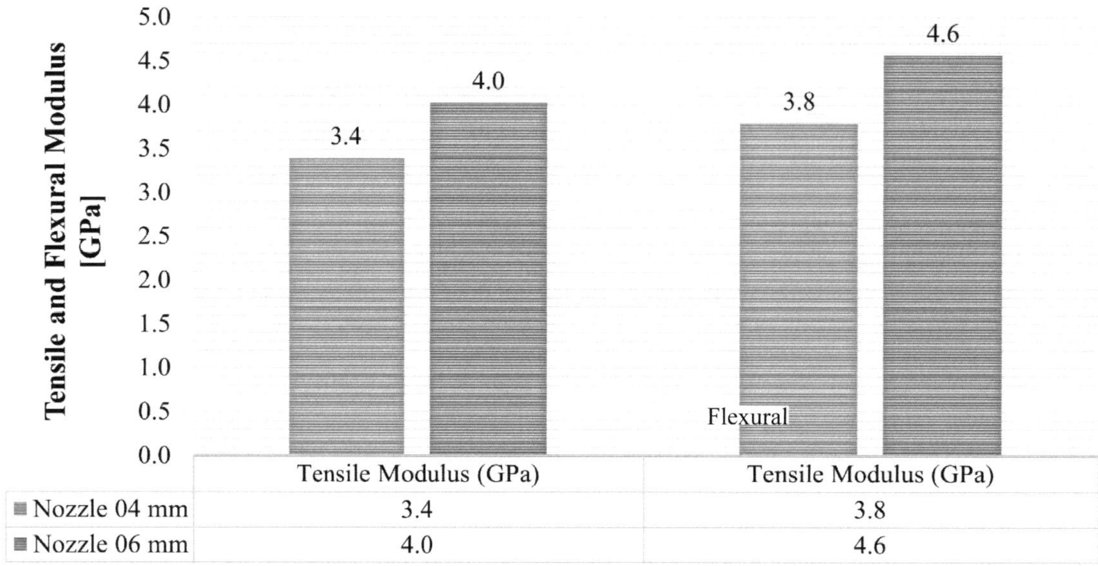

Fig. 7. Tensile and flexural modulus of PET-CF FDM 3D printed material with different nozzle diameter

Analyzing the modulus of elasticity during tensile and flexural tests, it was concluded that FDM 3D printing of PET-CF material with a larger nozzle diameter increases the modulus of elasticity (Fig. 7). The analysis shows that using a nozzle diameter of 0.6 mm compared to 0.4 mm increases the modulus of elasticity in tension up to 17.6%, and in flexural up to 21%.

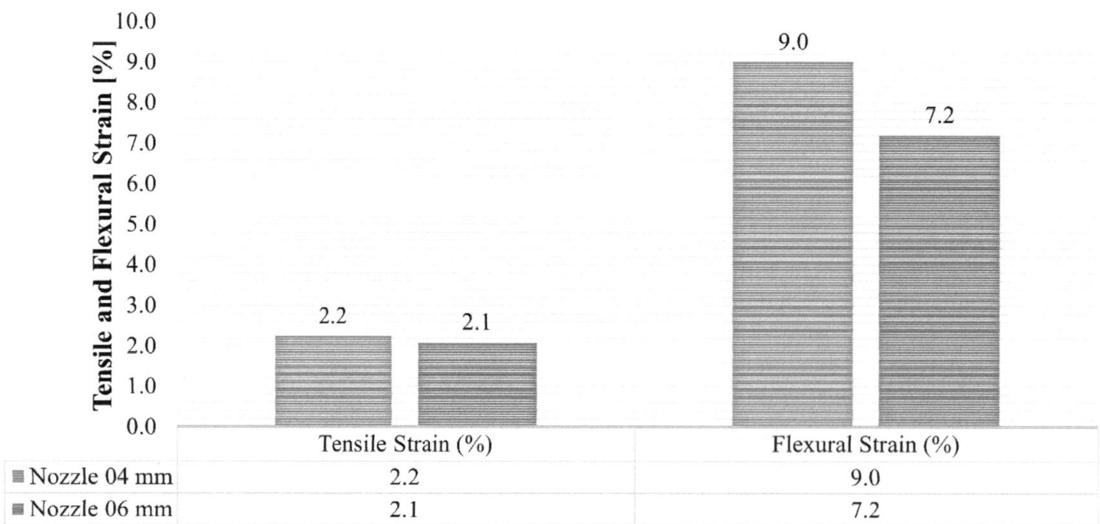

Fig. 8. Tensile and flexural strain before brake of PET-CF FDM 3D printed material with different nozzle diameter

In contrast to the effects on strength and modulus of elasticity, the influence of nozzle diameter on the strain before brake during tensile and flexural tests shows that using a larger nozzle diameter reduces the strain. Fig. 8. illustrates that the strain before brake in the tensile test was reduced by 4.5%, and in the flexural test, it decreased by up to 20% when the PET-CF material is FDM 3D printed with a 0.6 mm nozzle diameter. In addition to the mechanical properties, the influence of the nozzle diameter on the printing time was also analyzed, where the printing time was expressed as the time to 3D print one test sample for testing the tensile and flexural mechanical properties. As expected, using a 0.6 mm diameter nozzle for 3D printing testing samples reduces the printing time by up to 31% (Fig. 9).

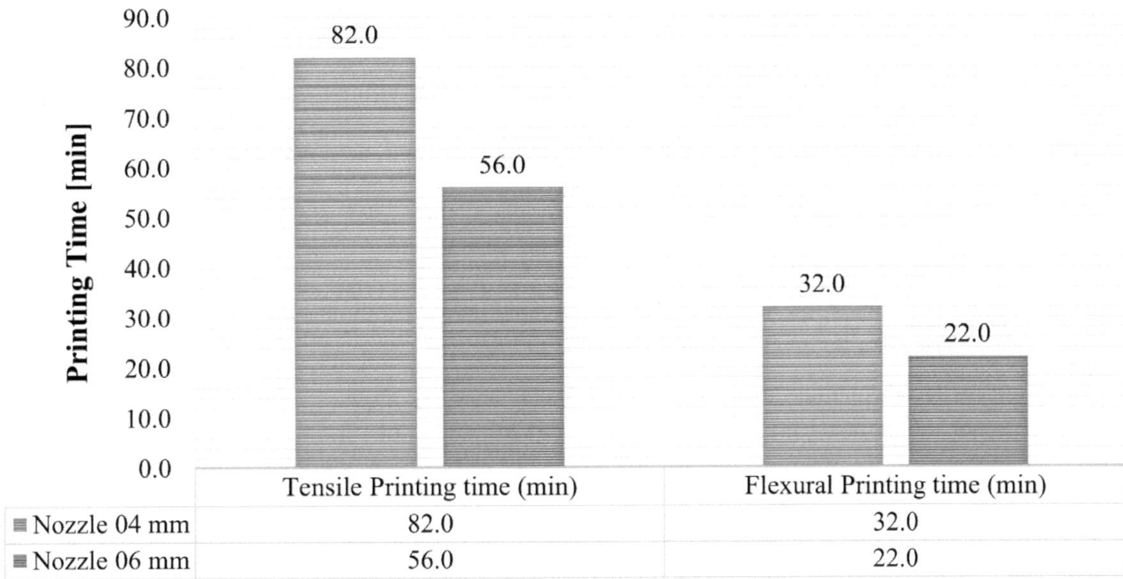

Fig. 9. Influence of nozzle diameter on printing time of tensile and flexural testing specimens

4. Conclusion

The study systematically evaluated the impact of nozzle diameter on the mechanical properties and printing efficiency of FDM 3D-printed PET-CF material. The findings can be summarized as follows:

- Strength: Nozzle diameter significantly influences the tensile and flexural strength of FDM 3D printed PET-CF material. Specifically, using a 0.6 mm nozzle diameter resulted in an increase in tensile strength by up to 21.6% and flexural strength by up to 14.8%.
- Modulus of Elasticity: The modulus of elasticity in both tension and flexural tests also showed improvement with a larger nozzle diameter. A 0.6 mm nozzle increased the modulus of elasticity by up to 17.6% in tension and up to 21% in flexural testing compared to a 0.4 mm nozzle.
- Strain Before Failure: Contrary to the observed enhancements in strength and modulus of elasticity, the strain before failure during tensile and flexural tests decreased with an increase in nozzle diameter. The strain before failure in tensile tests was reduced by 4.5%, and in flexural tests, it decreased by up to 20% when using a 0.6 mm nozzle diameter.
- Printing time: Using a nozzle with 0,6 mm diameter for 3D printing of PET-CF material reduces printing time by up to 31%, which finally affects the reduction of the price of the FDM 3D printed product itself.

These results indicate that while a larger nozzle diameter can enhance the strength and stiffness of FDM 3D-printed PET-CF materials, it may also lead to a reduction in ductility and speed up the process of 3D printing. Therefore, the selection of nozzle diameter should be carefully considered based on the specific requirements of the application, balancing the need for strength and stiffness against the desired level of material deformability. For future research, it is recommended to analyze how the nozzle diameter affects other mechanical properties such as compressive strength, hardness, and impact toughness. Additionally, extending the analysis beyond mechanical properties to investigate its effects on surface quality, dimensional accuracy, and other pertinent aspects in FDM 3D printed PET-CF materials is essential. This comprehensive approach will deepen our understanding of nozzle size's multifaceted influence and contribute to refining additive manufacturing processes for enhanced product performance and quality. Also, analyze other dimensions of the nozzle diameter to make optimizations.

5. Acknowledgments

Acknowledgments to UltiMaker company for their invaluable support not only for this research but also for advancing research in the broader field of FDM 3D printed materials.

6. References

[1] D. Hodzic and A. Pandzic. (2019). "Influence of carbon fibers on mechanical properties of materials in fdm technology," in 30th DAAAM International Symposium, pp. 334–342. doi: 10.2507/30th.daaam.proceedings.044.

[2] A. W. Gebisa and H. G. Lemu. (2018). "Investigating effects of Fused-deposition modeling (FDM) processing parameters on flexural properties of ULTEM 9085 using designed experiment," Materials, vol. 11, no. 4, doi: 10.3390/ma11040500.

[3] A. Pandžić, E. Kadric, and S. Kolesar. (2023). "Investigating The Influence Of Printing Orientation And Filament Drying On Tensile And Flexural Strength Of Fdm-Printed Carbon Fiber-Reinforced Polyamide Composites," in Annals of DAAAM and Proceedings of the International DAAAM Symposium, DAAAM International Vienna, pp. 146–153. doi: 10.2507/34th.daaam.proceedings.020.

[4] D. Hodzic and A. Pandzic. (2021). "Influence of Infill Design on Compressive and Flexural Mechanical Properties of FDM Printed PLA Material," in 32nd DAAAM International Symposium, pp. 0191–0199. doi: 10.2507/32nd.daaam.proceedings.029.

[5] T. S. Tamir, G. Xiong, Q. Fang, X. Dong, Z. Shen, and F. Y. Wang. (2022). "A feedback-based print quality improving strategy for FDM 3D printing: an optimal design approach," International Journal of Advanced Manufacturing Technology, vol. 120, no. 3–4, doi: 10.1007/s00170-021-08332-4.

[6] A. A. Al-Tamimi, A. Pandžić, and E. Kadrić. (2023). "Investigation and Prediction of Tensile, Flexural, and Compressive Properties of Tough PLA Material Using Definitive Screening Design," Polymers (Basel), vol. 15, no. 20, Oct. doi: 10.3390/polym15204169.

[7] A. Dadashi and M. Azadi. (2024). "Optimization of 3D printing parameters in polylactic acid bio-metamaterial under cyclic bending loading considering fracture features," Heliyon, vol. 10, no. 4, Feb. doi:10.1016/j.heliyon.2024.e26357.

[8] A. Pandzic and D. Hodzic. (2021). "Mechanical properties comparison of PLA, tough PLA and PC 3D printed materials with infill structure – Influence of infill pattern on tensile mechanical properties," in IOP Conference Series: Materials Science and Engineering, p. 012019. doi: 10.1088/1757-899x/1208/1/012019.

[9] D. Hodžić and A. Pandžić. (2022). "Influence Of Printhead Nozzle Diameter On Mechanical Properties Of Fdm Printed Pla Material And Printing Time," in Annals of DAAAM and Proceedings of the International DAAAM Symposium, DAAAM International Vienna, pp. 231–237. doi: 10.2507/33rd.daaam.proceedings.032.

[10] I. J. Solomon, P. Sevvel, and J. Gunasekaran. (2020). "A review on the various processing parameters in FDM," in Materials Today: Proceedings, Elsevier Ltd, pp. 509–514. doi: 10.1016/j.matpr.2020.05.484.

[11] W. Kiński and P. Pietkiewicz. (2020). "Influence of the Printing Nozzle Diameter on Tensile Strength of Produced 3D Models in FDM Technology," Agricultural Engineering, vol. 24, no. 3, pp. 31–38, doi: 10.1515/agriceng-2020-0024.

[12] F. He and M. Khan. (2021). "Effects of printing parameters on the fatigue behaviour of 3d-printed abs under dynamic thermo-mechanical loads," Polymers (Basel), vol. 13, no. 14, doi: 10.3390/polym13142362.

[13] P. Czyżewski, D. Marciniak, B. Nowinka, M. Borowiak, and M. Bieliński. (2022). "Influence of Extruder's Nozzle Diameter on the Improvement of Functional Properties of 3D-Printed PLA Products," Polymers (Basel), vol. 14, no. 2, doi: 10.3390/polym14020356.

[14] A. Smirnov, N. Nikitin, P. Peretyagin, R. Khmyrov, E. Kuznetsova, and N. W. Solis Pinargote. (2023). "Experimental and Statistical Modeling for Effect of Nozzle Diameter, Filling Pattern, and Layer Height of FDM-Printed Ceramic–Polymer Green Body on Biaxial Flexural Strength of Sintered Alumina Ceramic," Journal of Composites Science, vol. 7, no. 9, doi: 10.3390/jcs7090381.

[15] F. Kartal and A. Kaptan. (2023). "Investigating the Effect of Nozzle Diameter on Tensile Strength in 3D-Printed Printed Polylactic Acid Parts," Black Sea Journal of Engineering and Science, vol. 6, no. 3, pp. 276–287, doi: 10.34248/bsengineering.1287141.

[16] M. N. Sudin, N. M. Daud, F. R. Ramli, and M. A. Yusuff. (2023). "The Effect of Nozzle Size on the Tensile and Flexural Properties of PLA Parts Fabricated Via FDM," Science, Engineering and Technology, vol. 3, no. 1, doi: 10.54327/set2023/v3.i1.71.

[17] A. Dadashi and M. Azadi. (2024). "Optimization of 3D printing parameters in polylactic acid bio-metamaterial under cyclic bending loading considering fracture features," Heliyon, vol. 10, no. 4, doi: 10.1016/j.heliyon.2024.e26357.

[18] UltiMaker. (2024). "UltiMaker PET CF Technical data sheet General overview." Accessed: May 29, 2024. [Online]. Available: https://support.makerbot.com/s/article/000002928

DOI: 10.2507/35th.daaam.proceedings.018

MODELLING INFLUENCE OF HEAT INPUT ON GMAW FILLET WELD GEOMETRY

Petar Tasić, Ismar Hajro, Adi Pandžić & Almira Softić

This Publication has to be referred as: Tasic, P[etar]; Hajro, I[smar]; Pandzic, A[di] & Softic, A[lmira] (2024). Modelling Influence of Heat Input on GMAW Fillet Weld Geometry, Proceedings of the 35th DAAAM International Symposium, pp.0143-0149, B. Katalinic (Ed.), Published by DAAAM International, ISBN 978-3-902734-44-0, ISSN 1726-9679, Vienna, Austria

DOI: 10.2507/35th.daaam.proceedings.018

Abstract

One of the most often used welding processes in manufacturing components made of unalloyed structural steels is Gas Metal Arc Welding (GMAW). It is a relatively simple and cost-effective process providing high deposition rates and productivity. Moreover, it can be combined with other processes and/or fully robotized, both of which improve productivity even further. However, maintaining quality control of welds at these high productivity rates can be rather challenging task. There are standards describing the acceptance levels and corresponding weld quality, with some of the considered criteria based on the visible weld geometry. It would be beneficial to connect welding parameters with the weld geometry, so that deviation from defined geometry causes correction of welding parameters. This paper describes influence of heat input on leg size and penetration for GMAW fillet welds of unalloyed steel with thicknesses of 4 and 8 millimetres in horizontal position, with varied parameters being welding current and speed. Dependence is described through models based on linear regression analysis.

Keywords: GMAW; unalloyed steel; weld geometry; weld quality; model

1. Introduction

Gas Metal Arc Welding (GMAW) is one of the most extensively used joining processes in (pre)fabrication of various structures. It can provide considerably higher productivity than Gas Tungsten Arc Welding (GTAW) or Shielded Metal Arc Welding (SMAW), due to its high deposition rates. [1] Furthermore, GMAW process can be easily modified or adapted for automation or robotisation. [2] This means improvement of working conditions and safety for welders. [3] Despite this, GMAW is still in significant portion used as manual welding process, increasing the possibility of introduction of flaws and defects caused the effect of human factors. [4] On the other side, such high deposition rates and, consequently, fast production of welds pose challenge regarding weld quality check. The easiest way to do it would be reliable control of heat input, i.e. essential welding parameters (welding current, voltage and welding speed). [5] Regardless, to achieve full control of welding process and its outcome, non-essential parameters (e.g. gas flow, inclination, contact tip to work distance) should be controlled as well. In this way, the same weld bead geometry features can be achieved all along the seam. [6]

Controlling process parameters becomes particularly of interest when it comes to Wire Arc Additive Manufacturing (WAAM), which is executed by deposition of metal layers on top of each other (like cladding). In these cases, control of the bead geometry is in significant portion control of product geometry. [7] The idea of weld quality control through weld geometry control is not unknown, but it was used mostly for hybrid, beam and friction welding processes. As there are many recommendations for parameters selection for GMAW (and for arc processes in general), such relationships between the process parameters, the weld geometry and the weld quality are predominantly recently developed. [8] In many cases, driving forces are increased production volume and quality requirements on one side, and, on the other, reduction of manufacturing costs. Such concept of weld quality control relies on model, which has pivotal role in connecting welding process, heat input and weld geometry, as shown in Fig. 1.

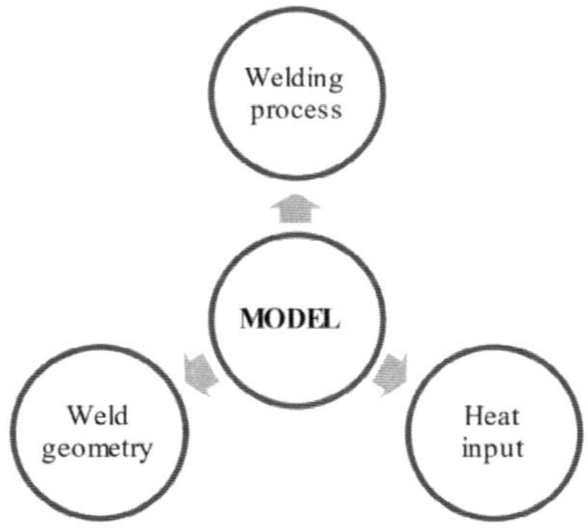

Fig. 1. Schematic representation of model connecting welding process, heat input and weld geometry [8]

Models describing weld geometry as function of heat input are relatively common. Some of the oldest models are based on corelation between leg size z (or weld thickness a) and heat input Q and are developed for welds of unalloyed structural steel made by arc processes, including GMAW. [9], [10] More recently, models are developed by using statistical tools (e.g. regression) for single process, one welding position and one type of steel, increasing model accuracy. [2], [11], [12] There are also doctoral theses dealing with prediction of geometry as function of heat input. [13], [14] Most recently, models are developed for GMAW using advanced statistical and mathematical tools, with strong potential for their implementation in manufacturing. The choice of process is driven by its widespread application in metalworking industry. [15], [16], [17] In some cases, models are coupled with real-time acquisition of weld geometry features by employing artificial vision, effectively establishing connection as shown in Fig. 1. [18], [19]

On the other side, models of heat input as function of weld geometry are scarce, but could be useful as well. For example, they could serve to enable on-line adjustment of parameters during welding to achieve certain weld geometry features, e.g. specified leg size z or weld thickness a, making such approach particularly useful in case of transitions between several welding position comprising single seam.

This study presents some of own experimental research conducted using GMAW process for single-pass fillet weld of non-alloyed structural steels with two thicknesses. Based on the results, the relationship between heat input and weld geometry has been established and evaluated by using statistical tools, with the purpose to see whether such relationship could be used in real manufacturing process.

2. Experimental research

To establish corelation between weld geometry features and heat input, a series of single-pass fillet welds on 4 mm and 8 mm thick plates of structural steels P355NL2 and S235JR (respectively) have been made using GMAW. Both steels are unalloyed low-carbon structural steels, with chemical composition given in Table 1 and mechanical properties given in Table 2.

	C	Mn	P	S	Si	Cu	N	Nb	Ni	Ti
S235JR	0.13	0.55	0.010	0.009	0.21	0.29	0.007	-	0.08	0.002
P355NL2	0.17	1.42	0.017	0.002	0.22	0.02	0.002	0.01	0.02	0.010

Table 1. Chemical composition (% wt.) of used steels [20], [21]

	Yield strength	Tensile strength	Elongation A_5	Impact toughness
S235JR	306 MPa	429 MPa	39%	149 J at -20 °C
P355NL2	435 MPa	555 MPa	32%	171 J at -20 °C

Table 2. Mechanical properties of used steels [20], [21]

Both steels have been chosen because of their widespread use for beams, supports, platforms, mine equipment, as well as in shipbuilding, and are known for excellent weldability and relatively low price. These thicknesses have been chosen as they are common for mentioned applications and can be used for single-pass fillet welds. All of them are made by using welding wire G 42 4 M21 3Si1 (EN ISO 14341-A) with diameter of 1.2 mm. Shielding gas in all experiments was pure CO_2, with flow varying from 13 to 17 litres/min, as recommended for GMAW. Contact Tip to Workplace Distance (CTWD) was kept between 12 and 15 mm, depending on current. All welds are made in flat position (PB), since it can support the biggest range of welding parameters.

There are three essential welding parameters for GMAW, namely welding current (I, A), arc voltage (U, V) and welding speed (w, cm/min). However, power source used in experiment was digitally controlled Daihen Welbee WB-P500L, with its electronics synergically corelating voltage with welding current, hence there were only two independent welding parameters, i.e. welding current and welding speed. Welding parameters were chosen to represent relatively wide range, with values measured during experiment given in Table 3.

	Welding current	**Arc voltage**	**Welding speed**	**Heat input**
4 mm	90 – 180 A	18.6 – 23.2 V	11 – 31 cm/min	0.33 – 1.05 kJ/mm
8 mm	165 – 190 A	22.2 – 28.5 V	14 – 61 cm/min	0.59 – 1.27 kJ/mm

Table 3. Welding parameters used in experiment

Heat input (Q, kJ/mm) is defined by (1), and for this experiment is calculated using average values of current and voltage measured by independently connected multimetar. Heat input efficiency η (also known as thermal efficiency) is usually taken based on recommendations, and in this case, it is η=0.8. [22], [23]

$$Q = \frac{I \cdot U}{w} \cdot \eta \tag{1}$$

All samples have been visually examined right after welding, with ones with significant visible imperfections being discarded and not considered in further analysis. Remaining samples have been cut, ground, polished and etched with nital. In that way, macro-sections of joints were prepared in accordance with EN ISO 17639. Afterwards, macro-sections were photographed, and geometry features analysed in AutoCAD, as shown in Fig. 2.

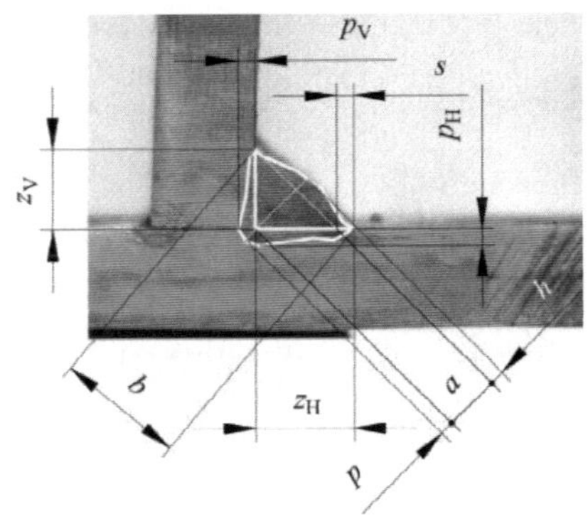

Fig. 2. Measurements and geometrical features visible at macro-section

Geometry features were measured in accordance with ISO/TR 25901-1, while weld imperfection tolerances for required quality levels were determined by ISO 5817. Additionally, two new geometry features are proposed. The one is average weld leg size z_a, described by (2) as average between leg size in horizontal (z_H) and vertical (z_V) direction. Similarly, average penetration p_a, described by (3), is average penetration between those in horizontal (p_H) and vertical (p_V) direction. The reason for introduction of average values is weld asymmetry. Ideally, there should be no asymmetry at all, but due to reality of manual welding process, it is always present, and it is considered as one of the representatives of weld quality. However, it is consequence of welder's manipulative abilities, and not heat input, and therefore omitted in this analysis.

$$z_a = \frac{z_H + z_V}{2} \tag{2}$$

$$p_a = \frac{p_H + p_V}{2} \tag{3}$$

3. Results and discussion

Based on the results of experiment, four models were developed for heat input, one as function of average leg size and other of average penetration, both for two thicknesses. Multiple linear regression has been used for this, as it was used in many previous research comparable to this one, and because problem is not too complex, including just few variables. Coefficients of regression models are obtained by the least squares method and are given in Table 4 and 5 for welds on 4 mm steel sheets, and Table 6 and 7 for 8 mm. Graphical representation of developed models are given.

	Coefficient	Standard error	*P*-value
Intercept	-0.41881	0.08814	$5.47\cdot10^{-05}$
z_a (mm)	0.17778	0.01437	$7.27\cdot10^{-13}$

Table 4. Coefficients of regression model $Q=f(z_a)$ for 4 mm sheets

	Coefficient	Standard error	*P*-value
Intercept	0.39433	0.07103	$6.15\cdot10^{-06}$
p_a (mm)	0.47230	0.11798	0.00042

Table 5. Coefficients of regression model $Q=f(p_a)$ for 4 mm sheets

	Coefficient	Standard error	*P*-value
Intercept	-0.04303	0.12076	0.72429
z_a (mm)	0.13444	0.01715	$1.53\cdot10^{-08}$

Table 6. Coefficients of regression model $Q=f(z_a)$ for 8 mm sheets

	Coefficient	Standard error	*P*-value
Intercept	0.74737	0.09075	$5.80\cdot10^{-09}$
p_a (mm)	0.11151	0.06562	0.10036

Table 7. Coefficients of regression model $Q=f(p_a)$ for 8 mm sheets

As possible to see, the *P*-values for intercept in Table 6 and p_a in Table 7 are bigger than 0.0500, meaning that intercept and p_a should have no effect on model's relevance, i.e. they are not statistically significant. Nevertheless, they are included in corresponding models due to physical nature of welding process being analysed. Basic data regarding regression analysis and developed models is given in Table 8.

	4 mm		**8 mm**	
	$Q=Q(z_a)$	$Q=Q(p_a)$	$Q=Q(z_a)$	$Q=Q(p_a)$
Coefficient of determination, R^2	0.84528	0.36401	0.68710	0.09348
Significance, *F*	$7.27\cdot10^{-13}$	0.00042	$1.53\cdot10^{-08}$	0.10036
Standard error	0.06669	0.13521	0.12712	0.21638
Number of experiments	30	30	30	30

Table 8. Results of regression models for two thicknesses

As possible to see from Table 8, significance *F* for model $Q=Q(p_a)$ for 8 mm has value of 0.10 (it is bigger than 0.05), what means that developed model is not sufficient to explain dependence of Q on p_a. Therefore, it will not be considered in further analysis. It is also possible to see that models based on average leg size have significantly bigger coefficient of determination (R^2) than those based on average penetration, meaning they are more reliable. Models for heat input as function of z_a and p_a are given with (4) and (5) for thickness of 4 mm, while (6) gives model of heat input as function of z_a for thickness of 8 mm.

$$Q=-0.41881+0.17778\cdot z_a \tag{4}$$

$$Q=0.39433+0.47230\cdot p_a \tag{5}$$

$$Q=-0.04303+0.13444\cdot z_a \tag{6}$$

Graphical interpretation of models given by (4), (5) and (6) is given in Fig. 3, 4 and 5, respectively. For comparison purposes, figures contain original experimental measurements as well.

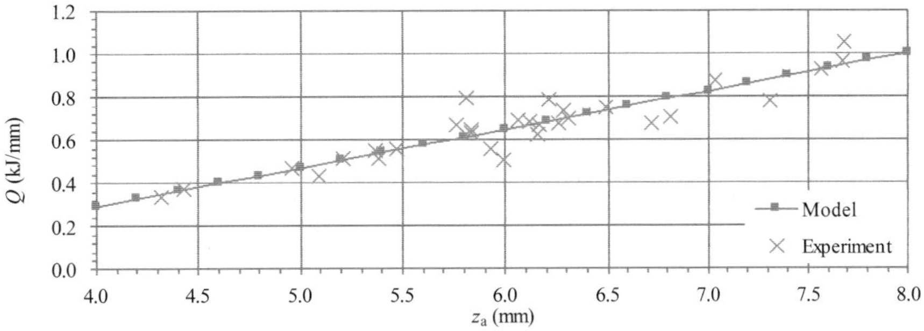

Fig. 3. Dependence of heat input on average leg size for thickness of 4 mm

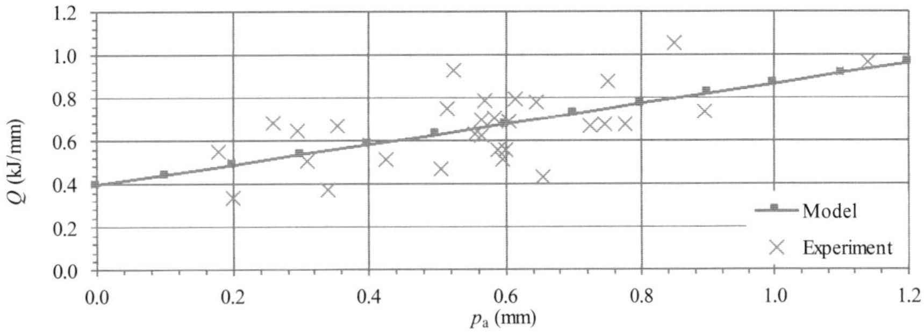

Fig. 4. Dependence of heat input on average penetration for thickness of 4 mm

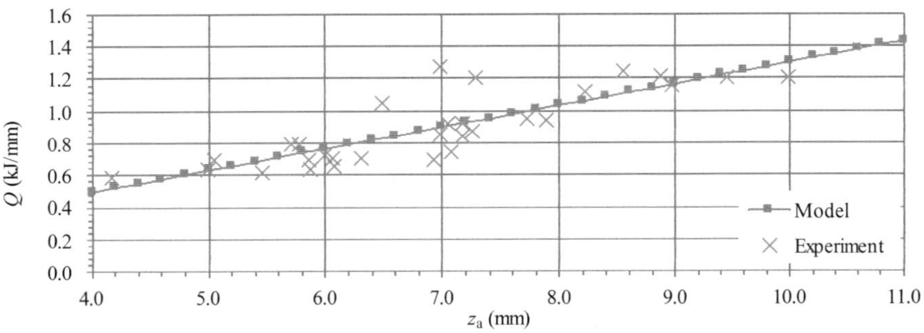

Fig. 5. Dependence of heat input on average leg size for thickness of 8 mm

Direct comparison between models for $Q=Q(z_a)$ for thicknesses 4 mm and 8 mm is given in Fig. 6.

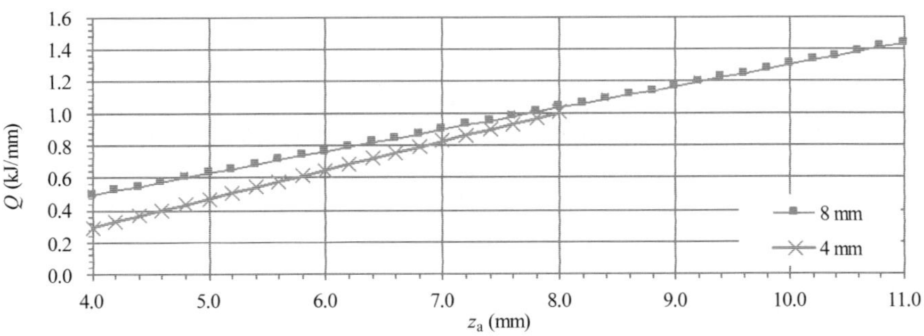

Fig. 6. Comparison of heat input models as function of average leg size for both thicknesses

It is noticeable that required heat input for 4 mm is lower than the one for 8 mm for the same average leg size, and that difference is more obvious as average leg size is smaller. Considering that the wire diameter is the same in both cases, and the difference between gas flows is insignificant, it is possible to conclude that the reason for difference lays in heat transfer in steel sheets. Due to difference in masses, thin sheet (4 mm) gets heated more rapidly than thick one (8 mm), hence more heat remains to melt filler metal, which, in turn, enables formation of weld with lower heat input.

According to models, for average leg size of approximately 8 mm, both models give the same heat input of about 1 kJ/mm. It should be noted that both current and welding speed affects heat input, and that the same value can be achieved with different combinations of them. However, from the technological point of view, having heat input of 1 kJ/mm on 4 mm thick steel sheet is rather extreme case, regardless the combination of current and welding speed, as it is on the verge of burning through the base metal.

4. Conclusion

Gas Metal Arc Welding is widely used process in making steel components and structures, primarily due to high productivity enabled by high deposition rates. Additionally, it is easily fully automated and/or coupled with robots, and it is also process that has many modifications. All this pushes productivity even further, requiring new ways to control welds. On the other side, it is process with no slag and with relatively small amount of fumes and dust, so it is suitable for coupling with artificial vision to scan welds and measure its geometry features in-situ in real time. That would maintain required weld quality level and increased productivity, leading to reduced costs. However, to implement this, the model connecting welding parameters (current, voltage and welding speed, heat input) with weld geometry and dimensions must be developed. In case such model is implemented in manufacturing process, parameters could be adjusted instantaneously during welding, enabling required weld profile along seams, even in cases of complex geometry and various welding positions.

This paper presents results of research on heat input dependence on two geometry features of welds. Experimental part of the study has been conducted making several series of fillet welds on unalloyed steel sheets with thicknesses of 4 mm and 8 mm, manually by using digitally controlled welding device and common filler metal and shielding gas. All welds were made in flat position, and afterward cut, polished and etched, so geometry features can be measured on prepared macrographs. For analysis, two of them were calculated, namely average leg size and average penetration. Each of these averaged values is calculated between values in horizontal and vertical direction. It is a new approach to define geometry features, as standards do not describe them, but rather consider welds with ideal geometry, i.e. symmetrical.

Based on experimental data, four models have been developed to describe heat input dependence on average leg size and penetration, for two thicknesses, with three of them proved to be statistically relevant. Models based on average leg size shown significantly higher coefficient of determination (R^2) than those based on average penetration. For both thicknesses, models shown that heat input does not strongly depend on average penetration, i.e. heat input has small or no influence on penetration. This is something that is opposite from real-life experience and physics of welding process, meaning it needs deeper and more detailed analysis. Models also shown that required average leg size can be achieved with smaller heat input on 4 mm than on 8 mm thick steel sheets, what is consequence of differences in heat transfer.

It would be possible to conclude that proposed approach of determining heat input as function of average leg size and average penetration has certain advantages, in particular if applied in manufacturing of steel structures, but still needs more data for verification and validation, and usage of improved or different statistical tools.

5. Further research

Further research have several possibilities, each of them with interesting perspective. On one side, it should focus on application of more advanced statistical tools that could provide better understanding of corelation between heat input and geometry features. On the other side, entire research can be reproduced, but this time using robot instead of welder. In this way, the welder's influence could be precisely determined. Further research could also include multi-pass welds, as well as welding stainless steels and aluminium alloys.

6. References

[1] Girón-Cruz, J. A., Pinto-Lopera, J. E. & Alfaro, S. C. A. (2022). Weld bead geometry real-time control in gas metal arc welding processes using intelligent systems, The International Journal of Advanced Manufacturing Technology, Vol. 123, 2022, pp. 3871-3884., DOI: 10.1007/s00170-022-10384-z

[2] Yaakub, M. Y., Tham, G., Rahim, W. M. A. W. A., Radzi, M. A. R. M. & Mahmud, A. (2013). Prediction of Welding Parameters and Weld Bead Geometry for GMAW Process in Overhead T-Fillet Welding Position(4F), Advanced Materials Research, Vol. 686, 2013, pp. 320-324., ISSN 1662-8985, DOI: 10.4028/www.scientific.net/AMR.686.320

[3] Popescu, M. (2010). Welding Ergonomics – Elements. Proceedings of the 21st International DAAAM Symposium, B. Katalinic (Ed.), Vol. 21, No. 1, ISSN 1726-9679

[4] Czifra, J., Vicikova, J. & Szabo, P. (2010). Automated production system for pipelines of vessels. Proceedings of the 21st International DAAAM Symposium, B. Katalinic (Ed.), Vol. 21, No. 1, ISSN 1726-9679

[5] Tham, G., Yaakub, M. Y., Abas, A. K., Manurung, Y. H. P. & Jalil, B. A. (2012). Predicting the GMAW 3F T-Fillet Geometry and Its Welding Parameter, Procedia Eng., Vol. 41, 2012, pp. 1794-1799., DOI: 10.1016/j.proeng.2012.07.385

[6] Filho, C. A. M. D., Sena, M. E. N., Neves, D. A., Brandi, S. D. & Neto, R. M. de A. C. (2023). Weld bead geometry prediction using artificial neural networks and genetic algorithm hybrid model in GMAW, Proceedings of 27th International Congress of Mechanical Engineering, 2023, DOI: 10.26678/ABCM.COBEM2023.COB2023-2140

[7] Maraki, M. R., Mahmoodi, M., Yousefieh M. & Tagimalek, H. (2023). Prediction and optimization of weld geometry in gas metal arc welding (GMAW) using least squares support vector machine, Journal of Welding Science and Technology of Iran, Vol. 8, 2023, ISSN: 2676-6787

[8] Tasic, P., Hajro, I., Hodzic, D. & Pandzic, A. (2020). Prediction of Geometry of Pulsed GMAW Fillet Welded Joints of Structural Steels. Proceedings of the 31st DAAAM International Symposium, pp.0444-0452, B. Katalinic (Ed.), ISBN 978-3-902734-29-7, ISSN 1726-9679, DOI: 10.2507/31st.daaam.proceedings.062

[9] Bosworth, M. R. (1991), Effective Heat Input in Pulsed Current Gas Metal Arc Welding with Solid Wire Electrodes, Welding Journal, Supplement, 1991, pp. 111-117

[10] Miller, D. K. & Funderburk R. S.(2001). Tools for reviewing Welding Procedures, Welding Journal, Vol. 80, No. 7, 2001, pp. 40-43., ISSN 0043-2296

[11] Gucwa, M. & Beczkowski, R. (2014). The Effect of Heat Input on the Geometric Properties of Welded Joints, Archivess of Foundry Engineering, Vol. 14, 2014, pp. 127-130., ISSN 1897-3310

[12] Ribeiro, R. A., Santos, E. B. F., Assunção, P. D. C., Maciel, R. R. & Braga E. M. (2015). Predicting Weld Bead Geometry in the Novel CWGMAW Process, Welding Journal, Vol. 94, No. 9, 2015, pp. 301-311., ISSN 0043-2296

[13] Khalaf, G. H. (1998). Neuro-fuzzy control modelling for gas metal arc welding process, Ph.D. Dissertation, Department of Manufacturing Engineering, Loughborough University, Loughborough, UK

[14] Benyounis, K. Y. (2006). Prediction and Optimization of Residual Stresses, Weld-Bead Profile and Mechanical Properties of Laser Welded Components, Ph.D. Dissertation, School of Mechanical and Manufacturing Engineering, Dublin City University, Dublin, Ireland

[15] Khrais, S., Al Hmoud, H., Al, A. A. & Darabseh, T. (2023). Impact of Gas Metal ArcWelding Parameters on Bead Geometry and Material Distortion of AISI 316L, Journal of Manufacturing and Materials Processing, Vol. 7, 2023, DOI: 10.3390/jmmp7040123

[16] He, Y., Li, D., Pan, Z., Ma, G. Yu, L., Yuan H. & Le, J. (2020). Dynamic Modeling ofWeld Bead Geometry Features in Thick Plate GMAW Based on Machine Vision and Learning, Sensors, Vol. 20, 2022, DOI: 10.3390/s20247104

[17] Biber, A., Sharma, R. & Reisgen, U. (2024). Robotic welding system for adaptive process control in gas metal arc Welding, Welding in the world, 2024, DOI: 10.1007/s40194-024-01756-y

[18] Lima, E. J., Castro, C., Bracarense, A. Q. & Campos M. M. (2005). Development of a model to predict weld bead geometry during the GMAW process using high speed camera. Proceedings of 7th International conference Trends in Welding Research, 2005

[19] Silva, R. H. G., Galeazzi, D., Schwedersky, M. B., Mendoça, F. K., Bonamigo, A. V. & Marques, C. (2021). An adaptive orbital system based on laser vision sensor for pipeline GMAW welding, Journal of the Brazilian Society of Mechanical Sciences and Engineering, Vol. 43, 2021, DOI: 10.1007/s40430-021-03081-7

[20] Inspection certificate 0083718076, dated 18.10.2023

[21] Inspection certificate 730/0045862, dated 29.04.2020

[22] Hajro, I. & Tasic, P. (2020). Influence of heat input and welding position on geometrical properties of GMAW fillet welds of unalloyed steel, TEM Journal, Vol. 9, No. 2, 2020, pp. 496-500., ISSN 2217-8309, DOI: 10.18421/TEM92-10

[23] Tasic, P.; Hajro, I. & Hodzic, D. (2017). Technological and Economical Aspects of Arc Welding of Al 5754, Proceedings of the 28th DAAAM International Symposium, pp.854-858, B. Katalinic (Ed.), Published by DAAAM International, ISBN 978-3-902734-11-2, ISSN 1726-9679, DOI: 10.2507/28th.daaam.proceedings.119

DOI: 10.2507/35th.daaam.proceedings.019

THE STUDY OF THE EFFECTIVENESS OF ENERGY AND INFORMATION TRANSFER THROUGH A LARGE REFLECTOR USING ITS CONDUCTIVE MATERIALS

Yuriy Konoplev, Fedor Mitin & Alexey Yuev

This Publication has to be referred as: Konoplev, Y[uriy]; Mitin, F[edor] & Yuev, A[lexey] (2024). The Study of the Effectiveness of Energy and Information Transfer through a Large Reflector using its Conductive Materials, Proceedings of the 35th DAAAM International Symposium, pp.0150-0156, B. Katalinic (Ed.), Published by DAAAM International, ISBN 978-3-902734-44-0, ISSN 1726-9679, Vienna, Austria
DOI: 10.2507/35th.daaam.proceedings.019

Abstract

The article discusses various methods for transmitting energy and information using the electrical conductivity of tungsten components within a large, transformable structure. It explores the principles of data transmission and examines the most common standards in this field. Additionally, it evaluates the feasibility of creating a new information transmission protocol that could be used in outer space applications. Due to the fact that these energy systems consider not only the shape and material composition of a structure, but also its ability to accurately transmit energy and signals, developing algorithms for both energy and data transmission presents a complex scientific and technical challenge. Technical solutions have been developed that allow for more efficient use of the design of large-sized space-based reflectors. The results of field tests have confirmed the benefits of the developed methods for transmitting energy and information using tungsten components in grid design.

Keywords: Large-sized reflector; Transformable structure; Electrical conductivity; Tungsten; Energy transfer; Information transfer.

1. Introduction

Currently, the development and advancement of technology for the production and operation of large-sized transformable structures represents an innovative solution to several challenges within space-based technologies. Antennas with large effective areas enable remote sensing of the Earth's surface and provide high-quality, noise-resistant communication. However, such structures present significant challenges in terms of their manufacturing and operational processes, particularly when it comes to deploying them into a functional position in outer space and adjusting the shape of the radio-reflecting mesh [1], [2], [3], [4].

As a prototype, let us consider a reflector with a controllable reflector shape (Fig. 1) [5] and [6]. The challenge in operating reflectors with a precisely controlled shape lies in the difficulty of launching large-scale structures into orbit from the ground in a functional state. This requires their deployment in outer space.

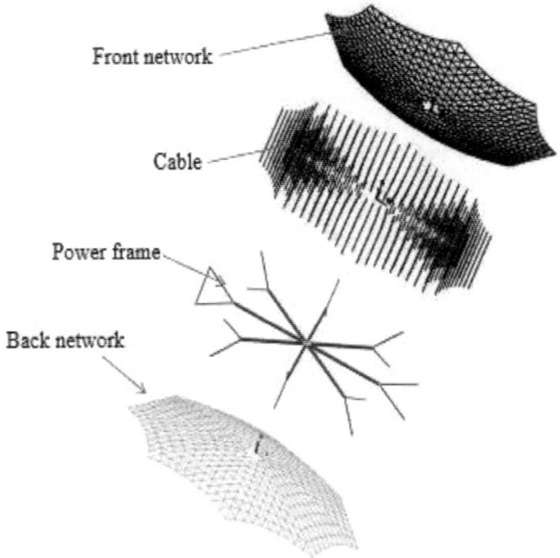

Fig. 1. Structure of a large-sized transformable reflector

A significant challenge in this regard is ensuring the accurate reproduction of the reflector's geometry in space, precisely as designed. Any deviation from the intended shape can result in a change to the radiation pattern, affecting the antenna's efficiency [7]. It is essential to deploy and subsequently adjust the specified reflector shape, as the structure may undergo changes due to external influences. To determine the current configuration of the antenna, a geometric scan using a total station (or laser tracker) is performed. Actuators are used to adjust the shape, which are located throughout the entire area of the reflector, by altering the lengths of the cables stretched between the front and back networks [8].

Actuation devices are operated via wired connections, which, in addition to adding significant additional mass to the system, also pose risks of entanglement and cable failure during deployment. Therefore, it is essential to develop alternative methods for transferring information and power to actuators [9], [10].

2. Analysis and selection of the energy transfer method

In order to operate the shape control system for a large-sized transformable structure, power must be supplied to the actuators. A major challenge in this regard is the use of redundant wired connections [11].

There have been studies in the field of energy transmission using an open optical channel. This approach has a relatively low efficiency, as optical receivers are not very efficient. In addition, while it offers the potential to eliminate wired connections, it requires the use of complex laser positioning systems for receivers and design of energy storage solutions, as it is not feasible to transmit energy continuously. This approach is challenging to implement in terrestrial environments due to atmospheric dispersion, which further reduces efficiency. Several countries are pursuing research in this area, but no commercialized space-based solution has yet been developed [12], [13]. Another potential method of energy transfer could be the use of microwaves. In this scenario, a significant amount of energy would be distributed through space past receiving devices.

In this paper, a system for transmitting energy using structural conductive elements in a grid is proposed. The radio-reflective net consists of tungsten, a good conductor, which allows it to be used entirely as a conductor. Because of the need to maintain almost the entire active area of the reflector energized, the efficiency of this circuit is lower than that of directly wired connections, although it is many times higher than that of an open optical channel.

This approach allows for the transmission of not only energy but also information over a network, which necessitates a number of modifications to the energy transmission structure, as it becomes necessary to transmit at different voltage levels. In this context, it is important to note that on larger structures, there will be a significant voltage drop due to tungsten resistance. This requires the use of multiple synchronized on-board power supplies or step-down devices for elements located closer to sources. Due to the need for different logic levels to be transferred, it is not possible to maintain a constant voltage. This issue is addressed by incorporating an energy storage system, based on either a battery or supercapacitor, into the actuator [13], [14]. In this scenario, information transmission is possible through simple presence/absence of voltage. However, this results in short-term disruptions to the power supply. Therefore, it is necessary to store energy in order to maintain operation during information transmission events. Another option is to utilize the conventional supply voltage of the control circuit as a reference zero. Alternatively, a higher voltage level, such as 1.5 times the standard voltage, can be used as a unit. In this instance, only overload protection would be required, allowing for a reduction in system weight and size.

3. Analysis and selection of the method of information transmission

To select an information transfer protocol, it is essential to consider the design characteristics of the network protocol. The reflector comprises two main components – the front and back network (Fig. 1). The rear side is used for grounding, hence only one line is available. This significantly reduces the number of potential methods for transmitting information. Under these circumstances, the UART interface appears to be the most suitable option. A common ground is provided by the backside of the reflector. Although two lines are typically required according to the UART specification, one can be omitted, as data transmission is only required in one direction.

The circuit includes a master microcontroller, which processes control signals and transmits them to a network of slave microcontrollers. Communication between the master and slave devices is established using the UART (Universal Asynchronous Receiver/Transmitter) protocol, with the TX (transmit) output of the master connected to the RX (receive) input of the slaves. This arrangement is dictated by the lack of a dedicated second channel for communication and the requirement to transmit data from the slaves to the master device.

To verify the functionality of this design, an experiment was carried out using tungsten as the data bus. Two microcontrollers were used in the experiment (Fig. 2). The research was conducted on models with significant size limitations. While this does not fully conform to the definition of a large-sized structure, it does allow for low-cost, in-kind validation of the viability of the proposed algorithms. At present, the practical implementation of the suggested methods on large-scale structures is achievable, although it has been little explored, which could potentially lead to additional challenges in their application to real-world scenarios.

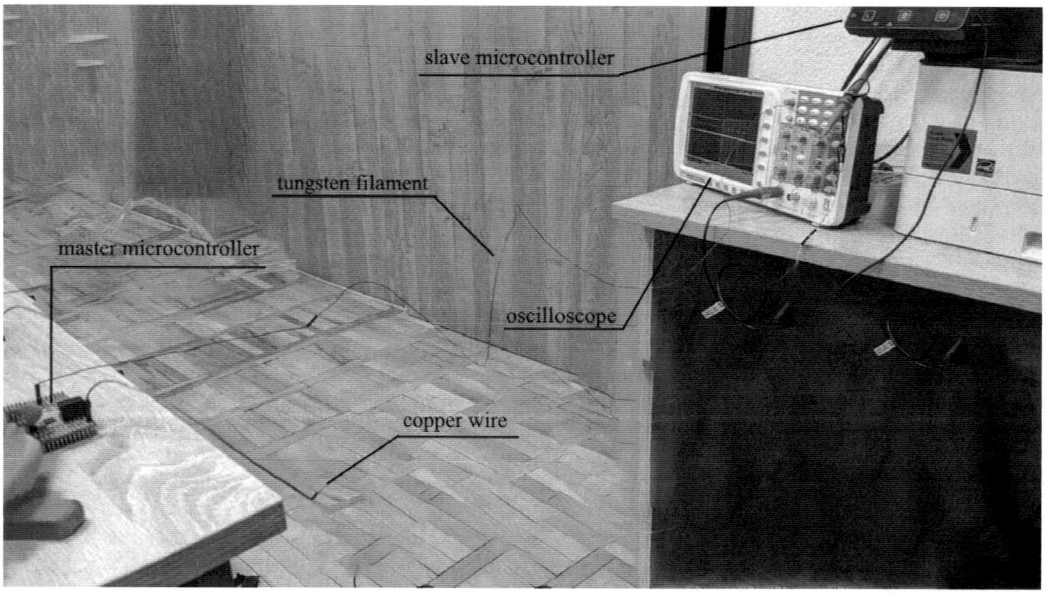

Fig. 2. Experimental setup

In the developed data transfer algorithm, each character is assigned a separate byte. This is an efficient approach that ensures stable data transmission, even if the packet needs to be expanded, and not just in extreme cases. The first two bytes indicate the slave device's number, for example 01, and the next byte indicates the direction of the device's rotation (0 for clockwise and 1 for counterclockwise). The next three bytes specify the time for the executing device to work. While this message does not provide all the information necessary for a real device's operation, it simulates the operation well enough. Initially, the experiment was performed on a 1-meter-long tungsten wire using two STM32F407 microcontrollers. A communication channel was established using tungsten and the ground was provided by a regular copper wire. The phrase "01 0 200" was transmitted between microcontrollers at a rate of once per second. The accuracy of the transmission was verified when sending a message from a subordinate microcontroller to a personal computer. Initially, the rate was set at 9,600 kilobits per second (kbit/s). There were no transmission errors for an hour. Next, the rate was increased to 14,400 bits per second (baud) and then adjusted according to specifications. Transmission errors started to occur at 56,000 baud. This indicates the possibility of increasing data transfer speed, but this can also lead to errors and reduced compatibility. A rate of 9,600 kbit/s is sufficient to transfer the necessary amount of data (one byte per millisecond). After replacing the tungsten wire with weave, general grounding was also performed using tungsten.
The experiment was repeated and the results were consistent. Therefore, the UART protocol ensures reliable signal transmission [14].

The developed algorithm for the system's operation, considering data transmission, involves the following steps:

- Determination of the current state of the network using a laser tracker.
- Calculation of the necessary offsets for actuation points.
- Transmission of these offsets to the master microcontroller.
- Generation of information packets by the master microcontroller.
- Transfer of these generated packets through the tungsten network elements using the UART (Universal Asynchronous Receiver/Transmitter) protocol.
- Reception and processing of the packets by slave microcontrollers.
- Execution of the corresponding control actions by executive devices.

One of the challenges associated with the use of the UART protocol lies in its application to reflectors with a larger radius. While reducing the data transmission speed can help increase the range, the use of tungsten has limitations that restrict the feasibility of this approach. To ensure the successful transmission of a signal over long distances, a method of separating receivers at the logical level is employed. This is because voltage drops over extended distances, and packet lengths must be reduced to account for this phenomenon. Additionally, addresses are not necessary, as there is no address system in place. Addresses are used to determine which device a command is directed towards, and this increases packet length. When using this protocol, losses can become significant, so it is important to reduce message lengths to minimize the risk of errors. Because it is not feasible for the slave microcontroller to report on the status of received packets, any error could disrupt device operation. Using accurate ADCs allows for more precise tracking of voltage changes. To confirm this, a current propagation model has been developed using the Matlab software package. This model shows voltage drops at specific points, with coordinates indicated in millimeters from the lower left corner of the antenna structure. The algorithm used is illustrated in Fig 3, and the simulation results are presented in Table 1.

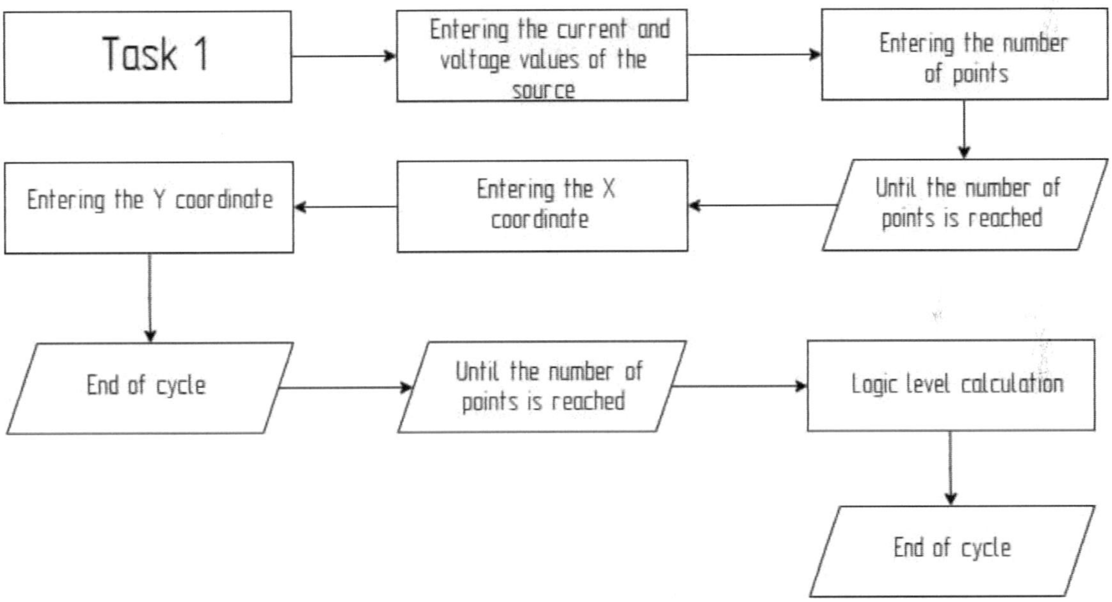

Fig. 3. The structure of the algorithm for modeling the distribution of stresses across network points

The address can only contain one bit, which is necessary if the network nodes are located at equal distances from a reference point. For instance, in the given table, this is the case for points 2 and 4, or 3 and 7. Due to the short distances in the layout, the voltage differences in the table are relatively small. Another possible solution could be to consider transferring the grounding to the spoke along which the actuator moves. This would allow for the rear part of the network to be freed, making it possible to utilize two transmission lines and providing access to several other protocols that require two lines, such as RS-485. The choice of this particular protocol avoids issues related to long-distance transmission of information. However, the physical implementation of this protocol is typically in the form of twisted pair, and it can be difficult to predict proper operation in scenarios involving lines spaced over a significant distance using rather unusual materials such as tungsten. As a result, testing of such a configuration is required. This approach has been identified as a key direction for further research in this field.

Capacities in the circuit elements arise when the weave is designed in such a manner that gaps occur. If air or vacuum appears in the interweaving, a planar capacitor may form. Its capacitance depends on the thickness of the resulting air or vacuum layer. Let us assume that this thickness is 1 millimeter. Then, $C = 0.058 * 10^{-12}$ F. This capacitance is quite small and therefore does not significantly affect the design. Therefore, it is impractical to calculate the capacitance taking into account the wire's geometry [15].

Point number	The X coordinate	The Y coordinate	Voltage value
1	0	0	10.0000
2	100	0	9.9900
3	200	0	9.9800
4	0	100	9.9900
5	100	100	9.9859
6	200	100	9.9776
7	0	200	9.9800
8	100	200	9.9776
9	200	200	9.9717

Table 1. The results of modeling the stress distribution at points in the tungsten network

4. Physical features of tungsten energy transfer

In the Ansys software, a mathematical representation of the grid system was created to verify voltage drops from the starting point. This model takes into account heating and current intensity. The findings are presented in Fig. 4 and Fig. 5.

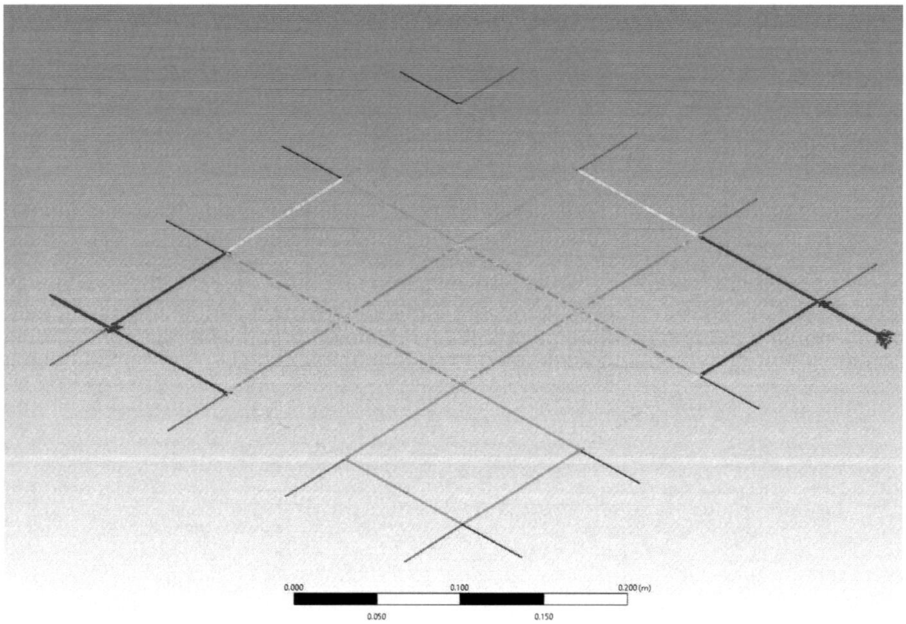

Fig. 4. Simulation of the current distribution in the ANSYS program

When moving from a grid node to its center, the number of nodes in the network increases, resulting in a decrease in the current strength within the segments between nodes. In the center of the network, current strength is at its minimum. As the current flows through the middle of the network, however, the number of nodes begins to decrease and the current strength begins to increase. This indicates that there are additional network parameters, in addition to resistance, that affect voltage and current strength. For instance, the shape of the network and the location of voltage application play a role. According to the Joule-Lenz Law, heating will occur unevenly due to high amperages, which can affect the geometry of the grid. These factors must be taken into consideration when designing a power supply system [16]. When transferring the developed techniques to a full-scale product, it is suggested to conduct additional testing that may reveal new opportunities to enhance the methods discussed in this paper [17].

5. The design of the layout facets of the reflector

To investigate the principles described, it is proposed to construct a mock-up simulation of a reflector array with a side length of 1 meter. The triangular frame will be constructed from aluminum tubes and the connections will be made from ABS plastic. The mesh will be completely composed of tungsten filaments, which will bring the scale of the model closer to that of the actual product. The small scale also allows for access to various testing chambers that can replicate space conditions. An example of a potential layout for investigating the processes under consideration is shown in Fig. 6.

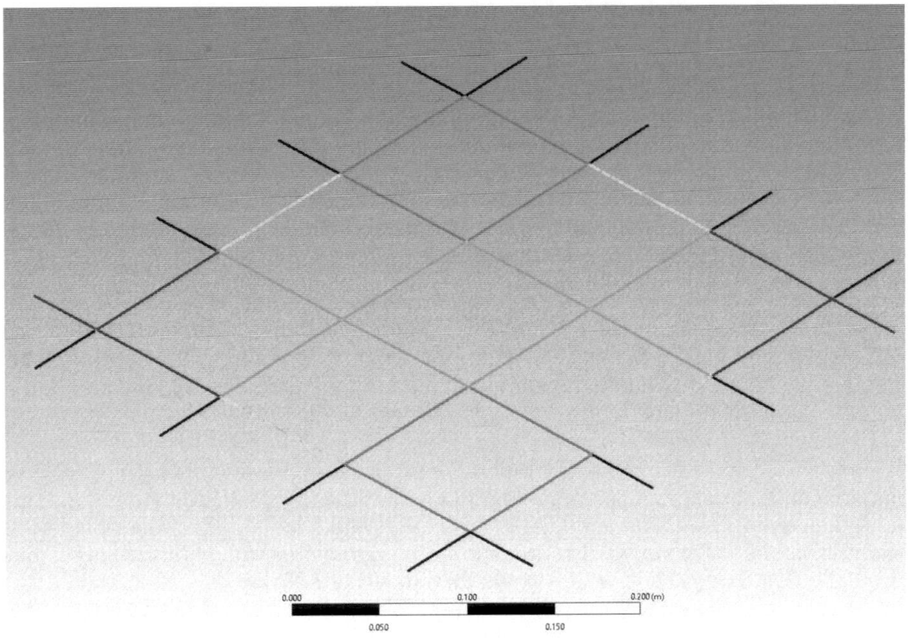

Fig. 5. Simulation of the heating of the grid in the ANSYS program

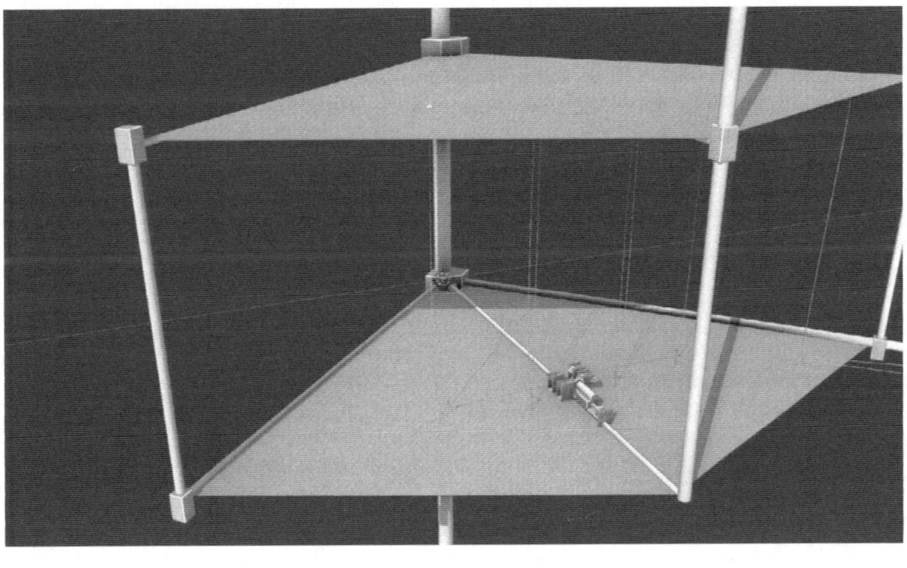

Fig. 6. The layout model in the Blender environment

6. Conclusion

The research presented in the article is designed to solve the problem of optimized energy information support in shape control systems of large-sized structures. The authors considered possible solutions to this problem and chose a method for transmitting energy and information through the elements of a tungsten network. During the research, results have been obtained demonstrating the potential for developing a system to transfer energy and information utilizing network elements. Techniques for physically constructing such a system have been proposed, taking into consideration the specific requirements of power supply for actuation points as well as operational algorithms. Recommendations have been provided for selecting methods for constructing energy supply, contingent on the design characteristics of the reflector, facet utilized, and weight and size constraints. Techniques to reduce the length of a message have been developed, enabling the transmission of less information and, consequently, enhancing the accuracy of data transfer. Additionally, considering the physical structure of a conductor opens up novel research objectives in this area, such as addressing uneven heating of the structure and optimizing stress application points.

In the future, the authors propose to conduct research on a more complex layout, which has a more accurate repetition of the geometry of the large-sized structures used. In addition, it is planned to work out algorithms over long distances, which will allow us to obtain new data and improve algorithms for energy information exchange.

7. Acknowledgments

The research was supported by grant No. 22-79-10112 from the Russian Science Foundation, https://rscf.ru/en/project/22-79-10112/.

8. References

[1] Xuelin Du, Jingli Du, Hong Bao, and Guohui Sun. (2018). Deployment Analysis of Deployable Antennas Considering Cable Net and Truss Flexibility. Aerospace Science and Technology, Vol. 82–83, pp. 557–565.

[2] Zhengrong Chu et al. (2014). Modeling and Analysis of a Large Deployable Antenna Structure. Acta Astronautica, Vol. 95, No. 1, pp. 51–60, ISSN 0094-5765, DOI 10.1016/j.actaastro.2013.10.015

[3] Guarriello, A., Loison, R., Bresciani, D., Legay H. & Goussetis, G. (2023). Structural and Radio Frequency Co-Design and Optimization of Large Deployable Reflectarrays for Space Missions. IEEE Transactions on Antennas and Propagation, Vol. 71, No. 5, pp. 3916-3927, doi: 10.1109/TAP.2023.3247940

[4] F.V. Mitin, A.D. Shirshov, Y.V. Konoplev, A.A. Yuev, A.I. Krivushov. (2024). Control system of the radio-reflective surface shape of a large-size space reflector using its current-conducting structural parts, Acta Astronautica Vol. 219, pp. 592-602, doi: https://doi.org/10.1016/j.actaastro.2024.03.061.

[5] Fedor Mitin, Alexey Krivushov (2022). Optimal Control of the DC Motor in the Tuning of the Space-Based Reflector of the Radar-Reflecting Netting, Proceedings of the 33rd DAAAM International Symposium, pp.0247-0253, B. Katalinic (Ed.), Published by DAAAM International, ISBN 978-3-902734-36-5, ISSN 1726-9679, Vienna, Austria DOI: 10.2507/33rd.daaam.proceedings.034

[6] Yuriy Konoplev, Alexey Krivushov, Fedor Mitin & Alexey Yuev (2023). Construction of The Mathematical Model of The Structure of The Actuator and Design of Control System, Proceedings of the 34th DAAAM International Symposium, pp.0183-0192, B. Katalinic (Ed.), Published by DAAAM International, ISBN 978-3-902734-41-9, ISSN 1726-9679, Vienna, Austria DOI: 10.2507/34th.daaam.proceedings.024

[7] Bakulin, V.N. and Borzykh, S.V. (2020) Modeling of the Deployment Process Dynamics for Large-Sized Transformable Space Structures. Izv. Vuz. Av. Tekhnika. Vol. 63, No. 4, pp. 50–56 [Russian Aeronautics (Engl. Transl.), vol. 63, no. 4, pp. 610–617].

[8] Konoplev, Yu. V., Shirshov A.D., Yuev A.A. (2023). Research and Prospects of Application of Current-Conducting Structural Parts of the Reflector for Optimization of the Energy-Information Support of the Radio-Reflective Surface Shape Control System. Russian Aeronautics. Vol. 66, No. 4. pp. 703-707. DOI 10.3103/s1068799823040086. EDN GUDQCN.

[9] Ponomarev S.V. (2011). Transformable reflectors of spacecraft antennas // Bulletin of the Tomsk State University. Mathematics and Mechanics. Vol. 16, pp. 110–119.

[10] Hanqing Deng, Tuanjie Li, and Zuowei Wang. (2015). Pretension Design for Space Deployable Mesh Reflectors under Multi-Uncertainty. Acta Astronautica, Vol. 115, pp. 270-276. DOI: 10.1016/j.actaastro.2015.05.038

[11] Matveev S.A., Strakhov S. Yu., . Khromikhin D.A. (2016). Organizing energy-and-information exchange between devices for controlling the shape of a transformable antenna, using fiber-optic technology. Journal of Optical Technology. Vol. 83, No. 11, pp. 703-707. DOI 10.1364/JOT.83.000703. EDN YNFYKD.

[12] Tishkov A. I., Konoplev Y. V., Yuev A. A., Shevtsov I. V. (2019) Development of a Combined Electric Power Storage Device. Russian Aeronautics. Vol. 62, No. 2, pp. 321-325. DOI 10.3103/S106879981902020X. EDN CIDXEJ

[13] Zhdanov, V.V., Koshtyal, Yu.M., and Kuznetsov, V.P. (2017). Analysis of the Principal Parameters of Modern Supercapacitors and Li-Ion Batteries with the Aim of Developing Heavy-Duty Power Supply Sources, St. Petersburg: FTI im. A.F. Ioffe, URL: http://www.rusbat.com/Interbat22032017/IoffeInterbat2017.pdf.

[14] Konoplev, Yu.V., Shirshov, A.D., and Yuev, A.A. (2023) Modernization of the Shape Control System for the Radio-Reflecting Surface of a Transformable Space Antenna. Izv. Vuz. Av. Tekhnika. Vol. 66. No. 2, pp. 156–159 [Russian Aeronautics (Engl. Transl.), 2023, vol. 66, no. 2, pp. 404–408].

[15] Ukhov, A.A., Bulat, M.P., Volobuev, I.A., Shapovalov, S.V., and Li, R.V. (2021) Wireless Power Transmission System for Powering Onboard Equipment of an Aircraft. Izv. Vuz. Av. Tekhnika. Vol. 64, No. 4, pp. 32–38 [Russian Aeronautics (Engl. Transl.), vol. 64, no. 4, pp. 613–619].

[16] Akira Meguro, Kyoji Shintate, Motofumi Usui, Akio Tsujihata. (2009) In-orbit deployment characteristics of large deployable antenna reflector onboard Engineering Test Satellite VIII. Acta Astronautica. Vol. 65, Ni 9–10, pp. 1306-1316, ISSN 0094-5765

[17] Rui Nie, Baiyan He, Shaoze Yan, Xiaofei Ma. (2019). Optimization design method for the cable network of mesh reflector antennas considering space thermal effects. Aerospace Science and Technology. Vol. 94, 105380, ISSN 1270-9638

DOI: 10.2507/35th.daaam.proceedings.020

HR-TECH AUTOMATION: A CASE STUDY OF RESUME DESIGN USING GENAI TECHNOLOGIES

Alexander Suleykin, Roman Babenko & Peter Panfilov

This Publication has to be referred as: Suleykin, A[lexander]; Babenko, R[oman] & Panfilov, P[eter] (2024). HR-Tech Automation: A Case Study of Resume Design using GenAI Technologies, Proceedings of the 35th DAAAM International Symposium, pp.0157-0164, B. Katalinic (Ed.), Published by DAAAM International, ISBN 978-3-902734-44-0, ISSN 1726-9679, Vienna, Austria
DOI: 10.2507/35th.daaam.proceedings.020

Abstract

With the development of technologies and the increase in the amount of data that companies work with, the use of large language models (LLM) is becoming an integral part of modern business processes. These models not only demonstrate impressive results in natural language processing, but also open up new horizons for automating routine tasks. In particular, in the field of HR-Tech, LLMs can significantly simplify the processes associated with resume processing, which makes them especially relevant for companies seeking to improve the efficiency of their HR departments working with a large amount of text data and resumes of candidates engaged in outstaffing both on the side of customers and contractors.

Keywords: Artificial intelligence; Generative AI; LLM; HR-automation.

1. Introduction

HR technology automation plays a key role in modern HR management, especially in the context of processing large amounts of data and increasing the efficiency of recruiting processes. One of the key aspects of automation is the processing of candidates' resumes, which includes their formatting and analysis for the subsequent selection of the most suitable specialists.

With the development of artificial intelligence (AI) and machine learning (ML) technologies, it has become possible to use these tools to automate the resume formatting process. Generative AI technology (GenAI) is one of the most promising areas in the field of HR automation, which can significantly increase the efficiency of HR specialists. The purpose of this article is to conduct a case study on the use of GenAI technologies to automate the resume formatting process. We will consider how these technologies can be used to speed up the process of CV formatting of random candidates to one specific unified format, improve the quality of resumes, speed up the process of their analysis and decision-making on candidate selection. In this paper, we consider different aspects of HR-processes automation using GenAI technologies, review difficulties of CV-formatting process, propose component architecture and Use Cases diagram, compare different open-source LLM-models, describe different data extraction pipeline options and give an estimation accuracy example. In addition, future work and other promising tasks in HR-tech industry with Gen AI technologies are described.

2. Literature review

Large Language Models (LLMs) are increasingly being utilized in Human Resources (HR) to streamline various processes, enhancing efficiency and effectiveness. Their applications range from automating resume screening to developing conversational agents for employee inquiries. Notably, the HR-MultiWOZ dataset provides a foundational resource for training LLMs in HR-specific dialogues, addressing privacy concerns while facilitating data generation (Xu et al., 2024). Additionally, LLMs have shown promise in automating repetitive tasks, such as handling time-off requests and medical claims (Afzal et al., 2024). Furthermore, LLMs can serve as zero-shot human models, improving interactions in HR-related scenarios by understanding context and planning responses ("Large Language Models as Zero-Shot Human Models for Human-Robot Interaction", 2023). Overall, the integration of LLMs in HR represents a significant advancement in optimizing workflows and enhancing employee engagement.

However, challenges remain, including the need for high-quality training data and the potential for biases in automated decision-making processes, which could impact fairness in HR practices (Zhang et al., 2023)(Salakar et al., 2023). Large Language Models (LLMs) are revolutionizing CV formatting and resume screening by enhancing the automation and efficiency of these processes. LLMs can effectively analyze unstructured data from resumes, capturing nuanced information that traditional methods often overlook. This capability is particularly beneficial in recruitment, where LLMs can summarize, grade, and even assist in decision-making regarding candidate selection. Key aspects include:

- **Enhanced Resume Screening.** LLMs automate the screening process, making it 11 times faster than manual methods (Gan et al., 2024). They improve the F1 score in resume classification, achieving 87.73% accuracy (Gan et al., 2024).
- **Multi-modal Understanding.** LLMs can integrate textual and visual data, enhancing the quality of CVs by generating descriptive content from images (Ma et al., 2024). This multi-modal capability allows for richer, more informative resumes that stand out to recruiters.
- **Addressing Unstructured Data.** LLMs excel at converting unstructured data into structured formats, ensuring critical information is not lost (Ghosh & Sadaphal, 2023).

While LLMs present significant advantages in CV formatting and recruitment, concerns about bias and transparency in automated systems remain critical issues that need addressing to ensure fair outcomes in hiring processes.

To sum it up, reviewed papers consider different HR-aspects automation, but there is no resume-formatting cases described in a detailed fashion. Thus, the novelty of the proposed paper focuses on the detailed resume-formatting Use Case based on different and most-advanced open-source LLMs. The paper provides a deep-dive into the process of converting CV of a random candidate to one specific unified format, describing architecture of proposed solution, difficulties, different techniques of interacting with LLM and the benchmarks, as well as the estimation of proposed solution and business value.

3. LLM for HR-Tech: Opportunities and prospects for use

LLMs open up a whole range of opportunities for HR specialists, allowing them to automate routine operations and improve work efficiency. For example, LLMs can:

- **Format resumes:** bring resumes to a single standard, which greatly simplifies their analysis and comparison;
- **Match candidates with vacancies:** analyze resumes and job requirements, identifying the most suitable candidates;
- **Generate Job Descriptions:** Create attractive and informative job descriptions that will be of interest to potential employees.

Automating such tasks with LM allows HR specialists to focus on more strategic issues such as talent selection, candidate interviews, optimization of internal HR processes and employee development.

A. The main difficulties of developing an AI-assistant for HR

Despite the promising prospects, the development of an AI-assistant faces a number of difficulties:

- **The specifics of HR cases:** the HR field requires a special approach that takes into account the variety of resume formats, the difference in the structure of resume writing among candidates and professional skills, which creates additional difficulties in the development;
- **Hallucinations:** LLMs can generate non-existent or implausible information, which can lead to errors in the selection of candidates or erroneous interpretation of candidates as not suitable for a specific vacancy;
- **The need to integrate external data and their constant relevance.** To achieve high results, integration with various systems and databases is required, which can be technically, financially and organizationally difficult. Additionally, it is necessary to ensure regular uninterrupted data updates, as the candidate database is constantly becoming outdated and not relevant.

4. Case Study: Automation of resume formatting

Our team was working on a project for a small IT company, the purpose of which was to speed up the resume processing process. At the entrance, we received resumes in various formats that needed standardization, and at the exit we received documents ready for consideration with a single structure and design.

4.1. Task description

It is necessary to process resume files automatically according to the specified template and requirements for structuring and formatting the text. The sources come in the following formats:

- .rtf format always looks the same,
- .doc/.docx - may look different,
- .pdf - always look different.

First name S.

Job title

Female – ж / Male - м, X years
Location: X city

Hard skills:

X, X1, X2

Experience — _ years _ months

The month and year of the start of work — The month and year of completion of work or present time	Job title - Duty 1; - Duty 2 etc.

Education:

- 20__, IHE
Faculty, specialty
- 20__, IHE
Faculty, specialty

About myself:

About myself; courses

Fig. 1. An example of a corporate template for structuring a resume

Additional requirements for the content of the converted resume:

- The resume must match the template, starting from the most recent experience to the very first (sometimes resumes are thrown off, where on the contrary people indicate the experience of old years first, and only the last one at the end);
- The functions performed must be written in a list, while in each sentence the subjects must be predicates (for example: “Module development...");
- The resume should not contain the names of companies and projects (it is necessary to depersonalize);
- The stack should not contain, for example, "English", etc., but only technologies;
- The point about education adapts to the candidate - not everyone has a higher education, when it is not necessary to remove the word "Higher" and leave just "Education". In the same place, if there are several of them - from the last to the old (year, name of the university/ college, etc., faculty and direction);
- If the resume contains completed courses and advanced training, this is indicated in the "About me" section.

4.2. Solution architecture

We have chosen DUC SmartSearch platform as the main platform for implementing the solution, as it was ideally suited for our task with minimal modifications. DUC SmartSearch is a digital ecosystem of AI assistants, its fork of the Danswer open-source system, which allows us to create add-ons on various LLM models, both proprietary (GigaChat, YandexGPT, ChatGPT, etc.) and Open-Source (LLAMA, GEMMA, etc.). A visual node pipeline editor has been added to the system, where you can independently create new individual LLM and AI agent orchestration pipelines for any subject area in low-code mode. The conceptual architecture of the solution is the following:

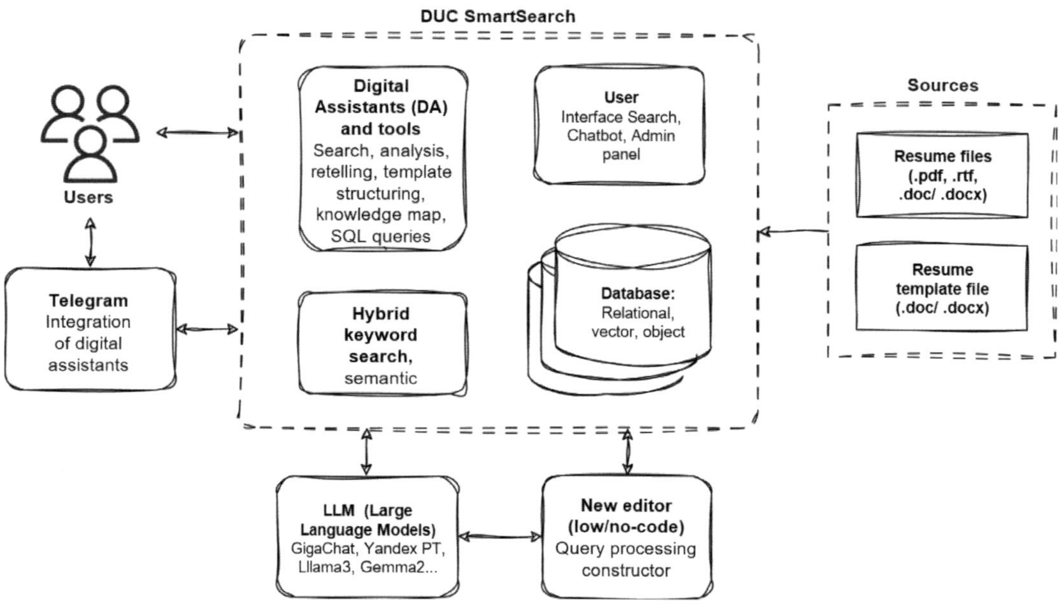

Fig. 2. A conceptual architecture of the resume formatting case implementation on the DUC SmartSearch platform

An additional digital assistant «HR assistant» was created in the system, and a special tool was developed and added - ResumeTool. This tool allows the digital assistant to work with files: upload the original resume, send it to the processing pipeline, fill in the template file with extracted data and save the formatted resume in the object storage. Users with the administrator role can upload their resume template or other documents in the tool settings. The following figure shows a diagram of the use cases of the developed resume formatting solution.

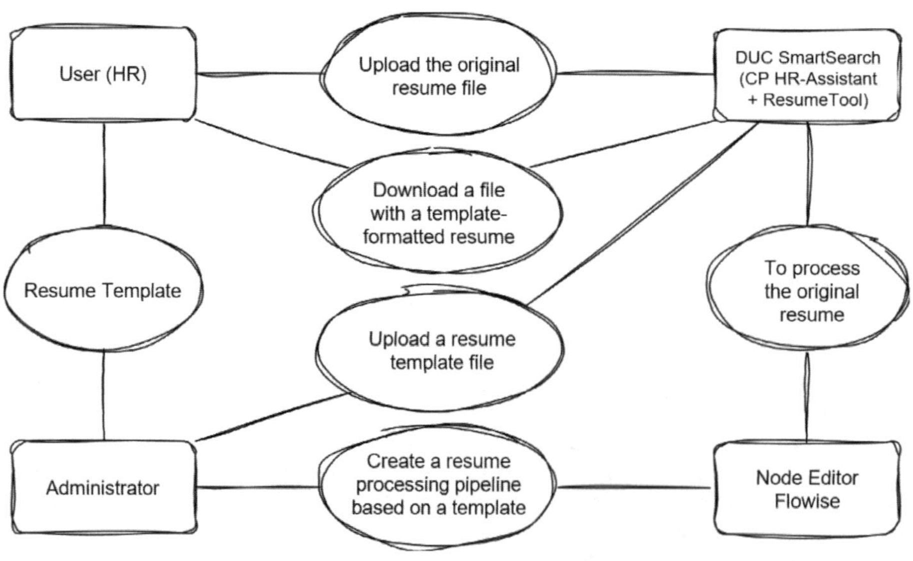

Fig. 3. Resume structuring according to a given template in DUC SmartSearch (use cases diagram)

5. Results: Smart Assistant for resume formatting

5.1. Resume processing in the DUC SmartSearch platform

The resume processing algorithm in the DUC SmartSearch system is as follows:

1. The administrator downloads the necessary resume template file on the settings page of the ResumeTool digital assistant "HR Assistant" tool;
2. Then the administrator sets up the pipeline for processing the initial resumes in the Flowise node editor (the stage of implementation and initial configuration of the system);
3. To process resume files according to a given template, an ordinary user goes to the "Chatbot" page, selects the configured digital assistant "HR assistant" and uploads the resume file for processing;
4. When the user sends a resume file on the Chatbot page, it is forwarded to the Flowise pipeline prepared, processed and the structured text of the resume is returned to the digital assistant. Next, the received text is processed into a specified template and the user is provided with a link to download the prepared file. The interface of the system is as follows:

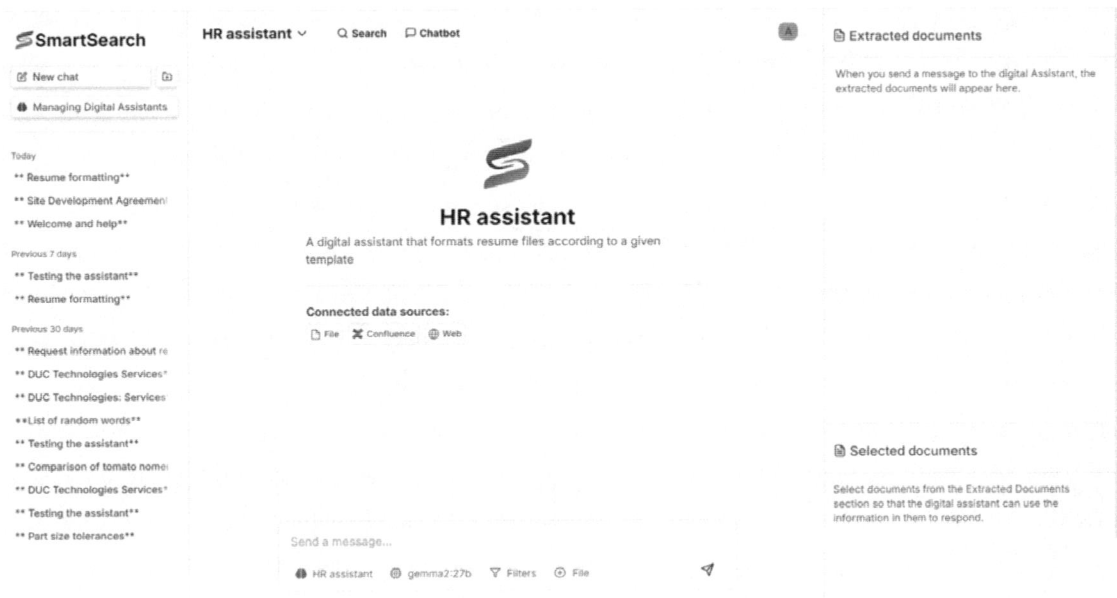

Fig. 4. Interface of the DUC SmartSearch Chatbot page

5.2. The development of the resume formatting pipeline

The solution for formatting the resume text was based on a data extraction approach similar to drawing up a Knowledge Maps – the approach, when all text is divided into N-topics, and for each topic a specific questions and typical answers are listed.

The main idea is to extract information in JSON format from unstructured data using LLM. During the development process, three pipeline schemes for information extraction were tested (Fig. 5):

In the first iteration of the developed pipeline, one common prompt was used to obtain information on all sections of the resume. The approach showed poor quality of the result - skipping data, ignoring instructions, and a high level of hallucinations (inventing). We believe that it is still difficult for language models to fulfill many text analysis requirements in a single query.

The second iteration of the pipeline was based on filling in the entire JSON structure at the first request and subsequent requeries to check and edit individual fields in the general JSON. The processing results became much better, and the solution was passed on to HR specialists for testing. After processing a couple of dozen resumes, shortcomings were identified, some of which were eliminated by tuning prompts.

Switching to the third version of the pipeline made it possible to eliminate the remaining comments, and also made it possible to flexibly configure each section individually. At the same time, the number of tokens transferred increased, since the entire text of the original summary was transferred to the model to extract each JSON field. This was not critical for us, because it did not significantly affect the processing speed, and the inference of the model was carried out locally on the GPU (without paying for tokens). In this project, the quality and stability of the processing result were prioritized over other non-functional requirements.

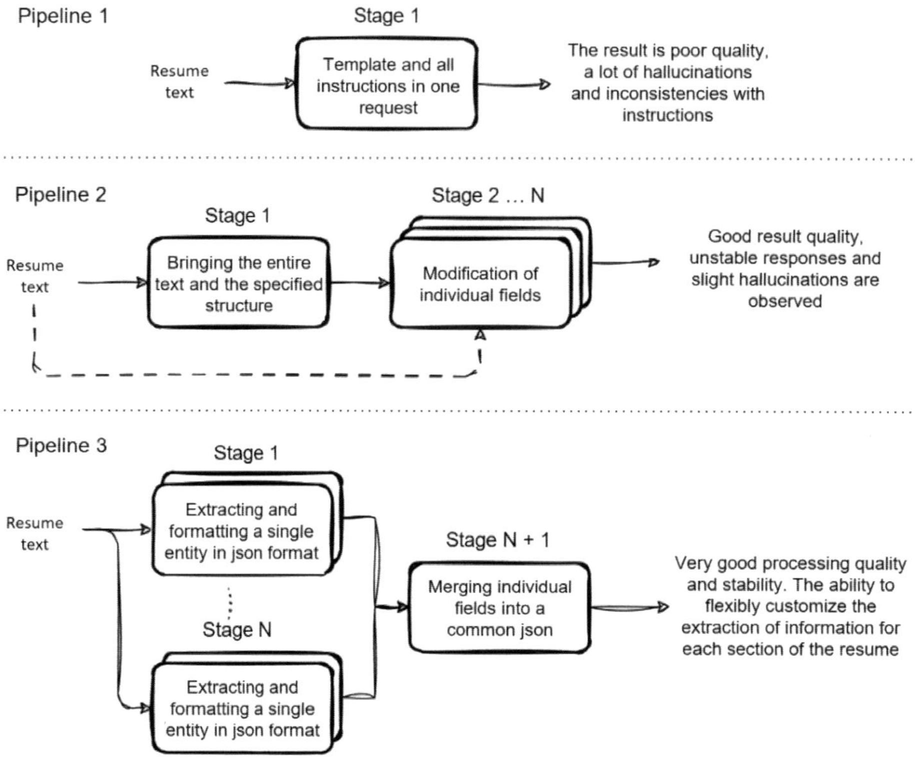

Fig. 5. Options for data extraction pipelines

5.3. Other tasks to be solved during the development process

Some .pdf summaries have a problem with incorrect encoding - when parsing text between letters in words, extra spaces appear. Such a "leaky" LLM text is very poorly processed or not processed at all:

```
Инстр уме нты  ▶ Jira;
▶ Redm ine;
▶ You Track;
▶ Activecollab;
▶ 1 C Bit rix 24;
▶ Bit rix: Уп р авлен и е сай том;
▶ Microsoft Office;
▶ Gantt Pro;
▶ Sm artsheet;
▶ Project Office;
▶  Fi rebase;
▶  Te st Flig h t .
 Ключевые навыки  ▶  У правление разработкой и внедрение
информационных проектов;
▶  Уп р а вл е н и е  б юд же то м  и  р и ска м и ;
```

Fig. 6. A fragment of the problematic text

Based on the results of experiments with flavors and various models, the following solution was found. At the beginning, we count the number of spaces in the text and calculate their percentage to the total number of characters in the text. If it is greater than some average value for normal resume texts, then remove all single spaces. This prepared text is sent to the preliminary pipeline, in which the model places the missing spaces. It does a good job with this task, unlike removing unnecessary spaces between letters in words. Next, the resulting text is processed according to the main pipeline.

6. Discussion

6.1. Project results

After the introduction of automation, the time for resume formatting was reduced by 6 times. The following shows how much time was spent on formatting before and after using the AI Assistant:

- Before automation: an average of 30 minutes for formatting one resume by an HR employee.
- After automation: 2 minutes for automatic formatting and 3 minutes for verification, a total of 5 minutes.

Taking into account the processing of an average of 100 resumes per month by one HR specialist, the implementation of this solution has made significant time savings (more than 40 hours of HR specialist work per month). Looking ahead, we note that the final solution works on the multilingual Gemma 2 27B model with Q4 quantization. These parameters allow us to run the model on a user segment video card with 24 GB of video memory.

6.2. Evaluation of results

As an alternative, we tested three more open-source local large language models: Mistral NeMo 12B Q8, Llama 3.1 8B FP16, Saiga-llama3 8B FP16. The quality of their processing turned out to be an order of magnitude worse than that of the selected model - inventing information, adding unspecified json fields, looping response generation, etc.

To evaluate the results of resume processing using the selected model, a testing methodology based on processing 10 random resumes with subsequent evaluation by three independent experts was applied. Points were awarded based on the quality of filling in each section of the resume separately on a 5-point scale:

1 - Fictional Information Is Present;
2 - The Information Was In The Text, But Was Not Extracted;
3 - Only Part Of The Information Has Been Extracted;
4 - All Information Has Been Extracted, But Formatting Requirements Have Not Been Met;
5 - All Information Is Extracted In The Required Form.

The average scores of the three experts and the final score are presented in the table below.

Summary section	**Average score**
Name and surname	4,8
Post	4,8
Gender	5
Age	5
City of residence	5
Technology stack	3,8
Full experience	5
Working periods	4,8
Functions	4,2
Education	4,6
About yourself (courses)	4,4
Final assessment	**4,7**

Table 1. Results of the evaluation of the formatting of resume files according to a given template

As it can be seen, the model does the worst job of filling the technology stack. The reason is that it does not interpret some technology names as development tools. This problem is solved by adding examples of technologies that the model skips to the prompts. Thus, during re-processing, these technologies will be extracted to the stack.

6.3. Scaling the solution

The integration of a node editor for configuring processing pipelines and downloading a template file allows SmartSearch administrators to configure a digital assistant to automatically process various types of documents according to specified requirements and template. Optionally, the implementation of document processing via Telegram app is available. The created solution is easily scaled to various resume formats. Since we used a modular approach, it only takes a few days to integrate the new functionality. This allows companies to quickly adapt to changing requirements and expand the capabilities of the HR assistant.

7. Conclusion

In this paper, we have developed AI-pipeline for CV-formatting. The proposed solution significantly automates the process of CV processing of job candidtates, saving up to 6 times of CV-processing time for HR-specialists. We have developed AI-pipeline using DUC SmartSearch platform, a fork of open-source Danswer system. We described AI-pipeline generation and defferent steps, data extraction and LLM communication steps, gave an overview of conceptual architecture, use cases diagram, described main difficulties of LLM-pipeline creation and developed a CV-template.

Additionally, we have tested developed AI-pipeline using different LLMs: Mistral NeMo 12B Q8, Llama 3.1 8B FP16, Saiga-llama3 8B FP16. Fort he best model, a testing methodology was proposed based on processing of 10 random resumes with subsequent evaluation by three independent experts.

8. References

[1] Weijie, Xu., Zicheng, Huang., Wenxiang, Hu., Xi, Fang., Rajesh, Kumar, Cherukuri., Naumaan, Nayyar., Lorenzo, Malandri., Srinivasan, H., Sengamedu. (2024). HR-MultiWOZ: A Task Oriented Dialogue (TOD) Dataset for HR LLM Agent. arXiv.org, abs/2402.01018 doi: 10.48550/arxiv.2402.01018.

[2] Anum, Afzal., Alexander, Kowsik., Rajna, Fani., Florian, Matthes. (2024). Towards Optimizing and Evaluating a Retrieval Augmented QA Chatbot using LLMs with Human-in-the-Loop. doi: 10.18653/v1/2024.dash-1.2

[3] (2023). Large Language Models as Zero-Shot Human Models for Human-Robot Interaction. doi: 10.48550/arxiv.2303.03548.

[4] Ceng, Zhang., Junxin, Chen., Jiatong, Li., Ze-bing, Mao. (2023). Large language models for human-robot interaction: A review. Biomimetic intelligence and robotics, doi: 10.1016/j.birob.2023.100131

[5] Esmail, Salakar., Jivitesh, Rai., Aayush, Salian., Yasha, Shah., Jyoti, Wadmare. (2023). Resume Screening Using Large Language Models. 494-499. doi: 10.1109/icast59062.2023.10454984

[6] Chengguang, Gan., Qinghao, Zhang., Tatsunori, Mori. (2024). Application of LLM Agents in Recruitment: A Novel Framework for Resume Screening. doi: 10.48550/arxiv.2401.08315

[7] Shuxiao, Ma., Linyuan, Wang., Senbao, Hou., Bin, Yan. (2024). Aligned with LLM: a new multi-modal training paradigm for encoding fMRI activity in visual cortex. arXiv.org, abs/2401.03851 doi: 10.48550/arxiv.2401.03851

[8] Preetam, Ghosh., Vaishali, Sadaphal. (2023). JobRecoGPT - Explainable job recommendations using LLMs. arXiv.org, abs/2309.11805 doi: 10.48550/arxiv.2309.11805.

DOI: 10.2507/35th.daaam.proceedings.021

INCREASING THE CONTROL RANGE OF AN AXIAL PISTON SWASH PLATE HYDRAULIC MACHINE

Stazhkov Sergey, Pryanichnikov Valentin, Elchinsky Victor,
Kuzmin Anton & Pham Tung Lam

This Publication has to be referred as: Stazhkov, S[ergey]; Pryanichnikov, V[alentin]; Elchinsky, V[iktor]; Kuzmin, A[nton] & Pham, T[ungLam] (2024). Increasing the Control Range of an Axial Piston Swash Plate Hydraulic Machine, Proceedings of the 35th DAAAM International Symposium, pp.0165-0170, B. Katalinic (Ed.), Published by DAAAM International, ISBN 978-3-902734-44-0, ISSN 1726-9679, Vienna, Austria
DOI: 10.2507/35th.daaam.proceedings.021

Abstract

Axial piston swash plate hydraulic machines are widely used in modern volumetric hydraulic drives, which is due to their high energy consumption and simplicity of design. A significant disadvantage of hydraulic machines of this type is the small control range caused by the large value of the minimum stable revolutions of the hydraulic drive shaft. In this research, the possibility of increasing the angle of inclination of the swash palte without increasing the friction forces in the piston- guide bushing pair to reduce the value of the minimum stable revolutions of the hydraulic drive shaft is considered.

Keywords: Hydraulic machines; Axial piston machines; Volumetric hydraulic drive; Control range, Tilt angle of the swash plate, Shaft torque.

1. Object of investigation

The object of research in this work is an Axial piston swash plate hydraulic machine (APGND), since it is widely used in modern hydraulic drives due to its high energy consumption and simplicity of design [1]. An increase in the angle of inclination of the hydraulic machine swash plate reduces the value of the minimum stable revolutions of the drive shaft of the APGND due to an increase in the useful component of the force applied from the shoe to the swash plate of the hydraulic machine [2]. However, an increase in the angle of inclination also leads to a greater misalignment of the piston in the guide sleeve, and accordingly to an increase in friction forces in this kinematic pair. Figures 1 and 2 show diagrams of forces in a piston pair of an APGND of standard and modified design [3].

Analyzing the force patterns, it can be seen that when the hydraulic machine is operating, the Fp force causes a reaction R in the piston joint C. This reaction decomposes into two components of the force: Ra and Rb [4]. The magnitude of these reactions affects the friction forces Ta and Tb, the combination of these two friction forces and fluid pressure will balance the horizontal component of the Ro force R. Similarly, the perpendicular component of the force Rn. Rn also balances the resultant force of the two pressures Ra and Rb.

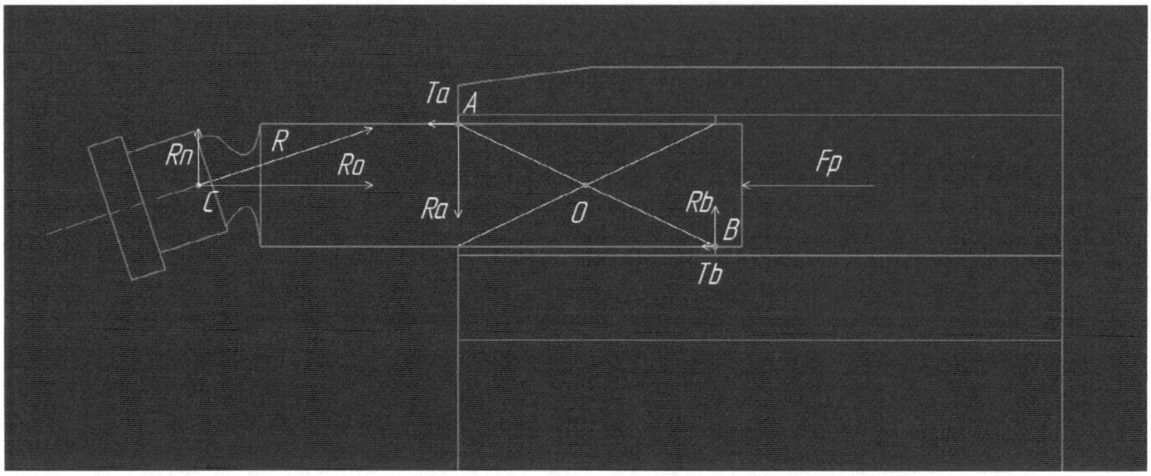

Fig. 1. The diagram of forces in the standard piston mechanism of the APGND

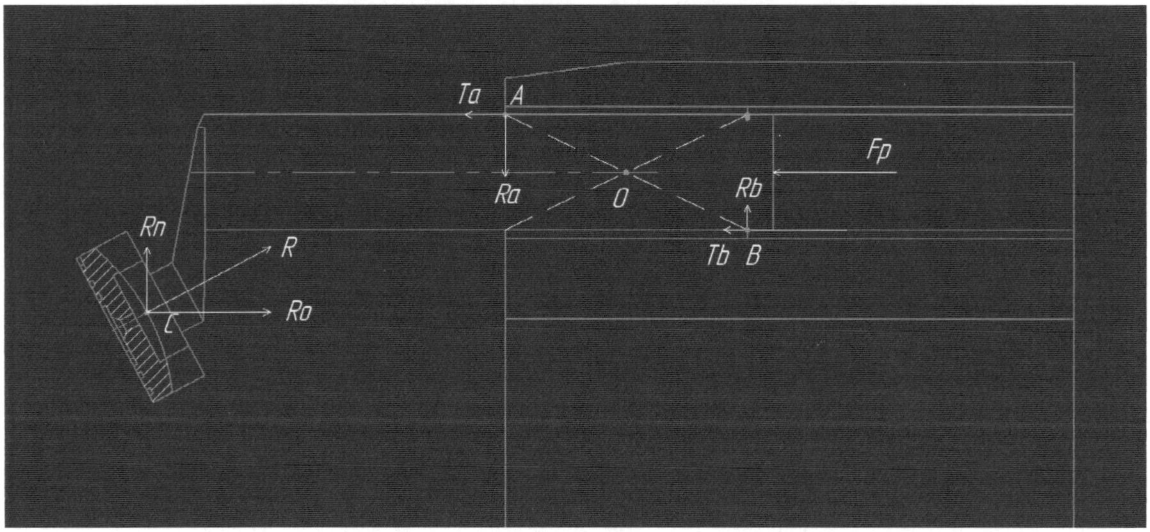

Fig. 2. The diagram of forces in the modified piston mechanism of the APGND

In nominal operation mode, the total torque and force of the entire system are balanced. Considering any power point (taking the point O as the center of the piston core), we have 3 cases depending on the position of the piston inside the guide sleeve [5]. The force diagram corresponding to each of these cases is shown in Figure 3.

I

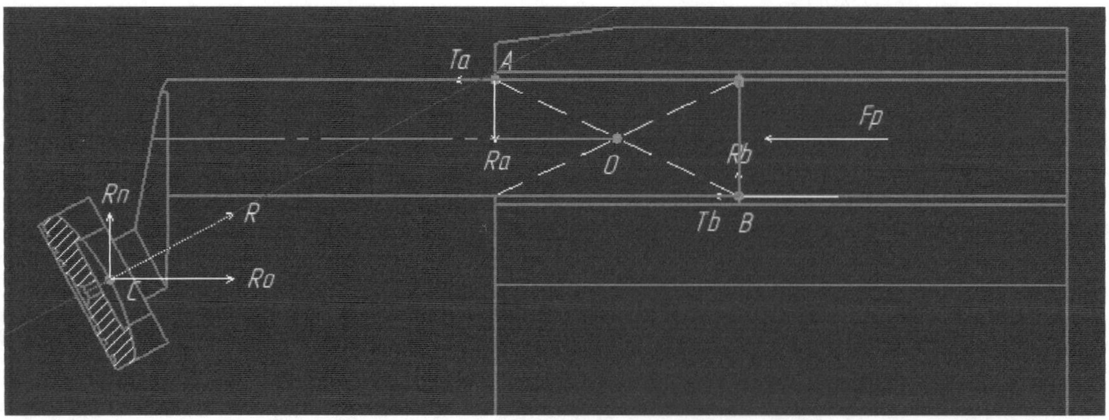

II

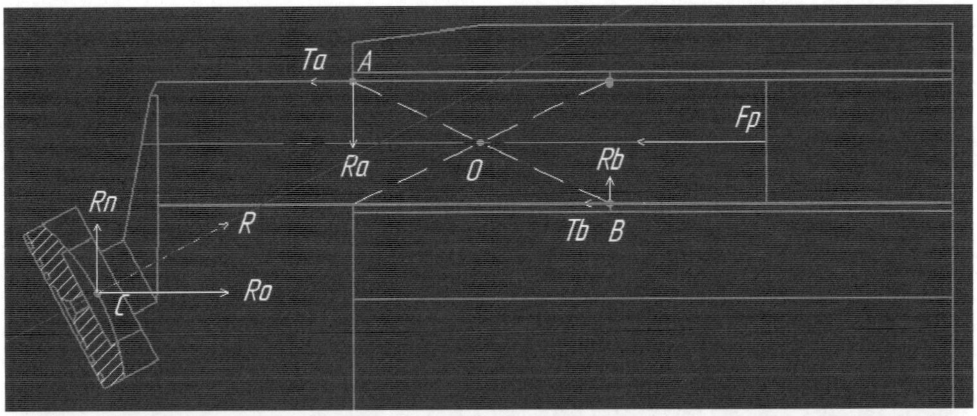

II

I

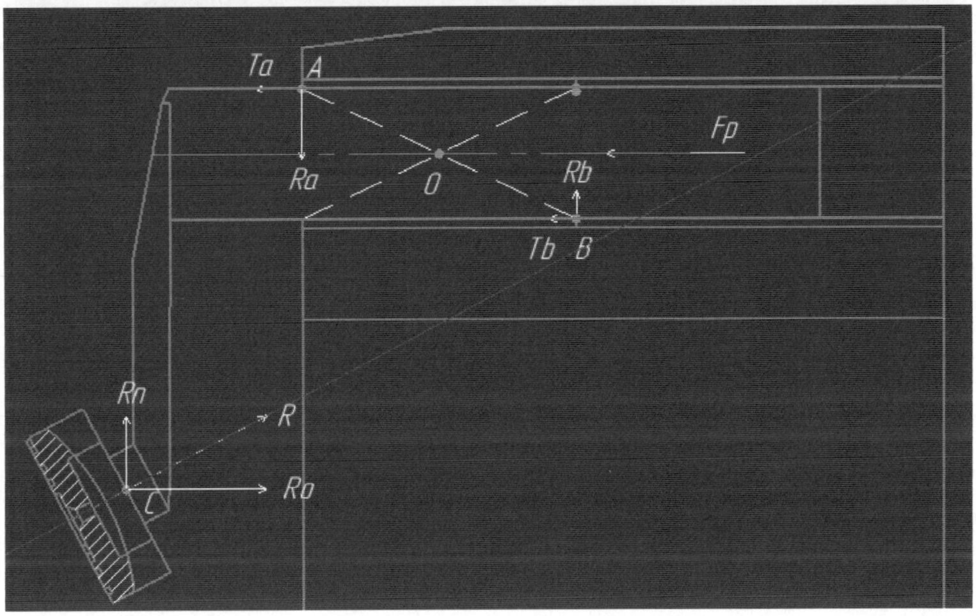

Fig. 3. Force diagram for 3 cases of piston position in a modified piston mechanism of the APGND

The equation of moments for each of these cases will take the form:

$$
\begin{aligned}
&M_{Rn} + M_{Ro} + M_{Ta} + M_{Tb} + M_{Ra} + M_{Rb} + M_{Fp} = 0 \\
&\Leftrightarrow Rn \cdot (l_{CO}) + Ro \cdot 0 - Ta \cdot r + Tb \cdot r - Ra \cdot l_{AO} - Rb \cdot l_{BO} + Fp \cdot 0 = 0 \text{ – I case} \\
&\Leftrightarrow Rn \cdot (l_{CA} + l_{AO}) - (Ta - Tb) \cdot r - (Ra + Rb) \cdot l_{AO} = 0 \text{ – II case} \\
&\Leftrightarrow Rn \cdot (l_{CA} + l_{AO}) - f \cdot (Ra - Rb) \cdot r - (Ra + Rb) \cdot l_{AO} = 0 \text{ – III case}
\end{aligned}
\quad (1)
$$

Where:
Fp – the pressure force of the working fluid on the end surface of the piston;
R– the reaction force of the swash plate;
Ro – the axial component of the reaction force of the swash plate;
Rn – the transverse component of the reaction force of the swash plate;
Ra – the reaction force of the outer part of the guide sleeve;
Rb – the reaction force of the inner component of the guide sleeve;
Ta – the friction force of the outer part of the guide sleeve;
Tb – the reaction force of the inner part of the guide sleeve;
γ – the angle of swash plate;
C – the center of the hinge;
d - the diameter of the piston;
r – the radius of the piston;
O – the geometric center of the guide sleeve;
l_{CA} – the amount of piston departure;
l_{AB} – the length of the guide sleeve;

l_{AO} – the axial distance from the edge to the center of the sleeve;
f – the coefficient of friction.

The strength of Rn is always proportional to the magnitude of γ. Consequently, by increasing the values of γ, we also increase the force Rn, which, while maintaining the friction force constant, will lead to a decrease in the torque of the hydraulic machine, and, as a result, to a decrease in the value of the minimum stable revolutions of the hydraulic drive shaft.

2. Investigation of the limiting angle swash plate of an axial piston hydraulic machine with a modified piston mechanism

The limit value for the friction force in the piston-guide sleeve pair is the value of this force in the standard piston mechanism of an axial piston hydraulic machine with an angle of swash plate γ > 18°. Operating pressure of the liquid P = 30 MПa. Then the force of the liquid pressure acting on the piston Fp is equal to:

$$Fp = P \cdot \pi \cdot r^2 = 30 \cdot 10^6 \cdot \pi \cdot 0.007^2 = 4618\ (H) \qquad (2)$$

All parameters were calculated from formula 1, taking into account the two moving piston points on the left and right borders:

	Ra	Rb	Ro	Rn	Ta	Tb	T
l_{CAmax}	3564	1736	5625	1827	677	330	1007
l_{CAmin}	2013	366	5070	1647	382	70	452

Table 1. Initial parameters

In the modified piston mechanism, rc = 12 mm is the maximum possible without changing the diameter of the hydraulic drive shaft. Taking into account the maximum values of all parameters obtained at different values of γ of the modification APGND, it was found that at a value of γ = 25 °, the values of the friction force T in the modified piston mechanism will be equal to the values of the friction forces in the standard piston mechanism with γ = 18 °.

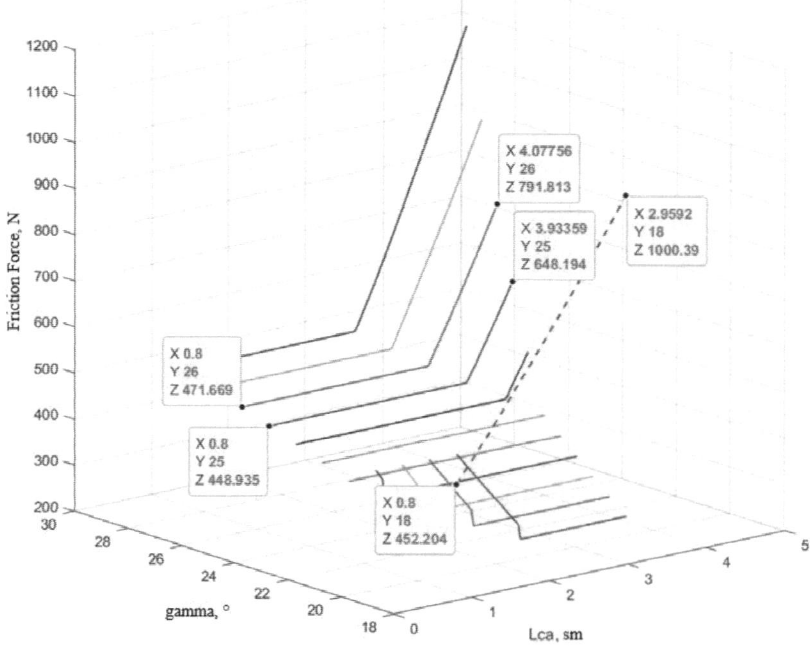

Fig. 4. The friction force in the piston-guide sleeve pair of the modified piston mechanism of the APGND at various γ at rc=12mm

If the value of γ is more than 25 degrees, the friction value will be greater than the standard mechanis,, which contradicts the criteria, and will not be used. Figure 4 shows a graph comparing the friction force between the piston and the guide sleeve when the piston is moving. The dotted line is a standard piston mechanism, the solid lines are values corresponding to different angles of inclination of modified piston mechanism with the same rc value (in the case of rc = 12mm on the graph). Due to the displacement of the rc piston hinge, a constant value of the friction force is observed for almost all the working stroke of the modified piston mechanism [6]. It can be seen from the graphs in Figure 4 that at an angle of inclination of 25°, the friction force generated by the modified piston is always less than for a standard piston when considering the same piston position. Figure 5 shows the torque from one piston acting on the main shaft on the work cycle.

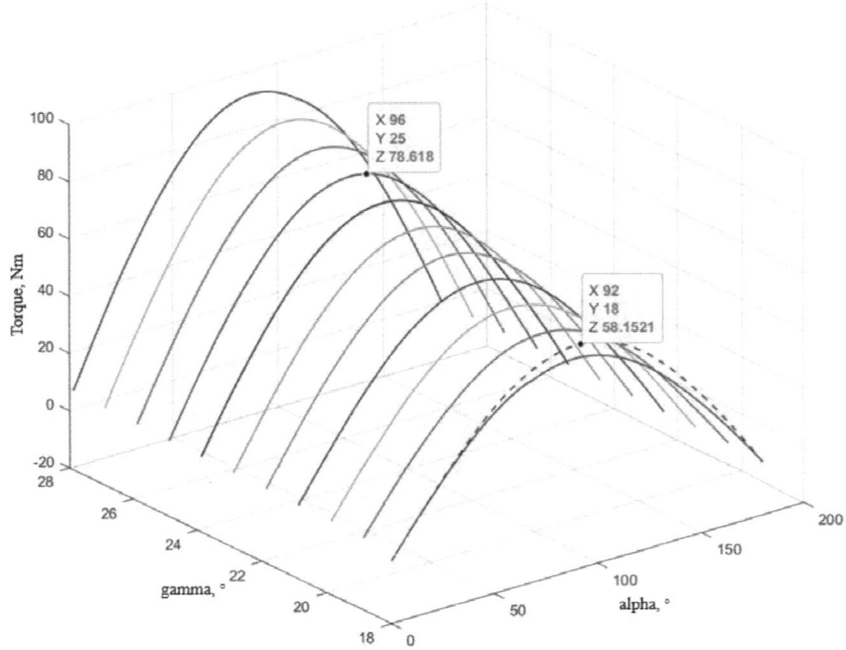

Fig. 5. The torque generated by a single piston mechanism of the APGND

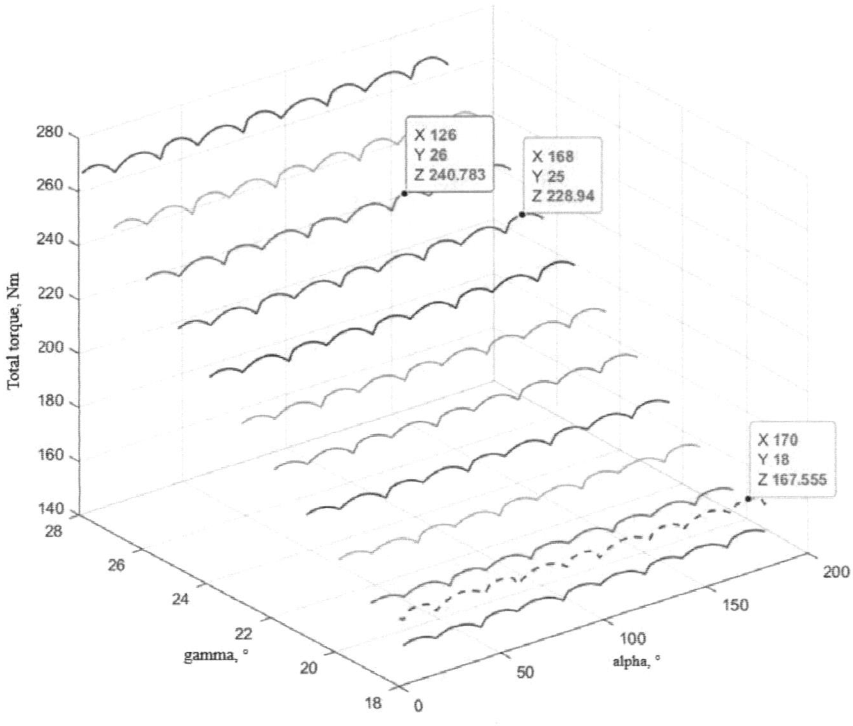

Fig. 6. The total torque generated by all (9) piston mechanisms of the APGND

Analyzing the obtained graphs, it can be concluded that when γ changes to 25°, the maximum torque compared to the initial case (γ = 18°) increased from 58 Nm to 78 Nm (35% higher). Figure 6 shows the total torque on the drive shaft of the APGND generated by all (9) piston mechanisms. The total torque also increased by 35% (from 167 Nm to 228 Nm).

3. Conclusion

From the results of the research, it can be concluded that with an increase in the γ to 25 °, the total torque on the drive shaft of an APGND with modified piston mechanisms increased by 35%, which in turn, with constant friction forces compared with piston mechanisms of standard design, will reduce the value of the minimum stable revolutions of the hydraulic drive shaft by approximately 30%. Due to the decrease in the value of the minimum stable revolutions of the hydraulic drive shaft, the control range of the APG with modified piston mechanisms and the γ = 25 ° will increase by 43% compared with the APGND with standard piston mechanisms and the γ = 18 °.

In the future, it is planned to develop and investigate the design of the entire undercarriage of the APGND with modified piston mechanisms and the γ = 25 °. Also, to develop a set of design documentation for the manufacture of a prototype of an APGND with a modified piston mechanism.

4. References

[1] V. A. Kamerlocher, V. V. Krasnokutsky, M. A. Rusanov, M. G. Shtyka (2018) / Hydraulic volume transmissions // Ministry of Education and Science of the Russian Federation, South Ural State University, Branch in Miass, Department of Automotive Engineering. – Chelyabinsk : SUSU Publishing Center. – 39 p.

[2] A.N. Dunaev (1985). Axial piston hydraulic machines with a swash plate. Manual. M.: MEI.

[3] Kuzmin, A[nton]; Popov, V[alery] & Stazhkov, S[ergey] (2016). Advanced Axial Piston Swash Plate Pump Parameters Recommendations, Proceedings of the 27th DAAAM International Symposium, pp.0556-0561, B. Katalinic (Ed.), Published by DAAAM International, ISBN 978-3-902734-08-2, ISSN 1726-9679, Vienna, Austria.

[4] T.M. Bashta (1974). Volumetric pumps and hydraulic motors, hydraulic systems. M.: Mechanical Engineering.

[5] A.O. Kuzmin (2020) / Improving the characteristics of adjustable axial piston swash palte hydraulic machines: specialty 05.02.02 "Mechanical engineering, drive systems and machine parts" : dissertation for the degree of Candidate of Technical Sciences - 135 p.

[6] S. Stazhkov, A. Kuzmin, V. Elchinskiy, N. Yakovenko (2020) / Tribological tests of the improved piston mechanism of the axial piston hydraulic machine // IOP Conference Series: Materials Science and Engineering : 15, Nizhny Tagil, 18–19 июня 2020 года. – Nizhny Tagil. – P. 012120. – DOI 10.1088/1757-899X/966/1/012120.

DOI: 10.2507/35th.daaam.proceedings.022

SOCIAL ENGINEERING AS A PROBLEM IN SMALL ORGANIZATIONS

Miroslav Tomšů, Veronika Rosíková & Matúš Madleňák

This Publication has to be referred as: Tomsu, M[iroslav]; Rosikova, V[eronika] & Madlenak, M[atus] (2024). Social Engineering as a Problem in Small Organizations, Proceedings of the 35th DAAAM International Symposium, pp.0171-0176, B. Katalinic (Ed.), Published by DAAAM International, ISBN 978-3-902734-44-0, ISSN 1726-9679, Vienna, Austria

DOI: 10.2507/35th.daaam.proceedings.022

Abstract

Digital information technology surrounds us at every step, and it is part of our everyday lives. This increased reliance on digital technology brings both convenience and risks. This paper explores the issue of social engineering in the context of its impact and consequences. In the context of small organizations, the risk of these attacks increases due to limited resources for cybersecurity and inadequate employee training. This paper discusses solutions to the issue of social engineering and suggests measures that small organizations can implement to reduce the risk of these attacks.

Keywords: social engineering; cyber security; cyber-crime; organization.

1. Introduction

The increasing interest in digital technology in organizations has heightened the need for security, especially in case of social engineering, which is considered one of the biggest threats. Many authors address the issue of social engineering from the perspective of its techniques and general prevention [1] [2], of social networks [3], of the cognitive aspect [4], of machine learning techniques [5,6], or of the COVID-19 pandemic when work in organizations shifted to home offices [7], but few authors discuss its implementation into organizational policies [8] [9]. This issue is addressed by a study that focuses solely on the banking sector [10] or healthcare organization [11]. This area should be more concerned with implementing [12] the security management system enshrined in European Union law [13]. The issue, however, is binding only for specific organizations to which the law applies [14]. However, more attention is paid to this issue, which leads to organizations needing to prepare for the threat of social engineering. This paper proposes specific security measures, methods, and tools that are not only against social engineering in small organizations.

2. Characteristics of social engineering

Social engineering is a form of attack that targets the weakest link in any security chain - a person. Moreover, people's actions are greatly influenced by their emotions, and this attack attempts to arouse the victims' positive or negative feelings. Most often, it is compassion, curiosity, shame or fear. When such emotions surge, the attackers rely on a judgment error that forces their victims to perform a task they would not normally do.

In short, social engineering can, therefore, be defined as a fraud based on psychological manipulation, the primary purpose of which is to lure the user into sensitive personal or payment information. It is much easier to fool the user and obtain data access than to circumvent the computer's security. In practice, the deceived person trusts the false information given to him via email, call or SMS. For this reason, he begins to act according to the intended instructions the attacker gave during this contact. These instructions most commonly include clicking a link, downloading an attachment or software to a computer, or transferring a specific amount of money [15].

2.1. The most common social engineering techniques

- **Phishing** - phishing is a form of social engineering. The attacker poses as a trustworthy entity and requests sensitive information from the victim under a seemingly legitimate pretext.
- **Spear Phishing** - is a targeted form aimed at a specific individual, organization, or company. Typical phishing campaigns do not target victims individually – they are not sent to hundreds or thousands of recipients. The attackers' tactic is to exploit acquired internal information or publicly available information, such as the names of executives, and then manipulate email addresses to resemble official ones.
- **Vishing** - a method similar to phishing emails, but instead of emails, it involves fraudulent phone calls. A cybercriminal disguises themselves, for example, as a bank or insurance company representative. Attackers benefit from making calls at night because their victims are less focused.
- **Smishing** - an attempt to obtain trustworthy information through SMS messages. Smishing's goal is often to redirect users to a website that collects sensitive data. Some messages may prompt users to send sensitive information even in direct replies to SMS.
- **Scareware** - malicious code that uses various techniques to induce anxiety and fear, indirectly forcing users to install additional malicious code on their devices. In the past, we have encountered fake antivirus products that warn about critical malware on the device. These programs are usually harmful themselves – they often display ads or collect data about the device or user.
- **Impersonation (Spoofing)** - the impersonation technique is the same in the physical world. Cybercriminals contact company employees, posing as, for example, the CEO, and attempt to manipulate the victim into taking specific actions, such as approving fraudulent transactions. [16]

2.2. Features of social engineering

Social engineering relies on weaknesses in the human psyche rather than gaps in technological solutions. It exploits people's trust, fear or inattention. An overly urgent message that tries to force the recipient to act without thinking or a non-standard request for sensitive data should raise suspicions of an attack. Reputable companies never ask for passwords or personal information via email or phone. The most common signs of social engineering include:

- **Time pressure**: All social engineering attacks are more successful if the victim believes it is necessary to make a quick decision.
- **Fear of missing out**: The fear of missing out phenomenon is behind the success of social networking. However, it also becomes an effective manipulation tactic if we believe the victim's inaction (refusal to disclose personal information) will lead to personal loss.
- **The fallacy**: Not all social engineering techniques are purely psychological. An attacker can use vulnerabilities in a website to redirect the victim to an unsafe site in the hope that they will enter personal or financial information.
- **Fake reviews**: This does not apply to companies that pay for store app reviews to make their product more appealing. Fake reviews can also make a website or service more credible and encourage the victim to enter personal information.
- **Fake identity**: This is an essential element of all phishing scams. Attackers present themselves as a trustworthy organization or individual to make the victim feel safe and provide sensitive information.
- **Incomplete information**: To be more convincing, hackers often use public or readily available information to encourage the victim to divulge even more. For example, scammers with an address, phone number, and the last four digits of a credit card can pose as an online merchant. [17]

3. Consequences of social engineering

The main risk is that the human factor always plays a significant role in social engineering techniques. While organizations may invest much money in security, it will only ever be as strong as its weakest link – the employees. Indeed, their mistakes are much more difficult to predict and prevent. For organisations, social engineering poses serious risks. Attackers can cause financial losses by convincing employees to transfer money into fake accounts, or they can obtain sensitive business information and data. It can cause not only the aforementioned financial losses but also a loss of credibility with clients and partners. Data breaches can result in serious legal consequences and penalties. If they do occur, other serious risks follow depending on the objectives of the specific attacker. The most common consequences of social engineering:

- Unauthorised access to the internal infrastructure of the company,
- installation of malware on a company computer (how to identify and fix a compromised computer),
- unauthorised entry into the workplace by an attacker after being let in by a deceived employee,
- extortion of sensitive access, personal or payment information,
- gathering sensitive information to launch the subsequent phases of the attack,
- direct transfer of funds by the defrauded person.

From this, we can easily deduce that the consequences for individuals and organisations can be devastating. Examples include loss of control of corporate systems, infected computers, empty bank accounts, encryption of critical corporate data or its complete loss. [18]

3.1. Features of social engineering

Social engineering is an integral part of cybercrime, and according to the IC3 and Verizon's 2023 report, social engineering is a growing problem. Verizon's 2023 report indicates that 74 % of data leaks were due to human error, and up to 50 % of social engineering attacks are based on pretexting. FBI, in turn submits that for the year 2023, there were:

- 12,5 billion USD stolen in U.S. cyberspace
- 521 652 victims were attacked in the U.S. (an additional 315 880 victims were uncovered across the 20 other countries examined).
- The most significant financial losses of up to 3,4 billion USD occurred among seniors over 60 years. To put this in perspective, the under-30 years population only lost about 401 million USD, nearly seven times the amount lost by the senior population. [12]

The 2022 State of the Nation's Cybersecurity Report from the National Cyber Security Council shows that social engineering is a pervasive and growing threat. The most common types of cyberattacks in 2022 included phishing, spear phishing, vishing, and phishing emails, all different social engineering techniques. The report also shows that attackers most often targeted the public, healthcare and private sectors. The National Cyber Security Index, which measures the preparedness of different countries to prevent cyber threats and their ability to manage cyber incidents that have already occurred, is very positive for the Czech Republic, as we are currently the safest of all countries, measured a score of 98,33. We must still be vigilant and prepared for the real threat of social engineering. [19]

4. Social engineering as a problem in the organization

In today's digital age, where both the private and public spheres increasingly rely on information systems and technology, cybersecurity is becoming a vital issue for organisations of all sizes and in all lines of business. In 2016, the European Union created a comprehensive document, NIS (Network Information Security), which aims to ensure a consistent and high level of security at the network and information systems level in EU member state organisations. Until now, national cybersecurity measures have varied considerably not only in form but also in practice. Only a few countries have had a legislative framework in this area, and not all have established top-level security groups such as CSIRTs/CERTs. The purpose of the Directive is to set out the basic requirements that all EU Member States must meet. [20] Many articles [21] [22] can be read about the term information security management system, but all describe organisations directly affected by the Directive and the law. Thus, they only apply to address some smaller organisations that are not required to implement cybersecurity laws and information security management systems.

5. Proposal of new measures

This paper proposes a methodology for managing a small organization in the private sector that will specify the basic principles of information security focused on social engineering. To do this, it is necessary to determine the importance organizations place on information security and whether they are adequately secured against attacks.

5.1. Research

Ten organisations with up to 50 employees, mainly in the private sector, were contacted. Semi-structured interviews were conducted with employees and employers. The interview lasted between an hour and an hour and a half, depending on the fluency of the interview and asking follow-up questions. The questions covered the following topics:

For employers:

- Is there cyber security training? If yes, how?
- Are employees allowed to use a private phone during work hours?
- Does each employee have unique PC login credentials? If so, are there specific password requirements? If yes, what are they?

- Do employees use work email for personal purposes as well?
- Are there company guidelines that specify rules of conduct for employees regarding social engineering threats and prevention of attacks?
- What types of documents do you attach to employee contracts?
- Do employees take company PCs home? If so, are there set rules for use outside of work? If so, what are they?

For employees:
- Have you been trained in social engineering (cybersecurity, information security)? If yes, how?
- Do you use a mobile phone during working hours?
- Do you have a unique password on your company PC, and have you never revealed the password to another person?
- Have you established company guidelines, for example, by collective bargaining agreement, which define your rules of conduct in information security?
- Do you take your company PC home? If so, do you have particular security rules, for example, in case of loss?
- Do you visit private email, social networking sites, etc., during working hours on a private phone?

The questions were asked individually or with follow-up questions as the conversation developed. The use of dynamic questions was minimal. Questions related to the main topic - social engineering - were asked first. The interviews concluded with a comparison of the state of the companies with the ideal state. The perfect state is defined as a state where the company has done everything possible to protect itself against social engineering attacks and sociotechnical manipulation techniques.

5.2. Evaluation

Organisations are very similar in terms of their approach to the issue. The data collected was very consistent; therefore, it was not difficult to isolate and describe the general characteristics of the group. Small companies have measures related to phishing emails; they take these as the biggest threat. Therefore, they mainly train employees to use PCs and apply basic restrictions on selected websites. Each employee has PC login credentials, but this is now the minimum-security standard. Some employees also own a company phone, but these devices do not have passwords or anti-virus systems. In general, small businesses do not have company guidelines that specify rules of conduct for employees regarding social engineering threats. Nor are such regulations used to regulate attack prevention. If an attack does occur, measures are created after the fact, i.e., when the attack happens. If the problem doesn't occur, it doesn't solve it, and they do not focus on prevention. They perceive the threats associated with social engineering as not very real. They have an advantage regarding the number of employees; their numbers are low, reducing the chances of personally carrying out a possible attack.

6. Information security methodology

This chapter aims to establish a basic guideline to consider during employee training. Another goal should be to create awareness among employees that they are the critical element of the organization's defence mechanism.

6.1. Information security guidelines for employees

The aim is to define precise rules and guidelines for employees to follow. The definitions must be accurate, and the organisation's people must understand their importance and respect them so that they become their business. In particular, the guidelines should cover the following areas: computer use, information handling, access data and passwords, internet and data storage, use of email clients, social networking, rules on telephone and mobile phone use, and document disposal.

6.2. Training and education

Every employee should receive basic security training. Initial training for new employees should be mandatory and access to computer technology should only be allowed after completion of this course. The training should be short so that it captures employees' attention, and they remember the important points. We should refrain from training sessions lasting several hours to a full day, where people are demotivated, and their attention span is zero. Emphasis should be on situations and cases that can harm the company, understanding their principle and providing specific instructions on corporate information security. Training programme should include the following points:
- the use of socio-techniques
- recognition of sociotechnical issues,
- procedures in case of suspicious requests for information,
- highlighting the need to authenticate persons requesting information,
- trusting others without adequate verification,
- handling sensitive information.

6.3. Audit

The employer should conduct regular checks to ensure employees adhere to the contractual security rules. Audits should be performed regularly and processed by an independent external provider.

6.4. Penetration testing

The penetration test mimics a hacker attack on network infrastructure or servers from an external or internal network. Penetration can also be conducted using the social engineering method, where the hacker uses non-technical tools such as exploiting people's trust and weaknesses. Attacks can help to obtain sensitive information or cause unavailability of services and unauthorised access to data.

7. Discussion

This paper has examined previous efforts to address the problem of social engineering, but none of them addresses the environment of small organisations in the private sector. While previous works have examined social engineering in terms of its techniques and prevention in general, social networks, cognitive and machine learning techniques, or the implementation of an information security management system in directly covered organisations, they have not addressed the problem of the impact of social engineering on small organisations. The findings of the brief research indicate that employees in organisations have only minimal security measures in place, perceive threats associated with social engineering as not very real, and have almost no corporate cybersecurity policy and rules of conduct regarding social engineering threats. Such rules are not even used to regulate the prevention of these attacks.

The main objective of this section is to critically evaluate the results of our work in a broader context. The employer must always provide a suitable environment for the employee by ensuring that all programs are up-to-date and that all devices have secure antivirus software. Many times, companies are attacked simply because of outdated software versions. These obsolete systems have vulnerabilities that even not-so-experienced hackers can find and get into the employee's PC. The advantage of having an in-house security solution provider is funding. For example, the task is assigned to an employee or performed by a department with expertise. Efficiency thus varies from case to case. Company connections often influence employees, and so the results can be manipulated. An external security investigator provides an independent view of the issue. The resolvers have a practical methodology and more experience with these types of audits. Therefore, they have no ties to the environment and staff and are not biased in their assessment. The disadvantages are the cost of acquiring an external firm, the disclosure of sensitive information, and the need to select an external firm.

Employee training methods are workable but need to be sufficiently updated and to reflect the dynamic environment in which most organisations find themselves today. Employees are not trained against all hazards. This can be done, for example, by creating an awareness campaign in which more emails are sent out to employees, explaining how to recognise a possible attack and what to do if an attack is suspected.

The issue of cyber security, in general, can be complicated for senior management in organisations to grasp due to traditional role divisions, with management focusing only on strategic management, difficult-to-explain (business) benefits, lack of understanding of cyber threats, vulnerabilities and measures, and failure to recognise the long-term benefits in protecting data and maintaining the company's reputation. The contribution of the proposed solution in this article is mainly the complexity and the level of measures leading to a secure cyber environment in organisations that do not address this issue. The benefit is not only to increase employee (user) awareness of potential threats in cyberspace, but it is also primarily focused on social engineering, but also, to protect the entire organisation and mitigate the impact of threats on organisations.

Further research should focus on improving efficiency and reducing costs associated with security measures. It may also include artificial intelligence and machine learning tools to facilitate specific processes. Long-term studies should also be conducted to evaluate the lasting impact of training on user security behaviour. Future research should include developing incident response plans that can be easily adapted to different types of organisations and threats.

8. Conclusion

This article looked at the impact of social engineering on small organizations. In comparison with the available literature, it was found that the issue still needs to be addressed at this level. The aim was to propose specific security measures, methods, and tools to counter social engineering in the particular environment of small organizations and enhance their cyber security. The research revealed that most of the employees of organizations are entirely under-protected. In the second part, specific security measures were proposed for managing small organizations, focusing on enhancing cybersecurity with a focus on social engineering. The discussion summarized the benefits of these solutions and the possible limitations they may pose and provide possible suggestions for future research.

9. Acknowledgments

This article was supported by grant project IGA/FAI/2023/002, „Protection of the population, soft targets and the information environment" solved in 2023-2024.

10. References

[1] Birthriya, S. K., Ahlawat, P. & Jain, A. K. (2024) 'A Comprehensive Survey of Social Engineering Attacks: Taxonomy of Attacks, Prevention, and Mitigation Strategies', Journal of Applied Security Research, pp. 1–49. doi: https://doi.org/10.1080/19361610.2024.2372986.

[2] Samad, A. (2024). Social Engineering in Cybersecurity: Prevention and Mitigation Strategies.

[3] Tomsu, M. Rosikova, V. Vojtek, P. & Hrabakova, M. (2023). Spear-phishing and Its Specifics Within Social Engineering, Proceedings of the 34th DAAAM; International Symposium, pp. 0379-0387, B. Katalinic (Ed.), Published by DAAAM International, ISBN 978-3-902734-41-9, ISSN 1726-9679, Vienna, Austria. doi: 10.2507/34th.daaam.proceedings.052

[4] Burda, P., Luca Allodi & Zannone, N. (2023). Cognition in Social Engineering Empirical Research: a Systematic Literature Review. ACM Transactions on Computer-Human Interaction. doi: https://doi.org/10.1145/3635149.

[5] Ramakrishnan, S., Malini Mittal Bishnoi, Shanmugan Joghee, S Jijitha and Kumar, A. (2024). Social Engineering:Role of Teachers in Cohabitation of AI with Education. pp. 1-6. doi: https://doi.org/10.1109/iccr61006.2024.10532897.

[6] D. V. Grbic & I. Dujlovic (2023). Social engineering with ChatGPT, 2023 22nd International Symposium INFOTEH-JAHORINA (INFOTEH), East Sarajevo, Bosnia and Herzegovina. pp. 1-5, doi: https://doi.org/10.1109/INFOTEH57020.2023.10094141.

[7] Sharma, V., Renu Saharan, Wilson, K., Sharma, D., Suresh Beniwal & Chander Parkash Dora (2022). Privacy and Security Challenges in the Era of the COVID-19 Pandemic. Advances in medical technologies and clinical practice book series, pp.287–308. doi: https://doi.org/10.4018/978-1-6684-5741-2.ch017.

[8] Salama, R. & Fadi Al-Turjman (2023). Cyber-Security Countermeasures and Vulnerabilities to Prevent Social-Engineering Attacks. CRC Press eBooks, pp.133–144. doi: https://doi.org/10.1201/9781003322887-7.

[9] Steinmetz, K.F. and Holt, T.J. (2022). Falling for Social Engineering: A Qualitative Analysis of Social Engineering Policy Recommendations. Social Science Computer Review. doi: https://doi.org/10.1177/08944393221117501.

[10] Momoh, I., Adelaja, G., & Ejiwumi, G. (2023). Analysis of the Human Factor in Cybersecurity: Identifying and Preventing Social Engineering Attacks in Financial Institution.

[11] Cartwright, A. J. (2023). The elephant in the room: cybersecurity in healthcare. Journal of Clinical Monitoring and Computing. ISSN 1123-1132.

[12] Farid, G., Warraich, N.F. & Iftikhar, S. (2023). Digital information security management policy in academic libraries: A systematic review (2010–2022). Journal of Information Science. doi: https://doi.org/10.1177/01655515231160026.

[13] Vandezande, N. (2024). Cybersecurity in the EU: How the NIS2-directive stacks up against its predecessor. Computer Law & Security Review. doi: https://doi.org/10.1016/j.clsr.2023.105890.

[14] Božić, V. (2023) Enhancing Hospital Cybersecurity: Implementing the NIS2 Directive for Resilience and Compliance. Journal of Health Care Management. pp. 220-235.

[15] Chetioui, K., Bah, B., Alami, A.O. & Bahnasse, A. (2022). Overview of Social Engineering Attacks on Social Networks. Procedia Computer Science, [online] 198(1877-0509), pp.656–661. doi: https://doi.org/10.1016/j.procs.2021.12.302.

[16] Aldawood, H., & Skinner, G. (2020). An advanced taxonomy for social engineering attacks. International Journal of Computer Applications, pp. 1-11. ISSN: 0975-8887

[17] Syafitri, W., Shukur, Z., Mokhtar, U.A., Sulaiman, R. & Ibrahim, M.A. (2022). Social Engineering Attacks Prevention: A Systematic Literature Review. IEEE Access, pp. 39325-39343. doi: https://doi.org/10.1109/ACCESS.2022.3162594.

[18] Wang, Z., Zhu, H. and Sun, L. (2021). Social Engineering in Cybersecurity: Effect Mechanisms, Human Vulnerabilities and Attack Methods. IEEE Access, pp.11895–11910. doi: https://doi.org/10.1109/access.2021.3051633.

[19] Report on the state of cyber security of the Czech Republic for 2022, Available: https://nukib.gov.cz/download/publikace/zpravy_o_stavu/Zprava_o_stavu_kyberneticke_bezpecnosti_CR_za_rok_2022.pdf

[20] Drivas, G., Chatzopoulou, A., Maglaras, L., Lambrinoudakis, C., Cook, A. & Janicke, H. (2020). A NIS Directive Compliant Cybersecurity Maturity Assessment Framework. IEEE Xplore. doi: https://doi.org/10.1109/COMPSAC48688.2020.00-20.

[21] Onyshchenko, S., Yanko, A., Hlushko, A., & Sivitska, S. (2020). Increasing information protection in the information security management system of the enterprise. In International Conference Building Innovations. pp. 725-738. Cham: Springer International Publishing.

[22] Alzahrani, L. & Seth, K.P. (2021). The Impact of Organizational Practices on the Information Security Management Performance. Information, 12(10), p. 398. doi: https://doi.org/10.3390/info12100398.

DOI: 10.2507/35th.daaam.proceedings.023

INFORMATION OVERLOAD AS A PROBLEM OF IT SPECIALISTS

Miroslav Tomšů

This Publication has to be referred as: Tomsu, M[iroslav] (2024). Information Overload as a Problem for IT Specialist, Proceedings of the 35th DAAAM International Symposium, pp.0177-0182, B. Katalinic (Ed.), Published by DAAAM International, ISBN 978-3-902734-44-0, ISSN 1726-9679, Vienna, Austria
DOI: 10.2507/35th.daaam.proceedings.023

Abstract

In today's information age, we are constantly witnessing significant technological advances that profoundly impact our daily lives. However, the constant advances and ever-increasing technologies and practices can result in too many demands and requirements on people. This paper aims to analyse how the problem of information overload manifests itself in IT specialists, how they notice it and how to prevent and, above all, eliminate information overload. The research is in the form of a questionnaire survey among IT specialists and a subsequent assessment of how severe and widespread the problem of information overload really is. Conclusion underlines that the chosen topic should be considered and that with the increasing advancement of technology in a fast-moving field like IT, these problems should be prevented.

Keywords: information overload; information technology; IT specialist; information environment; organization.

1. Introduction

Today, as we witness the rapid development of information and communication technologies, the amount of information is growing exponentially. This phenomenon, known as information overload, has become one of the significant problems for both individuals and organisations. Information overload is when an individual or group is overwhelmed by too much information, which negatively affects their ability to make decisions and process data effectively. This phenomenon was first described in the 1970s by Alvin Toffler in his book Future Shock [1]. Since then, this problem has dramatically intensified in the digital age.

Today, information overload is particularly evident in the work environment, where individuals face an enormous onslaught of emails, messages, work tasks and a constant influx of new data. According to surveys, up to 60% of workers feel overloaded by the information they must process daily [2]. This leads to decreased productivity, increased stress and poorer decision-making skills [3], [4]. As reported by [5], information overload can negatively affect individuals and entire teams, resulting in increased decision-making errors and reduced quality of work. The causes of information overload can be divided into several categories. The main factors include the increase in the volume of information available and the fragmentation and ambiguity of information, which is often dispersed between different sources and formats [6]. Another important aspect is the constant availability of digital devices, which brings the benefits of fast communication and increases the expectation of immediate responses and response times [7]. In response to this problem, various strategies and tools are being developed to eliminate or reduce information overload. These include automated information filtering systems, personalised search algorithms [8], efficient time management, and task prioritisation tools.

It is also essential to improve personal skills, such as managing information effectively [9], recognising essential resources, and minimising distractions [10]. However, this research has yet to focus on a specific industry, such as IT specialists. This paper aims to provide a comprehensive view of the issue of information overload among IT specialists and the working conditions that may also lead to the problem. It also aims to suggest improvements to working conditions, how to sort out unnecessary or irrelevant information, the mental health of IT specialists, and how physical health can affect information overload.

2. Information overload

The information overload on humanity continues to escalate. However, the term *information overload* was introduced to the broader public in 1970 by Alvin Toffler [11], who was an author, journalist and futurologist. This term was added to mean that when the information management required exceeded the capacity of the decision-making system, humanity creates much more information with time than humans can absorb, process, and study. Toffler was also instrumental in expanding the topic of digital revolution [12]. The sociologist Georg Simmel first became interested in the problem of information overload. Simmel was the first to notice the negative effect on the population, pointing out the information overload in modern urban lifestyles and observing the fatigue, aversion to knowledge and disinterest in responding to new situations in the population. Over the years, various psychologists have studied this topic, and it has been gradually observed that people have a limited capacity to absorb new information. During the evolution of humanity and the gradual acceleration of time due to information technology and technical developments, this topic has unfortunately become increasingly relevant, and more demands are being placed on people.

2.1. Explanation of the concept of information overload

Information overload is a massive problem in today's world, and with modernisation, this situation is occurring more often. Information overload can be defined as a condition in which a person is overwhelmed with too much information and data for a given decision but is also constantly inundated with information from many different sources. "Thousands of media and individuals spew out information today on social media, and we have reached a post-factual age where we are so overwhelmed with information that we may be on the verge of an information black hole." [13]

When a person experiences information overload, it can bring consequences in the form of decreased effectiveness, which also impacts poor decision-making in both personal life and work. When the problem of overload goes to the point where a person is unable to make decisions or shows no interest in the possibility of making decisions, so-called analysis paralysis can occur. Long-term persistent analysis paralysis is then called burnout syndrome.

Nowadays, there is so much information around us that there is no need to seek it out. On the contrary, information seeks us out [11]. At first glance, the abundance of information may seem optimistic. We have plenty of it around us, and we can choose; the more we know, the better. However, the human mind cannot be exposed to a constant intake of information. This phenomenon is called *information overload.* Simply put, it means that a person can no longer actively process this information, which can affect people's perception of the world around us. Beyond a certain point, when we are constantly exposed to new streams of information, the human mind can no longer sort through this information and assign any value, meaning, or significance to it.

Attentional fragmentation is another concept that goes hand in hand with information overload [14]. Nowadays, it is most often manifested by the fact that an individual who is focused on an activity such as work or study is easily tempted by mindlessly picking up the phone and, most often, browsing social networks. The gradual information overload in our brain causes more severe damage to human concentration, and the fragmentation of attention is still developing and will not leave a person calm.

2.2. Information overload in IT

Information technology is known for its speed of innovation and constant evolution. New technologies, programming languages, software tools and system updates appear almost daily. IT specialists must keep up with these changes and stay informed about the latest trends and practices. However, this can present challenges and lead to information overload.

One of the reasons for this is the vast amount of information available. Many online resources such as websites, blogs, articles, discussion forums, and social media offer information on new technologies. Keeping track of all these sources and filtering relevant information is challenging. IT specialists often must not only research new developments but also test and adapt them to be able to use new technologies in practice.

Another factor contributing to information overload in IT is the pressure for continuous learning. Stagnation in IT can mean obsolescence and loss of competitiveness [17]. Workers often must learn new languages, frameworks, and tools to meet the demands of the modern IT environment. This includes attending training courses and conferences and participating in regular study. In this effort to keep up-to-date, workers can feel pressure and stress that can lead to overwhelm. The rapid pace of development of IT projects must be noticed. Short deadlines and high client expectations often mean an increased workload and the need to adapt quickly to new requirements. This fast pace can lead to insufficient time to thoroughly process information and make decisions based on insufficient data. This increases the risk of errors and can negatively impact project outcomes.

There are several strategies to address information overload in IT. Staff should develop an effective system for filtering information and prioritizing tasks. This includes selecting the most important sources of information and using tools to automate and organize workflow. It is also essential to set clear goals and plan time to avoid distractions and inefficient multitasking [17]. In addition, IT specialists should seek support and collaboration with colleagues. Discussing and sharing experiences with the team can help filter and process information more efficiently. Setting aside regular time for learning and self-improvement is also essential [17]. Information overload is a real problem for IT specialists, resulting from the industry's speed of change and volume of information. However, with appropriate strategies for filtering, organizing and prioritizing information, collaborating with the team, and creating a balanced training plan, this problem can be managed, and greater productivity and efficiency at work can be achieved.

3. Methodology

3.1. Target group

This questionnaire was designed with IT specialists and individuals who work in the IT industry and deal with information related to the world of information technology in mind daily. The aim was to obtain responses from people with different specialisations and knowledge in the IT field. Professionals were approached, including programmers working with different programming languages, database specialists, software developers, theoretical computer scientists, analysts and hardware-oriented IT specialists. This ensured that the questions used encompassed different aspects of information technology and reflected the diverse experience and knowledge of people working in this field. The research was conducted in 30 organisations, with 1-2 IT specialists in each organisation being interviewed.

3.2. Questions used

This set of questions created a structured and comprehensive overview of the problem of information overload among IT specialists and was used for questionnaires, interviews or case studies. The questions were divided into several categories.

Extent and nature of information overload

- How often do you feel overwhelmed by the amount of information you work with daily?
- What types of information do you find most overwhelming (e-mails, documentation, technical specifications, messages from colleagues, etc.)?
- What volume of information do you typically process in a day?

Sources of information overload

- What are your work's main sources of information overload (internal communications, industry news, technical documentation, etc.)?
- How much influence do work tools (e.g. e-mail) have on your overload?

Impact on performance and productivity

- What impact does information overload have on your productivity and ability to concentrate?
- Do you feel that the quality of your work is decreasing due to the overload? If so, in what way?
- How does information overload affect your decision-making and problem-solving?

Managing information overload

- What strategies do you use to manage large amounts of information (time management, filtering, automation, delegation, etc.)?
- Do you use tools or techniques to streamline information handling (e.g. RSS feeds, task management applications)?
- How do you rate the effectiveness of these strategies in preventing or mitigating overload?

The role of the organisation and the working environment

- How does your organisation help you manage information overload (training, tools, policies to limit information flow)?
- Are there any practices or systems in your work environment that support information simplification?
- What role do colleagues and teamwork play in alleviating feelings of overload?

Psychological and physical impact

- What psychological or physical effects do you experience from information overload (e.g. stress, fatigue, burnout)?
- How often do you feel that information overload affects your personal life (spillover of work stress into leisure time)?
- What methods do you use to manage and prevent stress?

Future challenges and trends

- What changes do you expect about information overload in the next five years, given the evolution of technology and work practices?
- What innovations or tools could help IT specialists better manage information overload in the future?

3.3. Research results

A survey of information overload among IT specialists in 30 organisations found that most respondents (75 %) feel overloaded at least three times a week. The most overwhelming sources of information are emails and work tools. Overload significantly impacts their productivity, focus and decision-making processes, leading to more frequent errors and a slower work pace. Approximately 70 % of respondents confirmed that the amount of information decreases their work performance. IT specialists try to combat overload by using techniques such as blocking time to focus or automating routine tasks, but the success of these strategies is limited. In half of the cases, organisations offer training on managing overload, but many employees perceive that systematic procedures for managing information flow are lacking. The impact of information overload on mental health is also a significant issue. Approximately 35 % of respondents suffer from stress, with some reporting burnout syndrome. The research highlights the need for organisations to support their IT specialists better and to implement innovative tools to help reduce workload and improve the working environment.

4. Suggestions for improving conditions

If a company or organization has an IT sector in its structure, it should have a program to prevent information overload or even burnout syndrome in its employees. As mentioned in this article, the IT sector is an industry that is changing rapidly and constantly moving forward. New technologies are constantly emerging, resulting in constant employee perception and focus on these changes. The demands of other departments, supervisors, or customer requirements compound this.

4.1. Prioritisation of information

Using software tools to organise and manage tasks allows you to categorise tasks and resources according to priority and relevance. This allows employees to focus on the most important tasks and not be distracted by less relevant information.

The Scrum or Kanban methodology helps the IT team to define priorities and facilitate collaboration when distributing workloads.

4.2. Process automation

Automating repetitive tasks (e.g., by automating testing or scripting everyday operations) can save time for more important activities and reduce the cognitive load associated with manual processes. Automation eliminates the need to track every change manually, significantly reducing the amount of information that needs to be processed.

4.3. Information filtering and aggregation

Aggregation tools (for tracking news or monitoring system logs) allow you to filter and aggregate information, reducing the need to monitor multiple sources simultaneously. This allows IT specialists to get relevant data from one centralised source and minimise information overload from multiple channels.

4.4. Improving communication tools

Using centralised communication platforms such as Slack or Microsoft Teams, where conversations can be efficiently sorted by channel and topic, can reduce information chaos and distractions from the constant flow of messages and emails. Teams can take advantage of features to buffer non-urgent notifications and reduce interruptions.

4.5. Training and skills development

IT specialists should be trained in digital literacy and effective information retrieval. This includes the ability to identify trusted sources, distinguish relevant from irrelevant information, and the ability to delegate tasks or seek help within a team. Training to improve the ability to multitask and manage time properly can be helpful to minimise stress and improve concentration.

4.6. Deploy AI and analytics tools.

Artificial intelligence and machine learning can help predict what information or issues will take priority and automatically highlight relevant issues, making decision-making easier. For example, predictive analytics tools can monitor the performance of IT systems and highlight issues before they become critical.

4.7. Employer motivation

To provide valuable tools to relieve information overload and job frustration, employers (not just) can help IT specialists by focusing on their employees' intrinsic motivation and career growth. One critical approach is to invest in their professional development. In this way, organisations offer many opportunities, such as professional training, interactive workshops and access to the latest technologies. These activities allow IT specialists to expand their skills and knowledge and support their personal and professional growth. As a result, professionals gain greater confidence and competence to take on new tasks and responsibilities, which helps to reduce their information burden. One of the other essential aspects is the provision of greater responsibility.

4.8. Physical health

Maintaining good physical fitness and preventing information overload are critical factors in today's modern work environment to achieve optimal performance and maintain long-term interest in work. Strategies are needed to help keep fit and prevent burnout syndrome. One general recommendation is regular exercise, which is an integral part of a healthy lifestyle. The prolonged sitting and sedentary lifestyle typical of IT specialists can hurt physical health.

5. Discussion

Information overload is a significant issue for organisations in the digital era, especially in the rapidly changing IT industry. Several approaches can mitigate this phenomenon, each with its benefits and limitations. Process automation and information filtering are emerging as a critical strategy for reducing overload. Automating routine tasks using tools such as scripts or log management tools greatly relieves workers from repetitive tasks contributing to information chaos.

Automation also improves productivity and allows you to focus on higher value-added tasks. Research (19) shows that automation reduces information overload by eliminating redundant manual processes that lead to performance degradation. Introducing centralised communication tools like Microsoft Teams is vital in simplifying work processes. In IT environments requiring rapid communication, these tools can filter important messages and minimise information noise. However, excessive use of these tools without appropriate filtering can paradoxically lead to message overload, as confirmed by authors [19], who state that emails and chat platforms can be a source of stress if not managed properly. One of the biggest challenges of information overload is that the problem can be cognitive, and resource based.

The cognitive aspect focuses on the limited ability of individuals to process large amounts of information, which often overlaps with a lack of time or budget to process that information. For this reason, the right balance of information quantity and complexity must be struck so that these resources are used efficiently. Improving digital literacy and training on time management and multitasking can help workers better manage information overload. According to many authors [5], effective time management and identifying relevant information play a key role in how individuals manage the onslaught of data and information. Eliminating or mitigating information overload requires a multi-layered approach that includes not only technological solutions but also changes in the way of working and the development of individual skills.

Further research should consider how individual differences between IT specialists (e.g., experience, time management skills, tool usage preferences) and organisational cultures influence perceptions of information overload and its impacts. There is also a need to explore how this phenomenon can be minimised at the organisational level through education programmes, training or new approaches to project management. With the development of new tools and approaches to information management (e.g. intelligent data filtering and workflow automation), it is necessary to examine how these tools can reduce information overload. Research in this area should look at which technologies are most effective and how they can be implemented in the daily practice of IT specialists.

6. Conclusion

Information overload is a common problem affecting IT specialists in today's digital era. Data overload and a constant flow of information lead to a decline in performance, increased stress and reduced quality of decision-making processes.

This article addressed information overload in the specific environment of IT specialists. A short survey was conducted among IT specialists and found that most are affected by information overload. In the next section, the paper discussed recommendations for alleviating information overload. To effectively eliminate this phenomenon, various strategies need to be put in place, including both technological and organisational measures.

Process automation, data filtering and aggregation, and modern communication tools can help mitigate the impact of redundant information. Developing digital literacy and training in good time management also plays an important role, enabling staff to sort and process information more efficiently. Implementing these strategies leads to better use of resources, reduced cognitive load and improved work performance. The combination of these approaches can significantly improve the ability of organisations and individuals to meet the challenges of information overload and ensure optimal working conditions in a rapidly changing IT environment.

7. Acknowledgments

This article was supported by grant project IGA/FAI/2023/002, „Protection of the population, soft targets and the information environment" solved in 2023-2024.

8. References

[1] Toffler, A. (1970). Future shock. New York: Random House.

[2] Spira, B., & Goldes, M. (2007). "Information Overload: We have met the enemy and he is us", Basex Inc., Available at: http://freya.basex.com/web/webdownloads.nsf/e67dc0f5617d6e9c85256a99005ea0e7/ff10b594932040852570b9008022ce/$FILE/BasexInformationOverloadWhitePaper.pdf

[3] Eppler, M.J. & Mengis, J. (2004). The Concept of Information Overload: A Review of Literature from Organization Science, Accounting, Marketing, MIS, and Related Disciplines. The Information Society, 20 (5), pp. 325–344. DOI: https://doi.org/10.1080/01972240490507974.

[4] Bawden, D. & Robinson, L. (2009). The dark side of information: overload, anxiety and other paradoxes and pathologies. Journal of Information Science, 35 (2), pp. 180–191. DOI: https://doi.org/10.1177/0165551508095781.

[5] Jackson, T.W. & Farzaneh, P. (2012). Theory-based model of factors affecting information overload. International Journal of Information Management, 32 (6), pp. 523–532. DOI: https://doi.org/10.1016/j.ijinfomgt.2012.04.006.

[6] Hiltz, S.R. & Turoff, M. (1985). Structuring computer-mediated communication systems to avoid information overload. Communications of the ACM, 28 (7), pp. 680–689. DOI: https://doi.org/10.1145/3894.3895.

[7] Barley, S.R., Meyerson, D.E. & Grodal, S. (2011). E-mail as a Source and Symbol of Stress. Organization Science, 22 (4), pp. 887–906. DOI: https://doi.org/10.1287/orsc.1100.0573.

[8] Moe, S., Drüeke, T., Cunningham, J., Goodman, W., Martin, K., Olgaard, K., Ott, S., Sprague, S., Lameire, N. & Eknoyan, G. (2006). Definition, evaluation, and classification of renal osteodystrophy: A position statement from Kidney Disease: Improving Global Outcomes (KDIGO). Kidney International, 69 (11), pp. 1945–1953. DOI: https://doi.org/10.1038/sj.ki.5000414.

[9] Jackson, T.W. & Farzaneh, P. (2012). Theory-based model of factors affecting information overload. International Journal of Information Management, 32 (6), pp. 523–532. DOI: https://doi.org/10.1016/j.ijinfomgt.2012.04.006.

[10] van Knippenberg, D., Dahlander, L., Haas, M.R. & George, G. (2015). Information, Attention, and Decision Making. Academy of Management Journal, 58 (3), pp. 649–657. DOI: https://doi.org/10.5465/amj.2015.4003.

[11] Petr, M. (2022). Kyberkultura, hackeři a digitální revoluce. Grada Publishing a.s.

[12] Henry, D. (2016). Alvin Toffler, author of best-selling 'Future Shock' and 'The Third Wave,' dies at 87. [online] Washington Post. Available at: https://www.washingtonpost.com/business/alvin-toffler-author-of-best-selling-future-shock-and-the-third-wave-dies-at-87/2016/06/29/0d63748c-3e09-11e6-80bc-d06711fd2125_story.html.

[13] Vilém Sklenák (2001). Data, informace, znalosti a Internet. Nakladatelství C H Beck.

[14] Gazzaley, A. & Rosen, L.D. (2017). Distracted mind : Ancient brains in a high-tech world. MIT Press.

[15] Shenk, D. (1998). Data smog : surviving the information glut. San Francisco: Harperedge.

[16] Schick, A.G., Gordon, L.A. & Haka, S. (1990). Information overload: A temporal approach. Accounting, Organizations and Society, 15 (3), pp. 199–220. DOI: https://doi.org/10.1016/0361-3682(90)90005-f.

[17] Pink, D.H. (2018). Drive: The Surprising Truth about What Motivates Us. S.L.: Canongate Books Ltd.. ISBN 978-80-7554-104-8.

[18] Mandel, R. (2019). Global data shock : strategic ambiguity, deception, and surprise in an age of information overload. Stanford, California Stanford University Press.

[19] Tomsu, M., Valouch, J. & Vojtesek, J. (2023). Information Ecology, Information Hygiene, and Assessment of Information Vulnerabilities in Organization, Proceedings of the 34th DAAAM International Symposium, pp. 0388-0394, B. Katalinic (Ed.), Published by DAAAM International, ISBN 978-3-902734-41-9, ISSN 1726-9679, Vienna, Austria. DOI: 10.2507/34th.daaam.proceedings.053

DOI: 10.2507/35th.daaam.proceedings.024

DATA MASKING AND TOKENISATION

Maristela Bosnjak, Jasmin Redzepagic* & Luka Zgrablic

This Publication has to be referred as: Bosnjak, M[aristela]; Redzepagic, J[asmin] & Zgrablic, L[uka] (2024). Data Masking and Tokenization, Proceedings of the 35th DAAAM International Symposium, pp.0183-0190, B. Katalinic (Ed.), Published by DAAAM International, ISBN 978-3-902734-44-0, ISSN 1726-9679, Vienna, Austria
DOI: 10.2507/35th.daaam.proceedings.024

Abstract

This research paper examines mobile application security, focusing on data masking and tokenisation, which are essential for protecting sensitive data on Android and iOS platforms. It delves into platform-specific vulnerabilities and details techniques to safeguard data at rest and in transit. The study advocates for secure coding practices, advanced encryption, and regular security audits to enhance mobile app security. It emphasises the role of Mobile Device Management (MDM) solutions in strengthening security measures. To connect theory with practice, the paper includes real-world case studies that demonstrate the effectiveness of these security practices in various scenarios. These case studies provide actionable insights for developers, IT security practitioners, and professionals aiming to improve the development and management of secure mobile applications.

Keywords: mobile application security; data masking; secure coding practices; security audit.

1. Introduction

In the current era characterised by the widespread adoption of mobile applications, the overriding issue reverberating throughout the digital domain is the security of these pervasive technologies. As dependence on mobile apps grows, the urgency to address the increasing threats to data integrity and user privacy becomes paramount. This paper aims to dissect the complex structure of mobile application security, highlighting the critical importance of data masking and tokenisation. Amidst a constantly evolving technological environment, the vulnerabilities associated with Android and iOS platforms require detailed scrutiny.

This study investigates the substantial impacts of these vulnerabilities on user data and the possible consequences of security breakdowns, going beyond essential identification. This article examines many best practices, such as secure coding techniques, encryption tactics, and the critical function of regular security audits, considering that strong security measures are vital for safeguarding sensitive information.

Moreover, the discussion delves into Mobile Device Management (MDM) solutions, detailing their crucial contribution to bolstering mobile application security. Throughout this inquiry, real-world case studies and practical examples will shed light on the discussion, providing insights into practical implementations and emphasising the real-life impacts of embracing comprehensive security strategies. This exploration aims to identify and understand the security challenges in mobile application development and to offer practical solutions, laying a solid groundwork for developers, practitioners, and security experts.

This research paper addresses the primary problem of mobile application security, explicitly protecting sensitive data on Android and iOS platforms. This includes identifying platform-specific vulnerabilities and finding effective techniques to safeguard data at rest and in transit.

2. Mobile application security challenges

Mobile applications have become integral to modern life, offering unprecedented convenience and introducing significant security challenges. This chapter comprehensively explores these challenges, detailing the general concerns and the more specific threats related to data breaches and privacy vulnerabilities [1]. As mobile applications become increasingly embedded in everyday activities, the ecosystem becomes more dynamic and susceptible to various security threats. These range from unauthorised access to sensitive data to exploiting existing software vulnerabilities. This section thoroughly examines the complex interaction between user behaviours, app design choices, and an ever-evolving threat landscape, highlighting the persistent security issues prevalent across the mobile application domain.

A significant area of concern within this landscape is the risk of data breaches, which significantly threaten user privacy. Mobile users often store personal and sensitive information within applications, which can become targets for unauthorised access and data compromise. In this research, we will dig deeper into the implications of such breaches, utilising real-world examples to emphasise the critical need for stringent security measures. Recognising and understanding these security challenges is crucial for developing effective protective strategies. The subsequent sections of this text will delve into specific solutions to mitigate these risks. Discussions will focus on the critical roles of data masking, tokenisation, and mobile device management. By obscuring or anonymising sensitive information, data masking techniques help organisations comply with privacy regulations, mitigate the risk of data breaches, and protect customer confidentiality [1]. These strategies are essential for strengthening the security frameworks of mobile applications, addressing the identified challenges, and significantly enhancing the overall security posture of mobile environments. Furthermore, data anonymisation can strike an optimal balance between privacy protection and data utility, integrating techniques such as data masking [2]. This comprehensive approach ensures that mobile applications can continue to offer convenience without compromising the security and privacy of user data.

3. Platform-specific vulnerabilities

Let's now investigate platform-specific vulnerabilities that affect mobile operating systems, focusing on Android and iOS, which are predominant in the mobile industry. As mobile technology becomes increasingly widespread, it is imperative to understand the distinct security challenges each platform faces. We will delve into the unique vulnerabilities inherent to these systems and discuss how attackers can exploit them. By analysing the underlying architecture and security frameworks of Android and iOS, we aim to pinpoint their specific weaknesses to various attack vectors. Through comprehensive analysis, this chapter intends to provide developers, security professionals, and users with essential insights to better protect their devices from emerging threats. Gaining this knowledge is crucial for creating more secure applications and implementing practices that bolster the security of these widely used platforms.

3.1. Android vulnerabilities

While encouraging innovation, Android's open-source design poses specific concerns because of irregular security updates and fragmented software versions among various device makers. Comparing Android to iOS, which is an open-source operating system, iOS is typically less vulnerable to security flaws [3]. Common vulnerabilities in Android include:

- Permissions: Android applications are granted permission to access specific device resources. However, applications can request permissions they do not need or allow users to grant permissions to applications they do not trust accidentally.
- Insecure coding: Android applications are written in Java, a language known for its susceptibility to security vulnerabilities. Developers can make mistakes in their code that can allow attackers to exploit the application.
- Third-party libraries: Android applications often rely on third-party libraries to provide functionality. However, these libraries can also be vulnerable to security attacks.
- Rooted devices: Android devices can be rooted, meaning they have elevated privileges. This can make them more susceptible to attacks, as attackers can gain control of the device and install malicious software.
- Sideloading: Android devices allow users to install applications from sources other than the Google Play Store. This can introduce vulnerabilities, as users may not be able to verify the safety of these applications.

Some of the most significant attacks found in Android include [4]:

- Stagefright: Stagefright was a vulnerability in the Android multimedia framework that allowed attackers to execute code on devices remotely.
- Shellshock: Shellshock was a vulnerability in the Bash shell script interpreter that affected Android devices and other systems.

- Heartbleed: Heartbleed was a vulnerability in the OpenSSL cryptographic library that affected Android devices and other systems.
- Meltdown and Spectre: Meltdown and Spectre were vulnerabilities in Intel processors that affected Android devices and other systems.
- KRACK: KRACK was a vulnerability in the WPA2 wireless protocol that affected Android devices and other systems.
- QuadRooter: A set of four vulnerabilities affecting Qualcomm chipset software drivers that could give attackers root access to the device.
- Cloak and Dagger: A series of attacks exploiting Android's overlay and accessibility features, allowing attackers to hijack the user interface or steal information.
- Broadpwn: Broadcom's Wi-Fi chipset drivers could be vulnerable to remote code execution.
- StrandHogg: A vulnerability that allows malicious apps to masquerade as legitimate apps, enabling them to trick users into giving them sensitive permissions.
- Judy: A large-scale auto-clicking adware found in 41 apps that could potentially infect up to 36.5 million devices.

The vulnerabilities, such as Stagefright, BlueBorne, and QuadRooter, underscore significant security threats affecting millions of Android devices globally. These security flaws target everything from Bluetooth functionalities and multimedia processing to essential protocols like WPA2. Their discovery and exploitation highlight the ongoing struggle between system developers and cyber attackers.

In Q2 2023, the amount of malware, adware, or undesired software attacks on mobile devices increased. Throughout that time, 5,700,000 attacks were prevented. When a user attempted to pay for a transaction, the malware would steal their bank card credentials by inserting JavaScript code into the appearance of a well-known Asian online retailer. Among the other peculiar quarter finds was a movie-streaming app released on Google Play with a crypto miner [5]. Robust security mechanisms, regular system upgrades, and raising user knowledge are essential if Android continues leading the mobile operating system market. To safeguard their devices, users and developers must exercise caution, keep up with emerging security risks, and take preventative measures. A proactive approach is crucial to ensuring safe and dependable device management in our highly linked digital world.

3.2. iOS vulnerabilities

Because iOS is a closed-source operating system, security risks are lower than Android's. However, iOS is not immune to intrusions. Common vulnerabilities found in iOS include [6]:

- Application sandboxing: iOS applications are sandboxed, which means they are isolated from each other and the rest of the operating system. This can help to prevent attacks that target the operating system itself. However, attackers can still exploit vulnerabilities in individual applications.
- Insecure coding: iOS applications are written in Objective-C, a language known for susceptibility to security vulnerabilities. Developers can make mistakes in their code that can allow attackers to exploit the application.
- Rooted devices: iOS devices can be jailbroken and modified to gain root privileges. This can make them more susceptible to attacks, as attackers can gain control of the device and install malicious software.
- Sideloading: iOS devices do not allow users to install applications from sources other than the App Store, which can help prevent users from installing malicious applications. However, there are ways to sideload applications which can introduce vulnerabilities.

Some of the most critical iOS vulnerabilities include [7]:

- XcodeGhost: A malware strain propagated through compromised Xcode versions, Apple's official tool for developing iOS apps. Infected apps collect device information and upload it to servers controlled by attackers.
- Masque Attack - This attack exploited a flaw that allowed malicious apps to replace legitimate apps installed through the App Store. It was executed by luring users to install an app from a phishing link, replacing a genuine app without the user's knowledge.
- WireLurker: Discovered primarily targeting Chinese iOS users, this malware infiltrated iOS devices through malicious software on Macs and moved onto iOS devices via USB connections. It could steal information and install further malicious applications.
- Pegasus Spyware: A highly sophisticated spyware that exploits multiple zero-day vulnerabilities to hack iOS devices and spy on victims. It could read text messages, track calls, collect passwords, track location, access the target device's camera and microphone, and gather information from applications.
- KeyRaider: This malware targeted jailbroken iOS devices to steal account names, passwords, GUIDs, and Apple push notification service tokens. It led to unauthorised purchases and ransom demands.
- AceDeceiver: This malware exploited flaws in Apple's DRM protection mechanism (FairPlay) to install malicious apps on iOS devices, regardless of whether they were jailbroken.
- Checkm8: A boot rom exploit affected all iOS devices running on chips from A5 to A11. It allows permanent unmatchable jailbreaking of these devices, posing significant security risks.

- Trident: Used in combination with Pegasus spyware, the Trident exploit chain utilised three zero-day vulnerabilities to compromise iOS devices and install the spyware for surveillance purposes
- Ikee Worm: The first worm to target iOS, specifically jailbroken devices, which changed the device's wallpaper and attempted to propagate to other devices on the network. It primarily aimed to create botnets.
- Fake Jailbreak: Scams promising to jailbreak newer versions of iOS would instead install malware or adware on the devices of unsuspecting users, leading to privacy breaches and unauthorised use of the device resources.

The array of iOS security breaches, ranging from malware distributed through compromised versions of Xcode to advanced spyware like Pegasus, underscores the critical need for stringent security measures. These incidents highlight users' need to adhere to safe practices, such as restricting app downloads to trusted sources like the App Store, refraining from jailbreaking devices, and consistently updating software to protect against emerging vulnerabilities. Apple has issued patches to mitigate critical zero-day WebKit vulnerabilities for the latest iOS, tvOS, and macOS platforms. The patches will cover over 15 vulnerabilities that expose Apple devices to data leaks, arbitrary code execution, and denial-of-service attacks. The zero-day is the company's first for 2024 [8].

Moreover, these attacks underline the importance of ongoing user education in cybersecurity, as many exploits leverage user actions, like downloading a malicious app or clicking on a deceptive link. The continuous development of these threats points to a relentless challenge between cybercriminals and security professionals, driving forward the enhancement of security technologies and strategies to safeguard users in a digital age that is perpetually connected. That's also why we must develop new methodologies to mitigate these problems. For example, Magen supports the dynamic masking of payloads and the static masking of large data sets, offering format-preserving encryption and tokenisation [9].

4. Approaches to solving mobile application security problems

Data masking and tokenisation are critical methods for securing sensitive data when full disclosure is unnecessary. Data masking hides specific elements within a dataset to block unauthorised viewing of valid values, which is especially useful in testing, development, or third-party engagements. Masking can be static (permanent changes) or dynamic (real-time during data access). Tokenisation replaces sensitive data with non-sensitive tokens, which only a secure system can revert to the original data, making them useless to intruders without access. Both techniques reduce data breach risks and ensure compliance with strict privacy laws like GDPR and HIPAA, thus bolstering organisational data security frameworks and protecting sensitive data even if other security measures fail. We must remember that breaches happen daily, even in most regulated environments, such as the healthcare industry, which is the industry with the highest cost of data breaches. This further proves how important it is to develop secure and secure environments in which we will host these applications. Having in mind that most modern environments for application deployments include containers and CI/CD (Continuous Integration/Continuous Deployment) pipelines, it is essential to point out that the security of CI/CD toolsets is paramount as they directly affect the entire development and deployment pipeline, presenting a unique set of security challenges that must be rigorously managed to protect against both internal and external threats [10].

4.1. Data masking

Data masking is an essential security approach that shields confidential information by hiding certain parts of a dataset, allowing it to be used safely in settings where accurate data exposure could be dangerous, such as development, testing, or training. Creating a version that looks architecturally like the original but contains non-sensitive data makes it possible to use genuine datasets without disclosing sensitive information. There are various types of data masking, including dynamic data masking, which momentarily obscures data in real-time during query execution without changing the underlying database, and static data masking, which modifies data permanently before it is moved from a secure production environment. This approach is crucial for protecting against outside threats and complying with stringent data protection regulations, such as GDPR, which require careful data processing.

Data masking can protect sensitive information in mobile applications, such as credit card numbers, Social Security numbers, and customer addresses. In data masking, actual data is substituted with artificial or scrambled data, making it meaningless and less appealing to potential attackers. This reduces the value of the data and its usefulness if compromised. There are several techniques for masking data. A standard method involves replacing sensitive information with random characters; for example, a credit card number might be shown as "XXXX-XXXX-XXXX-XXXX." Alternatively, encryption can scramble the data, making it unreadable to anyone without the encryption key.

4.2. Tokenization

Tokenisation is a technique that replaces confidential information with distinct identifiers, or tokens, to strengthen application security. Tokenisation is necessary for protecting sensitive data in apps, such as credit card numbers, social security numbers, and personal health information. These tokens maintain essential details of the data without jeopardising its security. Applications handle less sensitive information directly by substituting tokens for sensitive data, which lowers the risk of data breaches and helps assure compliance with laws like PCI DSS and GDPR.

A real-world example would be a customer entering their credit card information on an online shopping application. Tokenisation substitutes a token for the actual credit card number. This token is kept in databases and used in the program procedures. The actual data is mapped back to the original data and is kept safe in a token vault concurrently. This vault's access is restricted to highly secure procedures, significantly reducing the methods by which a cybercriminal can obtain private information. As a result, even if an attacker gained access to the application's database, the extracted tokens would be useless without the tokenisation mechanism, guaranteeing the security of the sensitive data underneath.

5. Best practices for mobile app security

Mobile app security is increasingly critical in our technology-driven world, where mobile devices are ubiquitous in personal and professional settings. The challenge in securing mobile applications arises from the vast array of operating systems, various device types, and diverse user interactions. These apps are particularly vulnerable to security threats, leading to data theft, phishing schemes, and unauthorised device control. As reliance on mobile applications continues to surge for functions like financial transactions, shopping, and personal communication, it's imperative to safeguard them against emerging cyber threats.

Several best practices must be diligently followed to bolster mobile app security. Robust authentication processes, such as multi-factor authentication, should be implemented to minimise the risk of unauthorised access. Developers must continuously update and patch applications to address newly discovered security flaws. There are three main ways to achieve a better mobile app security level: secure coding, encryption, and audits, where appropriate. Secure coding techniques are essential to prevent vulnerabilities like SQL injections and cross-site scripting. There are multiple principles involved when trying to write secure code:

- Employ defensive coding practices: Developers should craft code that can withstand typical security threats, such as SQL injections, cross-site scripting, and buffer overflows.
- Implement robust encryption: Developers must utilise strong encryption protocols to safeguard sensitive information, including passwords and credit card details.
- Refrain from embedding sensitive data: Developers should avoid hardcoding sensitive information directly into their applications. Instead, sensitive data should be kept in secure storage options like databases or keychains.
- Utilize secure authentication methods: Developers are advised to implement secure authentication strategies, such as two-factor authentication, to protect sensitive data.

Encrypting data when transmitted and stored ensures that sensitive information remains protected. We can employ different techniques:

- Symmetric encryption: This method employs a single data encryption key. It's commonly used for file encryption and facilitating secure device communications.
- Asymmetric encryption: This method utilises a pair of keys—public and private—for data encryption and decryption. The public key is openly shareable, while the private key remains confidential. This encryption is often used in secure communications, such as online banking and email security.
- Hashing: A one-way encryption that transforms data into a distinct fingerprint. Hashes are primarily used to confirm data integrity, helping to ensure that files have not been altered during download.

Additionally, rigorous security auditing, testing, and vulnerability assessments should be conducted before app releases. These proactive measures help identify and rectify potential security weaknesses, enhancing the overall security posture of mobile applications and providing a safer environment for users. Again, we can do this in multiple different ways:

- Static security analysis: This process involves analysing an application's source code without executing it. It helps detect various vulnerabilities, such as buffer overflows, SQL injections, and cross-site scripting.
- Dynamic security analysis: While the application is active, this analysis can pinpoint issues that static security analysis might miss, including memory leaks and race conditions.
- Penetration testing: This method simulates attacks on an application to identify significant vulnerabilities that could be exploited.

If we look at mobile app security from the opposite direction, from the direction of end users, some of whom might be our employees, one of the main ways to protect them from malicious mobile applications is by using MDM (Mobile Device Management). These are some of the operational procedures that are possible if we use MDM:

- Enrolling, configuring, and deploying devices
- Managing users and groups
- Controlling device settings and security policies
- Remotely locking, erasing, or wiping devices
- Monitoring device activity and collecting data
- Protecting data at rest, in transit, and in use.

In terms of mobile device security, using MDM allows us to do the following:

- Device enrolment: MDM solutions can help to automate the enrolment process for devices, which helps to ensure that all devices are correctly configured and compliant with security policies
- Profile deployment: MDM solutions can be used to deploy profiles to devices, which can be used to configure the device settings, such as passcode requirements, app restrictions, and web browsing restrictions
- Remote management: MDM solutions can be used to remotely manage devices, which allows IT administrators to lock devices, wipe data, and install software updates
- Content filtering: MDM solutions can be used to filter content on devices, which helps to prevent users from accessing inappropriate or malicious content
- Data encryption: MDM solutions can be used to encrypt data on devices, which helps to protect data from unauthorised access
- Security auditing: MDM solutions can be used to audit device activity, which helps to identify and investigate security incidents.

From the aspect of corporate security policies, all these capabilities are a very welcome addition to corporate security. They've also become somewhat of a norm, especially in larger enterprises, where MDM has taken a significant foothold in the past decade. For example, we can count on the fact that there will always be employees who will lose their mobile devices somewhere, and we must have some remote management capability to deal with these situations.

6. Case-study: Data masking tool

Most companies use numerous databases and applications in their daily operations. These (production) databases are frequently copied to different environments for additional purposes, such as development, testing, acceptance, training, outsourcing, etc. Nevertheless, many of these systems include corporate vital and privacy-sensitive data and personally identifiable information (PII), which you are not permitted to utilise for such purposes.

Data masking, also known as anonymisation, is the process of modifying data so that it can still be used for testing and development purposes. Still, it makes it nearly difficult to identify an individual. For a case study, the data masking tool DATPROF [11] was used on a student information database containing real names, usernames, student classes, and other faculty information. The tool generated synthetic data to replace existing data.

The results showed how well the user-friendly interface of the Data Masking DATAPROF tool managed complex data relationships. This user-friendliness made complex data linkages easier to integrate and manage and drastically decreased the time and effort users need to put data masking methods into place. The tool's user-friendly interface ensured that confidential data was safeguarded without sacrificing the accuracy or functioning of the underlying data structures, even for users with differing degrees of technical proficiency. The tool's robust features also helped maintain data accuracy and consistency across various datasets, demonstrating its capacity to handle complex data masking problems in different application settings. Its features include support for XML and CSV files, integrated synthetic data generators, HTML audit and GDPR reporting, REST API testing for data automation, and an intuitive online interface.

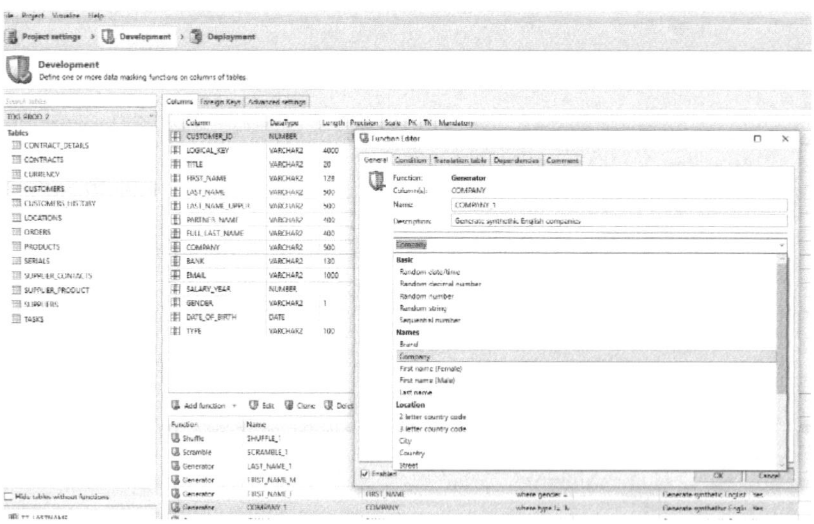

Fig. 1. DATPROF tool interface (picture from: https://www.datprof.com/. Accessed on 07.07.2024.)

Figure 1 shows the general layout of the DATPROF tool interface. Because of GPDR restrictions, the database used for the case study cannot be displayed.

7. Future research

Multiple promising areas for future research should be investigated. Investigating memory safety in mobile app development entails refining current programming languages or crafting new ones that automatically address prevalent issues such as buffer overflows and memory leaks. Compiler technology advancements might concentrate on sophisticated static analysis tools explicitly designed to identify memory-related vulnerabilities. This strategy allows developers to produce inherently safer code, thereby minimising the risks of exploitable vulnerabilities.

In automated security testing, the emphasis is on creating cutting-edge tools that meld smoothly with mobile development workflows. Enhancing dynamic analysis tools to replicate real-world attack scenarios more accurately and employing machine learning to foresee and adjust to emerging threats are critical aspects of this focus. Furthermore, masked token models introduce a sophisticated method of maintaining data privacy during the machine learning training phase [12]. Such improvements can drastically lower the time and resources required to pinpoint vulnerabilities, thus bolstering the security of apps before they are launched. New approaches are also being developed for masking private data using Named Entity Recognition and Deep Learning concepts to achieve data anonymity [13]. Furthermore, the NextGen Data Masking Engine introduces a versatile approach to handling complex data structures for adequate data security and compliance [14]. This topic is being actively researched as it can potentially change how we implement different security layers.

Moreover, customising threat modelling for the mobile sector involves adapting existing methods to the unique challenges the mobile environment poses, including aspects related to mobile operating systems, hardware, and communication networks [15][16]. By proactively identifying and addressing potential threats during the development phase, developers can craft safer mobile applications, significantly lowering the risks for users and developers in a world increasingly dominated by mobile technology. In the current era characterised by the widespread adoption of mobile applications, the overriding issue reverberating throughout the digital domain is the security of these pervasive technologies. As dependence on mobile apps grows, the urgency to address the increasing threats to data integrity and user privacy becomes paramount [17].

8. Conclusion

The paper's in-depth analysis of mobile application security—which mainly focuses on data masking and tokenisation—has shed light on the critical steps required to safeguard sensitive data in the context of the widespread use of mobile apps in today's world. The best methods for enhancing mobile app security were highlighted in this study, which also carefully investigated the security flaws and problems of the iOS and Android platforms. Data masking and tokenisation have become essential tactics to make sensitive data unreadable and provide a substantial layer of protection against unwanted access.

Our investigation also covered the difficulties that Android and iOS present, highlighting the significance of implementing platform-specific security controls. We also found that novel tactics, including data masking, are required to guarantee complete data protection in the cloud because traditional data security measures are insufficient. Data masking is an essential security measure that protects sensitive data by replacing it with realistic but not actual data, allowing organisations to use and share data without exposing personal or sensitive information.

The study examined a wide range of recommended procedures for protecting mobile applications, including doing in-depth security audits, utilising encryption, and using secure coding techniques. It highlighted how Mobile Device Management (MDM) solutions may improve security, demonstrating their importance in a comprehensive security plan. The case study showed how simple it is to perform data masking with a specialist tool for data masking like DATPROF.

In conclusion, the study not only determined the critical security controls for mobile applications but also offered developers and IT security professionals useful information and workable answers. The results highlight the importance of incorporating thorough security measures and employing cutting-edge security techniques to safeguard sensitive data in mobile applications.

9. References

[1] Badgujar, P. (2021). Implementing Data Masking Techniques for Privacy Protection. Journal of Technological Innovation, Vol.2 Issue 4, October-December 2021

[2] ŞAHİN, Y.; DOGRU, İ. (2023). An Enterprise Data Privacy Governance Model: Security-Centric Multi-Model Data Anonymization, Proceedings of Uluslararası Mühendislik Araştırma ve Geliştirme Dergisi, April 15, 2023, Uluslararası Mühendislik Araştırma ve Geliştirme Dergisi, DOI 10.29137/umagd.1272085

[3] Garg, S., & Baliyan, N. (2021). Comparative analysis of Android and iOS from security viewpoint. In Computer Science Review (Vol. 40, p. 100372). Elsevier BV. DOI 10.1016/j.cosrev.2021.100372

[4] Android Mobile Security Threats, Available from: https://www.kaspersky.com/resource-center/threats/android-mobile-threats Access: 2024-04-08

[5] IT threat evolution in Q2 2023. Mobile statistics, Available from: https://securelist.com/it-threat-evolution-q2-2023-mobile-statistics/110427/ Access: 2024-04-08

[6] What Are the iOS Security Vulnerabilities? Available from: https://www.preemptive.com/blog/what-are-the-ios-security-vulnerabilities/ Access: 2024-04-08
[7] Iphone Os : Security Vulnerabilities, CVEs. Available from: https://www.cvedetails.com/vulnerability-list/vendor_id-49/product_id-15556/Apple-Iphone-Os.html Access: 2024-04-08
[8] Apple Patches Critical Zero-Day Vulnerability for Macs and iPhones. Available from: https://www.spiceworks.com/it-security/vulnerability-management/news/apple-patches-critical-zero-day-vulnerability-macs-iphones/ Access: 2024-04-12
[9] Moffie, M.; Mor, D.; Asaf, S.; Farkash, A. (2022). Next Generation Data Masking Engine, Proceedings of Lecture Notes in Computer Science, 2022, pp. 152-160, Springer International Publishing, DOI 10.1007/978-3-030-93944-1_10
[10] Dakic, V.; Redzepagic, J.; Basic, M. (2022). CI/CD Toolset Security, Proceedings of DAAAM Proceedings, DAAAM International Vienna, 2022, pp. 0161–0164, DAAAM International Vienna, DOI 10.2507/33rd.daaam.proceedings.022
[11] DATAPROF tool for data masking. Available from: https://www.datprof.com/products/datprof-privacy/ Access: 2024-02-10
[12] Hou, L.; Geng, Y.; Han, L.; Yang, H.; Zheng, K.; Wang, X. (2022). Masked Token Enabled Pre-training: A Task-Agnostic Approach for Understanding Complex Traffic Flow, Proceedings of the Institute of Electrical and Electronics Engineers, February 10, 2022, Institute of Electrical and Electronics Engineers (IEEE), DOI 10.36227/techrxiv.19134854.v1
[13] Kodandaram, Satwik R.; Honnappa, K. & Soni K. (2021). Masking private user information using Natural Language Processing, International Journal of Advance Research, Ideas and Innovations in Technology, Volume 7, Issue 3 - V7I3-2048, ISSN: 2454-132X
[14] R. A.; Hegadi, R. S.; M. T. N. (2018). A Study on Big Data Privacy Protection Models using Data Masking Methods, Proceedings of the International Journal of Electrical and Computer Engineering (IJECE), Institute of Advanced Engineering and Science, October 01, 2018, p. 3976, DOI 10.11591/ijece.v8i5.pp3976-3983
[15] Aeri, M. (2020). Secure Data Sharing in Cloud Computing Systems: Techniques and Applications, Proceedings of the Turkish Journal of Computer and Mathematics Education (TURCOMAT), Ninety Nine Publication, December 15, 2020, pp. 2087–2094, DOI 10.17762/turcomat.v11i3.13606
[16] Stefanovic, D.; Nikolic, D.; Dakic, D.; Spasojevic, I.; Ristic, S. (2020). Static Code Analysis Tools: A Systematic Literature Review, Proceedings of DAAAM Proceedings, DAAAM International Vienna, 2020, pp. 0565–0573, DOI 10.2507/31st.daaam.proceedings.078
[17] Dakic, V.; Jakobovic, K.; Zgrablic, L. (2022). Linux Security in Physical, Virtual, and Cloud Environments, Proceedings of DAAAM Proceedings, DAAAM International Vienna, 2022, pp. 0151–0160, DOI 10.2507/33rd.daaam.proceedings.021

DOI: 10.2507/35th.daaam.proceedings.025

SECURE CODING GUIDELINES AND STANDARDS

Kristina Loncar, Jasmin Redzepagic* & Vedran Dakic

This Publication has to be referred as: Loncar, K[ristina]; Redzepagic, J[asmin] & Dakic, V[edran] (2024). Secure Coding Guidelines and Standards, Proceedings of the 35th DAAAM International Symposium, pp.0191-0198, B. Katalinic (Ed.), Published by DAAAM International, ISBN 978-3-902734-44-0, ISSN 1726-9679, Vienna, Austria
DOI: 10.2507/35th.daaam.proceedings.025

Abstract

This paper delves into fundamental secure coding techniques and guidelines to reduce cybersecurity risks during software development. By analysing past cyberattacks and current security weaknesses, the article underscores the importance of secure coding for maintaining the integrity of information. It details critical concepts from the Open Worldwide Application Security Project (OWASP), focusing on the OWASP Top 10 vulnerabilities list. The article also explores methods to address these vulnerabilities, including issues with access control, encryption errors, and injection flaws. The paper emphasises the necessity of adhering to secure coding protocols from organisations such as OWASP, SEI CERT, and ISO to bolster application security. Additionally, it addresses typical challenges developers encounter, such as inadequate security knowledge and limited resources. It offers strategies to cultivate an active security mindset among development teams, as recorded during the case study.

Keywords: secure coding practices; cybersecurity; OWASP; software vulnerabilities.

1. Introduction

In an era where digital transformation is integral to the success of organisations, the security of software systems has become critical. Secure coding standards and guidelines are essential for creating resilient apps to fend off hostile attacks and safeguard confidential information. These guidelines cover a wide range of recommended practices and regulations to handle the various security flaws that may occur while developing software. Through adherence to these standards, developers may effectively detect and address potential security vulnerabilities, guaranteeing that their software is not only operational but also resistant to external assaults. Secure coding techniques must be used to protect individual programs and the more extensive information technology ecosystem that contemporary businesses depend on.

Financial losses, legal repercussions, and reputational harm are just a few of the severe and far-reaching effects of ignoring secure coding procedures. Rigid adherence to secure coding standards could have prevented many undetected vulnerabilities that lead to data breaches and security incidents. Furthermore, these practices must be implemented to comply with regulatory requirements in several industries, highlighting their crucial significance in operational and legal frameworks. One might logically think that cyberattacks emerged alongside the advent of the first programming languages, computers, and networks. However, this isn't correct. Interestingly, the first instances of what could be considered cyberattacks took place in 19th-century France [1], where individuals used the telegraph system to pilfer financial market data - a concept almost inconceivable today.

Shifting to a more contemporary example, the conversation around cybersecurity was significantly shaped by an event in 1988 when Robert Morris, a Cornell University student, unintentionally unleashed a virus that rapidly spread across over six thousand MIT computers, causing millions of dollars in damages [2]. Despite his lack of malicious intent, Morris was prosecuted under the 1986 Computer Fraud and Abuse Act. Since that time, cybersecurity has become an increasingly critical and relevant field. With the evolution of the Internet and technological advancements, malicious actors have continuously exploited new vulnerabilities within software and tools. Understanding these significant attacks and ways to thwart them is essential in computer studies today.

2. OWASP top security risks

The Open Worldwide Application Security Project (OWASP) is a non-profit organisation with worldwide recognition which seeks to enhance software security. OWASP was founded in 2001 and offers abundant free and open tools, resources, and community-driven projects to assist corporations, security experts, and developers identify, comprehend, and mitigate web-based applications' most significant security risks. The group aims to increase software security awareness so that people and businesses can decide what constitutes a real danger to their software [3]. The following sub-chapters will present the most critical web application security risks.

2.1. Broken access control

A key component of information security is Access Control, which controls user access to resources inside an information system. It entails selective restriction of data, applications, and systems access to guarantee that only authorised users can carry out particular tasks [4].

If Access Control were compromised, we would take into account the following scenarios:

- Elevation of privilege or any violation of the Least Privilege principle
- Bypassing Access Control through the modification of the internal application state, HTML page, or URL
- Metadata manipulation, such as tampering with the JWT token.

The Open Web Application Security Project (OWASP) outlines several best practices for preventing and fixing issues related to broken access control. These best practices ensure that systems enforce appropriate permissions and restrict unauthorised access.

Here are some of the key recommendations:

- Least Privilege - Ensure that users are only granted the minimum access levels necessary for their roles. Regularly review and update permissions to ensure they are appropriate.
- Deny by Default - Implement an access control policy that denies all access by default, only allowing access to resources explicitly granted. This minimises the chance of unauthorised access due to misconfigurations or overlooked permissions [5].
- Access Control Mechanisms - Use role-based access control (RBAC) to enforce user roles and states in determining resource access. Where applicable, consider attribute-based access control (ABAC) for more granular control based on user attributes and resource tags.
- Secure Server-Side Enforcement—Ensure that access controls are enforced server-side and not reliant on the client-side. Client-side checks should only be used for user experience enhancements and must be duplicated on the server.
- Audit Logs and Monitoring—Implement comprehensive logging of all access control decisions and regularly monitor access patterns for anomalies. This helps detect and respond to unauthorised access attempts.
- Static and Dynamic Testing - Regularly test access controls statically and dynamically. Automated tools scan for common access control vulnerabilities and conduct periodic security audits. We can create tools that trace the impacts of code changes on the security of a given software and evaluate it [6]
- Segregation of Duties - Separate responsibilities and duties to reduce the risk of fraud or data tampering. This involves defining roles to limit the power of any single individual where sensitive transactions or data access are involved.
- Use of Secure Direct Object References - Ensure that when direct references to internal implementation objects such as files, directories, or database keys are exposed to users, they cannot manipulate them to gain unauthorised access to other objects.
- Token Management - Securely manage authentication tokens, session identifiers, and other credentials. Ensure they are not exposed in URLs, logs, or referrer headers.
- Regular Updates and Patch Management - Keep all systems updated with the latest security patches and updates. This helps mitigate vulnerabilities that could be exploited to bypass access controls.

By implementing these strategies, organisations can strengthen their security posture and reduce the risks of broken access control.

2.2. Cryptographic failures

Vulnerabilities resulting from poor use of cryptography to secure sensitive data are known as cryptographic failures in the OWASP Top Ten. This group of security hazards deals with how cryptographic safeguards might break, exposing or allowing unwanted access to confidential data. The brief overview of OWASP's best practices for preventing and resolving cryptographic failures [7] are:

- Insecure Algorithms and Practices: Cryptographic failures often stem from outdated or insecure algorithms, cypher modes, and hash functions. Additionally, problems may arise from poorly generated initialisation vectors or mismanaged cryptographic keys.
- Lack of Enforcement: Sometimes, encryption protocols are not adequately enforced due to missing directives, which leads to vulnerabilities in securing data.

To safeguard the confidentiality and integrity of data, OWASP advises:

- Robust Encryption: Encrypt all sensitive data, both at rest and in transit, with robust and vetted algorithms, cypher modes, and hash functions. It's also crucial to phase out outdated protocols like FTP and SMTP when transmitting sensitive information. Furthermore, new principles in cryptography might be used to enhance further encryption security - principles like quantum cryptography [8] or the application of fractal geometry in the field of cryptographic security [9]
- Manage Data Lifecycle: Proactively manage the storage and transmission of sensitive data. Disable data caching to prevent unintended data leaks and promptly delete unnecessary data to minimise risk exposure
- Secure Randomness: Ensure that cryptographic processes utilise true randomness. Avoid predictable seeding in cryptographic functions to prevent attackers from exploiting weak randomness in encryption, which could lead to data breaches.

These guidelines emphasise the importance of cryptographic solid practices and the need for a systematic approach to managing and securing data throughout its lifecycle, enhancing overall security measures against potential cryptographic failures.

2.3. Code Injection

Code injections occur when malicious code is introduced and executed in an application without proper security [10]. These vulnerabilities allow attackers to input code into areas where it is not intended to be, exploiting the system. Different types of code injections include:

- SQL Injections: Attackers insert SQL queries into inputs that directly interact with databases, allowing unauthorised access or operations
- Cross-Site Scripting (XSS): Malicious scripts are embedded into web pages through HTML script tags, affecting users who access these pages
- Command Injections: Commands are injected into an application, which then executes these commands at the operating system level
- Lightweight Directory Access Protocol (LDAP) Injections: Input parameters in LDAP queries are manipulated, altering directory services and information.

To effectively counter these injection threats, several defensive measures are recommended:

- Input Validation: Implement rigorous validation of all user inputs to detect and neutralise potentially malicious data. Use whitelisting approaches to only allow inputs that meet specific criteria.
- Use of Special Character Escaping: Escaping special characters in inputs can prevent interpreters from executing unwanted commands or queries.
- Client-side Script Management: Set the HttpOnly flag in cookies to prevent access to cookie data via client-side scripts, reducing the risk of XSS attacks.
- Parameterized Queries: Always use parameterised queries or prepared statements when accessing databases. This technique ensures that input data is treated as data rather than executable code, which helps prevent SQL injections.
- Content Security Policy (CSP): Implement a robust Content Security Policy to restrict the sources of executable scripts and mitigate XSS risks by specifying which scripts are allowed to run and from where.
- Regular Security Audits: Conduct regular security audits and use tools designed to detect injection vulnerabilities in code before production deployment.

Organisations can significantly reduce the risk of injection vulnerabilities by adopting these strategies, protecting their systems and user data from malicious activities.

2.4. Insecure design

Insecure design refers to vulnerabilities originating from fundamental system architecture flaws. It's crucial to understand the distinction between design and implementation: while a system with a robust design might function adequately even if not implemented perfectly, allowing room for future enhancements, a system with inherent design flaws cannot be rectified by flawless implementation, as it lacks essential security controls from the start [11]. Examples of such insecure designs include [12]:

- Storage of Non-Encrypted Sensitive Information: Unencrypted sensitive data exposes it to potential breaches and unauthorised access.
- Logs and Alerts Containing Exploitable Information: Logs that include sensitive data can provide attackers with the information they need to exploit the system further.
- Inadequate Isolation of Roles and Privileges: Poor segregation of duties can lead to excessive access rights, increasing the risk of insider threats or accidental misuse.
- Unrestricted File Uploads: Allowing files to be uploaded without restrictions or proper security checks can lead to the uploading of malicious files, potentially compromising the system.

To mitigate these risks, it is advisable to implement several strategic measures:

- Adopt a Security-by-Design Approach: Integrate security at the earliest design stages and throughout development. This includes using threat modelling to identify and address potential security issues before they become embedded in the system.
- Principle of Least Privilege: Ensure access controls are strictly enforced, granting users the minimum access necessary for their job functions.
- Encrypt Sensitive Data: Always encrypt sensitive data at rest and in transit to protect it from unauthorised access.
- Secure Logging Practices: Logs should not contain sensitive information. Implement proper log management and monitoring to prevent unauthorised access and manipulation.
- Robust File Handling Mechanisms: Implement strict validation and sanitisation processes for file uploads to prevent malicious files from being uploaded to the system.
- Regular Testing: Continuously perform unit and integration tests to identify and resolve vulnerabilities promptly. This proactive approach helps ensure security flaws are caught and remediated before the software reaches production.
- Feature Reduction: Remove any features that are not being used to reduce the attack surface. A streamlined codebase minimises potential entry points for attackers and simplifies security management.
- Adopt Microservices Architecture: Implement microservices to promote effective compartmentalisation of different application functions. This structure can isolate breaches to a single service without compromising the entire system.

Addressing these aspects of insecure design can significantly enhance a system's security posture, reducing vulnerabilities and strengthening its defences against potential attacks.

2.5. Security misconfiguration

Software engineers frequently use frameworks to develop applications [13], which reach their full potential only when their complex configuration files are correctly established and managed. While default configurations offer convenience, they are typically insecure and pose an easy attack target. Examples of security misconfiguration might include [14]:

- Inadequate Security Settings: Security settings on application servers, frameworks, libraries, and databases are not configured to secure values, or they remain unencrypted.
- Outdated Software: Security patches and updates are often disabled by default, making unpatched, outdated software vulnerable to exploits.
- Enabled Unnecessary Features: Unneeded ports, services, default accounts, and permissions are left enabled, increasing the attack surface unnecessarily.

To reduce the likelihood of configuration mismanagement, consider the following measures:

- Reduce Platform Complexity: Streamline platform functionality by removing or uninstalling unused tools and frameworks, which reduces potential vulnerabilities.
- Implement Automated Security Processes: Set up automated systems to regularly check and confirm the appropriateness of configuration settings across all environments, ensuring the latest security updates are consistently applied.
- Tighten Cloud Storage Settings: If using cloud storage, turn off unnecessary permissions and directory browsing to protect against unauthorised access and data breaches.

By diligently managing and securing configuration settings, organisations can significantly reduce the risks associated with security misconfigurations. Adopting these best practices secures applications and enhances overall system resilience against cyber threats. We must always be aware that even a minor and straightforward vulnerability can lead to extensive issues and substantial harm [15].

Addressing vulnerabilities in system architecture, configurations, and operational practices is essential for safeguarding sensitive data. Ignoring these security gaps can have disastrous consequences, enabling malicious actors to steal, modify, or delete critical information. Such breaches could allow attackers to forge identities and impersonate key figures, causing irreparable damage to an organisation's reputation and financial standing. Moreover, these vulnerabilities often remain undetected, particularly when they involve spying on unsecured communication channels or deliberately slowing operations by tampering with configuration files.

The long-term effects of these breaches can go beyond immediate operational and financial issues, potentially leading to regulatory compliance breaches. Such violations could result in substantial fines and a severe loss of client trust, which can be incredibly challenging to restore. In the most severe cases, inadequately designed and secured systems may suffer irreparable damage, leading to complete organisational failure.

One other method that we could employ is indeed using Trusted Execution Environments. The two most prominent technologies are Intel's SGX (Software Guard Extensions) and AMD SEV (Secure Encrypted Virtualization)[16]. It's a two-way relationship - secure coding practices become crucial when integrating hardware security technologies like Intel Software Guard Extensions (SGX) and AMD Secure Encrypted Virtualization (SEV). Intel SGX provides developers with the tools to create secure enclaves that shield code and data from external threats, even if there's a breach in the operating system. Proper secure coding techniques ensure that the data managed within these enclaves is secure, preserving its integrity. Conversely, AMD SEV secures data at the hypervisor level, encrypting virtual machines' memory, which is essential in environments where cloud resources are shared. Secure coding practices enhance the effectiveness of AMD SEV by ensuring that data is protected not only when it is stored but also while it is being transmitted or processed.

Thus, organisations must adopt stringent security measures recommended by frameworks like OWASP. Implementing strategies such as consistent security reviews, integrating security during the design phase, and enforcing rigorous access controls can significantly reduce these risks. By proactively tackling these vulnerabilities, organisations can shield themselves from the direct and collateral damages of cyberattacks, ensuring their long-term resilience and reliability in the digital landscape.

3. Secure Coding Standards

Coding standards consist of guidelines or best practices that help organisations minimise security vulnerabilities and programming errors, enhancing the security posture of their applications. These standards are universally relevant and provide a crucial defence against prevalent cyber threats for applications of any size. Notably, there is a significant educational gap in secure coding practices in academia; for example, among the top 24 undergraduate computer science programs in the US, only one requires a cybersecurity course [17]. This highlights the critical need for formalised secure coding standards in educational curriculums.

The absence of secure coding education underscores the importance of coding standards. These standards bridge the skills gap among developers, facilitating the efficient delivery of secure code with minimal disruptions. Implementing recognised coding standards optimises development time and offers cost savings by allowing developers to focus on enhancing other application features. Moreover, these standards bring a structured discipline to the codebase, helping developers navigate and strengthen code security and organisations comply with legislative and regulatory frameworks.

One of the most essential standards in this domain is the "OWASP Top 10", which identifies the most severe web application security risks. The Open Web Application Security Project (OWASP) also provides the Application Security Verification Standard (ASVS), which outlines methodologies for building, testing, and maintaining secure applications [18]. OWASP further supports developers through tools like the Zed Attack Proxy (ZAP), a free, open-source tool to discover vulnerabilities in web applications.

Beyond OWASP, the Software Engineering Institute (SEI) offers the SEI CERT Coding Standards, which address various programming languages, including C, C++, Java, Android, and iOS [19]. These standards offer detailed guidelines, examples, and explanations to help developers avoid errors such as undefined behaviours and vulnerabilities. OWASP and SEI CERT utilise the Common Weakness Enumeration (CWE) standard, which has been maintained by The MITRE Corporation since 2006. This standard provides a common language for describing software and hardware vulnerabilities [20], facilitating more transparent and more effective communication within the cybersecurity field.

The International Organisation for Standardization (ISO) is another key player, having developed over 24,000 standards across different sectors. The ISO 27000 series, particularly ISO 27034, focuses on information security and offers extensive guidelines on essential security measures such as authentication, encryption, and data validation to enhance application security. While not a direct coding standard, NIST SP 800-53 provides comprehensive security and privacy controls for federal information systems. Part of the more extensive NIST SP 800 series, these guidelines are available freely and can be adapted by organisations according to their specific security needs, unlike the certifiable ISO standards.

Adopting these different coding standards and security frameworks is crucial for developing secure, robust, and compliant software. By adhering to these guidelines, organisations can protect themselves against the dynamic landscape of cyber threats while ensuring their software meets global security standards.

4. Examination of real-world issues with applying security standards

A study was performed on how OWASP best practices are applied in real-time situations to understand better the difficulties and consequences of implementing these security standards. By examining real-world security breach scenarios and evaluating OWASP adherence, the study aims to identify frequent issues and offer suggestions for enhancing the adoption of these practices. Two basic scenarios were studied:

- Insecure Data Transmission in E-Commerce Platforms,
- Poor Authentication Mechanisms in Financial Applications Background

4.1. Scenario 1: Insecure Data Transmission

In the scenario, the e-commerce platform experienced a data breach where attackers intercepted unencrypted data transmitted between the website and its users. The investigation revealed that the platform was not enforcing HTTPS for all communications, contrary to OWASP's recommendation for secure transmission. This lapse in security allowed attackers to exploit the vulnerability and gain unauthorised access to sensitive customer information.

The OWASP Top Ten highlights the importance of encrypting data in transit over HTTPS to guarantee its integrity and secrecy. Despite recommendations, the e-commerce platform's HTTPS migration was challenging. Two of these difficulties were managing outdated systems that were difficult to update and a general lack of knowledge about the dangers of sending unencrypted data. In addition, the vulnerability was made worse by the platform's incorrect SSL/TLS setups and expired certificates, which exposed data to possible interception.

4.2. Scenario 2: Poor Authentication Mechanisms

Robust authentication procedures are necessary to confirm user identities and stop illegal access to private data. Effective authentication protects against cyber threats by limiting authorised users' access to protected resources. Even if attackers can get passwords through phishing or other methods, compromising user accounts becomes much more challenging when robust authentication measures, such as multi-factor authentication (MFA), are used.

The scenario in the study included a breach in which attackers gained unauthorised access to user accounts by taking advantage of weak authentication methods on a financial application. Because the application did not use multi-factor authentication (MFA) and only used password-based authentication, it was especially susceptible to brute force and credential-stuffing attacks. The attackers eventually compromised many user accounts and gained access to sensitive financial information by using automated tools to systematically test various password combinations and reusing credentials that they had acquired from previous breaches. This hack highlights how crucial it is to implement robust authentication procedures to safeguard user accounts and private information.

5. Analysis of the study

Because financial systems handle large amounts of valuable data, they are often the focus of cyberattacks. Robust authentication procedures are necessary to confirm user identities and stop unwanted access. The analysis of the real-world scenario focused on how people interacted with the security incident, looking at how human factors affected the breach and pinpointing the issues that were found. These problems—explained in more detail below—showcase how important human behaviour, consciousness, and judgment are regarding cybersecurity mishaps.

5.1. The people problem

In many development teams, security is often relegated to secondary importance, perceived as less exciting or immediately rewarding than developing new application features. This attitude can lead to a reactive security posture [21] spurred by miscommunication between developers and upper management. Even in teams that recognise the importance of security, non-technical stakeholders can pressure teams to prioritise short-term gains over long-term security, encouraging a culture resistant to adopting stringent security practices.

5.2. Inconsistency

Security practices must be integrated at every phase of the software development lifecycle—design, development, testing, deployment, and maintenance. Establishing a Secure Software Development Lifecycle (SSDLC) or implementing automated security scans can help enforce consistent and proactive security measures across all projects. Without such systems, security implementation can become sporadic and reactive, undermining the effectiveness of any security protocols.

5.3. Insufficient knowledge

It is difficult to apply security best practices since software engineers frequently lack a solid foundation in the "subject" when they first enter the industry. The situation is exacerbated when the required tools or frameworks are too complex or unwieldy for practical use. This knowledge gap can hinder the effective adoption of necessary security measures, leaving applications vulnerable.

5.4. Lack of resources

Beyond knowledge gaps, a sheer lack of resources can impede the enforcement of security measures. Financial constraints may prevent teams from acquiring the best tools for security implementation. Time constraints are another critical factor; teams may find themselves pressed for time, unable to adequately address security concerns within project timelines, or dealing with large legacy systems where implementing standards would require an impractical overhaul.

Addressing these challenges requires a multifaceted approach, including better education and training in secure coding practices, improved communication between technical and non-technical team members, and adequate resource allocation to security tasks. Organisations must also foster an environment where security is considered integral to the development process, not an optional add-on. By overcoming these obstacles, the tech industry can enhance its defence against the growing threat of cyberattacks, ensuring a safer digital environment for all users.

6. Future work

Future research in secure coding practices and standards offers numerous promising avenues that can significantly enhance cybersecurity measures within the software development community. One crucial area of focus is integrating secure coding education into the academic curriculum at all educational levels. It is essential to educate computer science students who will develop mobile apps to know how to create secure cypher applications [22]. Investigating effective methods to embed these standards from early education to university programs could help cultivate a generation of well-versed developers in cybersecurity principles, thus bridging the current educational gap. Awareness of the many security vulnerabilities and accounting for and handling all error states are crucial steps to writing secure code [23].

Developing automated security tools incorporating artificial intelligence and machine learning could also revolutionise how security threats are predicted, detected, and mitigated. Such advancements would enhance the capabilities of security tools like OWASP ZAP and alleviate the security management burden on developers, allowing them to concentrate more on innovative aspects of software development. Static and dynamic code of malware reverse engineering can also help here, and knowing the architecture of the attacked application might be of great help [24]. Exploring human factors influencing secure coding practices also presents a valuable research direction. This includes studying the psychological and organisational elements that impact the adoption and implementation of security measures. Insights gained could lead to more effective team structures and development processes that inherently promote better security practices. Another promising future research area involves advancing defensive programming techniques. Developing new frameworks and programming languages that integrate security as a core component could establish new benchmarks in software development, making software more secure from the outset. Furthermore, as technologies like IoT, blockchain, and quantum computing evolve, there is a critical need for specialised security standards that address the unique challenges posed by these technologies.

Future research could also explore the synthesis of security standards across various technology domains, such as web development, mobile applications, and cloud computing. This would help create a cohesive security strategy that minimises the inconsistencies in security implementations across different platforms. Additionally, longitudinal studies on the effectiveness of secure coding practices would provide valuable insights into their real-world efficacy and long-term benefits, enabling the refinement of current practices and validating their sustained impact on cybersecurity.

7. Conclusion

Cybersecurity is a critical issue in the technology sector, highlighting the need for rigorous security measures in all aspects of software development. While the industry recognises its importance, its integration into the computer science curriculum in schools and universities remains limited. Organisations like OWASP are at the forefront of addressing this gap by continuously identifying and evaluating common security threats and developing strategies to mitigate them. These efforts are supported by various security standards that help organisations implement secure practices efficiently and cost-effectively. The study of problems encountered when implementing OWASP best practices illustrates web application security's complex nature and importance. While OWASP offers thorough recommendations for guarding against typical security threats, implementation frequently encounters significant difficulties. However, adopting these best practices ultimately rests with the organisations themselves. It is imperative that the broader tech community, including educational institutions and businesses, understand the significance of cybersecurity and actively participate in creating a more secure digital landscape. As technology continues to evolve, so does the sophistication of cyber threats, making it crucial for us to stay ahead through proactive security practices.

By embracing these security measures and standards, we can combat potential threats and build a more secure future for cyberspace, protecting sensitive data and maintaining users' trust worldwide. This ongoing commitment to cybersecurity is essential for fostering a resilient digital environment that can withstand the challenges of tomorrow.

8. References

[1] https://arcticwolf.com/resources/blog/decade-of-cybercrime/, (2022). A Brief History of Cybercrime, Accessed on: 2023-12-06

[2] https://www.fbi.gov/history/famous-cases/morris-worm, (no date). Morris Worm, Accessed on: 2023-12-06

[3] https://owasp.org/Top10/A00_2021_Introduction/#how-the-categories-are-structured, (2021). OWASP Top 10 - 2021, Accessed on: 2023-12-06

[4] https://owasp.org/Top10/A01_2021-Broken_Access_Control/, (2021). A01:2021 – Broken Access Control, Accessed on: 2023-12-06

[5] https://owasp.org/www-project-proactive-controls/v3/en/c7-enforce-access-controls, (no date). C7: Enforce Access Controls, Accessed on: 2023-12-06

[6] Othmane, L. B., Jamil, A.-M., Abdelkhalek, M. (2019). Identification of the Impacts of Code Changes on the Security of Software, Proceedings of the 2019 IEEE 43rd Annual Computer Software and Applications Conference (COMPSAC), IEEE, ISSN 0730-3157, DOI 10.1109/COMPSAC.2019.10268

[7] https://owasp.org/Top10/A02_2021-Cryptographic_Failures/, (2021). A02:2021 – Cryptographic Failures, Accessed on: 2023-12-06

[8] Moric, Z.; Pinjuh, J. & Kurelic, S. (2021). Securing communications in chatter application. Proceedings of DAAAM, at Vienna, Austria, ISSN 1726-9679, DOI 10.2507/32nd.daaam.proceedings.013

[9] Blahova, M.; Mikulicova, M. & Hromada, M. (2020). Utilization of Fractal Geometry Possibilities for Information Systems Security. Proceedings of DAAAM, at Vienna, Austria, ISSN 1726-9679, DOI 10.2507/31st.daaam.proceedings.085

[10] https://brightsec.com/blog/code-injection/, (2022). Moradov, O.; Code Injection in Brief: Types, Examples, and Mitigation, Accessed on: 2023-12-06

[11] https://owasp.org/Top10/A04_2021-Insecure_Design/, (2021). A04:2021 – Insecure Design, Accessed on: 2023-12-07

[12] https://crashtest-security.com/insecure-design-vulnerability/, (2022). Sengupta, S.; Insecure Design, Accessed on: 2023-12-08

[13] https://www.aquasec.com/cloud-native-academy/supply-chain-security/security-misconfigurations/, (2023). Security Misconfiguration: Types, Examples & Prevention Tips, Accessed on: 2023-12-09

[14] https://owasp.org/Top10/A05_2021-Security_Misconfiguration/, (2021). A05:2021 – Security Misconfiguration, Accessed on: 2023-12-08

[15] Stanciu, A.-M.; Ciocarlie, H. (2023). Analyzing Code Security: Approaches and Tools for Effective Review and Analysis, Proceedings of the 2023 International Conference on Electrical, Computer and Energy Technologies, IEEE, ISBN:979-8-3503-2782-3, DOI 10.1109/icecet58911.2023.10389326

[16] Vukasovic, M.; Miladinovic, D.; Milakovic, A.; Vuletic, P.; Stanisavljevic, Z. (2022). Programming Applications Suitable for Secure Multiparty Computation Based on Trusted Execution Environments, Proceedings of the 2022 30th Telecommunications Forum (TELFOR), IEEE, DOI 10.1109/TELFOR56187.2022.9983726

[17] https://hbr.org/2019/08/every-computer-science-degree-should-require-a-course-in-cybersecurity, (2019). Cable, J.; Every Computer Science Degree Should Require a Course in Cybersecurity, Accessed on: 2023-12-13

[18] https://owasp.org/www-project-application-security-verification-standard/, (2022). OWASP Application Security Verification Standard, Accessed on: 2023-12-13

[19] https://www.linkedin.com/advice/0/what-most-effective-secure-coding-standards#cert-secure-coding-standards, (2023). What are the most effective secure coding standards for system development?, Accessed on: 2023-12-13

[20] https://cwe.mitre.org/about/index.html, (no date). About CWE, Accessed on: 2023-12-13

[21] https://snyk.io/learn/secure-coding-standards/, (no date). The Three Pillars for Implementing Secure Coding Standards, Accessed on: 2023-12-13

[22] Xu, J.; Yuan, X. (2015). Developing a course module for teaching cryptography programming on Android, Proceedings of the 2015 IEEE Frontiers in Education Conference (FIE), IEEE, DOI 10.1109/FIE.2015.7344086

[23] Yu, H.; Jones, N.; Bullock, G.; Yuan, X. (2011). Teaching secure software engineering: Writing secure code, Proceedings of the 2011 7th Central and Eastern European Software Engineering Conference (CEE-SECR), IEEE, ISBN:978-1-4799-8454-1, DOI 10.1109/CEE-SECR.2011.6188473

[24] Moric, Z.; Branstett, L.; Petrunic, R. (2022). Static-Analysis Techniques of Malware Reverse Engineering, Proceedings of DAAAM, Vienna, Austria, ISSN 1726-9679, DOI 10.2507/333rd.daaam.proceedings.024

DOI: 10.2507/35th.daaam.proceedings.026

QUALITY ASSESSMENT IN PETG 3D PRINTING

Edwing Isaac Rosales Molina, Emelly Adriana Barrera Domínguez,
Roxana Yamileth Cárcamo Alvarado & Diana Cruz

This Publication has to be referred as: Rosales Molina, E[dwing] I[saac]; Dominguez, E[melly] A[driana Barrera]; Alvarado, R[oxana] Y[amileth Cárcamo] & Cruz, D[iana] (2024). Quality Assessment in PETG 3D Printing, Proceedings of the 35th DAAAM International Symposium, pp.0199-0203, B. Katalinic (Ed.), Published by DAAAM International, ISBN 978-3-902734-44-0, ISSN 1726-9679, Vienna, Austria
DOI: 10.2507/35th.daaam.proceedings.026

Abstract

Nowadays, 3D printing has become more popular due to advancements in printing technology and materials. Nevertheless, it is necessary to continue expanding our understanding about it, as having a better comprehension of the relationship between printing parameters and part quality improves material use, energy consumption, and time efficiency. Furthermore, it can foster greater confidence in this manufacturing process. In this context, this study presents a quality assessment of PETG parts printed by Fused Filament Fabrication (FFF). The objective of this study is to contribute to this matter by describing the quality of PETG-printed parts produced with a cold bed printer under specific settings (temperature, speed, piece length, and infill percentage). It was identified that, under different conditions, warpage was the most prevalent defect in the experimental probes used. Qualitatively, it was observed that the surface finish was better at a lower speed (100 mm/min). In all conditions, parts with dimensions larger than the nominal dimensions were obtained. Comparatively, the smaller dimensions appear to be further away from the nominal value. The results observed imply that the CAD models must be adjusted to achieve a higher level of accuracy in the final dimensions of the product.

Keywords: 3D printing; defects; PETG; quality; FDM; FFF.

1. Introduction

At the present time, 3D printing is part of our life, as it has become more popular due to advancements in printing technology and materials. The market offers a wide range of prices for equipment and supplies, and there are technologies that do not require special installation or training for use. It is also worth noting that 3D printing has found applications in several different fields, including medicine, engineering, education [1] and more.

One of the most widely used technologies is Fused Deposition Modelling (FDM), a term trademarked by Stratasys [2], which is also known as Fused Filament Fabrication (FFF). This process entails passing a filament through an extruder, whereby the filament is softened to create the product layer by layer. The filament is typically composed of a polymeric material; however, composite filaments (fibre-reinforced polymers) are already commercially available. Research has been done on different aspects of the FFF process. Some studies have explored mechanical properties such as tensile strength [3], hardness and flexural strength.

In this respect, finding out how printing parameters affect these properties has been of particular interest [4], [5]. This has also included analysing post-processing operations [6], such as acetone chambering or heat treatment. In addition to the mechanical properties of the product, it has also been relevant to evaluate the extent to which the final geometry aligns with the desired design specifications. Accuracy [2], and precision [7] are two variables that have been also deemed important.

Other studies have been conducted with the vision of identifying new applications for the 3D printing process. It has been discovered, for instance, that the process can be used to create laminates comprising different materials [8], or to print on a substrate [9]. Nevertheless, it is necessary to continue improving the understanding about it. Good comprehension of the printing process improves part quality, material use, energy consumption, and time efficiency. Furthermore, it can foster greater confidence in this manufacturing process.

Although the use of heated bed printers is recommended for printing PETG parts, certain circumstances may motivate the use of cold bed technology, such as the unavailability of heated bed printers or the need to print larger components. Additionally, based on prior empirical experience, quality issues can occur when using this technology to print large parts with the same printer settings that do not produce defects in smaller parts. Therefore, it is important to understand the quality of parts produced in this context. The objective of this study is to contribute to this matter by describing the quality of PETG-printed parts produced with a cold bed printer under specific settings (temperature, speed, piece length, and infill percentage). In the context of the study, quality was understood as the absence of visual defects—such as warping, delamination, and poor adhesion in the top layer—as well as compliance with the expected dimensions of the printed pieces.

2. Experimental Procedure

A part was designed and printed to observe defects and measure different dimensions (Fig. 1). Sixteen pieces were made setting eight treatments (eight combinations of temperature, speed, and infill, as detailed in Table 1). Layer height was a held-constant variable (0.15 mm). The material used was a PETG filament (diameter 1.75 mm) from Zhuhai Sunlu Industrial Co. A BQ Witbox 2 3D printer with cold plate was used. Printing trajectories were programmed using Ultimaker Cura v.5.7.1. The part was oriented at 45° measured from the x-axis to obtain the longest possible travels (Fig. 2) as previous experience had shown that long paths led to defects.

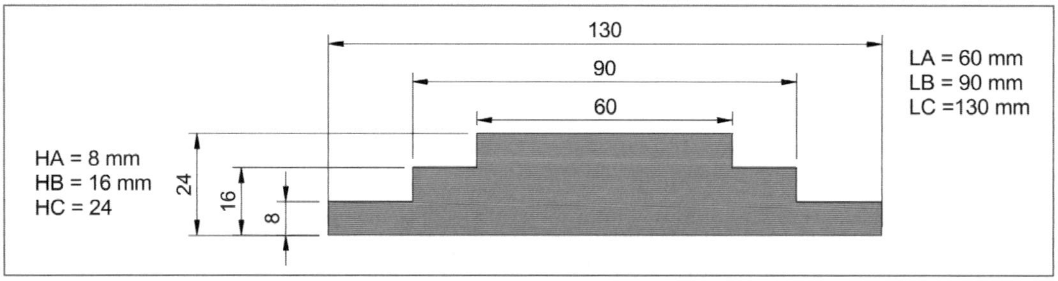

Fig. 1. Specimens dimensions.

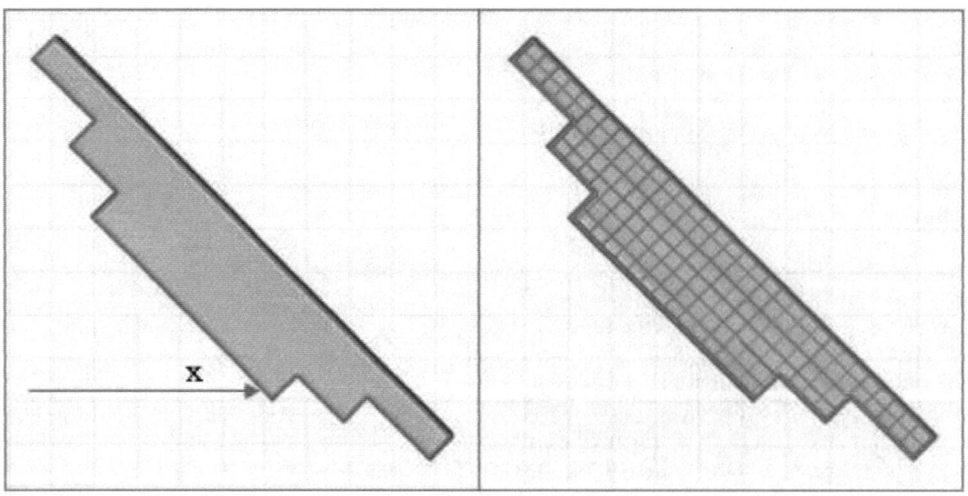

Fig. 2. Printing orientation and type of infill

Temperature (°C)	Print speed (mm /min)	Infill (%)
220	100	10
235	200	30

Table 1. Printing conditions

The printed parts were checked for defects and measured. Three heights and three lengths were measured for each part. The heights and lengths were plotted to identify trends. One limiting condition experienced was maintaining a uniform working temperature in the equipment. This was issue was particularly observed when operating near the maximum configurable printer limit (240 °C).

3. Results and Discussion

3.1. Qualitative analysis

Each specimen was subjected to a visual inspection to identify any defects (Fig. 3). It is important to consider that the visual inspection used to assess defects may have limitations in sensitivity or accuracy. The most prevalent defects observed were warping and warpage. Less frequently, delamination was also noted. In a few instances, waviness of the layers was also observed. Warpage was observed in every piece (Fig. 4).

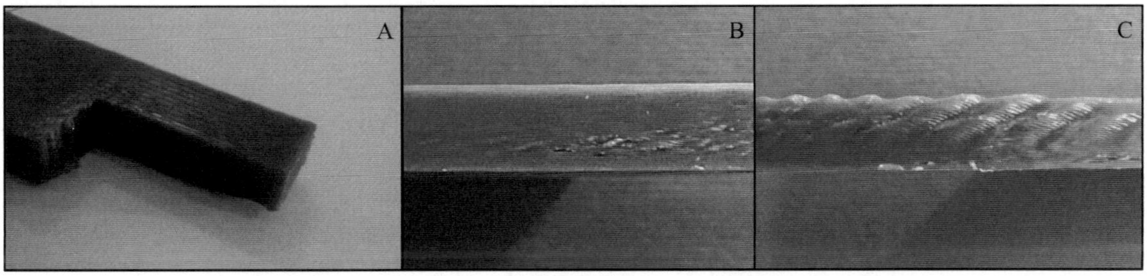

Fig. 3. Some defects identified: (A) warping, (B) delamination, and (C) waviness.

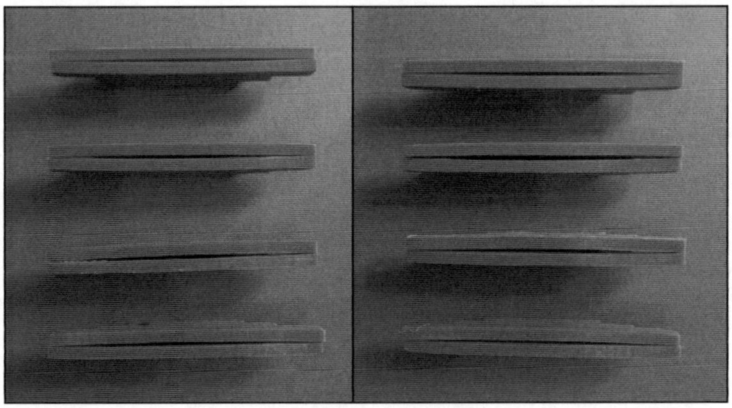

Fig. 4. Warpage in printed parts

Warpage is often attributed to residual stresses that are generated by different causes. In this instance, the deformation may be attributed to the discrepancy in cooling rates between the layers built on the bed and those deposited on top of other layers. In [10] the problem of these cooling rate differences is discussed extensively. In that study, it was observed that increasing the speed resulted in the bottom and top layers exhibiting a 15% and 13% decrease in residual stress, respectively; a drop-in warpage (~30%) was observed for both layers. In terms of surface finish, printing speed appears to be a key parameter. A better finish on the top layer was observed for parts produced at a lower speed (100 mm/min). This contrasts with the findings of the [11] study, which indicate that higher speeds (150 m/min) result in superior surface quality. It should be noted that in the case of [11], the material was PLA, whereas in the current study, it is PETG, and printing speed is considerably higher than it would be typically used for this material. It may be necessary to operate at this high speed to reduce the time required for printing and to exploit the properties of PETG. It is important to note that the decision to proceed under these conditions will inevitably result in a compromise of surface quality.

3.2. Quantitative analysis

To perform a quantitative assessment, the difference between the measured value and the nominal was calculated. Fig. 5 illustrates the magnitude of each discrepancy. The measured dimensions were larger than the nominal dimensions in all cases. It can also be seen that, with a few exceptions, the smaller dimensions (HA, HB, HC) are further away from their respective nominal values.

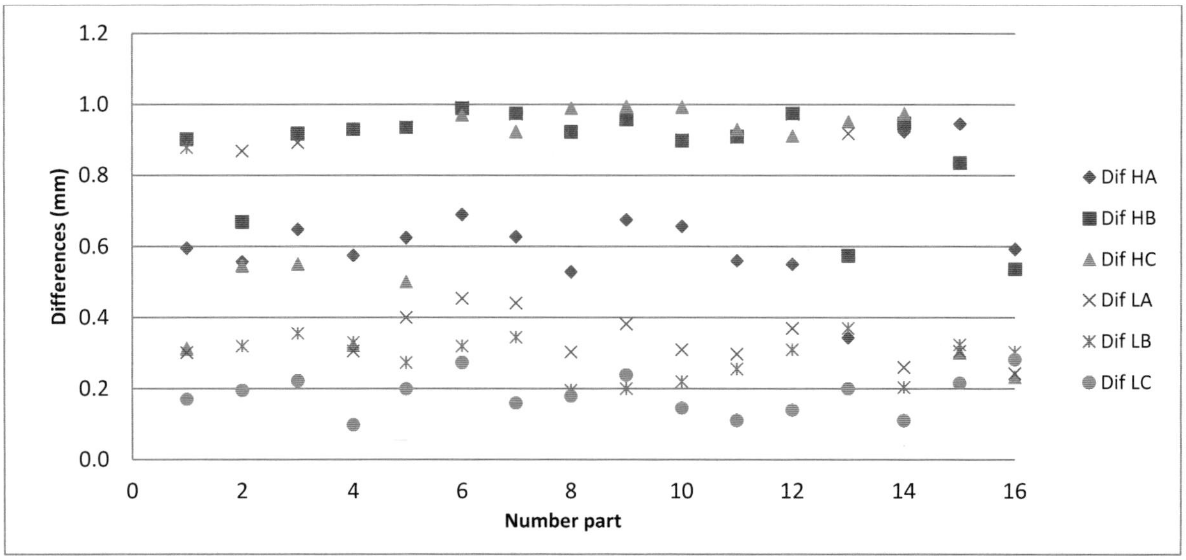

Fig. 5. Calculated deviations from nominal dimensions

4. Conclusions

This paper presents a quality assessment of 3D printed PETG parts under specific printing settings using cold bed technology in terms of the defects predominantly observed and conformance to designed dimensions. The results suggest that:

1. The most prevalent defects observed were warping and warpage. Less frequently, delamination was also noted. In a few instances, waviness of the layers was also observed.
2. Better surface finish was observed at the low printing speed (100 mm/min). It is possible to utilise a higher speed; however, this would have a detrimental impact on the quality of the workpiece.
3. In all instances, the dimensions exceed the nominal values of the model. This must be considered when generating three-dimensional models using computer-aided design (CAD) software. The largest discrepancies appear to occur for the smallest dimensions.
4. The results observed imply that the CAD models must be adjusted to achieve a higher level of accuracy in the final dimensions of the product, considering that under the analysed conditions, the process produces parts larger than expected.
5. Lack of adhesion was not a significant issue under the studied conditions, which would prevent the total rejection of the part as opposed to the effects of the other defects that occurred in the study.

This study reinforces the need to conduct further experimental research in order to have a better understanding of the factors that influence both the presence of visual defects and conformance to specified dimensions, considering the potential association between both quality issues. These efforts could also include the development of multiple regression models to estimate the dimensions of the 3D designs, with the objective of reducing the deviations from the nominal value along with the presence of shape defects. Other lines of research could include quality assessments regarding other dimensional characteristics of relevance based on the potential applications of the parts fabricated.

5. References

[1] Kroft L., Bicova K., & Milsimerova A. (2020). "Use of 3D printing technology in teaching the basics of milling technology," in Annals of DAAAM and Proceedings of the International DAAAM Symposium, DAAAM International Vienna, pp. 323–327, doi: 10.2507/31st.daaam.proceedings.045.

[2] Polak R., Sedlacek F., & Raz K. (2017). "Determination of FDM printer settings with regard to geometrical accuracy," in Annals of DAAAM and Proceedings of the International DAAAM Symposium, Danube Adria Association for Automation and Manufacturing, DAAAM, pp. 561–566, doi: 10.2507/28th.daaam.proceedings.079.

[3] Guessasma S., Belhabib S., & Nouri H., (2019). "Printability and tensile performance of 3D printed polyethylene terephthalate glycol using fused deposition modelling," Polymers (Basel), vol. 11, no. 7, doi: 10.3390/polym11071220.

[4] Loskot J. et al., (2023). "Influence of print speed on the microstructure, morphology, and mechanical properties of 3D-printed PETG products," Polym Test, vol. 123, doi: 10.1016/j.polymertesting.2023.108055.

[5] Shergill K., Chen Y., & Bull S., (2023). "What controls layer thickness effects on the mechanical properties of additive manufactured polymers," Surf Coat Technol, vol. 475, doi: 10.1016/j.surfcoat.2023.130131.

[6] Marković M. P., Cingesar I. K., & Vrsaljko D., (2023) "Maximizing Mechanical Performance of 3D Printed Parts Through Process Parameter Optimization," 3D Print Addit Manuf, doi: 10.1089/3dp.2023.0170.

[7] Bodur O., Stepanek V., Walcher E. M., & Durakbasa N. (2020). "Precision in additive manufacturing, optimization and evaluation of the accuracy of 3D printer based on GPS system," in Annals of DAAAM and Proceedings of the International DAAAM Symposium, DAAAM International Vienna, pp. 963–972, doi: 10.2507/31st.daaam.proceedings.134.

[8] Khan I. et al., (2023) "Parametric investigation and optimisation of mechanical properties of thick tri-material based composite of PLA-PETG-ABS 3D-printed using fused filament fabrication," Composites Part C: Open Access, vol. 12, doi: 10.1016/j.jcomc.2023.100392.

[9] Ražić S. E., Ludaš A., Kaurin T., & Zonjić T. (2023). "Applicability of polymers printed on textiles with a 3D printer for possible use in car interior," in IOP Conference Series: Earth and Environmental Science, Institute of Physics, doi: 10.1088/1755-1315/1128/1/012027.

[10] Samy A. A., Golbang A., Harkin-Jones E., Archer E., Tormey D., & McIlhagger A. (2021). "Finite element analysis of residual stress and warpage in a 3D printed semi-crystalline polymer: Effect of ambient temperature and nozzle speed," J Manuf Process, vol. 70, pp. 389–399, doi: 10.1016/J.JMAPRO.2021.08.054.

[11] Yadav A., Poorna Prakash B., Sai Dileep K., Arjuna Rao S., & Veeresh Kumar G. B. (2023). "An experimental examination on surface finish of FDM 3D printed parts," Mater Today Proc, doi: 10.1016/J.MATPR.2023.07.088.

DOI: 10.2507/35th.daaam.proceedings.027

FUZZY LOGIC-BASED SECURITY SYSTEM FOR REMOTE OUTDOOR CABINETS

Selahattin Mert Aydın, Emre Keklik, Anıl Akdoğan & Ali Serdar Vanlı

This Publication has to be referred as: Aydin, S[elahattin] M[ert]; Keklik, E[mre]; Akdogan, A[nil] & Vanli, A[li] S[erdar] (2024). Fuzzy Logic-Based Security System for Remote Outdoor Cabinets, Proceedings of the 35th DAAAM International Symposium, pp.0204-0209, B. Katalinic (Ed.), Published by DAAAM International, ISBN 978-3-902734-44-0, ISSN 1726-9679, Vienna, Austria
DOI: 10.2507/35th.daaam.proceedings.027

Abstract

Ensuring the security of cabinets located in areas far from physical access is of great importance. In such locations, various sensors are used to enhance the safety of these cabinets. These sensors monitor the surroundings of the cabinets and provide the necessary data to improve security. Fuzzy logic is employed to process this data effectively. With the advancement of technology, the importance of using fuzzy logic has significantly increased. In this study, a security system application for outdoor cabinets is designed using fuzzy logic. The primary goal of the study is to ensure the safety of cabinets in areas that are distant from human access. The details of this system are thoroughly explained in the following sections.

Keywords: Fuzzy Logic; Outdoor Cabinets; Security System; MATLAB; Fuzzy Logic Toolbox; Sensor

1. Introduction

Fuzzy logic enables the definition of variables and rules in a clear and flexible manner during modelling, though without requiring certainty. This method is particularly utilized in control engineering, especially in systems where traditional mathematical models are difficult to establish, or where known methods do not provide sufficient efficiency [1]. The advantages offered by fuzzy logic are as follows [2]:

- It closely aligns with human thinking processes and methods.
- Its application does not necessarily require a mathematical model.
- The simplicity of the software allows for the system to be established more economically.
- The use of membership degrees makes this technique more flexible than other control methods.
- It allows for the use of information that does not yield a definite outcome.
- It enables the modelling of nonlinear functions.
- A fuzzy logic-based model can be easily designed using only the experiences of experts.
- It is compatible with traditional control techniques [5].

Fuzzy logic is a mathematical system developed to handle uncertainties and imprecise information. Unlike traditional binary logic, this system allows for a more natural way of modeling uncertainties and vagueness present in the real world. The fundamental concepts of fuzzy logic include fuzzy sets, membership functions, if-then rules, fuzzification, inference, and defuzzification [3]. After deciding to design a fuzzy system, the first thing to do is to obtain "if-then" fuzzy rules. These rules are made with the help of experts. The fuzzy control chart is shown in Figure 1 below[4].

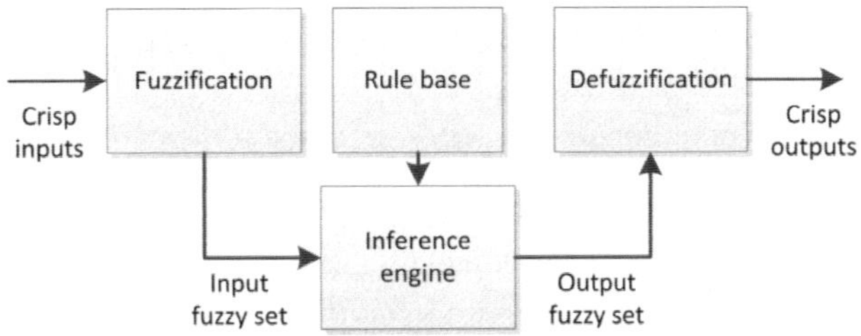

Fig. 1. Block Diagram of Fuzzy Logic

In the creation of the rule base; "Mamdani" method was used as fuzzy inference system. The Mamdani method is a system that transforms membership functions into fuzzy sets [10]. In the Mamdani system, the output of each rule is the element of the fuzzy set [11]. Outputs are obtained after the data taken as input are fuzzificated and evaluated in the database created by the expert[6]. In our study, these outputs were obtained using MATLAB application. Outdoor cabinets are used to store many electronic equipment, from battery packs to telecom equipment [7]. Currently, most outdoor cabins are installed on roadsides where environmental conditions are sometimes extremely harsh [8]. The safety of cabinets in places where human access is not easy is important. With this study, the security of outdoor cabinets has been tried to be increased. Although these cabinets are generally protected with locks, in remote areas, additional alert systems may be necessary to detect unauthorized access or potential damage [9].

With this study, it is aimed to increase the security of outdoor cabinets with a security system designed with the help of fuzzy logic. It is desired to design a security system that converts the data received from the sensors into an output with fuzzy logic for these cabinets located in places far from people.

2. Application

2.1. General Structure of the System

In this part of the study, the input values required for the control of the cabinet security system and the determination of the output values that will occur because of processing these values are shown [4]. In this study, we will examine the design of an expert system, which is controlled by fuzzy logic controllers with its general boundaries, makes the necessary controls for a safe cabin security system and can produce results accordingly. First, important, and relevant input and output values are determined. Then, the subset ranges are explained, and the shape and positions of the fuzzy subsets are determined. Then, fuzzy "if-then" rules are determined, and calculations are made. Here, we measure the force applied to the door and the smoke density in the cabin with the help of sensors we have placed at certain points on the door. In line with the values we have obtained, a possible door forcing, assault, etc. It is aimed to determine the danger by using fuzzy logic, with the warnings to be given to the environment with sound and light warnings and the warnings to be conveyed to the relevant units. In our study, a structure was considered to give warning from inside the cabin. According to the data to be obtained from the sensors, an alarm system will be used that will give loud noise and high flux light to the environment when necessary. By placing this system inside the cabin, it will give out sound and light through the spaces on the cabin and warn a living thing that may harm the cabin. This alarm system is shown in Figure-2.

Fig. 2. Outdoor Cabinet and Alarm System

2.2. Determining Input and Output Values

To prevent dangers that may occur because of violence or a fire that may occur in the cabinet, a system that gives an audible and visual alarm to the environment and sends the necessary warnings to the relevant units is controlled. In this direction, the values taken from the sensors placed in the cabinet were determined as input values, and the warning system level and Sound-Light alarm level were determined as output values.

Smoke(S) and Force(F) were determined as input parameters. The clusters of these values are shown in Figure 3.

S = { ZS, LS, MS, HS } (Zero Smoke, Low Smoke, Mid-Level Smoke, High Smoke)

F = { ZF, LF, MF, HF, UF } (Zero Smoke, Low Smoke, Mid-Level Force, High Force, Ultra Force)

Fig. 3. Input Values

The output parameters were determined as Warning System (W) and Light-Sound (VS). The clusters of these values are shown in Figure 4.

W = { NW, LW, MW, HW, UW } (No Warning, Low Warning, Mid-Level Warning, High Warning)

VS = { ZVS, LVS, HVS } (Zero Voice-Light, Low Voice-Light, High Voice-Light)

Fig. 4. Output Values

2.3. Creating the Fuzzy Rule Table

In this section, rules are determined by evaluating all the variables and possible differences. These rules have been designed under the conditions that may occur according to the values we have determined. The input to the system is defined in the "if-then" structure to be analysed on the basis of fuzzy logic, and the output data is processed. The rule table we created as a result of these data is shown in Figure 5.

		F				
		ZF	LF	MF	HF	UF
S	ZS	NW ZVS	LW LVS	MW LVS	HW HVS	HW HVS
	LS	LW LVS	LW LVS	MW LVS	HW HVS	HW HVS
	MS	MW LVS	MW LVS	HW HVS	HW HVS	HW HVS
	HS	HW HVS	HW HVS	HW HVS	HW HVS	HW HVS

Fig. 5. Rule Table

2.4. MATLAB Application

In this section, fuzzy logic inference of the system is made using the Fuzzy Logic Designer plug-in in MATLAB application in line with the rules we have determined in the previous sections. While making this design, the Mamdani method was chosen. The design prepared on MATLAB is shown in Figure 6.

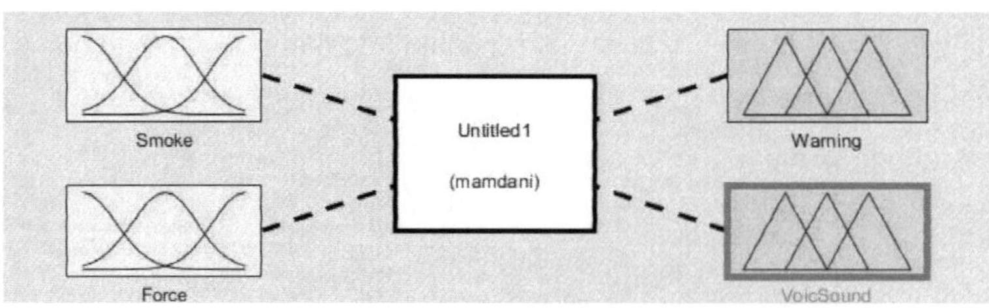

Fig. 6. Fuzzy Logic Design

After the rule table determined according to the inputs and outputs, we used in the design is prepared in the "if-then" structure. Membership Function Editor is used to determine membership functions on the application. With this editor, each input and output parameter is shown in graphically determined values. The graphs of the membership functions of the input and output parameters plotted on the application are shown in Figure 7, Figure 8, Figure 9 and Figure 10.

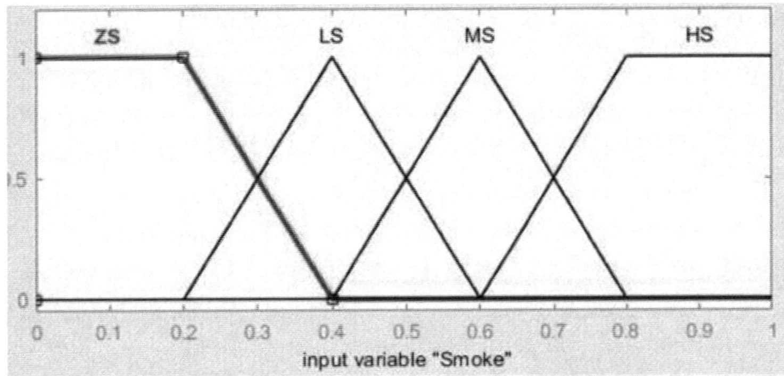

Fig.7. Membership Function Graph of Smoke Input

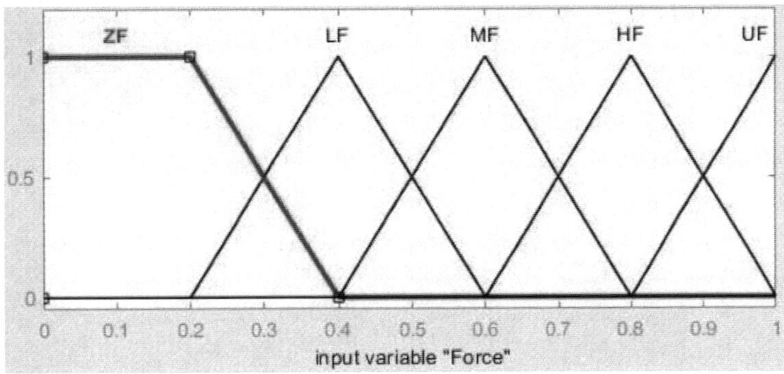

Fig. 8. Membership Function Graph of Force Input

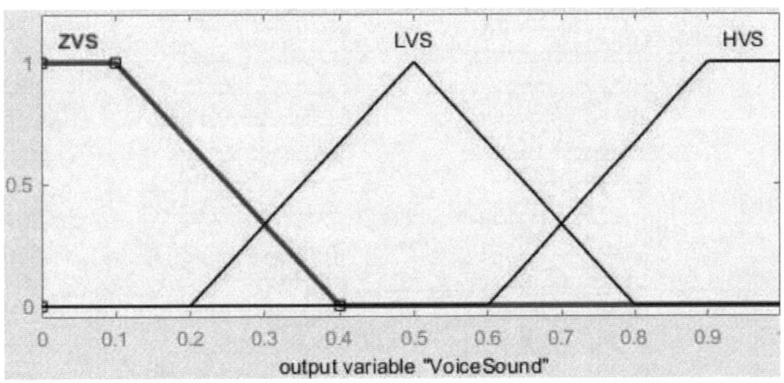

Fig. 9. Membership Function Graph of VoiceSound Output

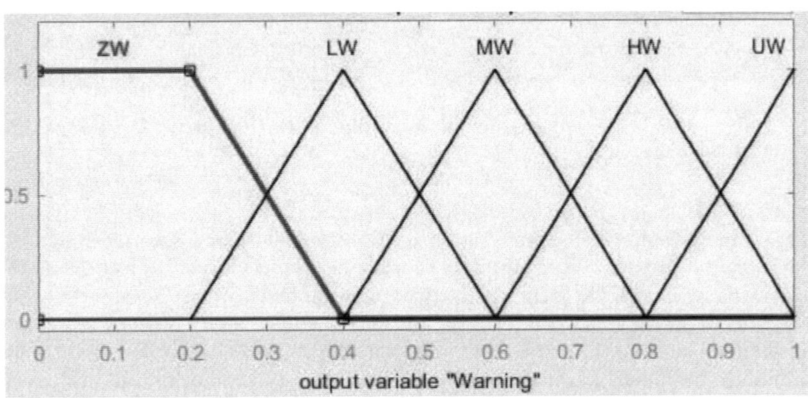

Fig. 10. Membership Function Graph of Warning Output

After the membership functions graphics were determined, results were prepared on the application. As a result of the design made with Fuzzy Logic Designer, the densities and 3D graphics of the outputs can be seen by changing the values of the inputs with the Rule Viewer window. The 3-dimensional graphics of the two outputs in the resulting state are shown in Figure 11. With these values, it can be observed by simulating the results that may occur according to the differences of the values coming from the real-life sensors of the system. The situations that may occur according to several different states of the input values are shown in Figure 12.

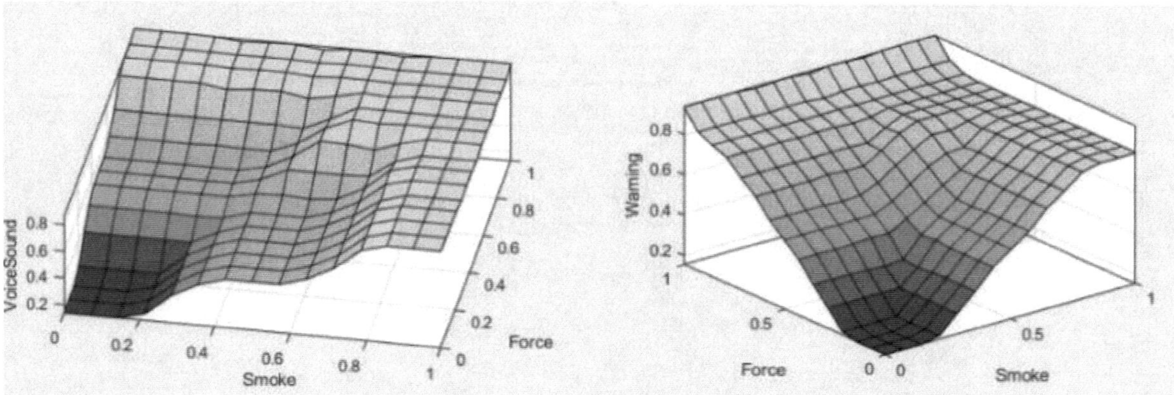

Fig. 11. 3D Graph of Outputs

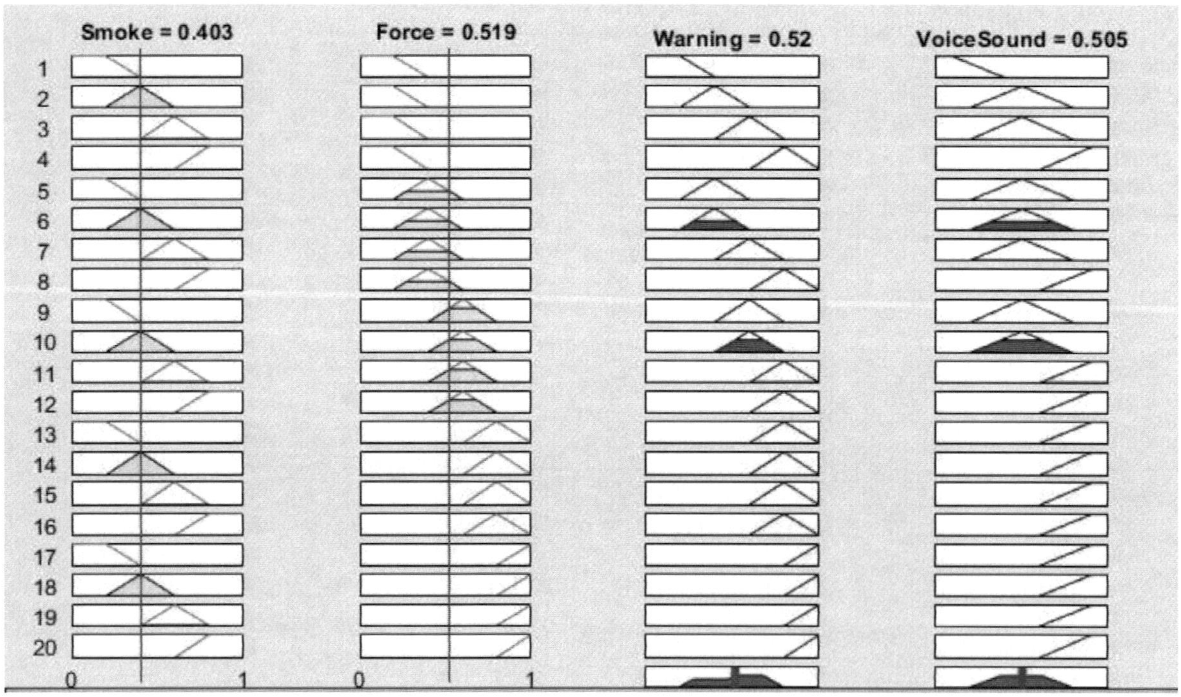

Fig. 12. Display of Inputs and Outputs on Rule Viewer

3. Conclusion

This study focused on the design and implementation of a fuzzy logic-based security system for remote outdoor cabinets. The primary problem addressed was ensuring the security of cabinets located in isolated areas, where traditional security measures may not be sufficient due to the difficulty of physical monitoring. To solve this problem, we designed a system that utilizes sensors to detect smoke and force applied to the cabinet, which are then processed using fuzzy logic to determine appropriate alarm responses, such as activating sound and light warnings. The fuzzy logic approach was chosen because of its ability to handle uncertain and imprecise data, which is common in real-world scenarios like security systems. By creating a set of "if-then" rules and employing the Mamdani method within the MATLAB environment, the system was able to effectively interpret sensor inputs and produce relevant outputs that trigger security alerts.

The results of the system showed that it could accurately assess potential dangers and respond with appropriate warnings, thereby enhancing the security of outdoor cabinets in remote locations. What is solved with this system is the ability to provide an automated and intelligent response to threats, such as forced entry or fire, without the need for constant human oversight. This solution reduces the risk of damage or unauthorized access to valuable equipment housed in these outdoor cabinets.

In terms of future work, this system can be expanded and improved by integrating more advanced sensors or incorporating additional environmental factors such as temperature and humidity. Moreover, future developments could involve connecting the system to a central monitoring unit for real-time alerts and interventions, making it even more responsive and effective in safeguarding remote infrastructures.

4. References

[1] Zadeh, L. A. (1965). "Information and control," Fuzzy Sets, vol. 8, no. 3, pp. 338-353, 1965.

[2] Hacımurtazaoğlu M. (2013), "Bulanık Mantık ile Manyetik Kilit Uygulaması", Akademik Bilişim– XV. Akademik Bilişim Konferansı Bildirileri

[3] Dubois D. & Prade H.M. (1980). Fuzzy Sets and Systems: Theory and Applications, Mathematics in Science and Engineering, v. 144. New York: Academic Press, pp. 9-34.

[4] Ödük, M. N. (2019). "Bulanık mantık yöntemi ve uygulamaları"

[5] Zadeh, L.A. (1965). "Fuzzy Logic," Computer, vol. 21, no. 4, pp. 83-93, 1988.

[6] Veselov, O[leg]; Veselov, A[lexandr] & Svetushenko, S[tanislav] (2019). The Use of Fuzzy Logic in Automatic Control Systems, Proceedings of the 30th DAAAM International Symposium, pp.0897-0903, B. Katalinic (Ed.), Published by DAAAM International, ISBN 978-3-902734-22-8, ISSN 1726-9679, Vienna,Austria

[7] Meng X.Z.; Lu Z.; Jin L.W.; Zhang L.Y.; Hu W.Y.;Wei L.C. & Chai J.C. (2015). Experimental and numerical investigation on thermal management of an outdoor battery cabinet, Applied Thermal Engineering 91, 210e224

[8] Delgado C. B.; Silva P. D.; Pires, L. C. & Gaspar, P. D.(2017). Experimental study and numerical simulation of the interior flow in a telecommunications cabinet, Energy Procedia 142, 3096–3101

[9] Aydin, S[elahattin] M[ert]; Keklik, E[mre]; Genes, E[ngin]; Akdogan, A[nil] & Vanli, A[li] S[erdar] (2022). Development of Pneumatic Tester for Life Testing of Industrial Electronic Rotary Latches, Proceedings of the 33rd DAAAM International Symposium, pp.0466-0470, B. Katalinic (Ed.), Published by DAAAM International, ISBN 978-3-902734-36-5, ISSN 1726-9679, Vienna, Austria

[10] Mamdani, E.H. & Assilian, S. (1975). "An experiment in linguistic synthesis with a fuzzy logic controller," International Journal of Man-Machine Studies, vol. 7, no. 1, pp. 1-13

[11] MathWorks Inc. (2021, October 24). Foundations of Fuzzy Logic [Online]. Available: https://www.mathworks.com/help/fuzzy/types-of-fuzzy-inference-systems.html

DOI: 10.2507/35th.daaam.proceedings.028

SECURITY PRINCIPLES IN CLOUD COMPUTING

Damir Josic, Matej Basic*, Luka Zgrablic

This Publication has to be referred as: Josic, D[amir]; Basic, M[atej] & Zgrablic, L[uka] (2024). Security Principles in Cloud Computing, Proceedings of the 35th DAAAM International Symposium, pp.0210-0215, B. Katalinic (Ed.), Published by DAAAM International, ISBN 978-3-902734-44-0, ISSN 1726-9679, Vienna, Austria
DOI: 10.2507/35th.daaam.proceedings.028

Abstract

This paper provides a comprehensive overview of cloud security principles, highlighting the critical areas of data confidentiality, identity security, access controls, and the role of monitoring and event logging in maintaining secure cloud environments. It addresses organizations' everyday challenges, such as improper identity and access management, insecure APIs, and potential unauthorized access risks. The discussion includes detailed strategies for mitigating these issues, such as implementing robust authentication protocols, network segmentation, continuous security assessments, and robust employee training programs. The paper also highlights the importance of multifactor authentication (MFA) and the innovative use of passwordless authentication to enhance security and user experience. It explores using SIEM (Security Information and Event Management) and SOAR (Security Orchestration Automation and Response) technologies to bolster cybersecurity through real-time analysis and automated incident response.

Keywords: cloud security; data confidentiality; Multi-Factor Authentication; identity security; Security Information and Event Management.

1. Introduction

Cloud computing has developed as a revolutionary technology, altering how corporations store, analyse, and manage data. It provides scalable resources, cost-effectiveness, and adaptability, enabling enterprises to adjust to evolving requirements. Nevertheless, these benefits are accompanied by considerable security challenges. The centralised structure of cloud services and the substantial volumes of data they manage render them a prime target for hackers. Maintaining data security, integrity, and availability is crucial for sustaining trust and promoting the broader adoption of cloud computing. This article examines essential security principles in cloud systems, concentrating on important aspects such as data encryption, identity and access management, and multi-factor authentication. Additionally, it analyses the significance of monitoring and event logging in improving cloud security. It addresses developing trends in applying Security Information and Event Management (SIEM) and Security Orchestration, Automation, and Response (SOAR) systems for real-time threat mitigation. This paper aims to present these challenges to readers and demonstrate ways to enhance their cloud security. It will explore strategies for managing access permissions, implementing advanced security tools, and the importance of real-time monitoring. The following sections will delve into critical areas such as Identity and Access Management (IAM), the importance of data privacy, and the necessity of monitoring security events as businesses navigate the complexities of cloud security.

2. Important Cloud Security Concepts

As cloud computing becomes increasingly popular for organisations, they face various challenges concerning data, identity, architecture security, and more. One of the primary challenges for organisations is taking accountability for their data. Users must manage access controls, implement best security practices, and monitor what is happening within their organisations and with their data. Many cloud security threats can be mitigated using proper configurations, correct security settings, cloud-native authentication protections, and various other security features [1].

Cloud computing security encompasses multiple layers, including infrastructure, data encryption, and identity management. A fundamental problem is data security, encompassing hazards such as data breaches and illegal access [2]. Numerous firms use encryption to guarantee data security, integrity, and availability. A multi-tiered encryption framework employing hybrid cryptographic methods, including RSA and Data Encryption Standard (DES), enhances security by implementing encryption at both user and server levels, substantially mitigating the risk of data breaches [3]. Alternative encryption techniques, such as attribute-based encryption (ABE) and proxy re-encryption (PRE), enhance the security of data exchange and storage across diverse settings [4]. A significant area for improvement in cloud security is data privacy, which is stored in shared environments. In public cloud installations with shared resources, safeguarding privacy becomes increasingly complex, necessitating sophisticated access controls and encryption technologies. Privacy-preserving methodologies, such as multi-factor authentication (MFA) and secure critical management systems, assist in mitigating concerns [5]. Cloud security also involves addressing vulnerabilities in Infrastructure-as-a-Service (IaaS), Platform-as-a-Service (PaaS), and Software-as-a-Service (SaaS) models. A prominent issue in IaaS and SaaS deployments is unauthorised access through weak authentication systems. Robust identity and access management solutions have been proposed to address these concerns, including federated identity systems and biometric authentication [6]. However, the challenge of securing virtual machines (VMs) against attacks like hyperjacking remains, as malicious users can exploit vulnerabilities in the hypervisor [7].

Implementing hybrid cloud models, which integrate public and private cloud environments, presents other security concerns, such as adhering to data protection rules and maintaining cross-cloud data integrity. Researchers have proposed alternatives to solve these difficulties, including end-to-end encryption and decentralised trust models [8]. Notwithstanding progress in cloud security technologies, human factors, including an inadequate understanding of security best practices, remain a substantial obstacle. Enhanced user knowledge, training, and improved governance norms can mitigate cloud security concerns [9].

This paper will focus on access controls, data security, and identity security in the following sections, using multifactor authentication to manage identity security. It will also explore the importance of event logging and how event-driven security can help detect, mitigate, and improve the overall security of cloud computing.

2.1. Access Controls and Identity Management in Cloud Environments

Access controls are implemented through an access management system that includes identity and credential management. These systems utilise various tools and policies to manage and define access to valuable resources such as data, systems, or physical assets. Effective access management is crucial to protecting these resources from unauthorised access. Familiar IAM (Identity and Access Management) challenges in cloud environments include improper provisioning or de-provisioning of services and users, maintaining inactive assigned users, managing numerous admin accounts, and users bypassing IAM controls. Implementing role-based access controls can also be challenging. Best practices to address IAM challenges in the cloud include developing governance strategies for identity management and using central directory services like Microsoft Active Directory to facilitate the provisioning, auditing, and de-provisioning of accounts.

All SaaS (Software as a Service) applications should support single sign-on (SSO) technology, requiring users to authenticate through SSO to ensure secure access. Organisations often need help safely and effectively providing employees access to systems and data. Access request management systems can help overcome these challenges, ensuring the organisation operates securely and efficiently [10][11]. Unauthorised access often occurs for several reasons, including weak authentication mechanisms, insider threats, insecure APIs, insufficient access controls, and social engineering or phishing attacks. To mitigate these and other unwanted scenarios, organisations should consider the following strategies:

- Strong Authentication and Access Controls: Implement robust authentication mechanisms and stringent access controls to ensure only authorised users can access cloud resources [10][11].
- Network Segmentation and Firewalls: Implement firewall controls for both inbound and outbound traffic, segment critical resources, and allow only authorised communication [10][11].
- Continuous Security Assessments: Perform regular security assessments to identify and address any improper access controls or configurations that could result in unauthorised access.
- Employee Training: Educate employees about security best practices, the risks associated with data breaches, and the importance of following security policies and procedures to prevent unauthorised access [10][11].

These strategies can help organisations protect their cloud resources from unauthorised access and other security threats. Effective access control and identity management are critical to maintaining the security and integrity of cloud environments.

2.2. *Identity Security and Multifactor Authentication*

Identity security in the cloud encompasses measures, policies, actions, and mechanisms to protect user and service account identities. This includes ensuring that only authorised users can access specific resources while maintaining the confidentiality and integrity of user and company data. Since cloud computing relies on the Internet to deliver services and resources, it is vulnerable to numerous security threats. Therefore, identity security in the cloud is essential for protecting users, company data, and other valuable resources.

To significantly reduce security risks, several policies can be implemented, including:

- Using strong and unique passwords for accounts and enabling multifactor authentication (MFA), especially for privileged accounts. This makes it significantly harder for attackers to gain unauthorised access [12].
- Allowing access to cloud accounts only through HTTPS and VPN. This ensures that data is encrypted during transmission, protecting it from interception [10].
- Regularly reviewing and updating security controls and settings in cloud accounts. This helps identify and mitigate vulnerabilities in real-time [13].
- Monitor cloud accounts for any unusual activity and investigate immediately if any occur. Continuous monitoring allows quick detection and response to potential security breaches [14].
- Staying informed about the latest security threats and best practices for managing identity security in the cloud. Keeping up-to-date with new threats and defences is crucial for maintaining robust security [15].

Multifactor authentication (MFA) is a core component of a strong identity and access management policy. MFA requires one or more verification factors, significantly decreasing the possibility of a successful cyber-attack. By requiring users to provide multiple verification factors, MFA minimises the potential for unauthorised access. Common authentication factors used in MFA include OTP (one-time passwords) generated by authenticator apps, software tokens, SMS-based passwords, physical USB tokens, and biometric authentication such as facial or fingerprint recognition. Less secure methods include personal security questions [15]. Implementing more secure methods and multifactor authentication makes it more challenging for unauthorised individuals to gain access to systems. MFA is also crucial in resisting brute-force attacks, phishing attacks, and other cyber-attacks. This extra layer of security operates on the principle that a user must present a username and password, approve the login using something they have (e.g., a phone), and provide something they are (e.g., fingerprint or facial recognition). Additional factors include time and location, where the time factor expects actions within an established time frame, and the location factor verifies if the user logs in from a known location [10]. Passwordless authentication is another method to enhance user security and simplify IT operations by eliminating the need to store, maintain, and rotate passwords. This method allows users to access applications or systems without entering a password, instead providing another form of evidence such as fingerprint, facial recognition, or hardware token code. It is often used alongside SSO to improve the user experience [5].

2.3. *Data Security*

As mentioned earlier, the Internet-based or public cloud model implies that data is transmitted over the Internet. Data losses or leakages can severely impact an organisation's reputation and lead to legal consequences. Therefore, data storage in remote data centres must be done with the utmost care. Key data security challenges include maintaining confidentiality, integrity, locality, etc. Classifying and treating confidential data appropriately is crucial to protect it from various attacks, such as cross-site scripting. Implementing robust encryption methods alongside access controls and monitoring suspicious activities is essential. Data should be encrypted both while stored ("data at rest") and while being transmitted ("data in transit"). Encrypting data during transmission prevents eavesdropping and unauthorised interception, while at rest, encryption ensures data protection even if physical storage media are compromised [10][12].

Maintaining data integrity is also vital to prevent data loss. Only authorised personnel should be able to modify the data, ensuring that unauthorised persons keep digital information unaltered. This guarantees that data is kept private, consistent, safe, and complete throughout its lifecycle [13][14]. Data locality is another significant challenge for organisations and cloud service providers. Data distributed across various regions can lead to compliance issues due to differing laws and regulations governing data. As data moves across different zones, the applicable laws may change, causing potential customer compliance issues. Data replication or backups for business continuity and disaster recovery can further complicate these issues [15].

2.4. *Monitoring and Event Logging*

Monitoring and event logging is crucial for IT teams to understand what is happening within their infrastructure. Significant actions or occurrences, known as events, can originate from networks, servers, firewalls, databases, operating systems, hardware infrastructures, or other sources. Event logging and monitoring are essential for auditing, compliance, security, troubleshooting, and alerting. Organisations can recognise unusual activity and improve security by monitoring environmental traffic.

Organisations address risks by setting up monitoring, logging, and alerting configurations to enhance cloud infrastructures' security, performance, and management, allowing security incidents to be detected before valuable data is stolen. Cloud monitoring encompasses a set of strategies and practices that will enable organisations to analyse, manage, and monitor the health, security, performance, and availability of their infrastructures and applications. It allows organisations to identify and address vulnerabilities or issues, preventing them from impacting business or end users. Regular audits through cloud monitoring ensure security standards and regulatory compliance. Modern cloud monitoring solutions include virtual network monitoring, machine monitoring, application performance monitoring, security and compliance monitoring, website monitoring, database monitoring, storage monitoring, and many others [1].

Some benefits of cloud monitoring include:

- Improving the security of cloud applications and networks.
- Enhancing service or application availability and performance due to quick issue reporting and solutions.
- Avoiding unexpected cloud costs.
- Usability on multiple devices.

SIEM (Security Information and Event Management) analyses security alerts generated by resources such as applications, networks, and servers [16]. Its essential functions are:

- Log collection – SIEM collects logs and event data from various sources.
- Data correlation and analysis – SIEM correlates data to identify patterns, anomalies, and security incidents.
- Alerts – SIEM generates alerts based on activities or events that may indicate a security threat.
- Compliance reporting – SIEM supports compliance efforts by providing reports based on security events and activities.

SIEM benefits include forensic analysis, threat detection, centralised visibility, and compliance management, making detecting, managing, and responding to security events easier [16]. SOAR (Security Orchestration, Automation, and Response) complements SIEM by focusing on task automation and incident response [16]. SOAR aims to automate security operations, identify specific events or threats in advance, and eliminate them through automated responses based on predefined workflows. This approach is known as a triggered outcome. SOAR can isolate infected systems, block malicious IP addresses, and deliver consistent and standardised responses [11]. SIEM and SOAR are often integrated to create comprehensive cybersecurity ecosystems [16]. SIEM allows for initial correlation and anomaly detection based on analysed data, while SOAR automates and orchestrates the response. This combined approach enhances infrastructure security by enabling quicker and more efficient incident response.

3. Future Works

Future research in cloud security is increasingly focused on advancing Zero Trust Architecture (ZTA), which guarantees that no user or device is automatically trusted, whether inside or external to the network. ZTA necessitates additional scrutiny to enhance authentication, access control, and the execution of detailed security policies. Integrating with sophisticated Security Information and Event Management and Security Orchestration, Automation, and Response systems is essential for improving real-time threat detection and automated incident response capabilities. These technologies facilitate expedited response times and enhance risk mitigation, yet their use in multi-cloud setups necessitates optimisation and scalability enhancements.

Artificial Intelligence (AI) and Machine Learning (ML) are essential in forecasting, identifying, and addressing cloud-based threats. Research should concentrate on creating more adaptive and intelligent models capable of detecting sophisticated attacks, including those employing adversarial strategies, while maintaining resource efficiency. Nonetheless, AI-driven solutions have problems regarding transparency and interpretability, requiring further research to ensure these systems are both successful and comprehensible.

Another area of research is the integration of homomorphic encryption and secure multi-party computation (SMPC), which could allow cloud services to process encrypted data without exposing sensitive information. These cryptographic methods hold promise but remain computationally expensive and challenging to scale for large cloud environments, requiring further optimisation.

Further research is needed in data privacy compliance, particularly in global cloud ecosystems where varying regional regulations, such as GDPR, CCPA, and new laws, continually evolve. More research is needed on developing privacy-preserving techniques that seamlessly comply with these regulations across borders, ensuring robust protection of user data while maintaining operational efficiency. Quantum computing is emerging as a potential threat to traditional encryption schemes, necessitating research into post-quantum cryptography. As quantum computers become more viable, securing cloud infrastructures against potential future quantum attacks is essential to avoid catastrophic data breaches.

Moreover, compliance with and enhancement of cloud security standards established by the National Institute of Standards and Technology (NIST) is essential. The NIST framework remains a fundamental resource, especially in risk management and incident response. Future research should investigate the alignment of upcoming technologies, such as AI, quantum computing, and edge computing, with NIST's growing standards to enhance cloud security.

4. Conclusion

One significant challenge for organisations is taking accountability for their data and configurations. Users must manage access controls, implement best security practices, and oversee their organisations' and information systems' activities. Access control is managed through systems that use identity and identification information, incorporating various tools and principles to safeguard critical resources, such as data, systems, and physical assets. A centralised directory service, like Microsoft Active Directory, is recommended to facilitate accounting, auditing, and de-provisioning, ensuring centralised control and visibility.

To mitigate undesirable scenarios, organisations should consider the following strategies:

- Strong Authentication and Access Control Measures: Ensure cloud resources are accessed only by authorised users.
- Network Segmentation and Firewalls: Implement firewalls for inbound and outbound traffic, segment critical resources, and permit only authorised communications.
- Continuous Security Assessments: Conduct security assessments regularly to identify and rectify any improper access control measures or configurations that could lead to unauthorised access.
- Employee Training: Educate employees on information security best practices, the risks of security breaches due to unauthorised access, their role in preventing breaches, and the importance of adhering to security policies and procedures.

Identity security in the cloud involves all measures, practices, operations, and mechanisms implemented to protect the identity of user and service accounts. Multifactor authentication (MFA) is a critical component of a strong identity and access control policy, requiring additional authentication factors that significantly reduce the likelihood of successful cyber-attacks. Implementing secure methods and MFA makes accessing systems more challenging for unauthorised individuals. MFA is also crucial in combating brute-force attacks, phishing attacks, and other cyber threats. Passwordless authentication can further enhance user security and streamline IT operations by eliminating the need for password management. Data loss or leakage can severely impact an organisation's reputation and have legal consequences. Data stored in remote data centres must be handled carefully, using robust encryption methods, access controls, and monitoring for suspicious activities. Data should be encrypted at rest and in transit to prevent unauthorised access and ensure data integrity. Only authorised personnel should be able to modify data to avoid data loss.

Monitoring and event logging are vital for IT teams to understand the activities within their infrastructure. Significant activities or events, considered as events, can originate from various sources, including networks, servers, firewalls, databases, operating systems, and hardware infrastructures. Cloud monitoring involves strategies and practices that allow organisations to analyse, manage, and monitor their infrastructures and applications' health, security, performance, and availability. SIEM systems collect, correlate, and analyse generated data and events, providing real-time analysis and alerts based on potential security threats. SOAR systems complement SIEM by focusing on task automation and incident response, aiming to eliminate threats through automated responses based on predefined workflows. By integrating SIEM and SOAR, organisations can create comprehensive cybersecurity ecosystems that enhance infrastructure security through quicker and more efficient incident response.

5. References

[1] Singh, U.; Tiwari, A. & Sharma, S. (2020). Data Security in Cloud Computing, International Journal of Engineering Applied Sciences and Technology, Published by IJEAST , Vol. 4, No. 10, pp.170-173., ISSN 2455-2143, Rajasthan, India. DOI: 10.33564/IJEAST.2020.v04i10.033, in press.

[2] Miryala, N. K.; & Gupta, D. (2019) Security in Cloud Computing, 2021 2nd International Conference on Intelligent Engineering and Management (ICIEM), pp. 599-604, IEEE, DOI: 10.17148/ijarcce.2019.81225.

[3] Kumar, S.; Karnani, G.; Gaur, M. S. & Mishra, A. (2021) Cloud Security using Hybrid Cryptography Algorithms, 2021 2nd International Conference on Intelligent Engineering and Management (ICIEM). IEEE, Apr. 28, 2021. doi: 10.1109/iciem51511.2021.9445377.

[4] Sun, P. (2020). Security and Privacy Protection in Cloud Computing: Discussions and Challenges. Journal of Network and Computer Applications, Vol. 160, pp. 102642. ISSN 1084-8045, DOI: 10.1016/j.jnca.2020.102642.

[5] Cherviakova, T. (2020). Information Security of Cloud Computing Technology. The National Transport University Bulletin, ISSN 2308-6645, DOI: 10.33744/2308-6645-2020-1-46-427-436.

[6] Singh, U., Tiwari, A. K., & Sharma, S. (2020). Data Security in Cloud Computing. International Journal of Engineering Applied Sciences and Technology, Vol. 4, No. 10, pp. 170-173. ISSN 2455-2143, DOI: 10.33564/ijeast.2020.v04i10.033.

[7] Zunnurhain, K., & Vrbsky, S. V. (2023). Security in Cloud Computing. International Research Journal of Modernization in Engineering Technology and Science. DOI: 10.56726/irjmets45986.

[8] Seth, B., Dalal, S., Jaglan, V., Le, D. N., Mohan, S., & Srivastava, G. (2020). Integrating Encryption Techniques for Secure Data Storage in the Cloud. Transactions on Emerging Telecommunications Technologies, Vol. 33, DOI: 10.1002/ett.4108.

[9] Alenezi, M. (2021). Safeguarding Cloud Computing Infrastructure: A Security Analysis. Computer Systems Science and Engineering, DOI: 10.32604/csse.2021.015282.

[10] Dakić, V., & Ribarić, S. (2020). Judicial and Technical Improvement of General Data Protection Regulation, Proceedings of the 31st International DAAAM Symposium 2020, pp.0189-0196, Published by DAAAM International, ISBN 978-3-902734-29-7, ISSN 1726-9679, Vienna, Austria. DOI: 10.2507/31st.daaam.proceedings.025.

[11] Dakić, V., Jakobović K., Žgrablić L. (2020). Linux Security in Physical, Virtual, and Cloud Environments, Proceedings of the 33rd International DAAAM Symposium 2022, pp.0151-0160, Published by DAAAM International, ISBN 978-3-902734-36-5, ISSN 1726-9679, Vienna, Austria. DOI: DOI: 10.2507/33rd.daaam.proceedings.021.

[12] Morić, Z., Branstett, L., & Petrunić, R. (2023). Rust and Webassembly for Fast, Secure, and Reliable Software. Proceedings of the 33rd DAAAM International Symposium, pp.0165-0171, Published by DAAAM International, ISBN 978-3-902734-36-5, ISSN 1726-9679, Vienna, Austria. DOI: 10.2507/33rd.daaam.proceedings.023.

[13] Reithner, I.; Papa, M.; Lueger, B. & Cato, M. (2020). Development and Implementation of a Secure Production Network, Proceedings of the 31st DAAAM International Symposium, pp.0736-0745, Published by DAAAM International, ISBN 978-3-902734-29-7, ISSN 1726-9679, Vienna, Austria. DOI: 10.2507/31st.daaam.proceedings.102.

[14] Blahová, M.; Mikuličová, M. & Hromada, M. (2020). Utilization of Fractal Geometry Possibilities for Information Systems Security, Proceedings of the 31st DAAAM International Symposium, pp.0619-0625, B. Katalinic (Ed.), Published by DAAAM International, ISBN 978-3-902734-29-7, ISSN 1726-9679, Vienna, Austria. DOI: 10.2507/31st.daaam.proceedings.085.

[15] Radinger, T.; Stuja, K.; Wölfel, W. & Markl, E. (2017). Functional Safety Concept for a Handling Robot Built on Optical Systems, Proceedings of the 28th DAAAM International Symposium, pp.0168-0172, B. Katalinic (Ed.), Published by DAAAM International, ISBN 978-3-902734-11-2, ISSN 1726-9679, Vienna, Austria. DOI: 10.2507/28th.daaam.proceedings.022.

[16] Babu, L.D.D.; Krishna, P.V.; Zayan, A.M. & Panda, V. (2011). An Analysis of Security Related Issues in Cloud Computing, Proceedings of the Contemporary Computing - 4th International Conference, IC3 2011, Communications in Computer and Information Science, Published by Springer, Vol. 168, ISBN 978-3-642-22606-9, Berlin, Germany. DOI: 10.1007/978-3-642-22606-9_21.

DOI: 10.2507/35th.daaam.proceedings.029

COMPARATIVE ANALYSIS OF ULTRASONIC TESTING SYSTEMS IN FLAW SIZE ESTIMATION

Maria Grozdanic, Morana Mihaljevic & Luka Cvetko

This Publication has to be referred as: Grozdanic, M[aria]; Mihaljevic, M[orana] & Cvetko, L[uka] (2024). Comparative Analysis of Ultrasonic Testing Systems in Flaw Size Estimation, Proceedings of the 35th DAAAM International Symposium, pp.0216-0223, B. Katalinic (Ed.), Published by DAAAM International, ISBN 978-3-902734-44-0, ISSN 1726-9679, Vienna, Austria
DOI: 10.2507/35th.daaam.proceedings.029

Abstract

One of the primary challenges in ultrasonic testing (UT) is accurately estimating the size of detected flaws, which is essential for evaluating the quality of mechanical parts and their compliance with acceptance criteria. Selecting an appropriate ultrasonic testing system is crucial to enhance measurement reliability and accuracy in flaw sizing, contributing to improved quality assurance. This research aims to evaluate the performance of various UT systems, focusing on different ultrasonic devices and probes, in estimating flaw sizes. The Distance-Gain-Size (DGS) technique with two sizing procedures – DGS mode and DGS diagram (graphically) – was used for flaw size estimation. Measurements were conducted on three calibration blocks with Flat-Bottomed Holes (FBH), representing ideal flaws considering shape and position relative to the sound beam. Based on the measurement results, both DGS procedures provided FBH sizes close to their real sizes when performed with appropriate UT systems. Varying ultrasonic devices in UT systems do not affect the FBH size, while different ultrasonic probes have a significant effect. A lower frequency probe is appropriate for thicker calibration blocks, and both frequency probes are suitable for thinner calibration blocks, with a slight advantage for the higher frequency.

Keywords: flaw sizing; distance-gain-size technique; ultrasonic testing system; ultrasonic device; flat-bottomed hole

1. Introduction

Flaw size estimation is one of the main challenges in ultrasonic testing (UT) [1]. Accurate sizing of detected flaws within a test object is a critical aspect of quality assessment and ensuring structural integrity [2]. When using normal probes, a Distance-Gain-Size (DGS) technique is commonly used for flaw sizing and determining flaw acceptability based on predefined acceptance criteria, which are usually expressed as an Equivalent Reflector Size (ERS). ERS is a clearly defined, repeatable, and reproducible parameter used in standards to establish acceptance criteria [3] against which flaw size is estimated. Therefore, the DGS technique plays a crucial role in adjusting UT systems and improving the accuracy of flaw detection and sizing.

Besides using appropriate sizing technique, reliable characterisation and successful estimation of the detected flaw are affected by several factors, including the UT technique, test part, personnel, and UT system. Therefore, it is important to understand how different factors affect the accuracy of ultrasonic measurements.

Testing techniques can significantly affect the reliability and accuracy of the measurements. Yadav et al. [4] examined the effectiveness and accuracy of pulse-echo and immersion technique on dimensional measurements. They concluded that the immersion technique provides higher precision in thickness measurements than the pulse-echo technique. The immersion technique is less influenced by surface roughness since the transducer is not in direct contact with the test object, while the pulse-echo technique is more practical for field applications. Rungrueng and Prateepasen [5] compared the effectiveness of Phased Array Ultrasonic Testing (PAUT) and Time of Flight Diffraction (TOFD) in detecting and sizing small planar defects in thick wall steel materials. Their study concluded that both techniques have advantages and limitations, and a combined approach utilizing both techniques is recommended for optimal results. The influence of the test part on the accuracy of ultrasonic measurement includes the influence of surface conditions, such as surface roughness [6], and various coatings [7]. Furthermore, the role of the testing personnel is significant as well. Bertovic [8] investigated the impact of human factors on the effectiveness and reliability of various NDT methods, including UT.

Selecting the appropriate UT system is crucial for ensuring the reliability and accuracy of measurements, thereby enhancing the accuracy of ERS estimation. The effectiveness of flaw detection and the accuracy in estimating their sizes are influenced by the components of the selected UT system, which includes the ultrasonic device, ultrasonic probe, couplant, and calibration standard.

Ultrasonic probes vary in frequencies, which is directly related to sensitivity and capability of detecting flaws. Higher frequencies provide better sensitivity, enabling the detection of smaller flaws, but they are more attenuated and less capable of penetrating deeper into the test part. On the contrary, lower frequencies, are less attenuated and capable of deeper penetration but provide lower sensitivity. Thus, selecting a probe requires a compromise between sensitivity and penetration. Yadav [9] investigated how different ultrasonic probe frequencies impact defect detection and characterisation accuracy. Experiments, comparing various probe frequencies, found that higher frequencies (> 10 MHz) are most effective for surface and subsurface defects, while lower frequencies (< 5 MHz) are more reliable for deeper defects. Frequencies between 5 MHz and 10 MHz offer a balance between sensitivity and penetration, making them suitable for general-purpose inspections. In her research, Mihaljević [10] determined that the choice of the ultrasonic probes, considering their parameters (nominal frequency and bandwidth), and the parameters of the selected ultrasonic device (voltage excitation pulse and damping resistance) influences the result of the ultrasonic thickness measurement. Kim and Song [11] studied the effects of couplant in contact ultrasonic testing and compared how different couplant types impact measurement results, noting that high-viscosity couplants generally provide better coupling, resulting in stronger and clearer signals. Using reference standards and calibration procedures helps ensure that ERS measurements are accurate and repeatable. Reference blocks with reflectors like FBHs are commonly used to establish baseline measurements for flaw sizing. The aforementioned studies underscore the importance of selecting the appropriate UT system to achieve accurate and reliable results. However, while the influences of different UT system components have been investigated, the impact of various ultrasonic devices, and their combination with various probes, on the accuracy of flaw sizing has not yet been investigated.

The main goal of this research is to compare flaw sizing performed with various UT systems, with emphasis on various ultrasonic devices and ultrasonic probes, to determine which UT system provides flaw size estimation closest to the real value. The DGS technique was used for size estimation because DGS provides repeatable and comparable results of flaw sizing, that is ERS. The sizing was performed with the DGS technique using two sizing procedures – DGS mode and DGS diagram. Measurements were performed on three calibration blocks with Flat-Bottomed holes (FBH), which represent flaws of ideal shape and position in relation to the sound beam, to be estimated using the DGS technique.

2. Methods

To evaluate the performance of various UT systems in flaw size estimation, the size of an FBH was estimated. The FBH is a manufacturable version of a theoretical Disc-Shaped Reflector (DSR) used for development of DGS diagrams, making it an ideal flaw for sizing with the DGS technique. A detailed mathematical approach to predict the echo signals from FBH for normal probes is presented in [12]. Estimation of FBH size, expressed as the ERS, is carried out by the HRN EN ISO 16811 standard [13] using the DGS technique. The Pulse-Echo technique was used for calibration and all measurements.

2.1. Ultrasonic testing systems

Eight various UT systems, combining four ultrasonic devices (Figure 1) and two normal ultrasonic probes, were used for estimating FBH sizes.

The UT systems that were used to conduct the measurements consisted of the following components:

- ultrasonic devices: USM 100, USM 36, USM GO and USN 60 (manufacturer: Krautkrämer)
- ultrasonic probes: MB4S and MB5S
- couplant: gel ZGS
- calibration blocks: 5 0088, 5 0100 and 2 0300

To neglect the influence of the ultrasonic device's parameters (energy, voltage, damping and PRF mode) on results, parameters were the same during all measurements. Also, to isolate the influence of probe frequency on FBH size estimation, probes with nominal frequencies of 4 MHz (MB4S) and 5 MHz (MB5S) were selected, having the same probe diameter and frequency bandwidth. Table 1. shows the characteristics of used probes.

Ultrasonic probe	Nominal frequency, MHz	Diameter, mm	Bandwidth, %	Near field, mm
MB4S	4	10	25	15.6
MB5S	5	10	25	20

Table 1. Characteristics of ultrasonic probes MB4S and MB5S

a) b) c) d)

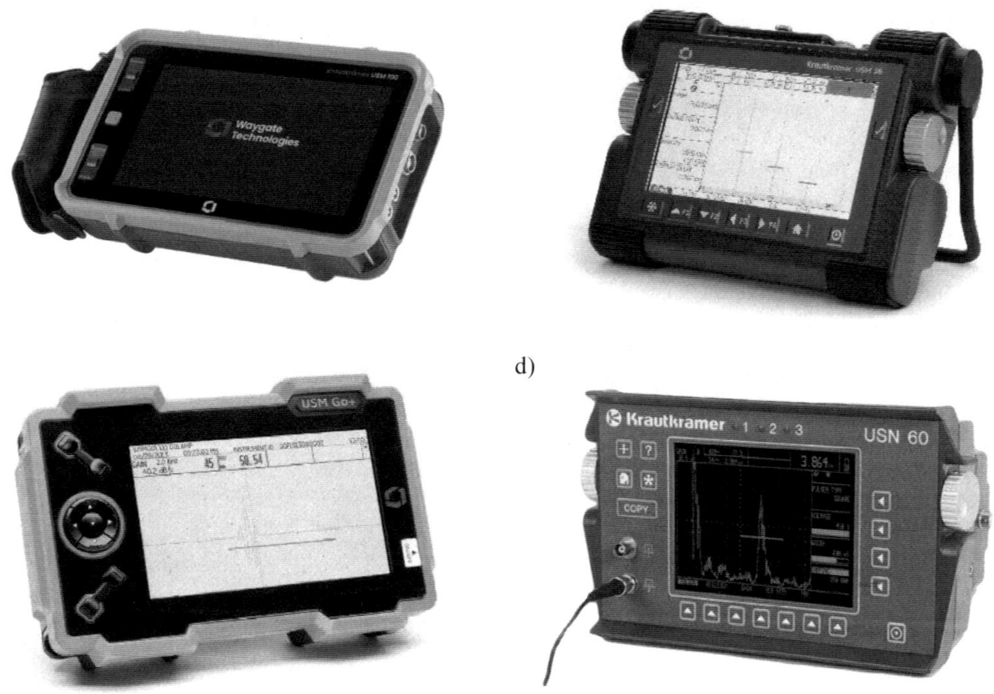

Fig. 1. Ultrasonic devices used for sizing FBHs [14]:
a) USM 100; b) USM 36; c) USM GO; d) USN 60

Figure 2 shows various ultrasonic devices used to investigate their accuracy in sizing FBHs in calibration blocks.

2.2. Calibration blocks with Flat-bottomed holes

Flaw size estimation was carried out on three calibration blocks with FBHs. The calibration blocks were made according to the ASTM E428-08 standard, and the data of the characteristic dimensions of the blocks are presented in Table 2.

Calibration block	Thickness, mm	Flat-bottomed hole	
		Diameter, mm	Depth, mm
5 0088	41.15	1.98	22.35
5 0100	44.20	1.98	25.40
2 0300	95.10	0.79	76.20

Table 2. Characteristics of calibration blocks with Flat-bottomed holes

The diameters of FBHs represent their sizes, which will be further used in comparison, as real values. Calibration blocks (Figure 2) were chosen to have FBHs positioned in the far field, that the size estimation using the DGS technique would be reliable.

Fig. 2. Calibration blocks with FBH (from left to right: 5 0088, 5 0100, 2 0300)

Figure 2 shows three calibration blocks with FBHs used in this research. Those blocks are part of a series of blocks containing FBHs of different sizes at different ultrasonic paths, representing reference reflectors, that are usually used to set testing sensitivity.

2.3. Ultrasonic testing systems calibration

For each UT system, a separate calibration was performed using the calibration block 5 0800.. Ultrasonic velocities and probe delays, used for further measurements, were determined. The obtained values were also used to carry out measurements on other calibration blocks (5 0100 and 2 0300), considering that all blocks are made of the same material. Table 3 presents ultrasonic velocities and probe delays determined using block 5 0800 and used for performing measurements with specific UT systems.

Ultrasonic device	**USM 100**		**USM 36**		**USM GO**		**USN 60**	
Ultrasonic probe	**MB4S**	**MB5S**	**MB4S**	**MB5S**	**MB4S**	**MB5S**	**MB4S**	**MB5S**
Ultrasonic velocity, m/s	5882	5879	5833	5879	5837	5879	5880	5876
Probe delay, µs	0.944	1.707	0.878	1.928	0.793	1.738	0.954	1.759

Table 3. Ultrasonic velocities and probe delays for calibration of different ultrasonic testing systems

After calibrating the UT system, the next step was to estimate the ERSs of FBHs. For this purpose, two measurement procedures of the DGS technique were used.

2.4. Flaw sizing using the Distance-Gain-Size (DGS) technique

The DGS technique can be performed either by using the DGS mode or by graphically reading from the DGS diagram.

Two sizing procedures that use the DGS diagram for flaw sizing are:
- DGS mode which uses DGS diagrams stored in the ultrasonic device and
- Manual use of DGS diagram for graphical estimation of ERS.

Both utilise the same DGS diagrams to estimate the ERS of detected flaws, but they differ in how ERS is determined. When using DGS mode, the ultrasonic device automatically displays flaw size by analysing received signals from flaws, whereas manual use of DGS diagrams involves graphically reading ERS from physical DGS diagrams. Both sizing procedures are shown in Figure 3.

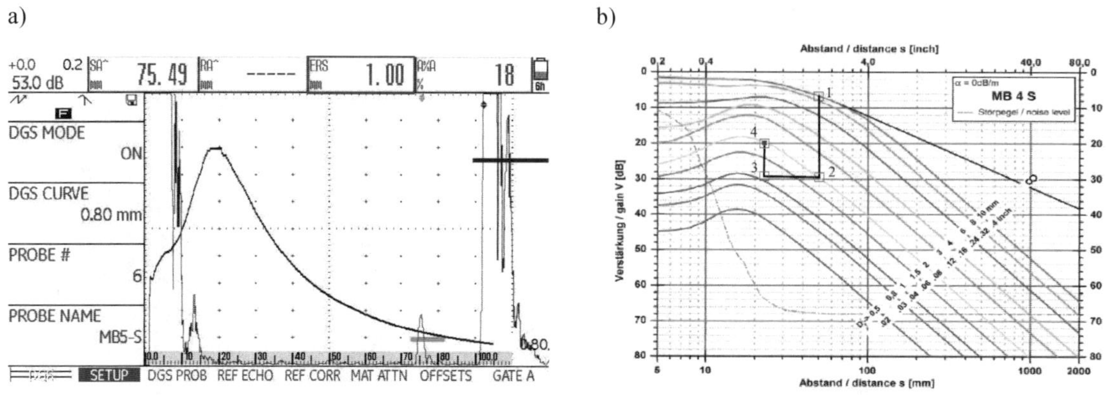

Fig. 3. ERS readings using different DGS technique procedures:
a) DGS mode; b) DGS diagram - graphically

Figure 3 shows the two sizing procedures that utilise the DGS diagram for flaw sizing. On the left (Figure 3a), an A-scan from the ultrasonic device USM 36 displays the ERS reading. On the right (Figure 3b), a specific DGS diagram for the normal probe MB4S outlines the steps for setting sensitivity (steps 1 and 2) and estimating ERS (steps 3 and 4), where step 4 presents estimated ERS.

DGS mode is a special function integrated into some ultrasonic devices. It utilises DGS diagrams stored in the ultrasonic device for ERS estimation by recording the reference echo (for normal probes is usually the back wall echo from the test object) and recalling the desired curve. Generally, each DGS diagram refers to a specific ultrasonic probe and is unique only to it due to its beam geometry and characteristic values of the ultrasound field. ERS of the detected flaw is read off the screen (Figure 3a). The DGS mode enables a more efficient evaluation of flaw sizes and acceptability. By reducing personnel subjectivity in estimation and improving the reproducibility of results, the process becomes faster and more cost-effective. This integration enhances the accuracy and reliability of flaw sizing and evaluation, providing more accurate measurements and better flaw characterization.

Manual use of DGS diagrams involves graphical estimation of ERS by first recording the reference echo from the back wall of the test object. A gain value (in dB) from the specific DGS diagram is then read and added to the reference gain to achieve the desired sensitivity. Next, the maximum flaw echo signal is identified. If the echo of the flaw is greater or lesser than 80% of the screen height, the gain difference must be added to or subtracted from the point on the diagram. In this diagram, the horizontal axis represents the ultrasound path where the flaw is located in the test object, and the vertical axis shows the working gain. This process identifies a diagram point representing the size of the reference reflector, from which the ERS is read (Point 4 in Figure 3b). If the point lies between two curves, interpolation is required. Manual ERS estimation may introduce errors due to the subjectivity of the testing personnel.

3. Results and Discussion

The sizes of FBHs, expressed as ERSs, were determined using DGS mode and graphically using the physical DGS diagram. Sizing was performed with four UT devices and two normal probes. The estimated ERS values are presented in Table 4 and Table 5. Table 4 displays the results obtained using the MB4S probe, while Table 5 presents the results from the MB5S probe. All values in both tables are expressed in millimetres.

Ultrasonic device	**USM 100**		**USM 36**		**USM GO**		**USN 60**		
Calibration block	**DGS mode**	**DGS diagram**	**DGS mode**	**DGS diagram**	**DGS mode**	**DGS diagram**	**DGS mode***	**DGS diagram**	**Real value**
5 0088	1.95	1.96	1.89	1.95	1.89	1.95	-	1.95	1.98
5 0100	1.90	1.95	1.90	1.94	1.90	1.91	-	1.95	1.98
2 0300	0.79	0.82	0.84	0.84	0.81	0.81	-	0.83	0.79

Table 4. Estimated ERS values in millimetres when testing with normal probe MB4S

*Ultrasonic device USN 60 has no DGS mode function, therefore only ERS values using the DGS diagram were obtained.

The agreement between the ERS values obtained by the DGS mode and the DGS diagram was analysed. For each calibration block, the deviations between the ERS obtained by the DGS mode and the DGS diagram for each combination of the ultrasonic probe and ultrasonic device are shown graphically in Figure 4.

Ultrasonic device	**USM 100**		**USM 36**		**USM GO**		**USN 60**		
Calibration block	**DGS mode**	**DGS diagram**	**DGS mode**	**DGS diagram**	**DGS mode**	**DGS diagram**	**DGS mode***	**DGS diagram**	**Real value**
5 0088	1.99	2.00	1.98	1.90	1.95	2.00	-	2.00	1.98
5 0100	1.97	2.00	1.97	1.95	1.97	2.05	-	2.00	1.98
2 0300	0.99	0.95	1.00	0.75	0.95	0.95	-	0.85	0.79

Table 5. Estimated ERS values in millimetres when testing with normal probe MB5S

a)

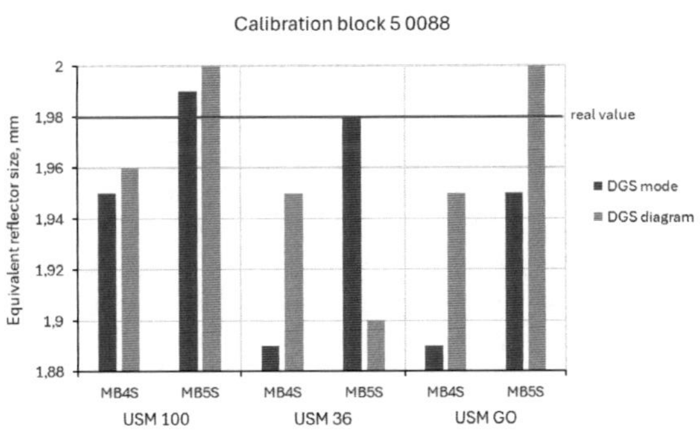

b)

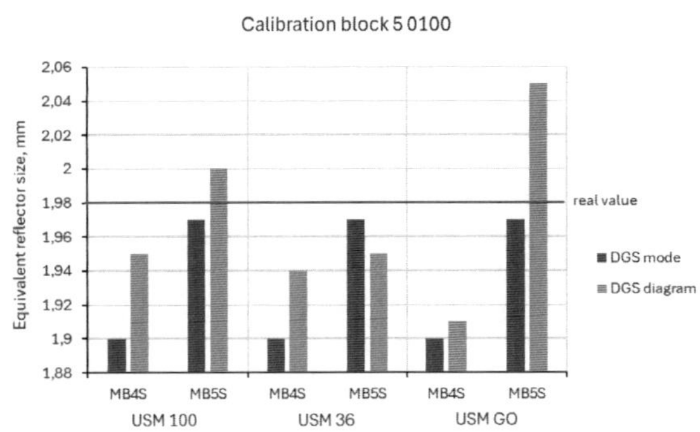

c)

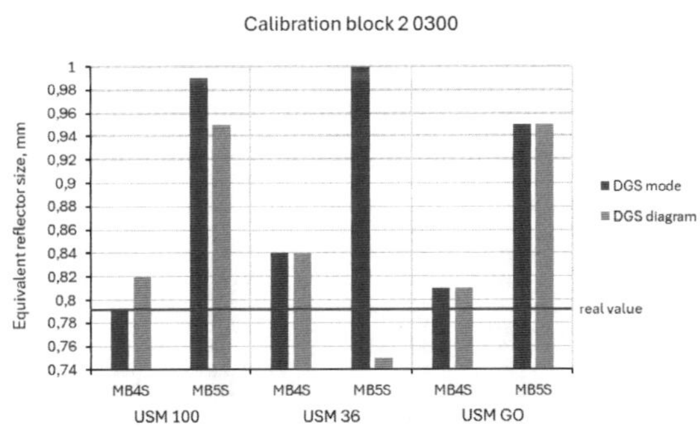

Fig. 4. Differences between ERS values obtained using DGS mode and DGS diagram for various UT systems for calibration blocks: a) 5 0088; b) 5 0100; c) 2 0300

The results in Figure 4 show that different UT systems exhibit varying levels of agreement between the DGS mode and the DGS diagram. Despite the influence of personnel subjectivity in estimating ERS using the DGS diagram, the differences are generally minor, with a maximum deviation of 0.08 mm (4%), excluding the UT system USM 36 – MB5S for calibration block 2 0300. The largest discrepancy between the DGS mode and the DGS diagram is observed when testing calibration block 2 0300 with the UT system USM 36 – MB5S. A deviation of 0.25 mm makes that UT system unreliable for accurate sizing.

For calibration block 5 0088 (Figure 4a) the best assessment between the DGS mode and DGS diagram is achieved with ultrasonic device USM 100 (deviation of 0.01 mm) for both probes. For calibration block 5 0100 (Figure 4b), the difference between results varies depending on the combination of the ultrasonic device and probe. For calibration block 2 0300 (Figure 4c), the differences are smaller compared to the other two blocks, but the results are not always close to the real value. Figure 4 also provides a visual representation of each measurement deviation from the real value, with more detailed deviation values shown in Figure 5. These deviations will be further discussed.

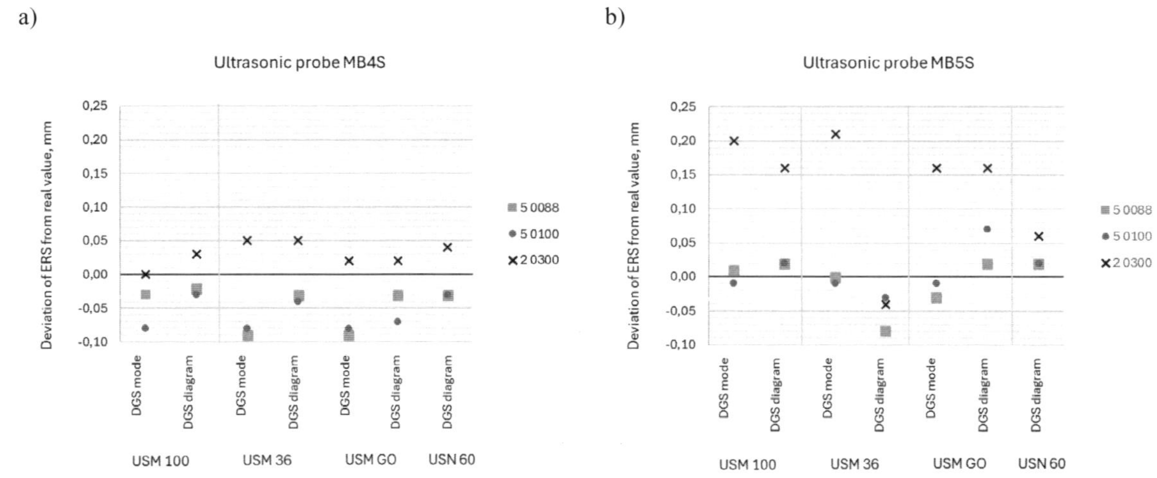

Fig. 5. Deviations of ERS values from real value testing with the ultrasonic probe:
a) MB4S; b) MB5S

Figure 5 provides a graphical representation of the deviation of estimated ERS values from real FBH sizes. Figure 5a shows that measurements using the MB4S probe exhibit minimal deviations using all UT systems, with a maximum deviation of 6.3%. Concretely, for calibration blocks 5 0088 and 5 0100, all ERS values are lower than the real values, whereas, for block 2 0300, nearly all results are higher than the real value. Generally, undersizing of flaws, as observed in blocks 5 088 and 5 0100, is more critical to safety than oversizing flaws. Furthermore, in blocks 5 0088 and 5 0100, manual estimation provides more accurate results than devices for all UT systems. Figure 5b indicates that when calibration blocks 5 0088 and 5 0100 are tested with the MB5S ultrasonic probe, estimated values are slightly closer to the real value compared with the values obtained testing with the MB4S ultrasonic probe, and UT system USM 100 – MB5S provides the best results. Significant deviations are visible when testing calibration block 2 0300. Figure 6 provides a more detailed presentation of the deviations from real values. Even though the ultrasonic device USN 60 does not have a DGS mode, it achieves excellent results in estimating ERS using a DGS diagram. For calibration blocks 5 0088 and 5 0100, the deviation from real values is up to 1.5% and for block 2 0300, it is 7.6%, which makes it the most suitable device for evaluating FBH in calibration block 2 0300 when testing with the MB5S ultrasonic probe.

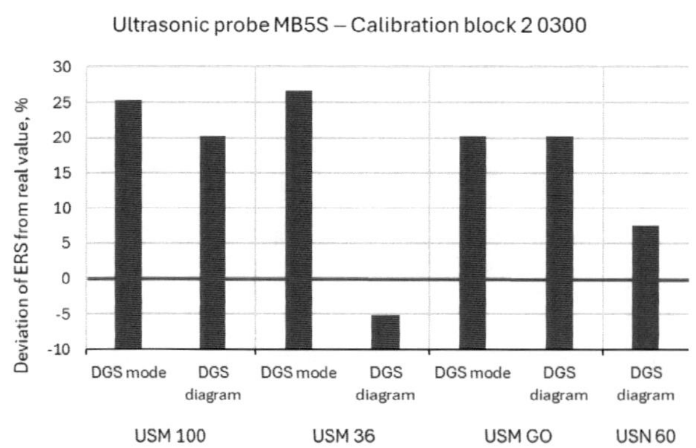

Fig. 6. Deviation of ERS estimated in calibration block 2 0300 testing with ultrasonic probe MB5S

Using the MB5S probe on calibration block 2 0300, regardless of the ultrasonic device and ERS estimation procedures, produces unreliable results with significant deviations. Estimated sizes are significantly larger than the real value. Generally, oversizing flaws may lead to unnecessary repairs or replacements of parts, whose estimated sizes are within the acceptable criteria, which causes unnecessary costs.

4. Conclusion

The influence of various ultrasonic devices (USM 100, USM 36, USM GO and USN 60) and ultrasonic probes (MB4S and MB5S) on the accuracy of flaw sizing using the DGS technique was investigated.

Two DGS procedures, DGS mode and DGS diagram, were used for sizing, and the results were compared. Sizing was performed on three calibration blocks, each containing a single FBH. The thinner calibration blocks, respectively 5 0088 and 5 0100, contained FBHs with diameters (sizes) of 1.98 mm at ultrasonic paths of 22.35 mm and 25.40 mm. The thicker calibration block, respectively 2 0300, contained the FBH with a diameter (size) of 0.79 mm at the ultrasonic path of 76.20 mm. The measurement results led to several findings.

For thinner calibration blocks the obtained sizes, expressed as ERS, closely match the real sizes for both sizing procedures and all UT systems. Generally, the maximum deviation was -0.09 mm (4.5%), but most results showed significantly smaller deviations from the real value, with slightly better estimation when using the MB5S probe. For the thicker calibration block, the results obtained when using the MB4S probe were accurate with minor deviations up to 0.05 mm (6.3%). The MB5S probe, however, resulted in significant oversizing with deviations up to 0.21 mm (26.6%).

Finally, it can be concluded that the DGS technique is generally accurate for estimating FBH sizes when used with the appropriate UT system. While the USM 100 device provided insignificant more accurate results, the overall choice of the ultrasonic device did not affect the ERS results. On the other hand, the influence of the probe frequency is significant and greatly affects the measurement results. For measurements on thinner blocks, the MB5S probe gives more accurate results. For measurements on thicker block, the MB4S probe provides results closer to the real value, while the ultrasonic probe MB5S should be avoided due to significant oversizing. Oversizing can be attributed to differences in ultrasonic frequencies; probes with higher frequencies have better sensitivity due to their smaller wavelength but are more attenuated over longer ultrasonic paths. Lower frequencies are more reliable for detecting and estimating sizes over longer ultrasonic paths. Additionally, the human factors in ultrasonic testing should not be overlooked, as the accuracy and interpretation of results highly depend on the skills and experience of the personnel conducting the measurements.

Further research will test additional calibration blocks with varying FBH sizes and ultrasonic paths to perform a more detailed analysis of the influence of UT systems on size estimation. That will help to refine the understanding of the impact of different ultrasonic devices and probe frequencies on flaw sizing accuracy using the DGS technique.

5. References

[1] Dwivedi, A.; Gupta, R. K.; Varma, R. K. & Narayanan, K. S. (2016). Effect of Processing and Microstructure on Ultrasonic Response of Aerospace Alloys: An Overview. Russian Journal of Nondestructive Testing, Vol. 52, No. 12, 703-709, ISSN 1061-8309

[2] Kim, H. J.; Song, S. J. & Kim, Y. H. (2003). Quantitative Approaches to Flaw Sizing Based-On Ultrasonic Testing Models, Proceedings of Review of Progress in Quantitative Nondestructive Evaluation, Thompson, D. O. and Chimenti, D. E. (Ed.), pp. 703-710, AIP Conference Proceedings, Bellingham, DOI: 10.1063/1.1570205

[3] Krautkramer, J: Ultrasonic Testing Level 2

[4] Yadav, K.; Yadav, S. & Dubey, P.K. (2021). A Comparative Study of Ultrasonic Contact and Immersion Method for Dimensional Measurements. MAPAN-Journal of Metrology Society of India, Vol. 36, No. 2, 319-324

[5] Rungrueng, C. & Prateepasen, A. (2019). Comparison of Detection and Height Sizing Ability of Planar Defect in Thick Wall Weld Steel Between PAUT and TOFD Techniques. International Journal of Recent Engineering Research and Development, Vol. 4, Issue 4, 21-38, ISSN 2455-8761

[6] Rodriguez, E. (2003). Effect of Surface Roughness in Ultrasonic Testing (Pulse-Echo by Direct Contact) in AISI/SAE 4340 Steel Samples, Available from: https://www.ndt.net/article/ecndt02/254/254.htm, Accessed: 2024-07-01

[7] Mihaljevic, M[orana]; Cajner, H[rvoje]; Markucic, D[amir] & Kozic, K[arlo] (2018). Coating Influence on Ultrasonic Thickness Measurement Result, Proceedings of the 29th DAAAM International Symposium, pp.0341-0346, B. Katalinic (Ed.), Published by DAAAM International, ISBN 978-3-902734-20-4, ISSN 1726-9679, Vienna, Austria DOI: 10.2507/29th.daaam.proceedings.049

[8] Bertovic, M. (2015). Human Factors in Non-Destructive Testing (NDT): Risks and Challenges of Mechanised NDT, Ph.D. Dissertation, Technical University Berlin, Berlin, Germany

[9] Yadav, K.; Dhiman, N.; Yadav, S. & Dubey, P. K. (2022). Effect of Probe Frequency on Defect Evaluation in Ultrasonic Non-Destructive Testing, New Delhi, Yadav. S. et al. (Ed.), pp. 14-17, New Delhi

[10] Mihaljevic, M. (2015). Measurement uncertainty estimation of the ultrasonic thickness measurement, Ph.D. Dissertation, Department of Quality, Doctoral thesis, University of Zagreb, Faculty of Mechanical Engineering and Naval Architecture, Zagreb, Croatia

[11] Kim, Y.; Song, S. J.; Lee, S. S.; Lee, J. K.; Hong, S. S. & Eom, H. S. (2002). A Study of the Couplant Effects on Contact Ultrasonic Testing. Journal of the Korean Society for Nondestructive Testing, Vol. 22, No. 6, 621-626

[12] Danilov, V.N. (2007). Calculation of Echo Signals from a Flat-Bottom Hole for Normal Probes. Russian Journal of Nondestructive Testing, Vol. 43, No. 11, 724-732, ISSN 1061-8309

[13] Non-destructive testing – Ultrasonic testing – Sensitivity and range setting (ISO 16811:2012), International Organization for Standardization, 2012

[14] https://www.bakerhughes.com/waygate-technologies, Accessed: 2024-07-04

DOI: 10.2507/35th.daaam.proceedings.030

METHODS FOR ESTIMATING THE ECONOMIC IMPACT MODEL OF PORTS

Astrida Rijkure

This Publication has to be referred as: Rijkure, A[strida] (2024). Methods for Estimating the Economic Impact Model of Ports, Proceedings of the 35th DAAAM International Symposium, pp.0224-0229, B. Katalinic (Ed.), Published by DAAAM International, ISBN 978-3-902734-44-0, ISSN 1726-9679, Vienna, Austria
DOI: 10.2507/35th.daaam.proceedings.030

Abstract

Port performance evaluation is becoming increasingly topical in modern port management, especially for comparing a port with competing ports. Port performance can be achieved by comparing the port to predefined standards of accuracy, completeness, cost, and speed. Some ports monitor their activities solely on the basis of functional characteristics, while others cover wider aspects such as administrative and financial performance. But the overall goal of any port is to maximise profits, develop and increase the volume of transhipment. The efficiency of ports varies considerably: not all ports function equally well, and in recent years, the gap between ports adapted to new logistical and economic requirements and those that have failed has widened. Differences in performance result in a significant diversion of traffic, with a negative impact on business opportunities in well-functioning ports. This threatens the efficiency and sustainability of the transport network and the competitiveness of the economy in general. This study will rank and group the criteria for port performance and when exiting these criteria, the methods for calculating these performance criteria will be derived.

Keywords: ports economic; port efficiency; blue economy; port performance indicators

1. Introduction

Port performance evaluation has become an increasingly important topic in contemporary port management, particularly in the context of comparing the efficiency and competitiveness of ports. As ports play a critical role in global trade and supply chains, their ability to meet evolving logistical, economic, and environmental requirements is paramount. Measuring port performance involves assessing key parameters such as accuracy, completeness, cost, and speed against predefined standards, which serve as benchmarks for efficiency and operational excellence. These parameters help identify performance gaps and areas for improvement, enabling ports to stay competitive in a rapidly changing global market.

Traditionally, many ports have focused on monitoring functional characteristics, such as cargo handling capacity, ship turnaround time, and berth occupancy rates. However, in recent years, port management has expanded to include broader aspects like administrative efficiency, financial performance, and environmental sustainability. The primary goal of any port remains maximizing profits, increasing transshipment volumes, and enhancing overall operational effectiveness.

The increasing complexity of global trade and the rising demand for seamless logistics networks have made these broader measures essential to evaluating port performance comprehensively. The efficiency of ports varies significantly, with some ports successfully adapting to new logistical and economic demands, while others have lagged behind. This divergence has led to increasing disparities in performance, which in turn impacts the global transport network and the competitiveness of various economies. Efficient ports that have embraced technological advancements and optimized their operations are capturing more business, while underperforming ports risk losing traffic, market share, and business opportunities. This dynamic has far-reaching implications, as the efficiency of ports is critical not only for the ports themselves but also for the sustainability and competitiveness of the entire transport network and the broader economy. This study aims to systematically rank and group the key criteria for port performance evaluation. By doing so, it will provide a comprehensive framework for assessing and comparing ports, taking into account functional, administrative, and financial aspects. Additionally, the study will derive methods for calculating these performance criteria, ensuring that the evaluation process is both rigorous and relevant to the modern context of port operations. The research will provide insights that can guide port authorities and stakeholders in making informed decisions about resource allocation, operational improvements, and strategic planning. By aligning the evaluation criteria with current industry standards and expectations, the study will contribute to enhancing port efficiency and competitiveness on a global scale.

The primary research question guiding this study is: *What are the most critical performance criteria for evaluating port efficiency, and how can these criteria be effectively measured and ranked to support decision-making in port management?* In answering this question, the study will offer a structured methodology for port performance evaluation, preparing the reader for the subsequent methods and results sections where the ranking, grouping, and calculation processes will be detailed.

The aim of the study is to compare port performance assessment models (Stochastic Frontier Analysis, Data Envelopment Analysis, Key Performance Indicators), identifying new aspects in port economic performance assessment and model development. The purpose of this study is to identify possible performance indicators (port equipment, cargo handling, capital equipment costs, staff costs, berth facility etc.) that would measure the impact of ports on the port economy.

2. Materials and methods

The principal materials used for the studies are as follows: various sources of literature, e.g. scholars' articles, research papers and the reports of institutions (European Commission, Ministry of Transport, World bank etc.). The following suitable qualitative research methods have been used: monographic; analysis and synthesis; grouping, logical and abstractive constructional (literature review). Due to the importance of maritime transport in countries' economies, in the literature, many studies for evaluating port efficiency have been carried out [7], [11], [14], [16].

3. Results and discussion

The result of the study is to assess the importance of the port industry and the need for specific performance indicators. This study has developed a system of port performance indicators that allows us to identify opportunities that generate the highest returns compared to other international standards in ports. It will also allow the impact of improvements made in ports to be assessed. Under conditions of competition, ports are not merely facing the issue of cost efficiency, but also the technical efficiency of services. Port efficiency is related to the efficiency of using available resources and the ability to provide high transshipment capacity services to their customers. Technical and cost efficiency must be in balance. Traditionally, port performance is evaluated in terms of actual and optimal throughput, measured in tonnes or number of containers processed and transshipped. A port can also use other performance indicators to evaluate the efficiency thereof.

To determine port capacity indicators, it is important to be aware of what the input data are. Usually, the principal port performance indicator is cargo turnover over a certain period of time. It is also the main parameter that is accessible for statistical analysis. However, the actual efficiency of a port is characterised by a range of other indicators, since, depending on the port administration model, the port administration itself is not involved in the reloading of cargo, which is performed by terminals that rent land from the port for the performance of cargo operations, while the responsibility of port administration is to ensure high-quality infrastructure to enable reloading of the particular cargo at the terminal of the port. Therefore, in the context of a port, cargo turnover at the port is an insufficient indicator. Due to the effectiveness of the programme, automated inspection has already found applications. The user merely needs to provide inputs to the software. [4]

In a global supply chain, customers expect services of intermediaries to be flexible and less expensive. In the context of maritime transportation, cargo owners expect delivery to be reliable, and non-costly. On this basis, Logistics services providers' focus on designing an efficient method in which ports advantageous to minimization of costs overall transport costs are preferred. [1] Various empirical studies have been conducted in recent years to examine the impact of economic efficiency on economic growth. In general, these studies reveal that measuring economic efficiency, especially overall economic performance, has a statistically significant, positive effect on economic growth rates [2], [3], [5], [9], [13]. The author offers to group port performance indicators into 3 large packages: general indicators, financial indicators and operational indicators. According to the author, this grouping of indicators enables one to realise how important the selection of the criteria that need to be assessed and calculated is.

Averages	Finance	Actions
waiting time	berth throughput	ship expenditure by GT
berthing time	berth handling volume per 1m	handling per ton operation profit
cost berthing time	cargo port time per tonn	gross profit margin
ship capacity per day	berth occupancy ratio %	
shipping turnover	berth usage ratio	

Table 1. Port performance indicators

Definitely, all aforementioned indicators must be financed, including the balancing of required administrative costs with financial indicators that are required to monitor transport infrastructure and the provision of transport services irrespective of the model that is used. The administration of the port must make decisions regarding how the volume of public or private resources will be divided among different tasks, as well as transport infrastructure and the provision of transport services. These indicators are directed towards the increase in port profits. These indicators can be mutually attributed to each other or by using additional values or criteria. World Bank (1993) defined the study of the Port productivity index that was divided into three types: by operational productivity, assets (equipment) productivity and financial productivity indicators. Each of these indicators can be calculated. Scientists have offered different combinations and formulae for measuring and calculating these criteria. One of the methods for indirectly determining unknown parameters is the use of computer models that can replace costly experimental tests. [6] The author has developed her own simplified table for the calculation of these criteria, which provides a formula and a concise explanation of the expected result for each criterion and the use thereof.

This table graphically demonstrates how simple, but important methods, by attributing respective criteria relative to other criteria, can be used to assess how efficient the particular port is, whether it is customer focused or profit orientated only, how cost efficient services provided by the port are, and how well equipped and efficient the overall operation of the port itself is. It is important to evaluate the throughput capacity of each berth within a certain time period – usually it is one year. The speed of loading and unloading of a ship characterises technical provisions, which are also important characteristics of port performance. This service provision indicator characterises the quality of service provision at the port; loading and unloading speed of the ship is a very important element in determining the costs of transport services. Usually, this indicator is evaluated in hours per unloaded tonnes. It is important to evaluate how large a team is attracted to unload one ship, which allows one to evaluate the productivity of work, as well as the automation of loading procedures and whether the time of return on investment may be evaluated, if the human factor is replaced with automation. How many accidents occur at the port during loading operations, whether these accidents can be prevented, if the processes were automated. But the present study addresses measurable and evaluable criteria. It is important to evaluate how intensively the berths of the port are loaded, what the business hours of the berth are and the number of personnel employed. The load of a berth actually points to the level of demand for the services of the port. This can be measured in time periods (a year, a month, a week) and is usually expressed in %.

Productivity indicators that point to cost efficiency are important; they involve the evaluation of costs or the reloading of one tonne of cargo. This indicator is very difficult to measure, since it involves sensitive company data. In the competition between port terminals, data on the costs of reloading one tonne of cargo are difficult to obtain, since they depend on the amount of cargo, intensity of deliveries and form of co-operation with the customer; therefore, reloading costs of the same type of cargo or reloading tariffs may differ for each of the customers.

Port effectiveness operating objectives will differ between privately-owned and government-owned ports. If the port is privately owned, its effectiveness economic operating objective might be to maximize profits or to maximize throughput subject to a minimum profit constraint. If the port is owned by government, its effectiveness economic operating objective might be to maximize throughput subject to a zero operating deficit (where port revenue equals cost) or subject to a maximum operating deficit (where port revenue is less than cost) that is to be subsidized by government. A port's economic production function represents the relationship between the port's maximum throughput and given levels of its productive resources, i.e.

$$\text{Maximum Port Throughput} = f\,(\text{Port Productive Resources})$$

where throughput may be the number of containers (measured in 20-foot equivalent units or TEUs) or tons of cargo handled and port productive resources include labor, immobile capital (e.g. berths and buildings), mobile capital (e.g. cranes and vehicles), fuel and ways (e.g. port roadways and railways). If the port achieves the maximum throughput for given levels of its resources, then it is technically efficient; otherwise it is technically inefficient. A port's economic cost function represents the relationship between the port's minimum costs to be incurred in handling a given level of throughput, i.e.

$$\text{Minimum Port Costs} = g\,(\text{Port Throughput})$$

,where the costs are those incurred in the use of the port's resources, e.g. wages paid to labor and vehicle fuel expenses.

	Criteria	Formula	Description
Averages	waiting time	$\frac{total\ ship\ waiting\ time}{number\ of\ ships}$	Usually, the average indicator value is calculated for a period of 1 year, but it can also be evaluated every month. This indicator demonstrates the speed of the performance of loading operations, and enables one to evaluate how efficient the technical equipment of the berths is. This indicator is critical for the customer, since time is money.
	berthing time	$\frac{total\ berthing\ time}{number\ of\ piers}$	The equipment of berths differs and this criterion allows one to evaluate which berths are efficient and which ones require upgrade.
	cost berthing time	$\frac{income\ from\ the\ docking\ charge}{total\ call\ number}$	The average indicator that not only points to the intensity, but also the adequacy of the determined docking charge.
	ship capacity per day	$\frac{total\ cargo\ volume}{number\ of\ ships\ x\ port\ time}$	This indicator enables the port to determine how high the load or intensity of the port is. The criterion is very important for the evaluation of the use of port capacity.
	shipping turnover	$\frac{port\ time}{number\ of\ ships}$	The indicator that characterises the load or intensity of the port. The indicator that shows how much time ships stay in this port on average.
	berth throughput	$\frac{total\ cargo\ volume}{number\ of\ piers}$	The average cargo turnover at the berth points to the load of berth, which is important when assessing capacity criteria.
Finance	berth handling volume per 1m	$\frac{total\ cargo\ volume}{Bert\ lenght\ (m)}$	The indicator points to the volume of cargo per 1 m of berth area. This criterion can also be used to determine the capacity of warehouse use, except that the unit of measure used is m², instead of m.
	cargo port time per tonn	$\frac{total\ cargo\ volume}{port\ time}$	The indicator determines how much time is required to process a particular cargo volume within a selected time period.
	berth occupancy ratio %	$\frac{ship\ berth\ time}{number\ of\ ships\ x\ 365\ day} X100$	The indicator demonstrates the percentage (%) of time the specific berth has been in use during a year, as well as the intensity of its use.
	berth usage ratio %	$\frac{handling\ berth\ time}{ship\ berthing\ time}$ x 100	This indicator demonstrates the share of loading operations within the total amount of time spent by the ship in the port. Respectively, the time that the ship spends in the port.
Actions	ship expenditure by GT	$\frac{total\ port\ income}{total\ GT}$	Gross tonnage is the criterion that determines the capacity of the port to accept ships of a particular size. The higher the GT, the more specific the requirements for certain port infrastructure.
	handling per ton operation profit	$\frac{port\ operation\ profit}{total\ handling\ tonn}$	The indicator determines the profit generated as a result of reloading one tonne of cargo.
	gross profit margin	$\frac{port\ operation\ profit}{operation\ income}$	Gross profit margin is a financial indicator used to evaluate the financial health and business model of the company that demonstrates the amount of money left over from product sales after subtracting the cost of goods sold. It can be calculated by dividing the gross profit by total income.
	capacity used by the port	$\frac{total\ port\ handling\ capacity}{total\ cargo\ volume} x\ 100$	It is very important to determine this criterion, since at every port the maximum cargo loading capacity must be calculated depending on its area, number of berths and equipment in order to evaluate the proportion of port capacities that are used every year, which enables decision making on whether an additional increase in port capacity is required.

Table 2. Methods for calculating port performance indicators

If the port provides throughput at a minimum cost (given the unit costs or resource prices to be paid), then it is cost efficient; otherwise it is cost inefficient. If a port has the effectiveness operating objective of maximizing profits and is cost inefficient, it can obtain greater profits for the same level of throughput service by lowering costs in becoming cost efficient. However, note that a port can be cost efficient without being effective. [15] Port users seek information that has a direct impact on their business and the commercial choices they make. They are interested in the performance of the port in terms of operational time factors such as vessel waiting or cargo dwell times, or indeed labour productivity. Both factors will affect cargo transit costs. [10] These are formulas for the reaching of the objectives of every port development programme, which enable evaluation of the development direction and targets to be strived for, ways of optimising port procedures and achieving maximum profit at minimum investment. However, there must be an awareness that the optimisation of resources is not always sufficient for the economy of ports, competitiveness must be ensured, market trends must be followed, since it is possible that, in the absence of investment in port development, your cargo shall move to a neighbouring port because you are no longer capable of serving ships of new fleets with larger displacement or ships that require more modern technologies to unload cargo or to fill up the ship with new generation fuel or other new technologies that are developing rapidly, and changes that occur particularly in the transport sector. This is a rapidly changing sector, it is syncing with time, developing new technologies and attempting to minimise the environmental impact. Port operation focuses on improving port sustainability in order to add value to the high profile of international ports and reduce environmental and social impacts. Together with this port operation supports maritime transport and the overall economic system. [8] At the same time, upgrading of processes can also have a negative effect on the development of the sector. Like energy efficiency measures delay the growth of the global oil industry and the development of the steel industry is delayed by the reduction of dimensions and mass of various products, the transport sector must take the fact that everything is becoming "more compact" into consideration. The process called dematerialisation of economy can also have a long-term negative effect. The upgrade of technologies reduces the mass of durable goods; the increase in energy efficiency reduces the mass of energy resources, and the fuels will be replaced by electricity in the future.
Nowadays port development is a rapid process. A couple of decades ago, ports were relatively isolated territories, where reloading of cargo took place, but nowadays ports represent a complex ecosystem that is closely integrated into the economy of the city or country, which unites several stakeholders and areas of operation. Contemporary ports are transferring from the simple reloading of cargo to "smart logistics". Therefore, cargo turnover at the port is not the only criterion to be analysed nowadays. Currently, the range of port services is very wide and everything needs to be analysed in conjunction, instead of evaluating through cargo turnover alone.

The choice on what port to use are made based on the criteria deemed relevant by the users which could vary between them. To an extent, the volume of cargo throughput in a port is dependent on the choice of the users on the port that offers the criteria. The frequency at which the ports are used by the users could as well be linked with the quality of services rendered at the port. It is therefore important for port authorities to know these factors so it can serve as a guide in policy formulation geared towards improving services at port and the market share. Policies formulated on these improves the quality of services rendered by the ports and increase the level of satisfaction derived by clients. [1], [12] The process of industrialisation in the countries of the Asian region is expected to have an equally important effect on the development of the sector. Along with common economic and global factors, future scenarios of the port and shipping sector are increasingly affected by the development of modern technologies – the sector of transport and logistics will be significantly affected by the regions, where robotics and 3D printing sectors are developed, since these will affect the volume of product and raw material flows, as well as the geography thereof.

The scenarios of port and shipping industry development will be affected by how the manufacturers of goods change, for instance, how their production volumes and capacity to respond flexibly to market trends change, how potential consolidation and mutual integration of manufacturing sector participants occur. The capacity to maintain high productivity, as well as predictable and stable chains of supply will definitely be important factors, as well as ecological aspects, which are becoming increasingly important in practically all sectors of the modern global economy. The sector of e-commerce is an important factor for the development scenarios of the future. It has seen a significant increase of volumes on a global level, thus considerably affecting the area of cargo transportation and logistics, as well as ports and maritime transport.

4. Conclusion

This study set out to evaluate the performance of ports by developing a comprehensive system of port performance indicators. The primary issue addressed was the need for accurate, standardized methods to assess port efficiency, especially in light of intense competition and evolving global logistical demands. Traditional metrics, such as cargo turnover, were found to be insufficient on their own, necessitating the inclusion of a wider range of indicators, including financial and operational metrics, to provide a more complete picture of port performance. The solution proposed by the study involved grouping port performance indicators into three major categories: general, financial, and operational. This approach enabled a more holistic evaluation of port efficiency by addressing both cost and technical efficiency, thereby balancing resource use with service provision. The study also introduced practical methods for calculating these indicators, which allow for easy benchmarking against international standards and the identification of areas for improvement. By using these indicators, ports can more accurately measure factors like berth throughput, handling capacity, and profitability, which are critical for maximizing returns and enhancing competitiveness.

The results demonstrate that effective port performance evaluation requires a multifaceted approach that includes not only operational efficiency but also financial viability and customer satisfaction. The study successfully developed a framework of indicators that can be applied across different ports to assess their relative performance, identify opportunities for growth, and optimize resource allocation. This system allows for the assessment of both immediate operational improvements and long-term strategic planning, making it a valuable tool for port authorities aiming to enhance their competitive edge. One key outcome of this study is the recognition that port efficiency cannot be solely measured by traditional cargo handling statistics. A broader view of efficiency, which includes the effectiveness of port infrastructure, financial health, and service quality, is essential for the modern port industry. By applying this comprehensive system of performance indicators, ports can better align themselves with global logistical demands, improve their technical and cost efficiency, and attract more business. Looking ahead, future research should focus on the development of automated systems and advanced technologies, such as AI and robotics, to further enhance the accuracy and efficiency of port operations. Additionally, as environmental sustainability becomes a more pressing concern, the integration of green technologies and practices into the performance evaluation framework will be critical. This will allow ports to not only remain competitive but also contribute to broader environmental and social objectives, ensuring their long-term sustainability in a rapidly evolving global economy.

5. Acknowledgment

ERDF project No. 1.1.1.2/VIAA/1/16/061 "Long-term impact of blue economy on the increase of the competitiveness of ports in the Baltic Sea region"

6. References

[1] Blessing A.S., (2019) Port selection criteria from logistics service providers' perspective: a case study of West Africa. World Maritime University Dissertations. 1221.

[2] Cebula R., Mixon J F., (2012) The Impact of Fiscal and Other Economic Freedoms on Economic Growth: An Empirical Analysis, International Advances in Economic Research, volume 18, pages 139–149

[3] Galindo M., Sotos F.E. Méndez M.T. (2007) Value at Risk and Economic Growth, International Advances in Economic Research, volume 13, pages 214–221

[4] Gheorghiță V., Geantă F.A. (2023) Design of an Experimental Model of a System for Product Sorting Through Image Processing, Proceedings of the 34th DAAAM International Symposium, pp.0203-0208, B. Katalinic (Ed.), Published by DAAAM International, ISBN 978-3-902734-41-9, ISSN 1726-9679, Vienna, Austria DOI: 10.2507/34th.daaam.proceedings.026

[5] Haralambides H.E., (2015) Competition, Excess Capacity and the Pricing of Port Infrastructure, Port Management pp 221-252

[6] Kozar I., Sulovsky T. (2023) Numerical Model for Identification of Fiber-Reinforced Concrete Parameters from the Crack-Mouth Opening Displacement, Proceedings of the 34th DAAAM International Symposium, pp.0544-0550, B. Katalinic (Ed.), Published by DAAAM International, ISBN 978-3-902734-41-9, ISSN 1726-9679, Vienna, Austria, DOI: 10.2507/34th.daaam.proceedings.072

[7] Magdiel A. Agüero-Tobar, Marcela C. González-Araya, Rosa G. González-Ramírez, (2023) Assessment of maritime operations efficiency and its economic impact based on data envelopment analysis: A case study of Chilean ports, Research in Transportation Business & Management, Volume 46, 100821

[8] Muangpan T., Suthiwartnarueput K.(2019) Key performance indicators of sustainable port: Case study of the eastern economic corridor in Thailand

[9] Park N., Yoon G. (2014) Port Capacity Evaluation Formula for General Cargo, The Asian Journal of Shipping and Logistics Volume 30, Issue 2, August 2014, pages 175-192

[10] Pazienza P, (2015) The Environmental Impact of the FDI Inflow in the Transport Sector of OECD Countries and Policy Implications, International Advances in Economic Research volume 21, pages 105–116

[11] Pérez I., González M.M, Trujillo L., (2020) Do specialisation and port size affect port efficiency? Evidence from cargo handling service in Spanish ports, Transportation Research Part A: Policy and Practice, Volume 138, Pages 234-249, ISSN 0965-8564

[12] Philipp R., (2020), Digital readiness index assessment towards smart port development, Sustainability Management Forum | NachhaltigkeitsManagementForum volume 28, pages 49–60

[13] Poufinas T., Galanos G., (2021) The Impact of Competitiveness on the Shadow Economy, International Advances in Economic Research, volume 27, pages 29–46

[14] Tongzon J., (2001) Efficiency measurement of selected Australian and other international ports using data envelopment analysis., Volume 35, Issue 2, 2001, Pages 107-122, ISSN 0965-8564

[15] Wayne K. Talley,(2006) Port Economics, PORT PERFORMANCE: AN ECONOMICS PERSPECTIVE, Research in Transportation Economics Volume 17, 2006, Pages 499-516

[16] Wilmsmeier G.,Tovar B., Sanchez R.J., (2013) The evolution of container terminal productivity and efficiency under changing economic environments, Research in Transportation Business & Management, Volume 8, Pages 50-66, ISSN 2210-5395

DOI: 10.2507/35th.daaam.proceedings.031

DEVELOPMENT OF A FRAMEWORK FOR ADAPTABLE AUGMENTED REALITY ASSISTANCE SYSTEMS FOR MANUAL ASSEMBLY TASKS

Kathrin Konkol, Nora Kießling, Niklas Engeländer & Alexander Schwarz

This Publication has to be referred as: Konkol, K[athrin]; Kiessling, N[ora]; Engelaender, N[iklas] & Schwarz, A[lexander] (2024). Development of a Framework for Adaptable Augmented Reality Assistance Systems for Manual Assembly Tasks, Proceedings of the 35th DAAAM International Symposium, pp.0230-0237, B. Katalinic (Ed.), Published by DAAAM International, ISBN 978-3-902734-44-0, ISSN 1726-9679, Vienna, Austria
DOI: 10.2507/35th.daaam.proceedings.031

Abstract

In recent years, the use of Augmented Reality applications in an industrial context has become more established, especially for user guidance during manual assembly tasks. Despite the value of such applications for the industry, their drawbacks, such as high development costs or the fact that AR applications are often tailored to one specific use case, slow down their industry adoption. To address this problem, the authors developed a framework that enables the creation of complex AR assistance systems by using a condition-based approach. It helps to describe the assembly process for each step of the task by offering building blocks in the virtual environment. It can therefore superimpose visual guidance during manual assembly tasks such as work instructions. The frameworks' goal is to improve the development of adaptable assistance systems but also decrease their overall development costs and therefore improve their acceptance for industry use cases. The authors developed the approach according to user requirements that they evaluated during prior research. To evaluate its efficacy, a software prototype was developed that guides workers during manual assembly tasks in line production for automobile components. The framework shows potential for quick setups and adaptability to changing work instructions through its modular structure.

Keywords: Augmented Reality; Manual Assembly; Framework; Assistance System

1. Introduction

Assistance systems in the context of manual assembly are already established in large parts of the industry. They are intended to shorten product development and working hours, reduce cognitive strain and the number of errors and thus improve the quality of work [1]. One technology for assistance systems, which is increasingly being integrated into industry procedures [2], is Augmented Reality (AR). It offers the possibility to display information directly in the worker's field of view on the object itself. AR assistance systems cover a wide range of industry-relevant scenarios. The most common scenarios of industrial AR and Virtual Reality (VR) applications in the industry include remote maintenance, virtual training and assistance during decision-making processes [3].

Due to their immense potential, many other industrial use cases were covered in the literature, such as the prototyping of low-cost automation solutions [4], and AR maintenance scenarios with or without an established connection to machine control [5]. The use of AR assistance systems supports users during their day-to-day work, which can reduce execution time and error occurrences without added cognitive load [6]. Despite their value for certain use cases, AR assistance systems also come with drawbacks such as high development costs for the AR solution [7] as well as their authoring tools [8], and the risk of adding cognitive load to workers, if not designed from a user-centric perspective [9], [10]. Furthermore, current AR solutions are implemented with differing hardware components, ranging from handheld devices to projectors, and therefore don't cover all requirements of individual workplaces [11], [12]. Existing AR assistance systems are therefore tailored to one specific, previously defined use case, which makes them rigid and leads to new costs whenever the production processes are changed.

A possible solution to these problems is the development of adaptable assistance systems [12], which can be supported by functionality that understands user cognition [8], and automation for authoring tools [13]. While some adaptable assistance systems have been developed [14], they can lack generalisation and applicability because they are developed and tested in strict laboratory scenarios and are therefore not close enough to actual industrial conditions [15].

This paper describes a framework that enables the creation of adaptable AR assistance systems by using a condition-based approach. It offers building blocks which describe the assembly scene, including individual assembly steps and components, and augments visual guidance accordingly. Further enhancements to analyse user cognition and automation of authoring tools are discussed.

2. Related work

2.1 AR Assistance Systems for Manual Assembly

Up until now, many different AR assistance systems to guide manual assembly were developed and are still in development, as each individual industry use case has its specific requirements that need to be met by the solution [7]. One example is the modernization of a control cabinet production process by Khokhlovsky et al. [16] in which a process prone to human error was improved by utilising AR technology as well as an assistant robot to improve assembly time and allow for better focus during the process. Another example was published by Mura et al. in 2016 [13]. Their optical see-through solution uses sensor data to provide visual guidance for correct component assembly, detecting mistakes and assisting operators. However, the complex setup and testing is time-consuming. A proposed solution is a self-learning procedure that analyses sensor data to automatically deduce assembly sequences.

Wang et al. [8] identify high time expenditure as a key disadvantage in AR assembly research, categorizing it into AR assembly guidance, training, and simulation/design. They focus on guidance and assistance systems, concluding that authoring processes and integration into enterprise workflows for AR guidance in complex assemblies are time-consuming. They suggest developing more intelligent AR assembly systems that adapt to the operator's cognitive status as a solution. Additionally, a user study comparing three different AR assistance methods for assembly guidance [7] concluded that there is not one best method for all factors but instead outlines advantages of each method depending on the individual use case it's being used for. The aim of the study was to assess the impact of the three AR assistance methods, namely a Head-Mounted Display, a smartphone, and a wristband with an external monitor, on performance, mental and physical workload, and acceptance. As there is not one AR solution that fits every use case, the authors of this paper assume that further systems will be developed in the future and existing ones will be adapted, leading to new costs every time the production processes are changed, or the product is modified. This could indicate a need for frameworks for assistance systems, which are designed with adaptability in mind, which is discussed in this paper.

2.2 Adaptable Assistance Systems

Adaptable assistance systems can reduce development costs by responding to context-specific information, such as user actions or errors. This is done by gathering context-based data—such as visual feeds, location data, or user posture—and comparing it to the planned assembly sequence. The results can then be used to implement adaptive system behaviours. In a survey of context-aware frameworks, Baldauf et al. [14] noticed their similarity regarding their layered structure, which separates the data acquisition from its use. This approach makes context sources reusable in varying scenarios. Standardised formats and protocols further improve context-aware systems, as resources can be focused on the development of the context service itself and not on the integration of the data.

Various approaches have been used to recognize user intentions. As early as 1990, a method was developed using postconditions and preconditions to define user actions, detailing the status before and after tasks along with action-specific parameters [17]. By analysing these parameters, the framework aimed to understand the context and generate context-sensitive instructions. The authors find this approach valuable as it breaks down assembly tasks in a machine-readable format, but it requires time-consuming, unique, hand-crafted conditions.

Similar approaches to interpret human behaviour in virtual space were presented in later years which either check if previously defined conditions, which can be stored as a Boolean operator are met [18], such as "is leg moving", or use checks for key poses of individual action sequences that define user actions [19]. In addition to these approaches, however, there are others that use AI to recognize user intention. In 2019, Qu et al. [20] used machine learning (ML) to extract service requesters' intentions from their text by analysing their sentence structure and keywords to make it easier for service staff to provide assistance. Although the feature extraction and performance analysis procedures depend on large datasets and focus on service requests for computer problems, they could be conceptually transferable to assembly tasks. One year earlier Ramamurthy et al. [21] performed a survey that investigated ML techniques for human activity recognition. They observed the trend to use deep learning architecture to mitigate traditional ML algorithm limitations. The approach could be adapted to industry use cases to enable comparison of current user action and the planned assembly sequence.

The adaptability of systems has also long been a topic of research. An approach to generate User Interfaces (UI) for assistance purposes based on virtual context was presented by Macik et al. in 2014 [22]. They distinguished between more and less experienced users, varying the level of visual assistance accordingly. This approach utilised a data structure that controlled elements of generated content, such as Input, Output, and action triggers. While the adaptable assistance system appears promising, the maintenance and development costs of the adaptive UI are significant, and the practical benefits seem insufficient. Recent approaches exhibit similar limitations. In 2023 Fisher and Fu [15] analysed object-related user intention in collaborative situations through line of sight. The proposed UI is adapted according to the recognized user intention with the goal of improving the efficiency of collaboration. Despite the innovative approach, the strict laboratory scenario leads to a lack of generalisation, which reflects the missing applicability for industry scenarios and illustrates the state of research in this domain. Work has also been done in the overlapping field of AI for Extended Reality (XR) applications. A literature review by Reiners et al. in 2021 [23] shows that AI has been leveraged in multiple areas, such as trainings, robotics and autonomous driving and concludes its increasing relevance. Its overlapping use has been classified into three main groups: interpretation of XR generated data, conferring intelligence on XR and Training AI. They propose guidelines to distinguish between two main objectives, one in which AI is serving and assisting XR, through pattern detection or improvement of XR user experience, or one in which XR is serving and assisting AI, through generating AI-ready data or to improve AI performance.

In conclusion, many approaches to use AI to understand user actions and improve assistance systems have been developed. Despite the amount of research in this field, the authors are not aware of current industry solutions, that offer these functionalities and are integrated into modern assembly lines. This paper further investigates current obstacles that hinder industry adaption of such systems.

2.3 Current Obstacles in the industry

Although the use of AR offers great potential for many assembly work situations, so far existing AR assistance systems have mostly been isolated solutions, each tailored to a specific use case; hence its integration into industry processes often fails to be realised. In a survey of 2020 Souza Cardoso et al. [12] investigated AR approaches across multiple domains, their use cases as well as used AR devices. They summarise the main challenges for AR technology in a production environment to be hardware concerns, which include the user's mobility and necessity to not occupy their hands, unfeasible tracking methods and low projection quality. They further found that of 120 studied AR applications, only 5% were implemented in a production environment and concluded that the industry does not prioritise developing their own applications but rather prefers to use a standard application that they can integrate into already existing processes. Adaptable systems that are tested in a production environment are one proposed solution to increase industry adaptation.

After reviewing recent research projects that focus on digital assistance systems, Apt et al. [7] summarise the reasons for the non-commissioning of such systems are the investment costs, legal framework conditions, context-related user-friendliness and ergonomics. They also mention the specific corporate culture and the associated expectations in terms of productivity increases, cost reductions or product and process innovations. In addition to functionality, it must be ensured that the capabilities of the system meet the requirements of the workplace and its environment. Utilising AR solutions on a day-to-day basis poses the risk of increasing cognitive load on operators and therefore worsens their performance instead of improving it [9]. For this reason, ease of use is of utmost importance to limit potential negative effects on user performance and health [10]. Even after initial development, using a system in ongoing operations incurs further costs due to the need to adapt functionality whenever production processes or products change. While incorporating AI into assistance systems could enhance adaptability and reduce some of these costs, it also presents an additional challenge. Despite growing interest in AI in the industry, there is reluctance to implement it in industrial operations, as developed solutions are often only tested in laboratory settings due to high integration costs [24]. AI has often not been incorporated into future corporate strategies, which is also due to a lack of suitable frameworks that support integration into existing business models [25]. Other reasons include the lack of suitable data preparation and the retrievability and reusability of developed digital models, which would support the development of frameworks [26].

3. Framework for adaptable AR Assistance Systems

The authors of this paper propose a framework that supports the implementation of adaptable AR assistance systems. In previous research the authors of this paper have conducted expert interviews to assess the requirements of such assistance systems [27]. They found that the experts expressed a need for informative and cognitive assistance features, which focus on the analysis of user action and usability improvements. The system should further be easily adaptable and cater to users' requirements without making the workers dependent on the system. The following concept deduces the experts' feedback, regarding the assistance systems' requirements.

3.1 Concept

After evaluating the work that has been done in the field, and considering current industry needs, the authors conclude that a framework for adaptable AR assistance systems must take the following points into account:

1. Adaptability through generalisation:
   Virtual content should be easily replaceable to keep maintenance costs low and enable a setup with many different use cases and engineering tasks in mind. It should promote versatility and ease of use, so the development costs can be kept low, and the setup can be adjusted to changes in production processes. It further enables the adaptation of the visual assistance to cater to the users' requirements.
2. Interchangeability of hardware:
   The framework should not be dependent on proprietary AR-technology, such as Eye Tracking or LIDAR sensors. By not relying on those, but rather by offering optional means of integration for them, the framework is compatible with more devices and therefore has a lower barrier of entry.
3. Automation through process integration:
   The framework needs to offer interfaces to standard industrial environments to ensure data flow and decrease time spent on data preparation. Providing means for process integration increases the acceptance of the solution across various industry use cases.

The designed assistance system framework consists of multiple components that support the setup process: a workflow, actions, conditions, and assistance features. Figure 1 visualises their structure and relation. A workflow is the sum of several actions and describes the assembly sequence. It describes the assembly process as accurately as possible with the data known to the maintainer of the software, with context that is perceivable by the sensor of the AR device. This is essential, as the system needs to be context-aware to check given conditions during the use of the application. Actions are a subset of a workflow and are defined by a set of conditions. Conditions are defined as preconditions and postconditions and are made of attributes that can be either a Boolean value or a numerical probability value.

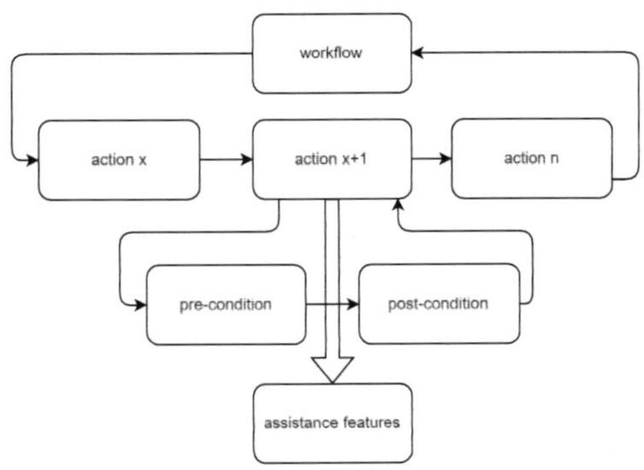

Fig. 1. Visualisation of assistance system framework

A precondition describes the status before its parent action is performed whereas a postcondition describes the status during the task. Any condition can be part of any action, but no two different actions may have identical conditions. For example, condition A may be a simple check to determine whether the user is interacting with a screwdriver, while another condition B could be whether the screwdriver is in contact with a specific screw. An action, "screwing" can now be defined with the two conditions A and B. Actions need to be unambiguous, so no further action can be defined by using only the conditions A and B. The final component is the assistance feature itself, which could be any information that guides the user, such as the description of the current task or the location of the components. It can be attached to a workflow, a task, or a condition, and therefore be adapted to user behaviour. It is necessary to provide the framework with a clear set of conditions that describe all relevant interactions within a given workflow.

Conditions are set up by the software maintainer and can be implemented either manually or via automated processes. As an example, the generic action of movement can be described by the transition of a component, such as a tool, from a fixed coordinate in an environment to another. This behaviour can be described by a single condition checking if the component has moved over time by monitoring its velocity. However, assembly tasks include complex actions that imply specific attributes and may require many conditions to be unique to the system. Examples of conditions can include the number of hands in motion, the velocity of an assembly component or the position of individual components during an assembly sequence.

3.2 Demonstrator

To evaluate the efficacy of the proposed concept, the authors developed an exemplary AR assistance system for manual assembly tasks for inline assembly of electric motor components. The development of the demonstrator is part of a publicly funded research project, which provides the industry use case, realistic components, and feedback of core users. A HoloLens 2 was used as an AR device. The framework itself and its virtual environment were developed using the real-time engine, Unity. All mentioned building blocks, such as the workflow, actions, conditions, and assistance features, were prepared as Prefabs, which are Unity components, and their logic was programmed with C#.

To set up the scene for the manual assembly task, the authors used a textual description of the assembly process, which was provided by the process designers of their research partners. The thereby deduced assembly task workflow consists of the following actions: 1. placing the lower lid, 2. fitting an inner ring, 3. placing a set of coils, 4. fitting an outer ring and 5. placing an upper lid. Figure 2 visualises the example workflow and action elements.

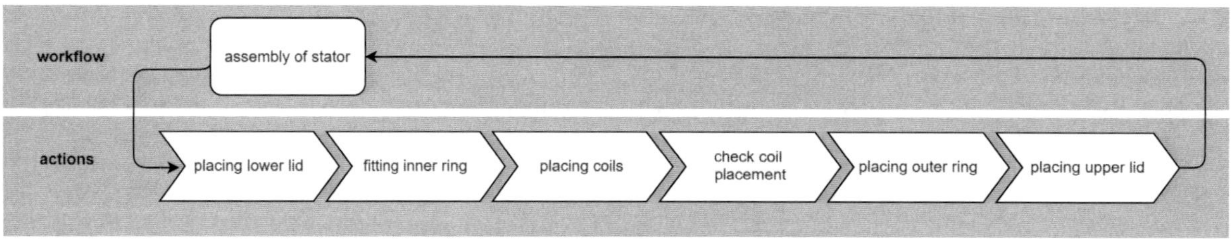

Fig. 2. Example workflow and actions

For each individual action, a set of simple conditions was defined. Once every action has been performed, the system expects the assembly process to start from the beginning. Figure 3 shows some exemplary conditions that were defined for the action "placing coils" and its attached assistance features. The following conditions were designed: 1. Coil was grabbed, 2. Coil is in the final assembly location, 3. A number of required coils have been placed. When the performance of the action is recognized by the system, the corresponding assistance features are shown to the user, which include a textual description of the current task, visual guidance on the coil storage location and visual guidance on coil assembly location.

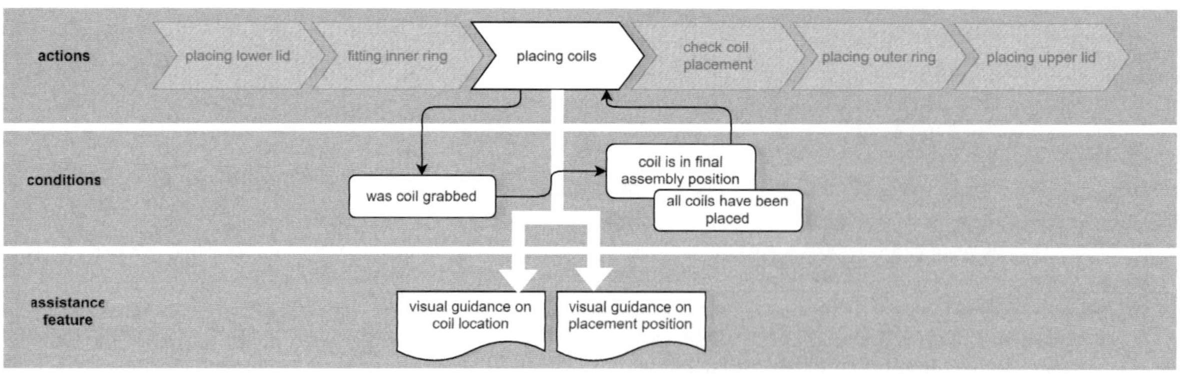

Fig. 3. Example conditions and assistance features

The AR application needs to be able to check whether the described conditions are true. To check if the user grabbed the coil or if the coil is in its assembly position, it is necessary to integrate some form of contextualisation. To achieve this, the authors used a combination of geometry tracking by using VisionLib[1] and QR code tracking that's natively

[1] https://visionlib.com/

supported by HoloLens 2. The demonstration is shown on a workbench that simulates the inline assembly environment, as shown in Figure 4. The assembly components are a simplified version of the real use case components and were 3D printed (see Figure 4, green box). As soon as the application runs, its assistance features are shown when either the respective QR Code or the assembly components are recognized by the system.

Fig. 4. Demonstrator-setup

While the user performs the assembly tasks, assistance features are visualised in their field of view. The following two assistance features were implemented:

1. Textual description of current assembly task and location of components (see Figure 5, left side)
2. Visual augmentation of final position of components (see Figure 5, right side)

Fig. 5. AR Assistance features

4. Discussion

The exemplary use of the framework for the development of the AR demonstrator gave the authors a clear structure on how to set up the virtual scene and how to describe the assembly tasks. The definitions of the conditions for each step of the task need to be set up, such as the placement of the coils shown in the example. In conclusion this indicates that the initial setup time of a virtual environment is currently not significantly decreased. The authors were already able to reuse created conditions within one assembly sequence, such as the storage location of coils, which indicates that future demonstrators can benefit from already created framework-components.

The developed system is a first building block that acts as a foundation for more elaborate and automated approaches to create workflows of complex assembly tasks. One reasonable approach would be to create a library of conditions for common assembly tasks and therefore shorten development times for setups, such as specific hand poses that are related to specific assembly tasks. Another approach could be to simplify or automate the creation of conditions. This could imply that the maintainer of the program performs the assembly tasks using the AR device to record the process.

Afterwards, the recorded steps need to be analysed, classified, and integrated into the framework. To achieve this, ML can be applied in finding definitions for conditions, for example, by analysing user movement to identify the beginning and end of a task or reading sensor data of involved tools. However, this would require a suitable model as well as proper training for the ML model. An alternative solution could be to create an interface to an existing software that describes work processes and use the existing description as content for the assembly task description and condition design.

Further important elements of the assistance system are the assistance features that guide the operator during the use of the application. Features are currently shown, whenever the preconditions of the actions currently performed are met. This simple approach should be enhanced by adding more variables which contextualise the users need for assistance. For example, the time it takes an operator to perform a specific assembly sequence can be compared. In case one task takes comparatively long, the system could suggest assistance that is suitable for the current assembly step. The condition-based approach could be enhanced for the application of assistance features to therefore react to users' requirements at runtime.

5. Next Steps

The authors are going to evaluate the developed AR demonstrator with a user study. During that study, they will evaluate whether the setup of the assembly task is feasible during the procedure by validating if the AR assistance features are shown at the right time. The users' UX and mental load during the use of the assistance system will be focus of the investigation. Based on the results, adjustments to the conditions and assistance features will be made accordingly. Another important step is to discuss the design of the framework with process designers and assistance system experts to assess its impact on their acceptance of AR assistance systems and possibly improve its design.

6. Conclusion

This paper presented the concept of a framework for AR assistance systems and the development of a demonstrator based on this concept that supports the inline assembly of electric motor components as an exemplary manual assembly task. The authors believe the presented approach to be well-suited for repetitive actions such as assembly tasks, as they are well defined by process designers and are often described in text form and therefore in a machine-readable way. The concept presented has the potential to decrease development costs of AR assistance systems for industry use cases, as it offers generalised building blocks and clear structure during the setup of the virtual environment. Its industry acceptance and efficacy remain to be verified.

7. Acknowledgments

This research is part of the Project 13IK022I "From conventional production plant to resilient competence plant through Industry 4.0"(German.: "Vom konventionellen Produktionswerk zum resilienten Kompetenz-Werk durch Industrie 4.0") and therefore partially funded by the Federal Ministry for Economic Affairs and Energy.

8. References

[1] Yang, X. & Plewe, D. A. (2016). Assistance Systems in Manufacturing: A Systematic Review, Advances in Ergonomics of Manufacturing: Managing the Enterprise of the Future. 978-3-319-41696-0, in: C. Schlick, S. Trzcieliński (Eds.), pp. 279–289, Springer International Publishing, Cham. https://doi.org/10.1007/978-3-319-41697-7_25.

[2] Dalle Mura, M. & Dini, G. (2021). Augmented Reality in Assembly Systems: State of the Art and Future Perspectives, Smart Technologies for Precision Assembly. 978-3-030-72631-7, in: S. Ratchev (Ed.), pp. 3–22, Springer International Publishing, Cham. https://doi.org/10.1007/978-3-030-72632-4_1.

[3] Damiani, L; Demartini, M; Guizzi, G; Revetria, R. & Tonelli, F. (2018). Augmented and virtual reality applications in industrial systems: A qualitative review towards the industry 4.0 era, IFAC-PapersOnLine, Vol. 51 No. 11 624–630. https://doi.org/10.1016/j.ifacol.2018.08.388.

[4] Manuri, F; Gravina, N; Sanna, A. & Brizzi, P. (2022). Prototyping industrial workstation in the Metaverse: a Low Cost Automation assembly use case, in: 2022 IEEE International Conference on Metrology for Extended Reality, Artificial Intelligence and Neural Engineering (MetroXRAINE), Rome, Italy. 978-1-6654-8574-6, IEEE, pp. 133–138. https://doi.org/10.1109/MetroXRAINE54828.2022.9967626.

[5] Kollatsch, C; Schumann, M; Klimant, P. & Lorenz, M. (2017). [POSTER] Industrial Augmented Reality: Transferring a Numerical Control Connected Augmented Realty System from Marketing to Maintenance, in: 2017 IEEE International Symposium on Mixed and Augmented Reality (ISMAR-Adjunct), Nantes, France. 978-0-7695-6327-5, IEEE, pp. 39–41. https://doi.org/10.1109/ISMAR-Adjunct.2017.27.

[6] Seth, A; Vance, J. M. & Oliver, J. H. (2011). Virtual reality for assembly methods prototyping: a review, Virtual Reality, Vol. 15 No. 1 5–20. https://doi.org/10.1007/s10055-009-0153-y.

[7] Apt, W; Schubert, M. & Wischmann, S. (2018). Digitale Assistenzsysteme: Perspektiven und Herausforderungen für den Einsatz in Industrie und Dienstleistungen, Institut für Innovation und Technik (iit), VDI/VDE Innovation + Technik GmbH. 978-3-89750-181-2, Berlin.

[8] Wang, X; Ong, S. K. & Nee, A. Y. C. (2016). A comprehensive survey of augmented reality assembly research, Adv. Manuf., Vol. 4 No. 1 1–22. https://doi.org/10.1007/s40436-015-0131-4.

[9] Cohen, Y; Faccio, M; Pilati, F. & Yao, X. (2019). Design and management of digital manufacturing and assembly systems in the Industry 4.0 era, Int J Adv Manuf Technol, Vol. 105 No. 9 3565–3577. https://doi.org/10.1007/s00170-019-04595-0.

[10] Di Pasquale, V; Simone, V. de; Miranda, S. & Riemma, S. (2022). Smart operators: How augmented and virtual technologies are affecting the worker's performance in manufacturing contexts, JIEM, Vol. 15 No. 2 233. https://doi.org/10.3926/jiem.3607.

[11] Deac, C. N; Popa, C. L; Ghinea, M. & Cotet, C. E. (2017). Using Augmented Reality in Smart Manufacturing, Proceedings of the 28th International DAAAM Symposium 2017. 9783902734112, in: B. Katalinic (Ed.), pp. 727–732, DAAAM International Vienna. https://doi.org/10.2507/28th.daaam.proceedings.102.

[12] Souza Cardoso, L. F. de; Mariano, F. C. M. Q. & Zorzal, E. R. (2020). A survey of industrial augmented reality, Computers & Industrial Engineering, Vol. 139 106159. https://doi.org/10.1016/j.cie.2019.106159.

[13] Mura, M. D; Dini, G. & Failli, F. (2016). An Integrated Environment Based on Augmented Reality and Sensing Device for Manual Assembly Workstations, Procedia CIRP, Vol. 41 340–345. https://doi.org/10.1016/j.procir.2015.12.128.

[14] Baldauf, M; Dustdar, S. & Rosenberg, F. (2007). A survey on context-aware systems, IJAHUC, Vol. 2 No. 4 263. https://doi.org/10.1504/IJAHUC.2007.014070.

[15] Fu, L.-C. & Fischer, R. (2023). Collaboration Assistance Through Object Based User Intent Detection Using Gaze Data, in: 2023 Symposium on Eye Tracking Research and Applications, Tubingen Germany. 9798400701504, ACM, pp. 1–2, New York, NY, USA. https://doi.org/10.1145/3588015.3590130.

[16] Khokhlovskiy, V; Oleynikov, V; Kostenko, D; Onufriev, V. & Potekhin, V. (2017). Modernisation of a Production Process Using Multicriteria Optimisation Logic and Augmented Reality, Proceedings of the 28th International DAAAM Symposium 2017. 9783902734112, in: B. Katalinic (Ed.), pp. 500–507, DAAAM International Vienna. https://doi.org/10.2507/30th.daaam.proceedings.067.

[17] Hudson, S. E. (Ed.) (1990). Proceedings of the 3rd annual ACM SIGGRAPH symposium on User interface software and technology - UIST '90, ACM Press. 0897914104, New York, New York, USA.

[18] Liu, J; Kuipers, B. & Savarese, S. (2011). Recognizing human actions by attributes, in: CVPR 2011, Colorado Springs, CO, USA. 978-1-4577-0394-2, IEEE, pp. 3337–3344. https://doi.org/10.1109/CVPR.2011.5995353.

[19] Baysal, S; Kurt, M. C. & Duygulu, P. (2010). Recognizing Human Actions Using Key Poses, in: 2010 20th International Conference on Pattern Recognition, Istanbul, Turkey. 978-1-4244-7542-1, IEEE, pp. 1727–1730. https://doi.org/10.1109/ICPR.2010.427.

[20] Qu, C; Yang, L; Croft, W. B; Zhang, Y; Trippas, J. R. & Qiu, M. (2019). User Intent Prediction in Information-seeking Conversations, in: Proceedings of the 2019 Conference on Human Information Interaction and Retrieval, Glasgow Scotland UK. 9781450360258, ACM, pp. 25–33, New York, NY, USA. https://doi.org/10.1145/3295750.3298924.

[21] Ramasamy Ramamurthy, S. & Roy, N. (2018). Recent trends in machine learning for human activity recognition—A survey, WIREs Data Mining Knowl Discov, Vol. 8 No. 4. https://doi.org/10.1002/widm.1254.

[22] Macik, M; Cerny, T. & Slavik, P. (2014). Context-sensitive, cross-platform user interface generation, J Multimodal User Interfaces, Vol. 8 No. 2 217–229. https://doi.org/10.1007/s12193-013-0141-0.

[23] Reiners, D; Davahli, M. R; Karwowski, W. & Cruz-Neira, C. (2021). The Combination of Artificial Intelligence and Extended Reality: A Systematic Review. Front. Virtual Real., Vol. 2. https://doi.org/10.3389/frvir.2021.721933.

[24] Arinez, J. F; Chang, Q; Gao, R. X; Xu, C. & Zhang, J. (2020). Artificial Intelligence in Advanced Manufacturing: Current Status and Future Outlook. Journal of Manufacturing Science and Engineering, Vol. 142 No. 11. https://doi.org/10.1115/1.4047855.

[25] Lee, J; Singh, J. & Azamfar, M. (2019). Industrial Artificial Intelligence. https://doi.org/10.48550/arXiv.1908.02150.

[26] Peres, R. S; Jia, X; Lee, J; Sun, K; Colombo, A. W. & Barata, J. (2020). Industrial Artificial Intelligence in Industry 4.0 - Systematic Review, Challenges and Outlook, IEEE Access, Vol. 8 220121–220139. https://doi.org/10.1109/ACCESS.2020.3042874.

[27] Konkol, K., Reusing, H. (2024). Requirements for an AI driven Extended Reality assistance system for assembly tasks. (In Press), 2024 the 10th International Conference on Virtual Reality (ICVR 2024). 979-8-3503-6422-4, in: pp. 381–389, IEEE, Bournemouth, United Kingdom.

DOI: 10.2507/35th.daaam.proceedings.032

OPTIMUM TILT ANGLE OF FLAT PLATE SOLAR COLLECTORS FOR DIFFERENT LOCATIONS IN BOSNIA AND HERZEGOVINA

Haris Lulic, Merima Gadara, Sadjit Metovic & Almira Softic

This Publication has to be referred as: Lulic, H[aris]; Gadara, M[erima]; Metovic, S[adjit] & Softic, A[lmira] (2024). Optimum Tilt Angle of Flat Plate Solar Collectors for Different Locations in Bosnia and Herzegovina, Proceedings of the 35th DAAAM International Symposium, pp.0238-0244, B. Katalinic (Ed.), Published by DAAAM International, ISBN 978-3-902734-44-0, ISSN 1726-9679, Vienna, Austria
DOI: 10.2507/35th.daaam.proceedings.032

Abstract

This paper shows the determination of the optimal angle of inclination for solar collector for different locations throughout Bosnia and Herzegovina over the course of the year. The research was conducted on an established solar system model designed for heating sanitary water for a hotel. It was found that, for one-position systems, the maximum solar thermal energy can be collected when the tilt angle of the collector is between 30° and 45°. For two-position systems, the optimal tilt angle is 30° during summer period and 60° for the rest of the year.

Keywords: Optimal tilt angle; flat plate water cooled solar collector; solar energy; function of the thermal efficiency of the solar collector; TRNSYS.

1. Introduction

In the era of globalization and excessive fossil fuel consumption, humanity faces the challenge of mitigating climate change by reducing greenhouse gas emissions and preventing the rise in the temperature of the Earth's atmosphere. Ahmović et al. [1] demonstrated the increase in average annual temperatures in continental climates attributed to climate change. One of the proposed solutions is to intensify the use of renewable energy sources. From an energy point of view, solar energy is an underutilized resource available globally and can be a qualitative and quantitative substitute for conventional energy sources [2]. One way to utilize solar energy for heating and hot sanitary water preparation is the use of active solar systems. To increase the efficiency of this type of solar system, it is necessary to increase the efficiency of solar thermal collectors. To increase the efficiency of flat solar thermal collectors, it is essential to optimize the angle of inclination of the collector relative to the horizontal plane and its orientation.

To collect the maximum daily amount of solar radiation that reaches the collector plane, dual-axis systems can be used to track the trajectory of the Sun depending on its position throughout the year. However, these systems increase the investment costs of the entire solar system, as well as the exploitation and maintenance costs. Additionally, since they rely on electricity for operation, they can reduce the overall efficiency of the solar system. For this reasons, in practice it is more common to design solar thermal collectors with fixed orientation and optimal angle of inclination relative to the horizontal plane. In his work, K. Skeiker [3] stated and explained the influence of several factors that directly affect the choice of the angle of inclination of solar collectors, namely:

- geographic latitude and altitude of the location,
- purpose of the solar system and period of use during the year,
- climatic conditions in the given locality.

In his work, V. H. Moncos [4] analyzed the correlation between the maximum solar irradiation on the collector plane and the angle of inclination relative to the horizontal plane for the site of Assuit, Egypt. Through the research, a mathematical model was developed that gives the possibility of determining the total solar radiation on an inclined surface of fixed orientation, for a certain day or period of the year. The results show that to maximize solar radiation collection on the plane of the solar thermal collector for a given location, its inclination angle relative to the horizontal plane should be adjusted eight times annually.

Similar research, using different methods of analysis, was carried out and presented in their work by Stanciu et al. [5]. Using the Hottel and Woertz model, the isotropic diffusion model and the HDKR model, they estimated the density of absorbed solar radiation on the collector plane. In the paper, they compared the values of the maximum incident solar radiation and the maximum absorbed solar radiation by the energy receiver. The results showed that at the optimal tilt angle, these values do not differ significantly.

D. Ibrahim [6] conducted a theoretical study of the optimal angle of inclination of flat solar thermal collectors relative to the horizontal plane for the area of Cyprus. During the research, the data, obtained by measurements, on the total solar radiation on the horizontal surface for the observed locality were used. The analysis included an assessment of the values of direct and diffuse radiation on the south-facing surface of the collector for different angles of inclination ranging from 0° to 90°. The results of the research are presented through the values of the optimal tilt angle of 48° for the winter period and 14° for the summer period.

Jafarkazemi et al. [7] investigated the influence of the angular deviation of the direction of the axis of the plane of solar thermal collectors from the south direction on the optimal tilt angle. During the research, data on total solar radiation at different angles of inclination of the collector plane and different orientations were used. The optimal tilt angle for the south-oriented collector field was determined to be 22°, which in value almost corresponds to the latitude of the location where the study was conducted. Also, the results show that the values of the optimal angle of inclination of solar thermal collectors, on a monthly and annual level, decrease with an increase in the azimuth angle of the collector (angular deviation from the south direction).

A mathematical model, using the software tool MATLAB, for determining the optimal angle of inclination of the collector relative to the horizontal plane was presented in their work by Pourfayaz et al. [8]. Using their model, it is possible to determine the tilt angles at which the maximum values of solar radiation on the surface of the collector are obtained for five cities in Iran at different latitudes. Similar research, in the same period, was conducted and the results were published in their work by Modarresi et al. [9]. The analysis they conducted was aimed at creating a model that predicts the optimal tilt angles of solar collectors for different locations in the southern and northern hemispheres. In doing so, it was shown that the value of the optimal angle of inclination at which the maximum value of solar radiation can be collected during the year does not depend on the longitude, but only on the latitude and the number of sunny days.

Analytical expressions for determining the optimal angle of inclination of solar thermal collectors for different geographical longitudes are given in the work of El-Kassaby et al. [10] and Aja et al. [11]. During the research, they used a fixed southern orientation of the collectors and hourly values of incident solar radiation.

Gunerhan et al. [12] in their work determined the daily, monthly and annual values of the optimal angle of inclination of the collector field at the site of Izmir, Turkey. The mathematical model used during the research also took into account the monthly level of atmospheric cleanliness in order to obtain the most realistic picture of the intensity of solar radiation that reaches the surface of the receiver. The results showed that the optimal tilt angle on an annual basis is 30.3°.

This paper presents the determination of the optimal tilt angle of collector for different locations in Bosnia and Herzegovina. The research was carried out on an already created model of solar system for the preparation of sanitary hot water for a hotel needs. The model was created in the TRNSYS software packages and the research was carried out over a year.

2. Potential of solar energy use in Bosnia and Herzegovina

Bosnia and Herzegovina has great potential for the use of solar and other energy sources and belongs to the countries of Europe with significant total solar radiation, which on an annual level ranges from 1250 kWh/m^2 in the north of the country and in mountainous areas up to 1600 kWh/m^2 in the south of the country.

If a comparison is made with the values of the total solar radiation on the horizontal surface in the countries of Central and Northern Europe, where the annual averages are 1150 and 1000 kWh/m^2year, respectively, the conclusion is reached that on average Bosnia and Herzegovina receives about 15% more solar energy compared to Central Europe, and 30% more than Northern Europe (Netherlands, Denmark, Great Britain). In the south of the country, in eastern and western Herzegovina, these percentages are even more pronounced, so that these regions lead in terms of the amount of available solar energy compared to central Europe by about 30% and compared to northern Europe by 50%. The natural potential of solar energy in Bosnia and Herzegovina amounts to 67.2 PWh per year, if every day of the year, an average of 3.6 kWh of radiant energy falls on each square meter of the horizontal surface. This value exceeds the total energy consumption in Bosnia and Herzegovina many times over.

Despite the fact that Bosnia and Herzegovina belongs to the countries of Europe with significant solar irradiation, the use of solar energy in this area can be considered insignificant. However, the level of use of renewable energy sources, especially solar energy, is still not at a satisfactory level. One of the reasons is the relatively low coefficient of usefulness of solar energy conversion into electricity. On the other hand, in solar applications used for heating purposes, the problem of energy accumulation is evident, i.e. the disproportion between the time of the most intense solar radiation (summer period) and the time when the need for heat is greatest (night and winter period). An adequate solution to the problem of energy accumulation results in an increase in investment costs, and thus the economic justification is questioned [13].

The application of low-temperature solar collectors for the purpose of preparation of sanitary hot water, heating of water in swimming pools, heating and air conditioning of spaces, development of solar technology in the territory of Bosnia and Herzegovina is currently the most realistic. The role and importance of the solar collectors in a solar installation indicates the fact that the operation of the entire system largely depends on the correct choice and installation of the collectors. An important factor for designing solar systems is determining the optimal tilt angle of collector, at which it is possible to collect the most of the solar thermal energy [14].

3. Determination of the optimal tilt angle of collector for different locations in the territory of Bosnia and Herzegovina

The optimal tilt angle of collector relative to the horizontal plane depends on the specific application, and parameters such as the seasonal variations and daily usage patterns of collected solar energy. To determine the optimal tilt angles for different locations in Bosnia and Herzegovina, a simple solar system for preparation of sanitary hot water for hotel needs was used. The system consists of flat-plate VITOSOL 100 collectors, a storage boiler, a solar pump, an expansion tank and regulation components. The system was modelled using the TRNSYS software package for location in Sarajevo, Bosnia and Herzegovina.

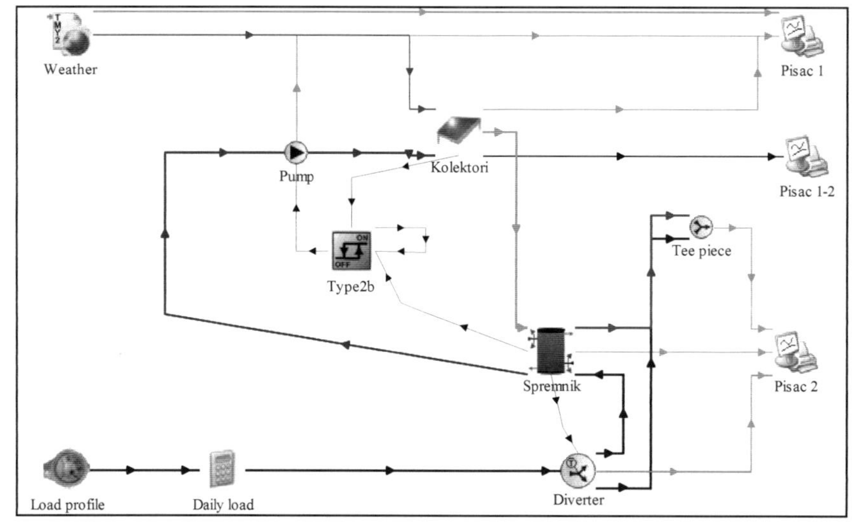

Fig. 1. Model of the solar thermal system for the preparation sanitary hot water for the hotel services [14]

When creating the model, already existing components owned by the library of the TRNSYS program package were used:

- Tank for the preparation of sanitary hot water "Type 4",
- Controller "Type 2b",
- Circulation pump "Type 3",
- Weather "Type 109",
- Diverter "Type 11",
- Printer "Type 65",

except for the collector component.

A new "Type 1b" collector component was created, whose program is written in the FORTRAN programming language. The optical efficiency function for the "Type 1b" collector is represented by a modified efficiency function depending on the tilt angle of collector for the incidence angle of solar radiation $\theta > 30°$, and by the correlation of the standard function for the incidence angle of solar radiation $\theta \leq 30°$.

The simulation time step used for this research is 15 minutes (from the offered values of the simulation time step, the minimum value was chosen, with the aim of achieving more accurate results) and was constant through the study [15]. The scheme of the model of the solar thermal system for the preparation of sanitary hot water for the needs of the hotel is shown in Fig. 1.

Determination of the optimal collector tilt angle was carried out for the following towns in Bosnia and Herzegovina: Sarajevo, Mostar, Zenica, Tuzla, Banja Luka and Livno. During research with the goal of maximizing solar thermal energy, the following tilt angles of collector φ were analyzed: 0°, 30°, 45°, 60°, 75°. The obtained results are presented in the form of a diagram (Fig. 2, Fig. 3, Fig. 4, Fig. 5, Fig. 6, Fig. 7).

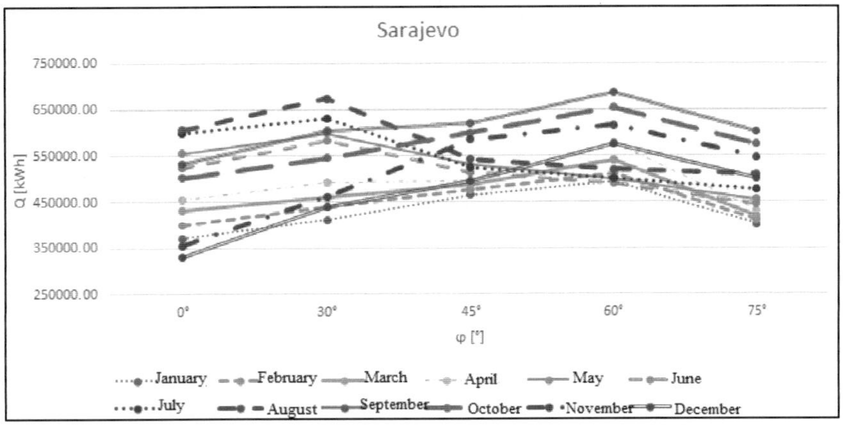

Fig. 2. Thermal energy collected for different tilt angle of collector, for the city of Sarajevo.

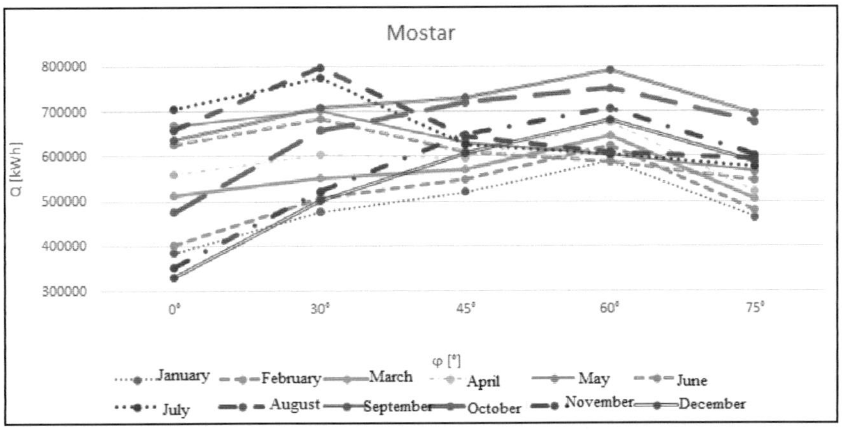

Fig. 3. Thermal energy collected for different tilt angle of collector, for the town of Mostar.

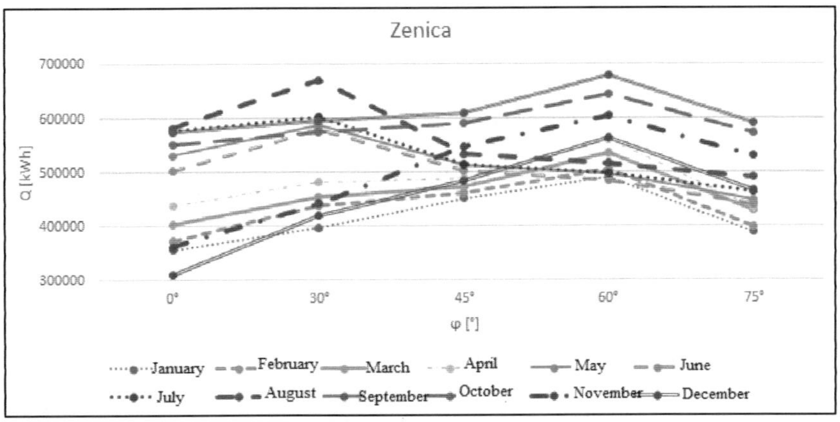

Fig. 4. Thermal energy collected for different tilt angle of collector, for the town of Zenica.

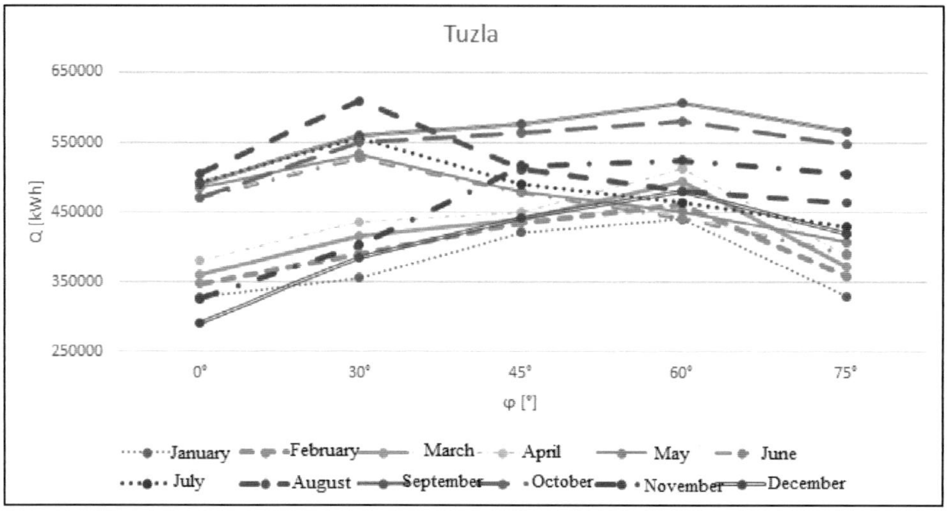

Fig. 5. Thermal energy collected for different tilt angle of collector, for the town of Tuzla.

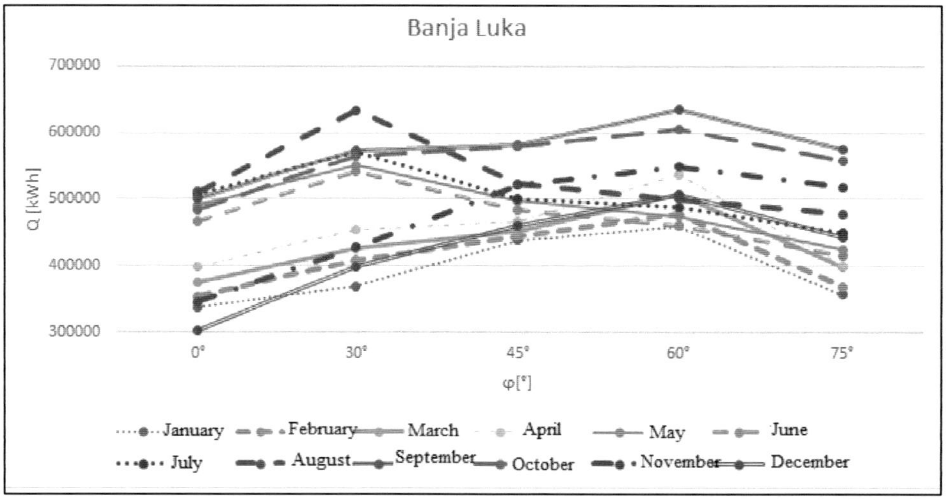

Fig. 6. Thermal energy collected for different tilt angle of collector, for the town of Banja Luka.

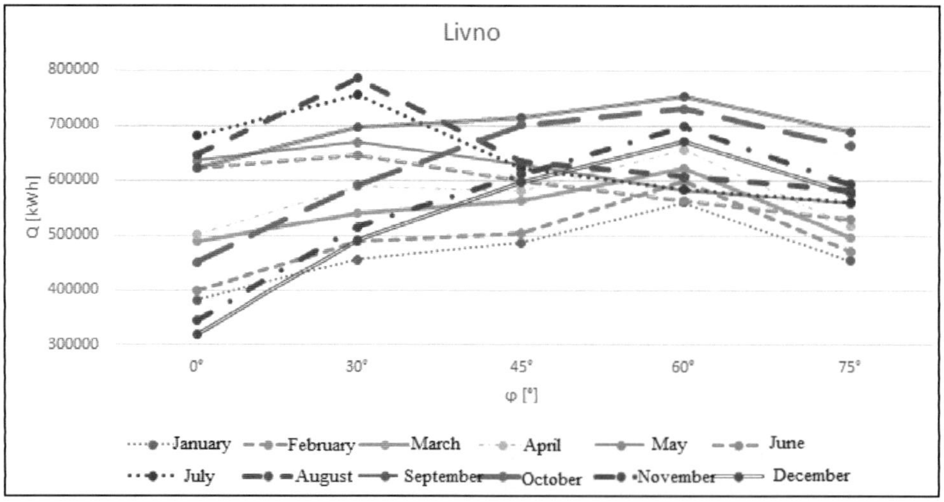

Fig. 7. Thermal energy collected for different tilt angle of collector, for town of Livno.

4. Discussion of the results

Based on the obtained diagrams (Fig. 2, Fig.3, Fig.4, Fig. 5, Fig.6, Fig.7) it is evident that the optimal slope in relation to the horizontal plane depends on the dynamics of the consumption of collected solar energy during the year. Solar applications that use solar energy in the months: May, June, July and August for the preparation of sanitary hot water in hotels, camps and other tourist facilities, the most of the solar energy can be collected at a tilt angle of collector $\varphi=30°$. If we are talking about solar applications that use solar energy in the months: January, February, March, April, September, October, November and December, the most collected solar energy can be achieved at a tilt angle of collector $\varphi=60°$.

Maximizing the collected solar energy for the preparation of sanitary hot water in hotels, camps, etc. in Bosnia and Herzegovina throughout the year can be achieved by using the so-called two-position collector tilt, where the collector tilt angle φ for the months: May, June, July and August is 30°, while for the months: January, February, March, April, September, October, November and December, the value of the collector tilt angle is 60°, as already stated.

For the practical application of solar energy, i.e. for designing and optimizing devices for using solar energy, it is necessary, in addition to the value of insolation on a horizontal surface, to know the amount of irradiated energy on inclined and vertical surfaces, as well as the radiation structure - the ratio of diffuse and direct radiation. For most applications, it is most advantageous to orient the solar radiation receivers (collectors) in the south direction, with the deviation from the azimuth angle of the collector ±10° (toward the east or west) slightly reduces the total energy arriving at the collector plane. Optimizing the slope results in an increase in the total irradiated solar energy, i.e. the influence of the direct component of solar radiation is intensified, while the diffuse component and the component of reflected radiation only add up.

The results of the conducted research show that the most of the solar energy throughout the year per unit of surface, can be obtained if the slope of the surface, facing south, is equal to a tilt angle of 30° to 45°.

5. Conclusion

This work presents the results of research on the optimal tilt angle of solar thermal collectors for various locations in Bosnia and Herzegovina, with a modified thermal efficiency function of the collector. Deviations of the standard instantaneous thermal efficiency function of solar thermal collectors from the efficiency function depending on different tilt angles of the collector relative to the horizontal plane, as presented in studies [13] and [15], range from 7-10%. This directly impacts the conversion of solar radiation into thermal energy and, consequently, the efficiency of the entire solar thermal system. All simulation models incorporate a standard instantaneous thermal efficiency function based on the type of collector and manufacturer, without correction for efficiency depending on the collector's tilt angle, which can lead to calculation discrepancies.

The aim of this study was to modify the existing component of a flat water-cooled solar collector in the TRNSYS software package, specifically the efficiency function for incident solar radiation angles less than and greater than 30°, and to simulate the operation of a simple solar water heating system for domestic hot water preparation for various locations in Bosnia and Herzegovina. The research results show the optimal tilt angles of south-oriented solar collectors relative to the horizontal plane at which maximum solar radiation conversion to thermal energy is achieved for different periods throughout the year.

The following conclusions can be drawn from this work:

- If it is a one-position system, the most solar energy can be collected, if the tilt angle of collector 30° and 45° and the surface of the slope facing south.
- If it is possible for the solar system to be in two positions during the period of use, then the optimal tilt angle of collector in the summer months is 30°, while in the rest of the period it is 60°.
- Optimizing the slope results in an increase in the total radiated solar energy, that is, the influence of the direct component of solar radiation is intensified.

Given that the installation of solar thermal systems depends on the slope and orientation of the roof surface and that, especially in cases of retrofit installations, it is sometimes impossible to achieve the ideal, southern orientation of the collectors, it may be interesting for future research to explore changes in the standard thermal efficiency function of the solar collector with changes in the azimuth angle.

6. References

[1] Ahmovic, I.; Dzaferovic, E. & Hadziahmetovic, H. (2018). Analysis of Temperature Data in the Area of the City of Sarajevo (Bjelave) in Period From 2000-2010, Proceedings of the 29th DAAAM International Symposium, pp.0260-0266, B. Katalinic (Ed.), Published by DAAAM International, ISBN 978-3-902734-20-4, ISSN 1726-9679, Vienna, Austria, DOI: 10.2507/29th.daaam.proceedings.037

[2] Lambić, M. (2013). Solar technologies, thermal and photovoltaic systems (orig. Serbian: Solarne tehnologije toplotni i fotonaponski sistemi), AGM, Belgrade

[3] Kamal Skeiker. (2009). Optimum tilt angle and orientation for solar collectors in Syria, Energy Conversion and Management 50, 2439–2448
[4] Moncos, V. H. (1994). Optimum tilt angle and orientation for solar collectors in Assiut, Egypt, Renewable Energy, Vol. 4, No. 3, 291-298
[5] Stanciu, C. & Stanciu, D. (2014). Optimum tilt angle for flat plate collectors all over the World – A declination dependence formula and comparisons of three solar radiation models Energy Conversion and Management Volume 81, 133-143
[6] Ibrahim, D. (1995). Optimum tilt angle for solar collectors used in Cyprus Renewable Energy, Vol. 6, No. 7, 813-819
[7] Jafarkazemi, F. & Saadabadi, S. A. (2013). Optimum tilt angle and orientation of solar surfaces in Abu Dhabi, Renewable Energy 56, 4-49
[8] Pourfayaz, F.; Shirmohammadi, R.; Maleki, A. & Kasaeian, A. (2020). Improvement of solar flat-plate collector performance by optimum tilt angle and minimizing top heat loss coefficient using particle swarm optimization, Energy Science and Engineering
[9] Modarresi, J. & Hosseinnia, H. (2020). Optimum Tilt Angle Model to Obtain Maximum Solar Energy, IETE Journal of Research, 549-557
[10] El-Kassaby, M. M. & Hassab, M. H. (1994). Investigation of a variable tilt angle Australian type solar collector. Renewable Energy, 327–332
[11] Aja, O. C.; Al-Kayem, H. & Karim, Z. A. A. (2013). Analytical investigation of collector optimum tilt angle at low latitude, Renewable Sustainable Energy 5
[12] Gunerhan, H. & Hepbasli, A. (2007). Determination of the optimum tilt angle of solar collectors for building applications, Building and Environment 42, 779–783
[13] Gadara, M. (2023). Experimental and numerical investigation of flat solar thermal collectors depending on the angle of inclination to the horizontal plane, PhD Dissertation, Faculty of Mechanical Engineering, Computing and Electrical Engineering, University of Mostar, Mostar, Bosnia and Herzegovina
[14] Gadara, M. & Lulić, H. (2021). Thermal efficiency of flat plate solar collectors and its improvement, In: 13th International Scientific Conference on Manufacturing Engineering Development and Modernization of the Manufacturing RIM, Proceediongs, pp. 90-96
[15] Gadara, M. & Lulić, H. (2023) Dynamic modelling of the solar system depending on the change in the tilt angle of the solar collector, In: 26th International Conference on Heating, Cooling and Air-Conditioning INTERKLIMA, University of Zagreb Faculty of Mechanical Engineering and Naval Architecture

DOI: 10.2507/35th.daaam.proceedings.033

RISK EVALUATION OF CZECH COMPANY WITH THE USE OF ROVS SOFTWARE

Jan Albrecht & Gabriela Králíčková

This Publication has to be referred as: Albrecht, J[an] & Kralickova, G[abriela] (2024). Risk Evaluation of Czech Company with the use of ROVS Software, Proceedings of the 35th DAAAM International Symposium, pp.0245-0251, B. Katalinic (Ed.), Published by DAAAM International, ISBN 978-3-902734-44-0, ISSN 1726-9679, Vienna, Austria
DOI: 10.2507/35th.daaam.proceedings.033

Abstract

This article examines the need for risk re-evaluation in a Czech organization to ensure that security measures remain relevant and effective in light of organizational changes. Whether the change is major or minor, reassessing potential risks is necessary to maintain adequate security standards. The paper describes the application of ROVS software for risk analysis, using the PIE analysis method for effective evaluation across various organizational fields. Following the thorough assessment, the article identifies key vulnerabilities and recommends targeted solutions to prevent or mitigate the impact of potential risks in case they occur. The paper demonstrates how the use of ROVS software can ensure a thorough evaluation of risks, followed by suggestions to improve the potential vulnerabilities. Furthermore, the work underscores the critical role of Knowledge Management in implementing effective security policies and enhancing overall risk management processes.

Keywords: Risk Assessment; Risk Analysis; PIE; Risk Management; Knowledge Management.

1. Introduction

Constant technological innovation is a very actual topic, and the rise is exponential, which many organizations need to work with and change some of their already implemented strategy policies [1]. Some could say that new technologies are an improvement on their side, so the security should be the same as before. With some of the technologies, that statement can be partially right; however, with many new technologies, it also brings new entry vectors for potential threats, and so they need to be included in the new security policies of an organization. Regardless of the range of changes made in a company, continuous re-assessment of potential risk is a mandatory process for effective risk management [1], [2].

For the re-evaluation to be effective, one must select a suitable method for the whole risk assessment process [3]. Thorough risk analysis should focus on targeting the most critical vulnerabilities and assign them adequate risk values. After the potential risks and their values are known, the analysis examines the already set security measures and looks for improvements [3], [4]. Based on those findings, the chosen analysis suggests suitable solutions for either the prevention of their occurrence or mitigation of their impact in case they appear [3], [4], [5], [6], [7], [8].

The ways of effective execution of a risk analysis are usually two, first option is to let an expert of risk management choose a specific risk analysis method that suits the best for the needs and which the expert knows the best. The second option could be to choose software that offers a thorough analysis with already implemented principles of some of the risk analysis methods. In the Czech Republic, there are only a few software options for such evaluations, one of them is the ROVS [9], which is widely used among organizations in this country and uses the principles of the PIE risk analysis method, known as one of the most common analysis methods [10]. The implementation offers a thorough examination of potential risks and has wide coverage of many fields that can be found in organizations.

A thorough risk analysis with a list of suitable suggestions is, without a doubt, one of the most important steps in risk management and dealing with potential threats. However, implementing those suggested measures is another significant factor in risk management. The data gathered from risk analysis is one thing, but how to work with them and effectively implement them in the security policies is sometimes overlooked, making the risk analysis process go partly to waste. One of the common mistakes can be an inadequate sharing of information about the explored potential risks among interested employees [11].

This gap can be filled with the use of Knowledge Management (KM), which focuses on the work with data and knowledge among the whole organization. To make the security measures effective and ensure that they will be followed correctly, the KM principles focus on the fact that everyone needs to be well informed not only about the potential risks but also about the impact they may cause once they occur. Once everyone knows the potential impacts and how disastrous they could be, there is a much higher possibility that they will follow the rules for their own safety and not only because it is something they need to follow [11], [12], [13], [14], [15], [16].

2. Methods

The risk re-evaluation is described on a fictional organization X, but its characterization is based on real data provided by a Czech company. To effectively execute the risk assessment, the work first focuses on the summarization of the organization fields that are later the targets of the risk analysis.

2.1. Characterization of the organization

The company works with various noble metals, specifically aluminium alloy and stainless steel. As the company grows, so does the number of its employees, which is now around one hundred. Since the work is highly concentrated on employees' work and is tightly related to security measures, the risk analysis of used methods and safety policies needs to be always up-to-date, and continuous re-evaluation should be a mandatory process of the organization.

During the last few years, several risk analyses were made, but as time goes on, so have the changes in the company, such as many technological innovations and the changes in some fields associated with new workplaces. Painting, varnishing, and welding of metal are some of the main jobs that the organization focuses on. As these fields are the major ones of the organization, the whole risk evaluation in this work is targeted mainly at them. However, the paper also includes some other areas closely related to the mentioned operations that were debated with the organization's leadership and needed to be re-evaluated.

The selected fields are listed below:

- Workspaces
- Storeroom
- Hand tools

2.2. Introduction of PIE analysis

Since the ROVS software works with the principles of PIE analysis, this section is to clarify its usage. PIE is an abbreviation where each letter represents a different category, "P" stands for probability of risk occurrence, "I" stands for possible impact of the risks (when they occur), and "E" can be described as the evaluator's opinion, on how the risk affects the overall safety for the selected area [9]. The values of each category can have different scaling meters, but the common scale is from 1 to 5, where the higher number usually means higher importance of the risks to be handled. The scale can be modified based on the needs of the evaluator and specific areas, but the paper will use the most common type, as the ROVS software also works with it.

$$P * I * E = \mathrm{R} \tag{1}$$

When the values of each category are assigned, the calculation of the final risk value, represented with the letter "R" is executed using the formula shown above (Eq. 1). Once the risk values are counted, the evaluator determines the importance of the risk and in which category of risks it is located based on the given table (Tab. 1).

Risk category	Determined risk value (R)	Description of the category
1.	>100	Very high risk, halt all operations
2.	100–50	High risk, immediate safety measures
3.	50–10	Risk, need for corrective actions
4.	10–3	Potential risk, increase of awareness
5.	3>0	Acceptable risk

Table 1. Risk categories based on their counted values.

2.3. Introduction of the ROVS software

The software is based on the principles of PIE risk analysis and can be nicely used for risk evaluation within any organization's risk management. Its usage is convenient not only because its risk analysis principles work in the easy-to-use software but also because it already meets the legislation requirements, which the organization needs to fulfil. The software is built on statistical data of the most frequent work accidents and related suitable measures. This paper works with the ROVS software for its wide database of potential work accidents, identification of potential risks that may occur, and overall wide coverage of many work fields, making the tool suitable for any organization that needs risk assessment, no matter the area of their focus [9]. ROVS has many prepared tables for a wide selection of fields, which can satisfy many types of organizations. The tables contain field specifications, a list of work operations under this field, and their related risk values. These risk values are counted using the PIE analysis principles (Eq. 1), as described in the chapter above. After each of the operations is assigned a corresponding risk value, the software suggests basic security measures based on the potential risks that the selected operation faces [9]. Thanks to the thorough and unified database, the ROVS software offers to evaluate many organizations with the same pre-set tables, and their principles are the same, always matching the legislative, no matter the field. This step assures that the evaluations will be at least a little bit objective and that the results of risk assessments are easily comparable with others, using the same principles. The fact that the tables are unified can help employees to keep their awareness of the security measures because of their similarity. Working with such software makes the risk assessment process easier and more efficient [9]. Despite all the positive sides of the ROVS software, it's important to stress that its effectiveness lies in the suitable choice of the tables and their values, making it greatly important to have well-experienced evaluators with a complex knowledge of the organization. Otherwise, it could downgrade the quality of the whole risk management.

3. Use of ROVS in selected fields

Based on selected field of organization, this section examines the potential risks of these areas and presents the data gathered using ROVS software. The paper shows a related table for each field with calculated values and their suggested security measures. These basic security measures are described more thoroughly in the next chapter of this paper.

3.1. Workspaces

In this field, the main focus is given to the operations executed there, such as painting, varnishing, and welding of noble metals. Since these activities are related to the work of people and organizations that need to keep the safety of employees at the highest possible level, their re-evaluation is very important. The selected operations to evaluate are pointed out in the table below (Tab. 2). For each identified risk, ROVS suggests security measures adequate to the operation to improve the overall security (Tab. 3).

#	Operation	Identification of risk - Workspaces	Risk values			
			P	I	E	R
#1	Electric arc and flame welding	- Burns to various parts of the body caused by hot sparks, droplets of molten metal and slag, and fragments of already solidified slag during its removal (a potential hazard is, for example, a hot particle falling into work boots); the risk is more serious during arc welding and gouging or hole cutting with oxygen.	2	3	1	**6**
#2	Electric arc and flame welding	- Burns to unprotected body parts (hand) from direct contact by the welder with a heated cut, cut metal material, and hot metal surfaces during heat transfer.	2	2	1	**4**
#3	Painting and varnishing	- Finger caught; struck of worker´s hand;	2	2	1	**4**
#4	Painting and varnishing	- Tripping; ankle sprain; hand injury caused by slipping; stumble;	2	2	1	**4**
#5	Painting and varnishing	- Worker collision and fall onto a vehicle, handling equipment, or stored items;	2	2	1	**4**

Table 2. Risk evaluation of workspaces using ROVS software (modified [9])

3.2. Storeroom

Almost every operation is somehow connected to the storage of materials needed for the work or for the actual storage of some completed products. No matter the purpose, correctly evaluating security measures in this field is another important part of the thorough re-evaluation. The risk values for adequate operations are shown in the table below (Tab. 4), and their related suggested security measures follow (Tab. 5).

#	Operation	Identification of risk - Storeroom	Risk values			
			P	I	E	R
#1	Storage racks	- Worker falls while servicing higher-level rack compartments.	1	2	1	**2**
#2	Storage racks	- Tripping or collision of a person with the rack structure and stored material.	2	2	1	**4**

Table 3. Risk evaluation of storeroom using ROVS software (modified [9])

3.3. Hand tools

Another closely related factor to be evaluated is the overall usage of hand tools. Regardless of the field in the organization, this also needs to be considered as one of the most important operations. Many employees in the workspaces use the tools daily, so the security measures should be well set to avoid needless accidents. A few selected operations are evaluated in the table below (Tab. 6). and their security measures are described in the next table (Tab. 7).

#	Operation	Identification of risk – Hand tools	Risk values			
			P	I	E	R
#1	Hand tools	- Worker struck by a loose tool, such as a hammer or head, coming off the handle.	3	2	1	**6**
#2	Hand tools	- Bruising of the left hand. - Hammer slipping out of the hand.	3	2	1	**6**
#3	Hand tools	- Abrasions and bruises on hands while working with tools in confined spaces, during repairs and maintenance.	2	2	1	**4**

Table 4. Risk evaluation of hand tools using ROVS software (modified [9])

4. Suggested security measures for selected fields

Using the ROVS tables, this part of the work shows suggested security measures related to the evaluated fields from the chapter above. The selected security measures, as were the identifications of risks, were also distributed to tables, so it is easier to understand them.

4.1. Workspace security measures

Both operations in this field are keen on safety while working; the main factors to keep the safety level high are to maintain order in the workplace, thorough and strict observance of given work processes, and use of adequate safety equipment, known as Personal Protective Equipment - PPE (gloves, glasses, etc.). Employees in this field should have corresponding education and, if needed, certified courses or licenses (for example, welding licenses), ensuring the worker's knowledge. Regardless of the employee's experience, frequent training should be provided, keeping the potential risks and consequences in the worker´s awareness. A thorough list of suggested measures by ROVS is shown in Table 5.

4.2. Storeroom security measures

Storage rooms usually have security issues based on their racks' construction, position, or overall room space. Manipulation with anything stored in these rooms should always be safe and easy, regardless of the position. Every person using the storerooms should be well informed of the room's safety measures and equipment. These measures need to be clearly depicted and described in the room itself and be included in the repetitive employee training to remind all the interested workers of potential risks and consequences they can go through in case they do not follow the security policies

#	Operation	Suggested security measures - Workspaces
#1	Electric arc and flame welding	• Proper welding practices and consistent use of PPE to protect eyes, face, and other parts of the body; • The hazards and precautions during oxygen cutting are similar to those of welding or flame cutting; increased risk arises from greater dust agitation and more extensive splattering of cut metal; Protection of the area below welding operations at height against hot splatter is essential.
#2	Electric arc and flame welding	• Use of PPE (gloves); • Adherence to correct work procedures.
#3	Painting and varnishing	• Items that are stored tightly together and lack parts for safe handling (such as loops or handles) should be placed on supports (avoid using rounded logs as supports); • When manually handling heavy objects, use appropriate aids, such as hand tools (e.g., trolley jacks).
#4	Painting and varnishing	• Flat, undamaged, and non-slip surfaces for floors, pathways, vehicle loading areas, and handling spaces.
#5	Painting and varnishing	• Maintaining order at the workplace and removal of protruding obstacles (e.g., raised covers, lids, mats, steps, thresholds, hoses, cables, flexible power cords, anchor bolts, etc.).

Table 5. Suggested security measures for workspaces using ROVS software (modified [9])

A thorough description of adequate security measures for the risks of storerooms is shown in Table 6.

#	Operation	Suggested security measures - Storeroom
#1	Storage racks	• Manual handling (loading and unloading of materials) from rack sections above 1.8 m is performed using safe equipment and aids (ladders, mobile steps, handling platforms, etc.); • Do not climb on the rack structure.
#2	Storage racks	• Maintain clear access or approach to the racks to avoid obstruction of loading and unloading of handling units and materials; • The width of aisles between racks and stacks corresponds to the method of material storage and is at least 0.8 m for manual handling; the width of the aisle for the passage of transport trolleys is at least 0.4 m wider than the largest width of the trolleys or loads.

Table 6. Suggested security measures for storeroom using ROVS software (modified [9])

4.3. Hand tools security measures

This category can be included in many fields across the organization since hand tools are part of many operations, and many workers use them daily. Employees have to be well informed about potential risks that any of the hand tools have and what the consequences are if the risks occur. The organization has to provide workers with adequate equipment (PPE) for operating such tools, and workers have to abide given security policies. As in other fields, continuous employee training must be set for effective risk management. A corresponding table (Tab. 7) with examples of suggested security measures describes their key point.

#	Operation	Suggested security measures – Hand tools
#1	Hand tools	• Do not use damaged tools (with a loose handle, deformed working part, etc.).
#2	Hand tools	• Focus while working, and if necessary, use hand protectors.
#3	Hand tools	• Adjust the workplace and organization to ensure that work with tools is performed in ergonomically suitable positions so the worker does not have to use tools overhead.

Table 7. Suggested security measures for hand tools using ROVS software (modified [9])

According to the analysis, the main benefits of introducing work with risks include the overall concept of risks that society faces or could face. At the same time, thanks to the use of this software, not only are the legislative requirements imposed on the employer met, but the requirements of client and certification audits are also met.

If the organization implements the suggested measures correctly, it will help not only to improve the safety and security of work fields across the organization but also to make the employees there feel safe and much more likely to adhere to the security policies. Sometimes the implementation of the suggested measures can fail, and the desired outcomes, which are improvement of safety and security policies, are not achieved, ending in a new re-evaluation process, making it not cost-efficient. At the same time, when a bad type of implementation is executed, such as too fast or without the right use of knowledge-sharing among an organization, employees can tend to be unmotivated to stick to the security policies, and the whole process of security measures implementation would go to waste. The disadvantages may also include the cost of the entire process, exacerbated by the fact that some work processes may be changed and the analysis needs to be repeated. At the same time, any company may need to introduce new workflows that need to be reacted to, which can bring the need for further updates and the associated labour time and costs.

5. Conclusion

The continuous re-evaluation of potential risk is a critical factor in effective risk management in any organization. Sometimes, this step of keeping the risk analysis up-to-date gets underestimated, and the data, thus the security measures themselves, are not relevant, which can create many potential vulnerabilities for the organization. The necessity for constant re-evaluation comes not only with the implementation of some changes in the company but also with the whole growth of the technologies around the organization, which is these days way faster and drastically changes the number and types of potential risks towards the company.

The paper focuses on the specific risk re-evaluation of an organization based on the data from a Czech company. Even though several risk analyses have already been executed, this mainly targets the selected fields of the company that are pointed out as the most important. The main goal of risk assessment is always the safety of employees and people in the organization area. For this re-evaluation, the work uses ROVS software, specifically made to evaluate the risks using PIE analysis principles, and follows the related legislation. After all the risks are evaluated and presented, the paper shows suggested security measures from the ROVS software, which are thoroughly described. Even when there are not some extremely critical risks that need to be dealt with, there are a few suggestions to make the work in the organization safe and ready to deal with potential risks.

The work suggests using Knowledge Management principles for effective risk management and implementing new suggested security measures. Such principles can help ensure a better understanding of the actual risks that the organization can face and also provide long-term observance of set measures, not only because the legislation and security policy say so but also because well-informed employees want to keep the safety of their work at the best possible level, now that they know of the potential consequences. One should not make risk assessments only because of changes or new implementations; effective risk management always focuses on continuous risk evaluation, so security measures are always up-to-date and the organization´s vulnerabilities are under control.

The limitation was evaluating only one organization, and the software does not have many competitors on the market. However, the aim was to show a way to effectively assess risks and the options for dealing with them. Furthermore, the organization can examine a different approach for the risk evaluation. For now, the risk assessment has been executed mainly to measure the security of workspaces and the safety of employees. As technological innovation rises, the company should also consider a thorough evaluation of its cybersecurity measures, where different work can be focused on how ROVS software can be used in this specific area or how Knowledge management helps the whole risk assessment process.

6. Acknowledgments

This work was supported by the Internal Grant Agency of Tomas Bata University - Faculty of Applied Informatics under project No. IGA/FAI/2023/003.

7. References

[1] O. V. Syuntyurenko (2022), Predicting Potential Threats and Megarisks in Information Technology Development, Scientific and Technical Information Processing, Vol. 49, no. 1, 48–59, doi: https://doi.org/10.3103/S0147688222010130

[2] D. F. Reding and J. Eaton (2020), Science & Technology Trends 2020–2040, https://www.nato.int/nato_static_fl2014/assets/pdf/2020/4/pdf/190422-ST_Tech_Trends_Report_2020-2040.pdf

[3] S. Haugen and S. Kristiansen (2022). Risk analysis methods, Maritime Transportation: Safety Management and Risk Analysis, (2nd ed., London, Routledge., 2022), pp. 175–251, doi: https://doi.org/10.4324/9781003055464

[4] V. Šilhánková and K. Netíková (2020). Komparace metod analýzy rizik pro hodnocení bezpečnostních hrozeb lokalit brownfields, 23rd International Colloquium on Regional Sciences, Vol. 1, Brno, 2020, pp. 489–497, doi: https://doi.org/10.5817/CZ.MUNI.P210-9610-2020-62

[5] IEC/ISO 31010:2019 (2019) Risk management – Risk assessment techniques, International Electrotechnical Commission, Geneva, (2019), preview: https://www.iso.org/obp/ui/#iso:std:iec:31010:ed-2:v1:en,fr

[6] Procházková, D. (2011). Analýza a řízení rizik. Praha: ČVUT. ISBN 978-80-01-04841-2

[7] Risk Analysis and Risk Management: Assessing and Managing Risks. Management Training and Leadership Training – Online [online]. Scotland (Edinburgh): Emerald Works Limited, © 2022 [cit. 2022-04-27]. Dostupné z: https://www.mindto-ols.com/pages/article/newTMC_07.htm

[8] Půček, Milan Jan (2020). Analýza Rizik. In: Terminál – interní prostředí [online]. Praha: Národní síť Zdravých měst, 2020, 2020 [cit. 2024-10-16]. Dostupné z: https://www.dataplan.info/img_upload/f96fc5d7def29509aeffc6784e61f65b/ana-lyza-rizik-text-metodika.pdf

[9] Rožnovský vzdělávací servis s.r.o. - Software. Online. Rožnovský vzdělávací servis. © 2001-2024. Dostupné z: https://www.rovs.cz/software. [cit. 2024-10-16].

[10] D. Kotkova, L. Kralik, L. Kotek and J. Valouch (2023), Multiple Criteria Decision-making: Risk Analyses for the Soft Target. ASTES Journal, Vol. 8, no. 2, 14–23, doi: http://dx.doi.org/10.25046/aj080202

[11] Albrecht, J[an]; Dostálová, P[etra]; Králík L[ukáš] & Králíčková, G[abriela] (2023). Benefits of Knowledge Management in Cybersecurity – Initial Study, Proceedings of the 34th DAAAM International Symposium, pp.xxxx-xxxx, B. Katalinic (Ed.), Published by DAAAM International, ISBN 978-3-902734-xx-x, ISSN 1726-9679, Vienna, Austria. DOI: 10.2507/34th.daaam.proceedings.069

[12] Paulova, I.; Vanova, J.; Rusko, M.; Hekelova, E. & Kralikova, R. (2017). Knowledge Managements for Improvement the Competitiveness of Organization, Proceedings of the 28th DAAAM International Symposium, pp.1221-1226, B. Katalinic (Ed.), Published by DAAAM International, ISBN 978-3-902734-11-2, ISSN 1726-9679, Vienna, Austria, DOI: 10.2507/28th.daaam.proceedings.170

[13] ISO 30401:2018 (2018) Knowledge Management Systems – Requirements, International Organization for Standardization, Geneva, preview: https://www.iso.org/obp/ui/en/#iso:std:68683:en

[14] Knoco Ltd (2021), Global Survey of Knowledge Management 2020-2021. Online. ©KNOCO LTD 2021. Knowledge management consultants. 2021. Accessed: https://www.knoco.com/knowledge-management-survey.htm.

[15] Milton, N. J. and Lambe P. (2020), The knowledge manager's handbook: a step-by-step guide to embedding effective knowledge management in your organization. Second edition. London, United Kingdom: Kogan Page Limited. ISBN 9780749484613. Accessed: http://search.ebscohost.com/login.aspx?direct=true&scope=site&db=nlebk&AN=2258716&authtype=ip,shib&custid=s3936755

[16] Talet, A. N. (2018), The Role of Knowledge Management with Risk Management for Information Technology Projects Risk Assessment. International Journal of Environment and Sustainability [online]. vol. 6, no. 2. ISSN 1927-9566.

DOI: 10.2507/35th.daaam.proceedings.034

FEASIBILITY STUDY AND OPTIMIZATION OF THE PRODUCTION OF WOOD JOINTS WITH AN INDUSTRIAL ROBOT

Stefan Machhammer, Kemajl Stuja, Erich Markl, Ali Aburaia & Arbnor Pajaziti

This Publication has to be referred as: Machhammer, S[tefan]; Stuja, K[emajl]; Markl, E[rich]; Aburaia, A[li] & Pajaziti, A[rbnor] (2024). Feasibility Study and Optimization of the Production of Wood Joints with an Industrial Robot, Proceedings of the 35th DAAAM International Symposium, pp.0252-0258, B. Katalinic (Ed.), Published by DAAAM International, ISBN 978-3-902734-44-0, ISSN 1726-9679, Vienna, Austria
DOI: 10.2507/35th.daaam.proceedings.034

Abstract

Wood is an important and environmentally friendly construction material. Currently, research in the use of wood as a construction material is increasing to improve the sustainability of buildings. The production of wooden components is currently carried out with special and expensive CNC machines (joinery machines) or by hand by skilled workers. A possible alternative solution is the use of industrial robots. However, there is only a limited amount of research and a lack of data in the field of wood machining with serial robots. In order to obtain further application possibilities and data, this paper investigates the feasibility of producing 33 classic wood joints with an industrial robot. For the feasibility study, simulations were carried out, as well as a total of 103 milling tests using a KUKA KR 16-2 with a spindle system. The feasibility study showed that the manufacturing of 29 classic wood joints with an industrial robot is possible and that tolerances of ±1 mm can be obtained. However, there are still numerous optimisation possibilities for the system and the process of wood machining with an industrial robot, which must be further analysed and optimised in future works.

Keywords: Industrial robots; woodworking; milling; wood joints; feasibility study

1. Introduction

Wood is a versatile and environmentally friendly building material and a current research topic in the construction industry to increase the sustainability of buildings [1]. One of these research topics are wood joints. Therefore, the wood industry is currently focusing on traditional and modern wood-wood techniques for joining different wood components and structures [2]. An important factor is the precision in producing wooden components, because a lack of precision in the manufacture of such connections can significantly influence the load-bearing behaviour and lead to damage to the components [3]. The production of the various wooden components in the wood industry is carried out either by expensive CNC machines (e.g. joinery machines) or manually by skilled workers, which can be labour-intensive. Although CNC machines are standard equipment for larger carpentry companies, they are often not affordable for small carpentry companies due to the high acquisition costs, as the corresponding required economic resources are not available or the risks of purchasing them are too high [4].

A possible alternative solution to produce wood joints and other wooden components are industrial robots, as they are associated with lower acquisition costs, allow a more flexible use, offer the opportunity to increase productivity in the woodworking sector and also enable the production of special wooden workpieces [5].

2. Motivation and aim

If the application area of industrial robots for the mechanical processing of wood is compared with other application areas such as aluminium processing, the comparison shows that there is only a small research field with a limited number of research papers in wood processing with industrial robots [4], [6]. In the field of processing aluminium with industrial robots, research papers employ various approaches to increase the accuracy and quality of the manufactured workpieces. For example, Huynh et al. [7] use multibody modelling, simulations and experimental tests to obtain data on cutting forces, vibrations and roughness during milling with an industrial robot. Another approach is to increase the pose-dependent stiffness of industrial robots for the machining tasks [8]. However, the papers in the field of wood processing with industrial robots focus mostly on domain-specific processing and manufacturing of specific components. For example, Geno et al. [9] investigated the processing of unshaped tree trunks and the associated process for automatic robotic manufacturing. For this, they examined the possibilities of connecting tree trunks and the challenges posed by the irregular geometry of tree trunks. In another work, Mesnil et al. [10] designed a robot cell to improve the flexibility in machining and manufacturing of different wooden workpieces for the prefabrication of a 50 m² wooden pavilion. They focused on the workflow and the productivity of the prefabrication. If we now look at the area of timber and the wood industry sector, Pancharowitsch et al. [4], [5], [11] carried out various studies on milling timber with an industrial robot. They conducted numerous experiments, tests and research work with the aim of obtaining objective data for wood processing with industrial robots and thus being able to make comparisons with CNC-controlled joinery machines. According to the authors, there are still large research gaps in the field of wood processing with industrial robots, where there is a lack of usable and comparable information, data, research results, general guidelines and approaches for the implementation and use of industrial robots for wood processing.

To obtain further application possibilities, approaches and data for the processing of wood by industrial robots, the aim of this work is to check the feasibility of producing different wood joints with an industrial robot. Therefore, the achievable accuracy and surface quality of the manufactured workpieces should be determined. Furthermore, a process from the design of the CAD models to the analysis should be created and the milling parameters and milling settings for machining should be determined.

3. Materials and methods

To achieve the goals, a methodology was created based on the approach of Pott and Dietz [12] for the conception and planning of industrial robot systems. In the first step, a potential analysis for the use of industrial robots in the wood industry was carried out and the classic wood joints according to DIN 1052 Section 15 and their subcategories in wood construction were analysed. In the second step, the requirements for the system for wood processing were worked out.

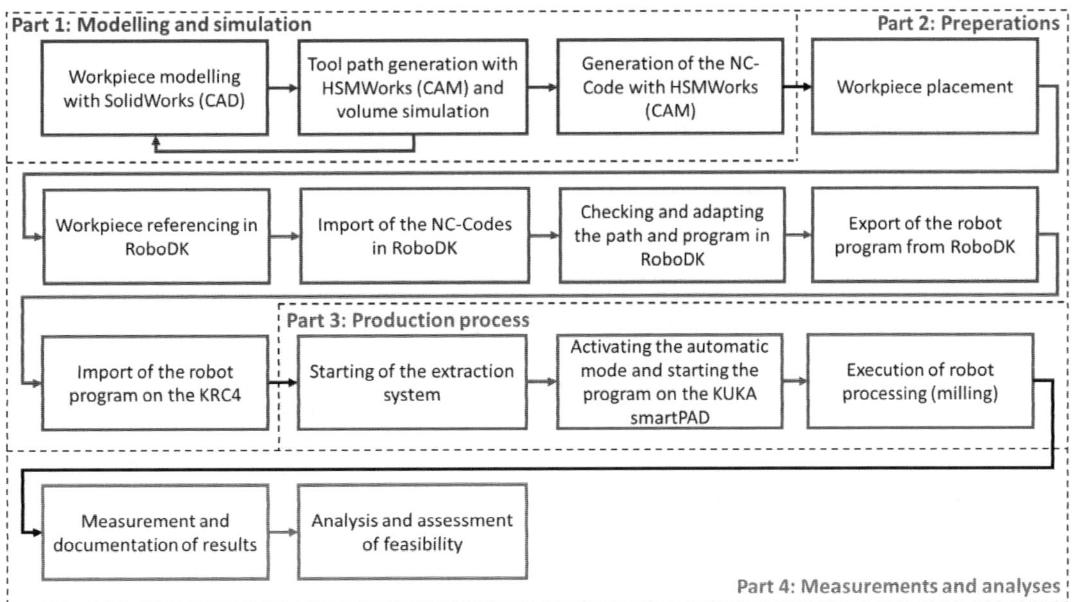

Fig. 1. Overall process from the design to the analysis of the measurement results for the milling tests

Fig. 2. Robot-Milling-Station of UAS Technikum Wien – 1) KUKA KR 16-2, 2) End effector system with asynchronous spindle, attachment changer and extraction system, 3) Control cabinet with frequency converter, 4) KUKA KR C4, 5) KUKA smartPAD, 6) Extraction system, 7) Clamping system with workpiece, 8) Tool magazine

In the third step, the feasibility study was carried out using simulations and milling tests. Therefore 33 selected wood joints from DIN 1052 were used as workpieces. These wood joints can be divided according to their geometric assignment into the five subcategories: lengthening joints, corner joints, framing joints, oblique framing joints and cross and halving joints. For the feasibility study, only those wood joints that can be modelled in CAD were used. Additional positional securing devices such as bolts, nails, dowels or wedges were not considered in this work. Wooden beams made of spruce or fir with dimensions of 250 mm x 78 mm x 78 mm were used as the workpieces. To carry out the simulation and milling tests, an overall process was created from the design to the analysis of the measurement results of the milling tests. The overall process is shown in Fig. 1.

For the robot simulations and milling tests, the robot milling station at the UAS Technikum Wien was used. This station is equipped with a KUKA KR 16-2, an asynchronous spindle (HFSAC-8022-24-ER16) and an attachment changer from Mechatron. The robot milling station at the UAS Technikum Wien is shown in Fig. 2. Furthermore, various wood milling cutters (with and without chip breaker) with diameters between 8 mm and 12 mm and a dovetail bit for dovetail joints with a diameter of 19 mm were used as milling tools. For the CAD modelling the software SolidWorks was used. As CAM software and for the volume simulations the additional application HSMWorks for SolidWorks was used. Moreover, the simulation and offline programming software RoboDK was used for the robot simulations and robot program generation. With the simulations the possibility of the production of the selected wood joints was checked.

For the milling tests, selected wood joints and their connectors were manufactured with the robot milling station. In addition, for the results of the milling tests a modified version of the criteria for assessing by Pantscharowitsch et al. [4] was used. The modified version includes the two main criteria accuracy (several geometric measurements) and surface quality (optical and haptic assessments for the surfaces and contours). For the accuracy, the mean value of the depth and contour measurements is calculated to compare the results. The measurements were carried out with a digital calliper from Lux with a resolution of 0.01 mm. The surface quality was rated with a value from 4 to 32. Where four is the worst value and 32 is the best value. The feasibility was then assessed based on the simulation and milling results. For this purpose, the tolerance value of ±1 mm was set for the geometric measurements of the milling tests, and the limit value of 16 points was set for the surface quality. The specified tolerance value of ±1 mm lies within all tolerance values of the standards ÖNORM B 1995-1-1:2023 [13] and ÖNORM EN 336:2013 [14]. In the fourth step, an analysis of the problems and the possible optimization options for the system, the overall process and the used hardware and software was carried out.

	Skilled workers	CNC machines	Industrial robots
Strengthen	Very high flexibility	High dimensional accuracy	High flexibility and usability
	Improvisational skills	Standardised and proven machines	Expandable workspace
	Learning skills	High production speeds	High customizability
	Experience	Proven programming	Low acquisition costs
	Adaptability	Efficient production times	Sufficient precision and dimensional accuracy
	Production of new and innovative workpieces		Production of new and innovative workpieces
Weaken	Fatigue behaviour	Limited flexibility	High mechanical compliance
	Low production speed	Limited working space	Complex process
	Limited strength/endurance	High acquisition costs	Missing data and information
	Distractibility	Large space requirement	Low production speed
	Low precision	Limited workpiece placement	Limited number of production system
	Shortage of skilled workers	Limited production	

Table 1. Comparison of the strengths and weaknesses of skilled workers, CNC machines and industrial robots for general woodworking based on the state of the art and the currently possible uses (procedure based on [15])

4. Implementation

4.1. Potential analyses

The potential analysis is based on a comparison of the strengths and weaknesses of skilled workers, CNC machines and industrial robots for general woodworking based on the process of [15] and is shown in Table 1. This comparison shows that skilled workers have a high flexibility and adaptability due to their experience which they can also use for new, complex and demanding processing tasks, such as the manufacture of special wood joints. However, skilled workers have a lower production speed and a lower level of accuracy compared to CNC machines. In addition, CNC machines have a high rigidity and precision which enable high levels of accuracy, repeatability and production speeds.

Although such CNC machines incur very high acquisition costs and require a lot of space. Furthermore, not all wood joints can be produced by standardized CNC machines due to limitations in the kinematics, tools, workspace, processing methods or control system. Compared to CNC machines, industrial robots have lower acquisition costs. Furthermore, they are significantly more flexible, and their workspace can also be expanded. However, there is a lack of experience, data and information in wood processing for industrial robots. This leads to complex processes and currently low production speeds. Nevertheless, industrial robots and their kinematic structure can produce special and innovative workpieces that could otherwise only be manufactured by hand, which is labour-intensive and time-consuming.

Due to the strengths of industrial robots, which are shown in Table 1, industrial robots can be used for numerous tasks in industrial wood processing and manufacturing. They offer potential applications in areas such as handling, testing or measuring, but also in mechanical processing such as milling, grinding, drilling or cutting. These include, for example, the production of wood joints, the production of chipboard panels, the assembly of wooden workpieces, or as part of production lines in wood processing

4.2. Feasibility study

As a first step in the feasibility study, various milling tests were carried out to determine the milling parameters and settings based on the overall process from Fig. 1. These tests were carried out for different milling tools and milling operations, whereby the milling parameters cutting speed, tooth feed and depth feed as well as the position of the workpiece and the robot configuration were varied. For this purpose, wood milling in the form of pockets and grooves was used, with a total of 89 milling tests being carried out and evaluated.

The milling tests for determining the wood milling parameters for the robot milling station showed that for the clearing operations in HSMWorks the "pocket clearing" operation achieved better surface quality with less marks than the "adaptive clearing" operation. Moreover, the cutting speed of 500 m/min and a value of 0.1 mm per tooth for the tooth feed proved sufficient to ensure the highest possible quality for surfaces with this setup. This results in a spindle speed of 15923.57 rpm. For the maximum depth feed, a value of half the cutter diameter is selected to minimize the forces from the milling process and to reduce breakouts and chatter marks. In order to maintain the surface quality, the "feed optimization" function is activated during the milling operation and the "corner deviation" is set to 0.0 mm. Furthermore, the stock-to-leave is selected as 0.5 mm. For the finishing operations of the milling tests, the cutting speed of 600 m/min as well as a tooth feed of 0.08 mm per tooth produced the best results. This results in a spindle speed of 19108.28 rpm. The "feed optimization" is also used for finishing operations and the “corner deviation” is also set to 0.0 mm.

Name of the wood joint	Manufactured connection parts	Manufactured wood joint
Tenon joint		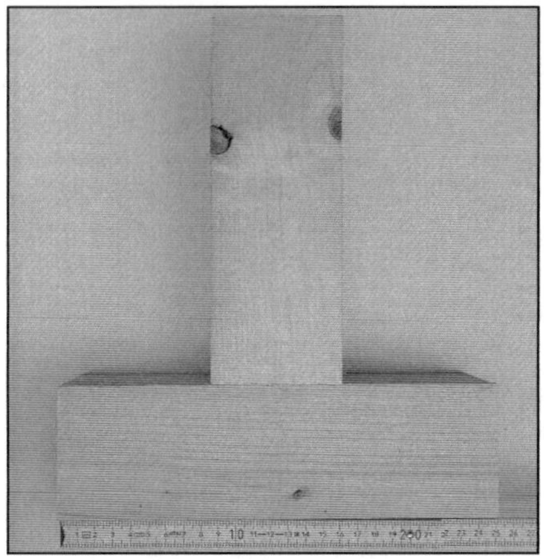
Oblique thrust joint		

Table 2. Two examples for the manufactured wood joints with pictures of the manufactured connection parts

In addition, the position of the workpiece is chosen close to the robot and centrally in front of the robot according to [16], to improve the stiffness of the system. Furthermore, the CAD models of the selected 33 wood joints were created in SolidWorks. Then the CAM operations were created in HSMWorks. Thus, the determined milling parameters were used for the milling settings of the different operations. With these models and the milling settings from the milling tests above, the volume simulations of the connectors were carried out in HSMWorks. Moreover, adjustments were made to the CAD models to enable processing with the available tools. Then the generated NC-codes from HSMworks were used to create the robot paths in RoboDK. In RoboDK the manufacturable connectors were then simulated again to check the reachability and the production with the industrial robot. After the simulations, seven wood joints were manufactured using the robot milling station. The produced joints were assessed according to the specified criteria. Two of the seven manufactured wood joints can be seen in Table 2.

5. Results

The potential analysis showed that there are numerous possible uses for robots in the wood industry as alternatives or additions to CNC machines and skilled workers. The biggest weaknesses of industrial robots are currently the lack of proven applications in the wood industry and the associated lack of data and information, as well as the complex processes for implementation.

The simulations of the feasibility study showed that the production of 29 of the 33 wood joints is possible by adapting the individual connecting parts to the available tools. Four of the wood joints failed the volume simulation in HSMWorks. The reason for three of the failed simulation was residual materials in the connectors, which could not be removed by the milling operations. For one wood joint, the total length of the milling cutter was not sufficient, and the tool holder collided with the workpiece.

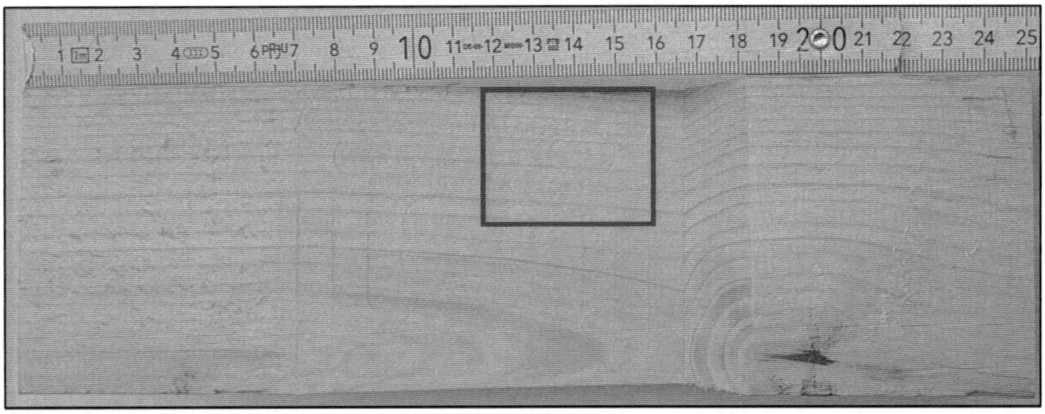

Fig. 3. Occurred chatter marks caused by vibrations and oscillations of the system

Furthermore, the results of the simulation tests in RoboDK showed that all wood joints that passed the volume simulation with HSMWorks also passed the simulation tests in RoboDK. In order to be able to manufacture the four connections that could not be produced, further tests with additional tools and further adaptations to the connections are necessary.

The milling tests of the connectors showed that in 13 of 14 milling tests, accuracies between -0.020 mm and 0.378 mm were achieved. One milling test achieved an average deviation of 1.325 mm due to a work object measurement error, which was outside the specified tolerance of ±1 mm. However, the tolerance could be maintained for the identically constructed connecting piece without any work object measurement errors. All surface qualities of the milling tests were above the specified limit of 16 points. Surface quality ratings between 20 and 28 out of a total of 32 possible points were achieved. The reasons for the low ratings were chatter marks on the surfaces, which can be seen in Fig. 3. These machining marks were caused by vibrations and oscillations of the system. To further reduce the chatter marks, further optimizations and tests must be carried out.

The simulation and milling tests showed that there are still challenges and optimization options for the manufacturing of wooden components with a KUKA KR 16-2 and the selected process to improve the quality of the production. In addition to the already mentioned chatter marks, the multi-sided machining of the workpiece is a challenge. At the moment, the component has to be moved by hand and referenced again each time. This is time-consuming and also leads to inaccuracies due to referencing errors that add up. A possible optimization options here is an additional clamping system that can rotate the workpiece to enable multi-sided machining of the workpiece in one operation. Another inaccuracy factor is the referencing and calibration of work objects and tools. The referencing of the work object is currently carried out by hand using the 3-point method from RoboDK via XYZ points, which can lead to inaccuracies during machining. Moreover, the TCP calibration functions from RoboDK is labour-intensive and error-prone for milling tools, which leads to additional work for the tool measurement process and to inaccuracies for the TCPs.

6. Conclusion

To sum up, it can be said that woodworking offers great potential for the use of industrial robots with numerous possible uses in the wood industry, such as the production of wood joints. However, there is a lack of information, experience and data for the use of industrial robots in the manufacturing of wood, which need to be further researched to enable the correct use. For this a feasibility study was carried out, to generate data und to verify the production of wood joints with an industrial robot.

The feasibility study, with various simulations and milling tests showed that the production of wood joints is possible using an industrial robot for milling. However, there are limitations regarding to the used hardware and software. For example, with the used milling tools and the CAM software HSMWorks four of the 33 wood joints could not be manufactured. To be able to manufacture these joints with an industrial robot for milling, further tests with additional tools are necessary. Furthermore, other CAM software should also be considered to check the manufacturing of the wood joints and to obtain additional results. Moreover, the milling tests showed that accuracies within the tolerance range of ±1 mm can be achieved if precise work object measurement is ensured. The results also show that the achieved accuracies are sufficient for processing of wood joints and construction timber. The evaluations of the surface quality ranged between 20 and 28 out of a total of 32 points, which is sufficient for the production of wood joints. The reasons for the low ratings are chatter marks on the surfaces. These chatter marks are visually recognizable, but do not affect the function of the joints. Thus, while the surface qualities are sufficient, it still offers potential for further optimization.

There are also other optimization possibilities for such a woodworking system with industrial robots, which must be analysed and addressed in further works. This includes for example, optimizing the overall process, which has numerous different interfaces. These interfaces generate additional effort during the process implementation.

An example for this is the additional creation of the robot programs in RoboDK for the KR C4 and the following transfer from RoboDK to the KR C4 for each milling process. Other optimization options are the improving of the stability of the system to reduce the chatter marks and optimizing the complex and error-prone measurements for tools and work objects, which are needed to ensure a flawless production of the joints and other wooden components.

7. References

[1] Ilgın, H. (2024). High-Rise Residential Timber Buildings: Emerging Architectural and Structural Design Trends, Buildings, Vol. 14, No. 1, 25, doi: 10.3390/buildings14010025.

[2] van Nimwegen, S.E. & Latteur, P. (2023). A state-of-the-art review of carpentry connections: From traditional designs to emerging trends in wood-wood structural joints, Journal of Building Engineering, Vol. 78, 107089, doi: 10.1016/j.jobe.2023.107089.

[3] Braun, M. & Kromoser, B. (2022). The influence of inaccuracies in the production process on the load-bearing behaviour of timber step joints, Construction and Building Materials, Vol. 330, 127285, doi: 10.1016/j.conbuildmat.2022.127285.

[4] Pantscharowitsch, M. & Kromoser, B. (2023). Automated Subtractive Timber Manufacturing—Joinery Machines Versus Industrial Robots, Journal of Manufacturing Science and Engineering, Vol. 145, No. 6, doi: 10.1115/1.4056924.

[5] Pantscharowitsch, M. & Kromoser, B. (2022). Investigation of industrial robots vs joinery machines for milling pockets in glulam: Comparison based on surface quality and 3D scans, Civil Engineering Design, Vol. 4, 1-3, 25–34, doi: 10.1002/cend.202100049.

[6] Makulavičius, M.; Petkevičius, S.; Rožėnė, J.; Dzedzickis, A. & Bučinskas, V. (2023). Industrial Robots in Mechanical Machining: Perspectives and Limitations, Robotics, Vol. 12, No. 6, 160, doi: 10.3390/robotics12060160.

[7] Huynh, H.N.; Verlinden, O. & Riviere-Lorphevre, E. (2017). Robotic Machining Simulation using a Simplified Multibody Model, In: Katalinic, B. (ed.), Proceedings of the 28th International DAAAM Symposium 2017, DAAAM International Vienna, 885–894, doi: 10.2507/28th.daaam.proceedings.123.

[8] Gotlih, J.; Karner, T.; Gotlih, K. & Brezocnik, M. (2018). Experiment Based Structural Stiffness Calibration of a Virtual Robot Model, In: Katalinic, B., Katalinic, B. (eds.), DAAAM International Scientific Book 2018, DAAAM International Vienna, 131–140, doi: 10.2507/daaam.scibook.2018.12.

[9] Geno, J.; Goosse, J.; van Nimwegen, S. & Latteur, P. (2022). Parametric design and robotic fabrication of whole timber reciprocal structures, Automation in Construction, Vol. 138, 104198, doi: 10.1016/j.autcon.2022.104198.

[10] Mesnil, R.; Gobin, T.; Demont, L.; Margerit, P.; Ducoulombier, N.; Douthe, C. & Caron, J.-F. (2023). Flexible digital manufacturing of timber construction: the design and fabrication of a free-form nexorade, Construction Robotics, Vol. 7, No. 2, 193–212, doi: 10.1007/s41693-023-00105-7.

[11] Pantscharowitsch, M.; Moser, L. & Kromoser, B. (2024). A study of the accuracy of industrial robots and laser-tracking for timber machining across the workspace, Wood Material Science & Engineering, 1–19, doi: 10.1080/17480272.2024.2324437.

[12] Pott, A. & Dietz, T. (2019). Industrielle Robotersysteme, Springer Fachmedien Wiesbaden, Wiesbaden, doi: 10.1007/978-3-658-25345-5.

[13] Standards, A. ÖNORM B 1995-1-1:2023 08 15, from https://www.austrian-standards.at/de/shop/onorm-b-1995-1-1-2023-08-15~p2670783, accessed May 3, 2024.

[14] Standards, A. ÖNORM EN 336:2013 11 15, from https://www.austrian-standards.at/de/shop/onorm-en-336-2013-11-15~p2042481, accessed March 22, 2024.

[15] Petzoldt, C.; Keiser, D.; Siesenis, H.; Beinke, T. & Freitag, M. (2021). Ermittlung und bewertung von einsatzpotenzialen der mensch-roboter-kollaboration, Zeitschrift für wirtschaftlichen Fabrikbetrieb, Vol. 116, 1-2, 8–15, doi: 10.1515/zwf-2021-0002.

[16] Klimchik, A.; Ambiehl, A.; Garnier, S.; Furet, B. & Pashkevich, A. (2016). Experimental study of robotic-based machining, IFAC-PapersOnLine, Vol. 49, No. 12, 174–179, doi: 10.1016/j.ifacol.2016.07.591.

DOI: 10.2507/35th.daaam.proceedings.035

EXPERIMENTAL INVESTIGATION OF VIBRATION DAMPING ABILITY FOR DIFFERENT SIMPLE CUBIC PERIODIC LATTICE STRUCTURES MORPHOLOGIES

Mario Soldo & Davorka Saravanja

This Publication has to be referred as: Soldo, M[ario] & Saravanja, D[avorka] (2024). Experimental Investigation of Vibration Damping Ability for Different Simple Cubic Periodic Lattice Structures Morphologies, Proceedings of the 35th DAAAM International Symposium, pp.0259-0267, B. Katalinic (Ed.), Published by DAAAM International, ISBN 978-3-902734-44-0, ISSN 1726-9679, Vienna, Austria
DOI: 10.2507/35th.daaam.proceedings.035

Abstract

Lattice structures have been the object of research by an increasing number of scientists in recent years following the rapid development of additive manufacturing. A series of improved mechanical properties have enabled the wide application of lattice structures in the automotive industry, aerospace industry, vibration damping, sports equipment, and medicine. In this paper, the impact of the morphology of additively manufactured PLA periodic Simple Cubic lattice structures on improving mechanical properties during vibration testing is investigated. It has been demonstrated that structures optimizing the morphological distribution of Simple Cubic unit cells offer better mechanical performance in terms of damping of vibration. Vibration testing has shown enhanced damping properties of vibrations with reduction of unit cell height in morphologically optimized lattice structure.

Keywords: lattice structures; Simple Cubic unit cell; additive manufacturing; PLA material; morphological distribution; vibration damping.

1. Introduction

Lattice structures are considered cellular structures (with honeycomb and foam structures) and were defined as a three-dimensional porous spatial structure formed by unit cells with different topologies that are spatially tessellated. An example of a lattice structure with the components of a unit cell (struts and nodes) is shown in Fig. 1. Depending on the arrangement of the unit cell, lattice structures can be periodic (if they belong to the group of non-stochastic cell structures) or random (if they belong to the group of stochastic cell structures) [1].

The topology of the unit cell refers to the type of unit cell, that is, the spatial position of the nodes that connect the struts. Spatial positions of the unit cell nodes are determined by the type of unit cell. The morphology of the unit cell is characterized by the shape and geometry of the struts and nodes. The morphology of the lattice structure refers to the spatial distribution and orientation of the unit cells within the lattice structure. Relative density is one of the crucial properties of lattice structures and represents the ratio of lattice structure volume to solid volume.

The main parameters that generally affect the mechanical properties (therefore the vibration damping properties) of lattice structures at the macro level are:

- topology of the unit cell;
- morphology of the unit cell;
- lattice structure morphology;
- relative density of the lattice structure and
- mechanical properties of the building material [3].

The growth in the use of additive manufacturing for final products has been visible last few years and the best indicator of growth is the increase in the use of additive manufacturing in the creation of patents. Additively obtained structures are widely used in the aviation industry [4]. The possibility of additive manufacturing of high-quality, complex specific components and the possibility of manufacturing components with lattice structures whose stiffness is lower than bone stiffness make lattice structures ideal for medical implants [5]. In recent years, a particularly large application of additively obtained plastic rigid lattice structures is in the production of sport and sports protective equipment [6]. A great potential that could be capitalized in the future is the application of lattice structures for acoustics and vibration damping, but it should be considered that additively obtained products are particularly sensitive to small irregularities during the manufacturing process [7]. The experimental investigation of the vibration damping ability of additively obtained plastic periodic lattice structures with different morphological distributions of unit cells (while the relative density is constant) is the subject of this research.

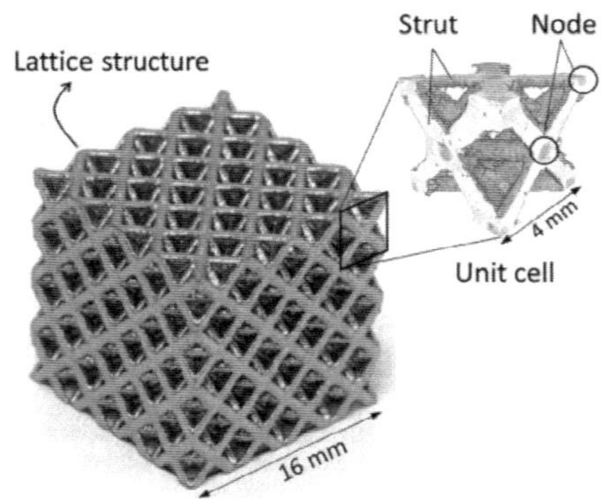

Fig. 1. Lattice structure formed by a repeating arrangement of octet unit cells made up of struts and nodes [2]

2. Previous research

In researching the ability of vibration damping of additively obtained lattice structures, several significant studies stand out. In a review paper by Dalela et al. [8], through a series of research papers mentioned, state the positive effect of the mechanical properties of metamaterials for the purpose of vibration control. However, due to the closedness and non-uniformity of the structure of porous materials, it is concluded that it is difficult to accurately predict the absorption of vibration energy and the behavior of oscillations. This is where additive manufacturing comes into play, which enables the production of uniform, periodic, open unit cells, where lattice structures show a good ability to oscillation wave damping and vibration behavior control, with generally good mechanical properties.

Chiocca et al. [9] investigate the occurrence of fatigue under displacement control of coated PLA lattice structures with FCC (Face – Centered Cubic) unit cells, obtained by extruding the material, at two different frequencies (3 Hz and 20 Hz). They proved that the coating process increased the fatigue strength for both frequency cases, and interestingly, they managed to model a sample that is suitable for both tensile and compressive testing. Wang et al. [10] examine the natural frequencies of 3D printed plastic samples with homogeneous and variable honeycomb unit cells. They vary the arrangement of the relative density of the test sample from a uniform to an optimized distribution of the relative density of unit cells. They experimentally prove that with this approach to the arrangement of unit cells with different relative densities it is possible to optimize the own frequency of the test samples.

Kayiran and Yaman [11] numerically and experimentally investigate the natural frequency for two different optimization approaches of a beam with cubic unit cells with equal value of volume fraction. The results show that the optimized beam has a 73% higher natural frequency (the first natural frequency is 325.3 Hz) compared to a uniform beam with the same volume fraction of 0.6 (the first natural frequency is 199.6 Hz). They successfully improved the vibration response without disturbing the external appearance (contour) of the beam. Elmadih et al. [12] investigate the influence of the size of the unit cell and the number of layers within the lattice structure (obtained by extrusion of plastic) on the natural frequency results in dynamic analysis. They conclude that the natural frequency of the lattice structure can be reduced by increasing the size of the unit cells and increasing the number of layers in the lattice structure.

Azmi et al. [13] investigate the influence of full and almost full BCC (Body – Centered Cubic) unit cells obtained by the process of extruding plastic material on the vibration characteristics of the struts and conclude that a full cross-section of the struts gives higher natural frequency values, and that the stiffness component of the lattice structure is much more significant compared to the mass component. Syam et al. [14] propose several design parameters (which directly affect stiffness and natural frequency) for the comparison of lattice structures and experimentally analyse natural frequency and compression strength. The main goal of the research was to achieve design parameters of lattice structures that provide high efficiency of vibration isolation along with good ability to withstand external compression loads. Also, they conclude that the defined parameters for the design of lattice structures are only suitable for experimental analysis, while they are not suitable for FE analysis since the simulation calculation is extensive due to the complex lattice structure geometry.

Monkova et al. [15] experimentally investigate the influence of geometrical parameters of BCC unit cells with different relative densities (through changes in BCC cell sizes and strut radius) on vibration damping and compression properties. The results of the experimental test showed that the ability of plastic lattice structures to dampen mechanical vibrations increases with a decrease in the volume ratio and with an increase in the cell size. Some research on periodic lattice structures was devoted to numerical and experimental analysis and showed examples of the application of lattice structures in gears, passenger seats, and as a vibration isolator. When a vibrational wave passes through lattice structures, it is attenuated based on the topology and geometry of the lattice, thus reducing the associated amplitudes. Although there is a large amount of research regarding the influence of lattice structure parameters on vibrations, there is not enough knowledge on how to improve damping efficiency by applying lattice structures based on different topologies, using functionally graded lattice structures, and changing geometric parameters such as cross-sectional area [16]. In this research, vibration damping abilities were tested on the different morphological arrangements of Simpe Cubic unit cells in lattice structures while the relative density of lattice structure was constant.

3. Modeling of different morphologies Simple Cubic periodic lattice structures

To define different morphological distributions of unit cells the lattice structures, the Simple Cubic (SC) unit cell was chosen as a type of unit cell, which is one of the most representative types of unit cells. The wireframe is modeled in the Rhinoceros program, and the dimensions of the volume for the wireframe model are 10 mm x 10 mm x 10 mm. It is an empirical assumption that those dimensions of the volume, in which the unit cell is modeled, are suitable for the material extrusion additive process. The SC unit cell is made with 12 struts, 8 of them are in horizontal position (where 4 struts are spatially parallel to each other) and 4 of them are in vertical position (which are also spatially parallel to each other), as shown in Fig. 2. The length of each SC unit cell strut within the previously defined volume is equal to 10 mm. The place where 3 struts join is called a node. SC unit cell has 8 nodes. By defining the spatial position and orientation of the struts and nodes, the topology (type) of the unit station is defined. By adding the cross section of the struts and nodes (in this case, a circular cross-section with a radius of 1.43 mm), the result is a defined unit cell morphology. When the morphology of the unit cell is defined, it is also possible to determine the relative density or porosity of the unit cell. In this case, the relative density of this SC unit cell with struts radius of 1.43 mm is 0.2, so porosity is 80%.

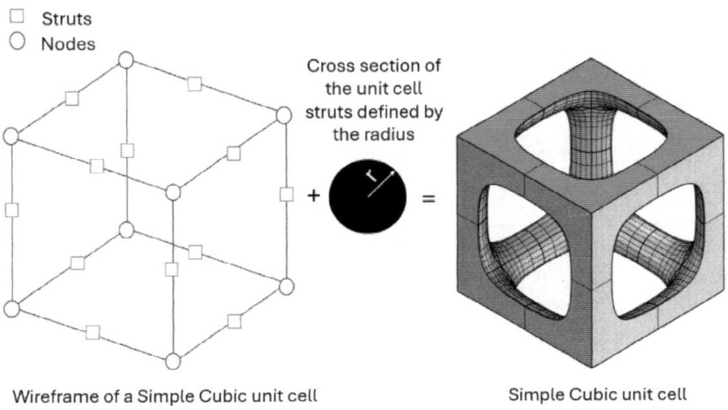

Fig. 2. Wireframe of a Simple Cubic unit cell (left) and Simple Cubic unit cells with defined struts radius (right)

In the Rhinoceros program, with the plug-in Grasshopper, an algorithm is defined for modeling three different morphologies of SC periodic lattice structures that, by changing the value of the radius of the struts and nodes, will have the same value of relative density (0.2), thus maintaining the same consumption of materials during the additive manufacturing process. The first series is designated as SC – A and has a rectangular spatial arrangement of SC unit cells within a lattice structure. The second series is designated as SC – B and has a spatial arrangement of SC unit cells along concave parabolas within a lattice structure. The third series is designated as SC – C and has a spatial arrangement of SC unit cells along convex parabolas within a lattice structure.

The SC – B and SC – C series differ, in addition to the parabolic spatial arrangement of the SC unit cells within the lattice structures, by the reduced heights of the unit cells in the zones closer to the place of assembly of the test specimens and the place of measurement of the vibration response of the lattice structures (upper and lower lattice layers), as it is shown on Fig. 3. A relative density of 0.2 was chosen for all three series, and the radii of the struts and nodes were assigned according to the relative density. Dimensions of samples for lattice structures with different morphology distributions of unit cells are 50 mm x 50 mm x 50 mm. The multiplication array of SC unit cell for SC – A series is 5 (length) x 5 (width) x 5 (height) and for SC – B and SC – C series, it is 5 (length) x 5 (width) x 7 (height).

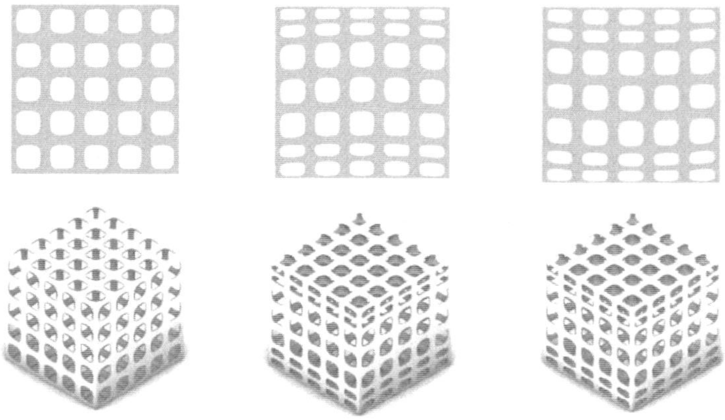

Fig. 3. Front (top) and isometric (bottom) view of models of lattice structures series SC – A (left), SC – B (middle) and SC – C (right)

4. Lattice structures additive manufacturing

Additive manufacturing of lattice structures was done on a Zortrax M200 Plus printer (material extrusion printer) with the support of the Z-Suite slicer program. The PLA (polylactic acid) material was selected with the printing parameters listed in Table 1. Material and the printer were selected according to earlier research on this topic [17]. Three identical samples of all series were printed at the same time to maximally reduce the influence of external factors on the quality of the test results. Table 2 shows the time spent on printing one series of lattice structures with three samples and the total material consumption (in grams and meters of used raw PLA material). The values from Table 2 proved the same consumption of material for all test sample production, so the accuracy of the defined algorithm in the plugin Grasshopper and program Rhinoceros for determining the same relative density over the radius of the struts and nodes was confirmed. Fig. 4 shows printed samples for experimental testing.

Selected material	Extruder temperature	Platform temperature	Layer thickness	Extruder diameter	Model infill
PLA	210°C	30°C	0.19 mm	0.4 mm	Solid (100%)

Table 1. Zortrax M200 Plus printing parameters

Series	Printing time	Material used
SC – A	21 h 42 min	102 g; 35.94 m
SC – B	21 h 45 min	101 g; 35.54 m
SC – C	21 h 38 min	101 g; 35.45 m

Table 2. Printing time and material used for sample production

Fig. 4. Printed samples of lattice structures for experimental vibration testing

5. Lattice structures vibration damping testing

The vibration testing was performed on a mechanical model adapted for the needs of vibration research on lattice structures. The vibration test, for all three series of morphologically different lattice structures, was carried out in three key steps: mounting the samples on the vibration platform, vibration excitation of the lattice structures via the vibration platform and saving the output data for validation of the experimental results. Vibration testing of additively manufactured lattice structures was performed using excitation and measurement equipment consisting of a vibration platform Model 25 – VP – T, measuring device VIBROSTORE 100 and appropriate program support for processing the results in the form of the ReO program (Report & Route Manager Software B&K Vibro – Brüel & Kjær).

A vibration platform Model 25 – VP – T was used to create vibration excitation in the vertical direction (perpendicular to the printing direction) of the testing samples. During the operation of the vibration platform, the frequency can be changed manually from 5 to 100 Hz (cycles per second CPS = Hz). The power of the electric motor of the vibrating platform is 1 KW. The motor must be adjusted in such a way that the mechanism rotates in a clockwise direction. The selection of the acceleration range is determined by a selector that changes the vibration speed over the selected frequency range automatically or manually. On this way, it is possible to choose the required frequency within the limits of 5 Hz to 100 Hz by applying appropriate limiters that correspond to the lower or upper limit frequency. The samples are tested in the frequency range from 10 Hz to 30 Hz.

The size of the amplitude and frequency of the vibration platform was determined with appropriate instruments for measuring vibrations (VIBROSTORE 100 with its associated sensors and ReO software). A vibration meter is an electronic device that processes vibration signals that it receives using a sensor that collects vibrations translated into electrical signals. After placing the magnetic part with the vibration sensor at the exact measuring point of the testing samples, the vibrations of the sample are converted into electrical signals that are transmitted to the measuring instrument via a cable. The signal is processed into vibration values that can be read on the screen of the measuring instrument. Acceleration or speed values are displayed on the screen of the vibrating device. It is important to place the magnet with the sensor always in the same location to achieve reparability of the test under the same conditions. For this reason, it is desirable to clean the surface where the magnet will adhere to remove unwanted impurities (oil grease, rust, dust, or residual paint) that can adversely affect the measurement results. The main screen of the VIBROSTORE 100 measuring device shows the state of the test sample in BDU (damage unit) and total g (or rms acceleration). Rms vibration (Root Mean Square Vibration) is one of the practical measures of vibration speed. It is often used to describe the vibration signal and represents the mean power of the measured velocity. It is obtained by measuring the peak amplitudes and multiplying by $1/\sqrt{2}$, which is approximately equal to 0. 707. Logically, dividing the rms value by 0.707 gives the peak amplitude value. VIBROSTORE 100 can work independently or with Brüel & Kjær software Vibro Report & Route Manager (ReO). Report & Route Manager is a powerful and highly functional vibration analysis editor that results in high quality reports. First, a route with selected names is defined in the ReO software, and the names are loaded into the VIBROSTORE 100. After testing, all recorded data from the VIBROSTORE 100 is loaded into the ReO program. Data in ReO can be displayed in many ways, including vibration frequency spectrums, ISO graphs, and "waterfall" graphs. Also, output reports can be generated automatically or adjusted manually.

Fig. 5. Samples of series SC – A, SC – B and SC – C glued on steel plates

The vibration test sensor of the VIBROSTORE 100 device can only be stably positioned on samples made of magnetic materials. Since the lattice structures are made of PLA material, it is necessary to connect two steel plates to the lattice structures sufficiently rigidly (one for mounting the lattice structure on the platform, and the other on the lattice structure for sensor positioning).

On the lower surface of the lattice structures (the surface that is printed first in the 3D printing process), a larger steel plate is glued. The board has 4 holes drilled in it for proper mounting on the platform of the vibration device. Three of the four holes have a diameter of 11 mm (for the passage of M12 screws), and one hole (upper left) has a diameter of 6.6 mm. The exact values of the distance between the holes are determined according to the distance of the threaded holes on the vibration platform to achieve tight clamping. For all the test samples to be successfully mounted at the same position on the platform, the position of the glued center of the sample is measured on the larger steel plate exactly in the middle of the plate. The longer and shorter sides of the plate are divided in half, and the center of the lower surface of the lattice structure is mounted in the center of the intersection. A smaller plate is rigidly attached to the upper surface of the testing samples (the surface that was printed last in 3D printing). While making smaller and larger plates, the requirements were fine processing to create a solid connection when gluing. Before gluing the plates, the contact surfaces of the plates with the sample, the vibration platform, and the magnetic sensor are cleaned. Testing samples of all three series (SC – A, SC – B and SC – C) of lattice structures with glued steel plates are shown in Fig. 5.

Fig. 6. Position of the test samples on the platform (left) and clamping the samples (right)

Experimental testing of three series of lattice structures with different morphological distributions of SC unit cells was performed at a temperature of 25°C (in one day) to avoid unwanted influence of external factors (temperature change, humidity) on the response of structures subjected to vibration excitation. Each test specimen is mounted in the same place on the vibration platform, as shown in Fig. 6. (left). Each sample is mounted with three M12 screws (ISO 4014) and one M8 screw (ISO 4014) and their corresponding nuts (ISO 4032) and washers (ISO 7091) as shown in Fig. 6. (right). Before the start of the vibration testing, a vibration sensor with a magnet was placed on the upper, smaller plate (exactly in the middle of the plate) and connected by a cable to the VIBROSTORE 100 device, as shown on Fig. 7. (left). Through the routes with the ReO program, each lattice structure of each series and each excitation frequency is assigned a suitable name. The structures were subjected to three different excitation frequencies of 10 Hz, 20 Hz, and 30 Hz via a manually adjusted frequency on the excitation device, as shown in Fig. 7. (right).

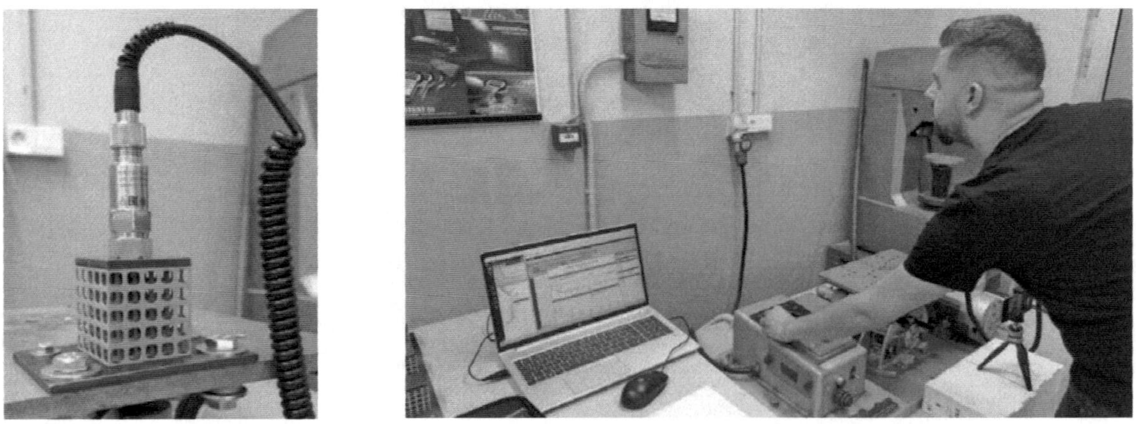

Fig. 7. Vibration sensor positioning (left) and manually adjusted excitation frequency (right)

During the test, rms values from the VIBROSTORE 100 measuring device were recorded for three different excitation vibration values. The obtained rms values (expressed in mm/s) of all test samples of lattice structures (SC – A, SC – B and SC – C) for all excitation frequency of 10 Hz, 20 Hz and 30 Hz are shown in Table 3.

Excitation frequency 10 Hz		Excitation frequency 20 Hz		Excitation frequency 30 Hz	
Sample	**rms value**	**Sample**	**rms value**	**Sample**	**rms value**
SC – A – 1 – 10	14.76 mm/s	SC – A – 1 – 20	28.66 mm/s	SC – A – 1 – 30	43.06 mm/s
SC – A – 2 – 10	14.18 mm/s	SC – A – 2 – 20	27.80 mm/s	SC – A – 2 – 30	41.59 mm/s
SC – A – 3 – 10	13.88 mm/s	SC – A – 3 – 20	28.13 mm/s	SC – A – 3 – 30	41.87 mm/s
SC – B – 1 – 10	14.36 mm/s	SC – B – 1 – 20	27.78 mm/s	SC – B – 1 – 30	42.04 mm/s
SC – B – 2 – 10	14.76 mm/s	SC – B – 2 – 20	27.90 mm/s	SC – B – 2 – 30	42.01 mm/s
SC – B – 3 – 10	14.44 mm/s	SC – B – 3 – 20	28.13 mm/s	SC – B – 3 – 30	42.47 mm/s
SC – C – 1 – 10	14.39 mm/s	SC – C – 1 – 20	27.52 mm/s	SC – C – 1 – 30	42.18 mm/s
SC – C – 2 – 10	14.73 mm/s	SC – C – 2 – 20	27.43 mm/s	SC – C – 2 – 30	41.68 mm/s
SC – C – 3 – 10	14.57 mm/s	SC – C – 3 – 20	27.79 mm/s	SC – C – 3 – 30	41.94 mm/s

Table 3. Measured rms values for three different series and three different excitation frequencies

6. Experimental results validation

In Table 4, average rms value and standard deviation of three different series (SC – A, SC – B and SC – C) for three different excitation frequencies (10 Hz, 20 Hz and 30 Hz) are listed.

Sample	average rms value	standard deviation
SC – A for 10 Hz excitation frequency	14.27 mm/s	0.37
SC – B for 10 Hz excitation frequency	14.42 mm/s	0.27
SC – C for 10 Hz excitation frequency	14.56 mm/s	0.14
SC – A for 20 Hz excitation frequency	28.20 mm/s	0.35
SC – B for 20 Hz excitation frequency	27.94 mm/s	0.15
SC – C for 20 Hz excitation frequency	27.58 mm/s	0.15
SC – A for 30 Hz excitation frequency	42.17 mm/s	0.64
SC – B for 30 Hz excitation frequency	42.17 mm/s	0.21
SC – C for 30 Hz excitation frequency	41.93 mm/s	0.20

Table 4. The average rms value and standard deviation of three different series for three different excitation frequencies

Series SC – A is taken as reference for comparing the results, since there are no morphological changes in the spatial arrangement of SC unit cells in lattice structures. At an excitation frequency of 10 Hz, the SC – A series gives an average value of rms response of the lattice structure of 14.27 mm/s. The SC – B series, at the excitation of 10 Hz gives a rms response of 14.52 mm/s, which is 1.7% higher than the rms value for the SC – A series. The SC – C series at the excitation frequency of 10 Hz gives rms response of 14.56 mm/s, which is 2% higher than the rms value for the SC – A series.

The value of the standard deviation, i.e., the deviation from the mean rms values of the test samples, shows that the smallest deviations among the samples are in the SC – C series with a deviation of 0.14, while the deviations in the SC – B series (0.27) and SC – A (0.37) higher. It can be concluded that at a low excitation frequency of 10 Hz, the SC – A series gives the lowest rms value, which makes it a better vibration dampener than the other two series, while at the same time, the SC – A series is the one that gives the worst results in terms of uniformity of results among the test samples. The best uniformity of results is shown by the SC – C series. The reason for this may lie in the stiffness of the SC – A series, which has large unit cells (10 mm), so even the smallest irregularity during printing can harm the quality of the test sample (also stiffness) and the unevenness of the vibration test results. According to the obtained average rms values, it can be concluded that at a low excitation frequency of 10 Hz, the defined morphological distribution of SC unit cells within the lattice structure gives a worse vibration response and lower damping ability than the rectangular spatial distribution of SC unit cells, but higher repeatability of the results.

While observing the rms value corresponding to the mentioned three series for an excitation frequency of 20 Hz, the following results are obtained. SC – A gives rms response of 28.2 mm/s, while for the SC – B series the rms response is 27.94 mm/s, which is 1% lower than the rms value of the SC – A series. For the SC – C series, the value rms is even lower than the SC – A series, by 2.2% and it is 27.58 mm/s. It can be concluded that the defined morphological distribution of SC unit cells within the lattice structure for excitation frequencies of 20 Hz gives a better vibration response and enables greater damping of excitation vibrations. Densification and reduction of unit cells in places close to mounting and testing lattice structures, resulting in dispersion of excitation vibrations from the excitation point to the measurement point. Thus, reducing the size of a SC unit cell has a positive effect on vibration damping at a higher excitation frequency of 20 Hz.

Standard deviation in the testing samples for an excitation frequency of 20 Hz, again, the SC – A series gives higher deviations among the results (on average 0.35), which makes the SC – B and SC – C series with a standard deviation of 0.15 more uniform.

So, at frequencies of 20 Hz, the conclusion can be that the SC – C series provides the best damping of vibrations measured vertically relative to the excitation, and the SC – C series is the one that gives the highest repeatability in results, with an average standard deviation of 0.15. For an excitation frequency of 30 Hz, the following rms values are obtained: SC – A 42.17 mm/s, SC – B value is equal to 42.17 mm/s, while for the SC – C series the average value is 41.93 mm /s. It is obvious that series SC – A and SC – B give the same rms value. If the standard deviations for the mentioned two series (0.64 and 0.21) are observed, it can be concluded that the SC – B series is the one that gives more uniform results during the vibration test. SC – A has higher deviations in the results. The lowest rms value and, at the same time, the most uniform test results are given by the SC – C series, which makes it optimal for use at excitation frequencies of 30 Hz.

The influence on the uniformity of the results and the reduction of the rms value is again influenced by the defined morphological distribution (by parabolas) of SC unit cells within lattice structures.

7. Conclusion and future research

This paper presents the modeling procedure of different morphologies of the SC periodic lattice structure that keep the same relative density of 0.2, the additive production of PLA lattice structures, and the examination of vibration damping ability of lattice structures through comparative rms values for excitation frequencies of 10 Hz, 20 Hz and 30 Hz. Experimental tests have shown that changing the morphological arrangement of SC unit cells within lattice structures gives different responses in terms of vibration damping in vertically excitation of lattice structures. The rms values obtained experimentally showed that at increased values of the excitation frequencies of 20 Hz and 30 Hz, the parabolic distribution of SC unit cells within lattice structures gives better possibilities of damping excitation vibrations. The reason for this lies in the height reduction of unit cells in the areas that are near the place of mounting the lattice structure and the place of measuring the rms response of the lattice structure. For a lower excitation frequency of 10 Hz, it is good to leave the morphological distribution of SC unit cells in the lattice structure in a rectangular distribution. Parabolic morphological distribution of SC unit cells in lattice structure for the specified frequency of 10 Hz does not give an optimal rms response.

Since the dimensions of the SC unit cell within the lattice structure are relatively small, the differences in the measured rms values are also reduced, but they are significant, and they should not be ignored. Future research could refer to the larger dimensions of the unit cells within the test samples of lattice structures. Furthermore, in this research, the shapes of the horizontal struts (straight or curved) were only changed, so that future research can also refer to the change in the morphological arrangement of the unit cells, which will affect the change of the vertical struts as well. Also, since there is not a large amount of research on the numerical analysis of vibration damping of lattice structures, it would be desirable to develop a model that will enable the numerical prediction of the behavior of plastic periodic lattice structures with different morphologies as vibration dampers.

8. Acknowledgments

Thank the Federal Ministry of Education and Science of the Federation of Bosnia and Herzegovina, who saw the potential in research of the ability for vibration damping of additive manufactured plastic periodic lattice structures and for financing the material for additive manufactured testing samples and the publication of this scientific research through a scientific project.

9. References

[1] Pan, C., Han, Y. & Lu, J. (2020). "Design and optimization of lattice structures: A review", Appl. Sci., Vol. 10, No. 18, pp. 1-36., DOI: 10.3390/app10186374.

[2] Babamiri, B. B., Barnes, B., Soltani-Tehrani, A., Shamsaei, N. & Hazeli, K. (2021). "Designing additively manufactured lattice structures based on deformation mechanisms", Addit. Manuf., Vol. 46, p. 102143., DOI: 10.1016/j.addma.2021.102143.

[3] Intrigila, C., Nodargi, N. A. & Bisegna, P. (2022). "The compressive response of additively-manufactured hollow truss lattices: an experimental investigation", Int. J. Adv. Manuf. Technol., Vol. 120, No. 5-6, pp. 3529–3541., DOI: 10.1007/s00170-022-08716-0.

[4] Maconachie, T., Leary, M., Lozanovski, B., Zhang, X., Qian, M., Faruque, O. & Brandt, M. (2019). "SLM lattice structures: Properties, performance, applications and challenges", Mater. Des., Vol. 183, p. 108137., DOI: 10.1016/j.matdes.2019.108137.

[5] Hulme, J., Sakhaei, A. H. & Shafiee, M. (2023). "Mechanical analysis and additive manufacturing of 3D-printed lattice materials for bone scaffolds", Mater. Today Proc., DOI: 10.1016/j.matpr.2023.02.278.

[6] Namvar, N., Moloukzadeh, I., Zolfagharian, A., Demoly, F. & Bodaghi, M. (2023). "Bio-inspired design, modeling, and 3D printing of lattice-based scale model scooter decks", Int. J. Adv. Manuf. Technol., Vol. 126, No. 7-8, pp. 2887–2903., DOI: 10.1007/s00170-023-11185-8.

[7] Ramadani, R., Pal, S., Kegl, M., Predan, J., Drstvenšek, I., Pehan, S. & Belšak, A. (2021). "Topology optimization and additive manufacturing in producing lightweight and low vibration gear body", Int. J. Adv. Manuf. Technol., Vol. 113, No. 11-12, pp. 3389–3399., DOI: 10.1007/s00170-021-06841-w.

[8] Dalela, S., Balaji, P. S. & Jena, D. P. (2022). "A review on application of mechanical metamaterials for vibration control", Mech. Adv. Mater. Struct., Vol. 29, No. 22, pp. 3237–3262., DOI: 10.1080/15376494.2021.1892244.

[9] Chiocca. A., Tamburrino, F., Frendo, F. & Paoli, A. (2022). "Effects of coating on the fatigue endurance of FDM lattice structures", Procedia Struct. Integr., Vol. 42, No. 2019, pp. 799–805., DOI: 10.1016/j.prostr.2022.12.101.

[10] Wang, X., Zhang, P., Ludwick, S., Belski, E. & To, A. C. (2018). "Natural frequency optimization of 3D printed variable-density honeycomb structure via a homogenization-based approach", Addit. Manuf., Vol. 20, pp. 189–198., DOI: 10.1016/j.addma.2017.10.001.

[11] Kayiran, A. M. & Yaman, U. (2019). "Effect of lattice structures on natural frequency of SLA fabricated parts", Procedia Manuf., Vol. 39, No. 2019, pp. 533–538., DOI: 10.1016/j.promfg.2020.01.413.

[12] Elmadih, W. (2020). Additively manufactured lattice structures for vibration attenuation Additively manufactured lattice structures for vibration attenuation, Ph. D. Dissertation, University of Nottingham, Nottingham, United Kingdom.

[13] Azmi, M. S., Ismail, R., Hasan, R., Alkahari, M. R. & Tokoroyama, T. (2017). "Vibration analysis of FDM printed lattice structure bar", Proc. Sakura Symp. Mech. Sci. Eng., No. November, pp. 33–35. Available from: https://www.researchgate.net/publication/321159615_Vibration_analysis_of_FDM_printed_lattice_structure_bar (2024-08-08)

[14] Syam, W. P., Jianwei W., Zhao, B., Maskery, I., Elmadih, W., & Leach. R. (2018). "Design and analysis of strut-based lattice structures for vibration isolation,", Precis. Eng., Vol. 52, No. August 2017, pp. 494–506., DOI: 10.1016/j.precisioneng.2017.09.010.

[15] Monkova, K., Vasina, M., Zaludek, M., Monka, P. P., & Tkac. J. (2021). "Mechanical vibration damping and compression properties of a lattice structure", Materials (Basel), Vol. 14, No. 6, pp. 1–16., DOI: 10.3390/ma14061502.

[16] Al Rifaie, M., Abdulhadi, H. & Mian, A. (2022). "Advances in mechanical metamaterials for vibration isolation: A review", Adv. Mech. Eng., Vol. 14, No. 3, pp. 1–20., DOI: 10.1177/16878132221082872.

[17] Šaravanja D. & Soldo, M. (2023). "Modeling of the Periodic Lattice Structure for Experimental, Numeric and Vibration Analysis", Ann. DAAAM Proc. Int. DAAAM Symp., Vol. 34, No. 1, pp. 26–34., DOI: 10.2507/34th.daaam.proceedings.004.

DOI: 10.2507/35th.daaam.proceedings.036

NUMERICAL SIMULATION OF FUSED FILAMENT FABRICATION PROCESS AND TENSILE TESTS

Mahmoud M. Farh, Mohamed Abdelkader & Viktor Gribniak

This Publication has to be referred as: Farh, M[ahmoud]; Abdelkader, M[ohamed] & Gribniak, V[iktor] (2024). Numerical Simulation of Fused Filament Fabrication Process and Tensile Tests, Proceedings of the 35th DAAAM International Symposium, pp.0268-0272, B. Katalinic (Ed.), Published by DAAAM International, ISBN 978-3-902734-44-0, ISSN 1726-9679, Vienna, Austria
DOI: 10.2507/35th.daaam.proceedings.036

Abstract

Fused Filament Fabrication (FFF) is a prevalent additive manufacturing process that uses thermoplastic materials such as polylactic acid (PLA) in filaments deposited layer-by-layer. This study focuses on the fabrication of components withstanding mechanical loads. However, the FFF parts are subjected to a time-varying thermal profile during manufacturing. Thus, these parts are subjected to geometry distortions and residual stresses due to temperature gradients and varying hardening degrees in adjacent layers, impairing the mechanical performance of 3D-printed structures. Rapid heating and cooling of the polymer feedstock further contribute to non-uniform internal stresses. These residual stresses and geometry distortion determine the research object. Finite element (FE) software ABAQUS is employed to simulate the 3D printing process and tensile test. The FE modelling includes a coupled thermo-mechanical analysis, and a tensile test simulated in two stages. The first analysis stage estimates the residual stresses' effects on mechanical performance through heat transfer, followed by static structural analysis, generating nodal displacement, stresses, and strains. The second stage uses this calculated stress distribution over time as a predefined stress field to define residual stresses acting in the 3D-printed samples for tensile tests. The experimental results are used to validate the tensile model predictions.

Keywords: material extrusion; fused filament fabrication; finite element simulation; thermo-mechanical analysis; residual stresses.

1. Introduction

Additive manufacturing (AM) has emerged as a transformative process in modern production technologies [1], with Fused Filament Fabrication (FFF) being one of the most prevalent methods for producing thermoplastic components. The increasing demand for parts that can withstand mechanical stresses, especially in industries like automotive and aerospace, highlights the need for a thorough understanding of the material properties and the process parameters involved in FFF. The FFF process manufactured parts experience a time-varying thermal profile, which introduces complexities such as geometry distortion and the development of residual stresses due to the polymer evolution from a semi-molten to a solid state during the manufacturing process [2]. These stresses have a detrimental effect on the mechanical performance of the produced parts, making this study area highly relevant.

Several studies have been conducted on the effects of thermal profiles in 3D printing processes, particularly focusing on the mechanical properties of printed parts [3], [4]. Recent advancements in numerical modelling have enabled researchers to simulate printed components' thermal and mechanical behaviours, providing insights into how different process parameters, such as printing speed, nozzle temperature, nozzle velocity, and layer thickness, influence the resulting properties [5]. Moreover, Finite Element Method (FEM) simulations have proven their efficiency in predicting mechanical deformations and stress distributions, furthering the understanding of the 3D printing process's intricacies. The physical optimization of FFF process parameters is costly, requiring time-intensive trials [6]. Therefore, Predictive simulation tools offer the potential to model the process, enabling parameter optimization to minimize part distortion and achieve the desired shape on the first trial. The FFF process is characterized by complex multiphysical phenomena, including solidification, heat transfer, and mechanical loads, which are closely interrelated with the process parameters and significantly influence the final part performance [7].

A comprehensive simulation procedure requires implementing a constitutive thermo-mechanical model for the extruded material, an accurate representation of process parameters, and experimental validation of predictions at each analysis stage. The FEM, coupled with element progressive activation, facilitates the simulation of the FFF process by utilizing various constitutive models of polymers and incorporating key process parameters such as extrusion temperature, tool-path patterns, nozzle velocity, layer thickness, filament width, etc. [8]. Zhang et al. [9] developed one of the earliest full 3D FEM models for the FFF process simulation, investigating the tool-path pattern on the FFF process through a thermo-mechanical analysis to predict the formation of residual stresses in a regular Acrylonitrile Butadiene Styrene (ABS) plate. The used material modelling was assumed to be linear elastic with temperature-dependent thermal properties. The study demonstrated that the tool-path pattern significantly influences both the distribution and magnitude of the computed residual stresses. The same simulation framework was later utilized in [10] to predict the impact of other process parameters, such as layer thickness, nozzle velocity, and filament width, on the final residual stresses and part distortion. The results demonstrated that the nozzle velocity is the most influential parameter affecting part distortion, followed by the layer thickness. Cattenone et al. [11] conducted simulations to investigate the influence of constitutive models on the FFF process for an ABS bridge-like part. The findings indicated that incorporating a temperature-dependent Young's modulus and yield stress, alongside constant thermal properties, resulted in a mean difference of 12% between the measured and predicted distortions.

This study simulated the FFF process and subsequent tensile tests using Finite Element (FE) modelling. ABAQUS software is employed to conduct a coupled thermo-mechanical analysis, focusing on the residual stresses and their effects on the mechanical performance of the printed components. Experimental tensile tests validate the simulation to ensure the accuracy of the results. Our findings contribute to the ongoing research by providing a deeper understanding of how process parameters affect residual stresses and mechanical performance, with implications for optimizing FFF processes for high-performance applications.

2. Simulation workflow

The FFF simulation workflow was proposed in [11],[12],[13] and illustrated in the schematic Fig. 1. In the preliminary stage, a CAD model defines the part's geometry; after that, the G-code outlines the FFF process parameters. The G-code is analysed to extract the time-dependent filament centrelines of the cross-sectional area of deposited material to compute the activation intervals of each finite element (FE) defining the Event-series data in ABAQUS. The numerical modelling of the FFF process is conducted using thermo-mechanical simulations with the activation intervals using element progressive activation approach, which are summarized in [14].

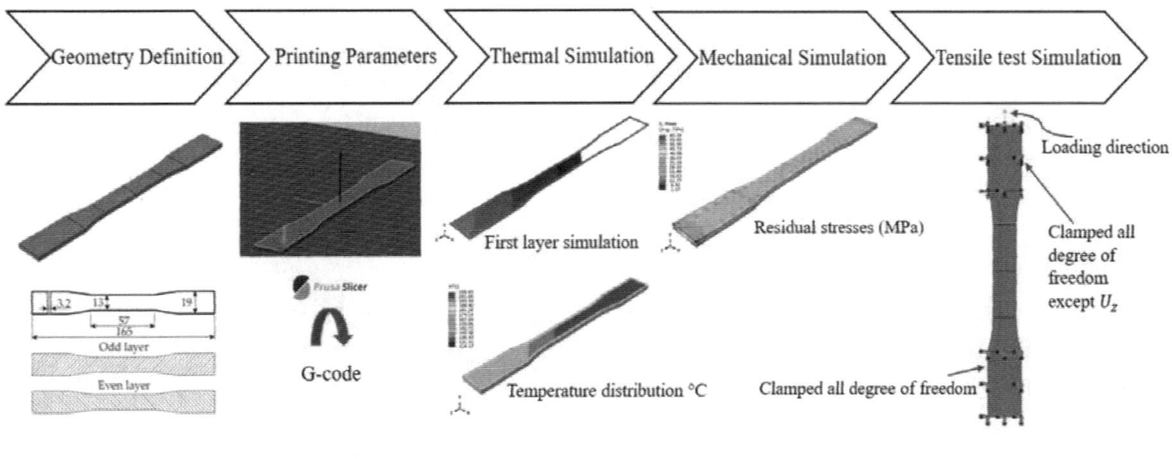

Fig. 1. Schematic workflow of the proposed study.

The Element progressive activation approach defines the printing path through time-dependent coordinates of the filament centreline processed from the G-code. The elliptical cross-section of the filament is assumed to remain constant and is represented by a rectangle. If the centre of the element lies within the defined rectangular boundaries, the element is activated and will be included in the analysis. The thermal model is first used to evaluate the temperature distribution as a function of time. The time-dependent temperature distribution is introduced as a solution-independent boundary condition into the mechanical model to estimate the residual stresses and part deformations. In the final stage, the residual stresses and geometry deformation are incorporated into the tensile test FE model to evaluate its influence on mechanical performance. Simultaneously, the experimental results in [15] were used to validate the model predictions.

3. Numerical simulation of the FFF process

Following schematic workflow Fig.1 of the simulation of the FFF process, the CAD model of the ASTM standard D638-14 determines the part's geometry. afterward, PrusaSlicer 2.3.3 slicing software is used to generate the G-code that outlines FFF process parameters in Table 1. A Python script is developed to re-elaborate the G-code, extracting the activation time of each FE and using it as input to define the event series data in ABAQUS. Full activation of elements was assumed during the element progressive activation while the specific boundary conditions were applied only for the activated elements. PLA is used as feedstock material in this study. Material characterization is performed experimentally to assess post-manufacturing material properties, including manufacturing defects. The temperature dependent material properties were demonstrated in [16],[17]. The selected element type was DC3D8: 8-node linear heat transfer brick and C3D20R:20-node quadratic brick for thermal and mechanical analysis, respectively. The element size along (x, z) axes was set to equal the nozzle diameter and along (y) axe was equal to layer thickness.

Printing Parameters	Typical Value
Extrusion Temperature	210 °C
Build Platform Temperature	60 °C
Print Speed	28 mm/sec
Infill Density	100%
Nozzle Diameter	0.4 mm
Number of layers	11

Table 1. Printing Parameters

4. Results

The proposed thermo-mechanical model predicted the temperature distribution Fig.2 for the test specimen. The temperature history is introduced as a solution independent boundary condition into the mechanical model to estimate the residual stresses and part deformations which are generated within the 3D printed specimen due to thermal cycling, cooling and process constraints. Higher residual stresses were observed along the specimen edges compared to the other locations. The formation of residual stresses was significant during the cooling process Fig.3. The thermo-mechanical analysis was followed by subsequent tensile test simulation. The first Tensile test simulation was performed excluding the manufacturing defects of FFF process. The second Tensile test model includes the residual stresses and geometry distortion. The later model indicates the impact of residual stresses on the mechanical performance of the 3D printed components. At the same time, the results were validated with the physical test achieving acceptable accuracy Fig.4. Average values of force-displacement results explained in [15] were considered during the validation of the FE model. The physical test demonstrated tensile load-bearing capacity of 1.7 kN at displacement 2.8 mm. The first tensile test model demonstrated higher predictions than the physical test. However, the second model demonstrated a reduction of tensile load-bearing capacity due to the incorporation of residual stresses and geometry deformation of the test specimen.

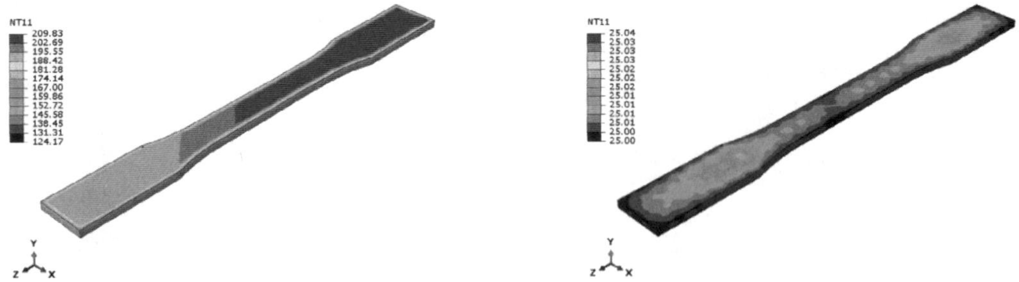

(a) (b)

Fig. 2. Temperature distribution (a) temperature distribution of test specimen at the end of extrusion, (b) at the end of cooling, where NT11 represents the nodal temperature (°C).

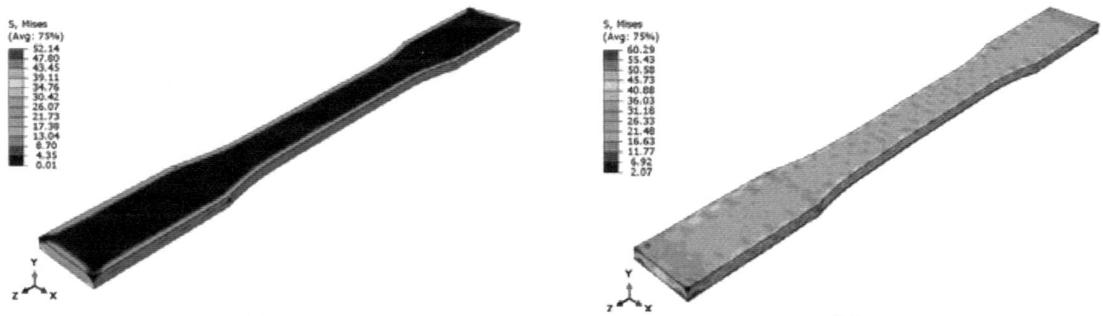

(a) (b)

Fig. 3. Residual stresses of test specimen (a) at the end of extrusion, (b) at the end of cooling, where S is in (MPa).

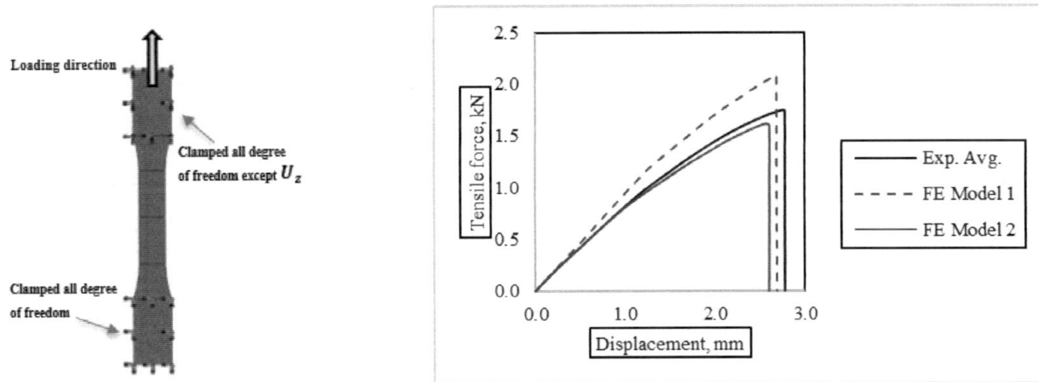

(a) (b)

Fig. 4. (a) Tensile test simulation setup, (b) Force-displacement curve where Exp. Avg. Represents the average typical values obtained from the physical tests [15]. FE Model 1, FE Model 2 represent the first and second tensile test simulations respectively.

5. Conclusion

This study discussed a numerical approach to predict the temperature distribution and residual stresses in PLA parts manufactured using the FFF process. The results demonstrated the potential influence of residual stresses and geometry deformation on the mechanical performance of the final component. To achieve enhanced accuracy, a detailed temperature-dependent material characterization should be considered. This could include the complex thermal and mechanical behaviour of the material at different stages, thus improving the simulation reliability. It was found that the computational efficiency of FE simulation depends highly on proper meshing strategies. In addition, the effects of radiation should be considered during the thermal simulation. However, the simulation approach of the FFF process faces some limitations as follows:

1. Thermal and Residual Stress Modelling: The simulation approach doesn't consider the full complexity of thermal gradients, residual stresses, and solidification effects during and after filament deposition. For intense solidification and volume changes, they are not adequately modelled, leading to incomplete predictions of distortions.
2. Material Defects: The current simulation approach does not fully capture the presence of voids and porosity in FFF parts, which significantly impacts their mechanical performance.
3. Layer Adhesion and Bonding: The interlayer bonding is a critical aspect of FFF part strength. The simulation approach doesn't capture the complexities of interlayer bonding, such as the temperature-dependent adhesion between layers, leading to inaccurate predictions.
4. Viscoelasticity and Melt Flow: The FFF process involves the extrusion of molten thermoplastics that exhibit viscoelastic behaviour. The simulation failed to accurately capture the flow dynamics of the melted filament and the solidification process.

Future research should focus on the effect of porosity and the air gap on the mechanical properties, incorporate realistic boundary conditions and environmental parameters, introduce composite materials such as reinforced thermoplastics, and consider the topology optimization of the FFF parts. Furthermore, incorporating multi-material printing strategies may enhance the performance of components subjected to complex loading conditions, offering new possibilities for advanced applications in additive manufacturing.

6. Acknowledgments

This work was supported by the Student Grant Competition (SGS) of the Technical University of Liberec, under project no. SGS-2023-5321.

7. References

[1] Bodur, O.; Stepanek, V.; Walcher, E. M. & Durakbasa, N. (2020). Precision in additive manufacturing, optimization and evaluation of the accuracy of 3D printer based on GPS system, Annals of DAAAM & Proceedings.

[2] Sheoran, A. J. & Kumar, H. (2020). Fused deposition modeling process parameters optimization and effect on mechanical properties and part quality: Review and reflection on present research, Materials Today: Proceedings, Vol. 21, pp. 1659-1672.

[3] Hodžić, D. & Pandžić, A. (2022). Influence of printhead nozzle diameter on mechanical properties of FDM printed PLA material and printing time, Proceedings of the 33rd DAAAM International Symposium, Vienna, Austria.

[4] Pandžić, A.; Kadric, E. & Kolesar, S. (2023). Investigating the influence of printing orientation and filament drying on tensile and flexural strength of FDM-printed carbon fiber-reinforced polyamide composites, Annals of DAAAM & Proceedings, Vol. 34.

[5] Behseresht, S.; Park, Y. H.; Love, A. & Pastrana, O. A. V. (2024). Application of numerical modeling and finite element analysis in fused filament fabrication: A review, Materials, Vol. 17, No. 17, pp. 4185.

[6] Hu, Z. & Mahadevan, S. (2017). Uncertainty quantification and management in additive manufacturing: current status, needs, and opportunities, The International Journal of Advanced Manufacturing Technology, Vol. 93, pp. 2855-2874.

[7] Ning, F.; Cong, W.; Hu, Y. & Wang, H. (2017). Additive manufacturing of carbon fiber-reinforced plastic composites using fused deposition modeling: Effects of process parameters on tensile properties, Journal of Composite Materials, Vol. 51, No. 4, pp. 451-462.

[8] Brenken, B.; Barocio, E.; Favaloro, A.; Kunc, V. & Pipes, R. B. (2018). Fused filament fabrication of fiber-reinforced polymers: A review, Additive Manufacturing, Vol. 21, pp. 1-16.

[9] Zhang, Y. & Chou, Y. K. (2006). Three-dimensional finite element analysis simulations of the fused deposition modelling process, Proceedings of the Institution of Mechanical Engineers, Part B: Journal of Engineering Manufacture, Vol. 220, No. 10, pp. 1663-1671.

[10] Zhang, Y. & Chou, K. (2008). A parametric study of part distortions in fused deposition modelling using three-dimensional finite element analysis, Proceedings of the Institution of Mechanical Engineers, Part B: Journal of Engineering Manufacture, Vol. 222, No. 8, pp. 959-968.

[11] Cattenone, A.; Morganti, S.; Alaimo, G. & Auricchio, F. (2019). Finite element analysis of additive manufacturing based on fused deposition modeling: Distortions prediction and comparison with experimental data, Journal of Manufacturing Science and Engineering, Vol. 141, No. 1, pp. 011010.

[12] Courter, B.; Savane, V.; Bi, J.; Dev, S. & Hansen, C. J. (2017). Finite element simulation of the fused deposition modelling process, Proceedings of the NAFEMS World Congress, pp. 11-14.

[13] Barocio, E.; Brenken, B.; Favaloro, A.; Bogdanor, M. & Pipes, R. B. (2020). Extrusion deposition additive manufacturing with fiber-reinforced thermoplastic polymers, In: Structure and properties of additive manufactured polymer components, pp. 191-219, Woodhead Publishing.

[14] Favaloro, A. J.; Brenken, B.; Barocio, E. & Pipes, R. B. (2017). Simulation of polymeric composites additive manufacturing using Abaqus, Science in the Age of Experience, pp. 103-114.

[15] Rimkus, A.; Farh, M. M. & Gribniak, V. (2022). Continuously reinforced polymeric composite for additive manufacturing—development and efficiency analysis, Polymers, Vol. 14, No. 17, pp. 3471.

[16] Espinach, F. X.; Boufi, S.; Delgado-Aguilar, M.; Julián, F.; Mutjé, P. & Méndez, J. A. (2018). Composites from poly (lactic acid) and bleached chemical fibres: Thermal properties, Composites Part B: Engineering, Vol. 134, pp. 169-176.

[17] Farah, S.; Anderson, D. G. & Langer, R. (2016). Physical and mechanical properties of PLA, and their functions in widespread applications—A comprehensive review, Advanced Drug Delivery Reviews, Vol. 107, pp. 367-392.

DOI: 10.2507/35th.daaam.proceedings.037

SDN and Network Security

Adriano Bubnjek, Damir Regvart* & Karlo Josic

This Publication has to be referred as: Bubnjek, A[driano]; Regvart, D[amir] & Josic, K[arlo] (2024). SDN and Network Security, Proceedings of the 35th DAAAM International Symposium, pp.0273-0278, B. Katalinic (Ed.), Published by DAAAM International, ISBN 978-3-902734-44-0, ISSN 1726-9679, Vienna, Austria
DOI: 10.2507/35th.daaam.proceedings.037

Abstract

This paper explores Software Defined Networking (SDN), detailing its architecture, components, and advantages over traditional network infrastructures. It also examines the security vulnerabilities inherent in SDN systems and strategies for their mitigation. The initial section of the paper covers SDN's essential features, such as scalability, enhanced visibility, and centralized control. It discusses its capability to operate independently of hardware while still utilizing it for traffic forwarding. The advantages and limitations of the SDN model are also analyzed. Subsequently, the paper categorizes various SDN security threats and attacks based on its unique architecture. It outlines preventive measures and responses to potential cyberattacks, including DDoS, ARP spoofing, and resource depletion.

Keywords: Software-Defined Network (SDN); network security architecture, network virtualization.

1. Introduction

Software Defined Networking (SDN) is an approach that allows network engineers and others to control underlying hardware infrastructure, which spans data centres, both private and public clouds, and application frameworks, using software-defined controllers or APIs (Application Programming Interfaces) from a centralized console or platform [2]. Essentially, SDN offers a programming-centric way to configure network and data traffic flow [4]. One of the notable advantages of SDN is enhanced network security [3], a primary focus of this seminar. Given the inherent complexity and layered nature of SDN, which is a critical component in many of today's enterprise infrastructures, this paper will briefly touch on certain SDN features and concepts without going into extensive detail. A fundamental distinction between SDN and traditional hardware-based networking is that SDN's control plane makes infrastructures more scalable, flexible, and manageable [1].

Furthermore, SDN differs significantly from conventional networking in terms of security implementations. This is partly due to its compatibility with Network Function Virtualization (NFV), which separates essential network security functions such as firewalls and load balancers (WAF) from dedicated hardware, instead hosting them in a software-defined instance that secures virtualized, containerized, or bare metal workloads. This paper will explore SDN architecture, its components, and core features in subsequent sections. It will also discuss how businesses can benefit from adopting a software-defined model, consider important security principles during the design phase, and address common threats, attacks, and vulnerabilities associated with SDN.

2. SDN Architecture and Key Features

As introduced earlier in this paper, Software Defined Networking (SDN) is a forward-thinking approach that leverages virtualization to separate the primary components and roles typically housed within traditional networking devices such as switches and routers [2]. Below is a figure depicting the SDN architecture and some of its integral components.

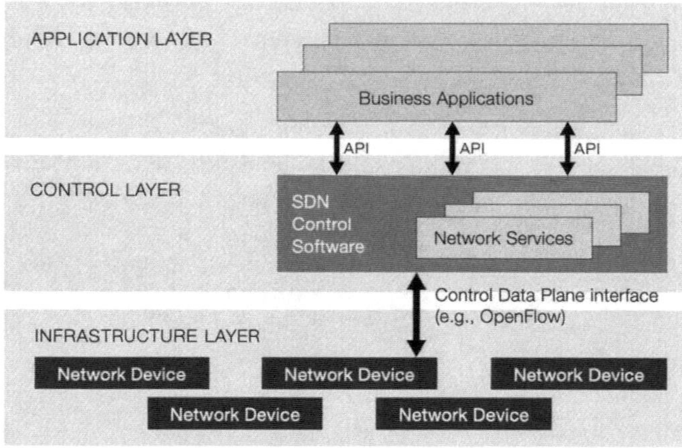

Fig. 1. Components and Architecture of SDN, https://www.sdxcentral.com/networking/sdn/definitions/what-the-definition-of-software-defined-networking-sdn/, accessed on 2024-05-05.

One of the principal distinctions in SDN is the division between the controller or control plane and the data plane [3]. The control plane creates network topologies or tables and sets packet handling policies. In contrast, the data plane handles the intensive tasks of forwarding packets and segmenting and reassembling data. These are standard functions within any conventional networking device. However, such a distributed networking framework does not scale well and lacks the speed required to keep pace with the rapid expansion of business development across various industries. Moreover, modifying or enforcing existing security policies within this traditional model is time-consuming, especially without third-party automation solutions like Ansible, Puppet, and Terraform, which may be excluded due to complexity or skill deficits [4]. SDN addresses these issues by centralizing the brain of the operation, the control plane, typically within a console or interface. In contrast, the data plane performs the primary network device functions. Other SDN components include application planes that support routing, load balancers, firewalls, etc., Northbound and Southbound APIs which facilitate connections between the SDN application layer, SDN controller, and networking devices, and OpenFlow, a protocol used to manage the southbound interface connecting the controller to network devices [4].

The SDN architecture is segmented into three layers: the application layer, the controller layer, and the networking devices layer [2]. The application layer contains programs communicating with the controller via API calls, such as load balancers and firewalls. The controller serves as the centralized element of SDN technology, acting as middleware between the SDN application and networking devices layers, facilitating data transfer between them. Lastly, the networking devices layer includes the hardware that hosts the data plane component. One might question the rationale and benefits of integrating such a sophisticated and costly model into an existing environment that seemingly meets all current business application and operational needs. A major issue in IT today is documentation, particularly of the network layer and its topology. All may seem well until a problem or legitimate threat arises. At this point, engineers often spend significant time manually investigating their own or their client's environments before they can even begin to address the issue. This process detracts from their primary responsibilities. This challenge becomes more manageable over time within the same company or with the same clients due to familiarity with the environment. However, this is neither sustainable nor scalable, particularly in large enterprises experiencing frequent changes [3].

In the below figure, Fig. 2., the SDN model proves its value, especially regarding networking. One of the most powerful features of a robust SDN model is its visibility and the ability to automatically generate network topologies from a centralized graphical console interface [4]. While this may not seem groundbreaking, it is a fundamental feature expected in any high-quality SDN product, such as NSX-T from VMware, ACI, and Meraki by Cisco, Dell SDN, Juniper SDN, and others. It is not uncommon for companies to rely on or complement their documentation with topologies created by SDN.

Beyond visibility and scalability, SDN facilitates the easy implementation of Quality of Service (QoS) for VoIP or video streaming, optimizes applications and services based on network metrics collected by SDN, reduces operational costs by minimizing the time required to troubleshoot and resolve issues, abstracts cloud resources, and simplifies the enforcement and implementation of security policies and models, which will be discussed in the following chapter. In summary, SDN offers significant advantages for organizations seeking flexibility, scalability, visibility, modernity, and security. However, SDN is not without its drawbacks and limitations.

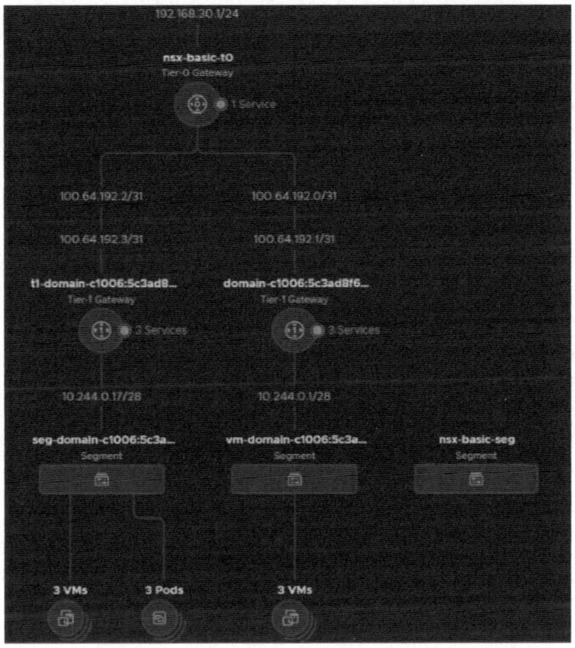

Fig. 2. Network Topology of vSphere with Tanzu Created by NSX-T

Its complexity necessitates reconfiguring existing network infrastructure, a daunting task that could span years depending on the size of the infrastructure and its business impact. Additionally, SDN demands highly trained and skilled employees due to its steep learning curve, which can prolong the implementation phase and complicate system maintenance once completed. This often leads companies to outsource, potentially resulting in vendor lock-ins and dissatisfaction among existing staff. Moreover, certain security limitations within SDN may be exploited by attackers, a topic that will be explored in the next chapter.

3. SDN Architecture and Component Attacks

When discussing attacks and vulnerabilities in Software Defined Networking (SDN), they can be broadly categorized into two groups: those targeting the SDN architecture and those targeting its components. Attacks on the SDN architecture typically focus on the application, control, and data plane layers, along with their respective interfaces [5]. In the application layer and its interfaces, three primary categories of attacks are prevalent: unauthorized access, malicious applications, and configuration issues. Specific examples within these categories include the lack of TLS implementation, insecure provisioning, and unauthorized application attacks. The security challenges posed by the separation of control and data planes in SDN architectures and the necessity of integrating security measures at each layer [5]. The control layer is susceptible to additional categories of attacks, such as data leakage, modification, denial of service, and broader system-level security issues. Examples include communication flooding between controllers and network devices and controller hijacking, which can have severe consequences [6]. The attacks on the data layer and its interfaces mirror those on the control layer, except for malicious applications, as the data layer does not directly interact with the application layer. Furthermore, attacks targeting SDN components are classified based on the affected component (Data Plane, Control Plane, Application Plane, Northbound API, Southbound API). The Application Plane is vulnerable to four major types of attacks: storage attacks, control message attacks, resource attacks, and access control attacks. There are techniques for enhancing the security of southbound infrastructure in SDN, which supports the need for robust security protocols and monitoring systems that could prevent such attacks [6]. A successful storage attack can lead to data corruption or theft from the shared storage used by SDN applications. Control message attacks can disrupt or manipulate communications between controllers and networking devices, affecting traffic management. Resource attacks aim to deplete memory or CPU resources on applications and controller components, causing disruption and performance degradation. Lastly, access control attacks target the controller, often due to misconfigured authentication, authorization, and accountability settings.

Three attack types are noted regarding the Data Plane: device, protocol, and side-channel attacks. Device attacks search for vulnerabilities in the software or firmware of SDN switches or routers. The utilization of multi-site solutions to mitigate potential network attacks offers practical approaches to securing SDN environments that could counter device attacks [7]. Protocol attacks target weaknesses in the OpenFlow protocol used to manage network traffic. In contrast, side-channel attacks analyze CPU consumption, power usage, and traffic flow to deduce patterns that may reveal information about specific services or applications [5].

For the Control Plane, potential attacks include manipulation, availability, and software hack attacks. Manipulation attacks target the controller's algorithms and understanding of the Data Plane, potentially leading to erroneous decisions [5]. Availability attacks aim to temporarily disable the controller, isolating parts of, or the entire, network through methods like resource depletion, flooding, and specific network traffic congestion [5]. Software hack attacks compromise the underlying system on which the controller operates, with lesser-known or non-enterprise-grade operating systems being particularly vulnerable.

A security-hardened, enterprise-grade OS like Photon OS can significantly reduce these risks. Northbound and Southbound APIs face three similar attack types: interception, eavesdropping, and availability attacks. For example, availability attacks might involve denial of service through excessive traffic; interception attacks could manipulate the flow table, resulting in incorrect packet forwarding; and eavesdropping attacks might involve control message spoofing, potentially leading to unauthorized control over network devices.

4. SDN Security Measures and Mitigations

Before delving into specific mitigation strategies for attacks, it's crucial to outline some general principles and security measures for hardening SDN protocols, components, and interfaces to minimize threats. A fundamental and critical security measure is the establishment of a robust framework for authentication and authorization [8]. This framework hinders unauthorized access to your SDN infrastructure and restricts internal access to approved entities. Additionally, implementing comprehensive auditing policies is vital, not only for compliance with standards like ISO 27001 but also for the overarching benefit of the organization. There should also be an effective method for evaluating new controls and their impact on the network's confidentiality, integrity, and availability. Moreover, a defense-in-depth strategy should be adopted to deter attackers and protect the system from various angles.

Specific security measures can be tailored according to the SDN layer involved. For the Application Plane, security can be enhanced by integrating protective measures directly into applications to monitor and actively defend against threats. This proactive security approach is crucial for managing potential security incidents. Proper authentication and encryption must be established to secure communications between the application and controller [8]. Additionally, publicly exposed web applications should be safeguarded with tools such as Web Application Firewalls (WAFs), standard firewalls, and proxies. Implementing TLS to encrypt traffic between networking devices and the SDN controller for the Data Plane is essential for preventing spoofing and eavesdropping. Remote access connections should use SSH instead of Telnet, and SNMPv3 should replace SNMPv2 for secure network device management over IP. Furthermore, tunnel endpoints must be secured with strong passwords. The SDN controller should be configured for high availability to eliminate potential single points of failure caused by attackers or system failures [9]. For example, deploying at least three controller instances across three virtualization hosts in a geographically distributed cluster can enhance system resilience. Role-Based Access Control (RBAC) policies must be enforced to ensure that only authorized personnel can make configuration changes. Additionally, logging and auditing mechanisms should be in place to detect unauthorized access and modifications. The following table outlines common attacks on SDN and corresponding mitigation steps:

Attack	Mitigation Steps
Password guessing/brute force	Enforce frequent password changes, use complex and multi-language passwords with special characters, change default settings, and avoid reusing passwords [8].
API exploitation	Regularly update and patch servers and SDN components.
Traffic sniffing	Encrypt traffic using robust protocols (e.g., TLS instead of SSL).
Side channel attack	Utilize robust encryption algorithms like AES and ECC to secure SDN components and network elements.
ARP spoofing	Implement network segmentation (VLAN), Dynamic ARP Inspection (DAI), and switch port security.
Network manipulation attack	Design a redundant SDN Control Plane layer with multiple controller instances and secure communications.
Misconfiguration attack	Implement RBAC, automated auditing, continuous monitoring with intrusion alerts, and automation tools for repetitive tasks. Provide security training and adopt a zero-trust networking approach.
DDoS attack	Use a high-availability (HA) controller architecture, rate limits, and packet-dropping techniques at the control plane, and monitor network traffic with an SDN-distributed firewall [9].
Resource attack	Monitor system metrics and implement rate limits and resource quotas where possible.
DNS spoofing	Implement DNSSEC to secure DNS communications.
Rogue SDN controller	Establish a robust authentication framework, use strong encryption for component communications, and implement RBAC, continuous auditing, and Multi-Factor Authentication (MFA).

Table 1. SDN Attacks and Mitigations

Although the strategies above aim to secure SDN from various threats and minimize the likelihood of attacks, complete protection against cyberattacks is not feasible. Therefore, having a mitigation strategy and countermeasures ready for well-known attacks is imperative.

5. Future Works

One significant area for future research involves the integration of artificial intelligence (AI) and machine learning (ML) into SDN. These technologies could revolutionize how networks predict, detect, and respond to anomalies and security threats. Exploring how AI can optimize network traffic flow and security protocols could lead to more autonomous network systems that reduce human error and enhance security efficacy. This advancement could usher in a new era of intelligent networking where systems are capable of self-management and real-time security enforcement. Another promising research trajectory is the development of advanced security protocols tailored explicitly for SDN environments. This includes creating dynamic security policies that adapt in real-time to emerging threats. Leveraging the centralized control inherent in SDN to implement rapid security updates across the network could significantly enhance the responsiveness and adaptability of network defenses. Additionally, as quantum computing poses new risks to traditional encryption methods, future studies should consider the implications of quantum technologies on SDN security, particularly concerning data encryption and secure communications. The role of SDN in emerging networking paradigms such as 5G, Internet of Things (IoT), and edge computing offers ground for research. These areas, characterized by their vast scale and complexity, could greatly benefit from SDN's capabilities in managing large arrays of devices and data flows. Investigating how SDN can provide dynamic, efficient solutions to manage and secure these expanding networks will be crucial as these technologies continue to grow and intersect.

6. Conclusion

This paper has thoroughly explored the multifaceted domain of Software Defined Networking (SDN), delineating its architecture, components, and intrinsic advantages over traditional network frameworks. The paper highlighted SDN's transformative potential in modern network management and security by unpacking critical features such as enhanced scalability, visibility, and centralized control. A focal point of the paper was the detailed examination of SDN's security landscape, where various threats and vulnerabilities specific to its unique architecture were categorized and analyzed. The paper highlighted prevalent security risks such as DDoS attacks, ARP spoofing, and resource depletion and articulated strategic mitigation measures to fortify SDN environments against such threats. Emphasizing the importance of robust authentication, authorization frameworks, and comprehensive auditing protocols, the paper advocated for a proactive security posture that aligns with contemporary compliance and operational standards.

Looking beyond the current applications of SDN, the paper proposed several promising directions for future research. Integrating artificial intelligence and machine learning is a revolutionary next step, potentially enabling more autonomous and dynamically secure networks. The call for enhanced interoperability and the development of standardized security protocols tailored to SDN environments further indicates this technology's ongoing evolution and maturation. Moreover, exploring SDN's applicability in burgeoning fields like IoT and 5G illustrates its pivotal role in the next generation of network technology.

7. References

[1] Jimenez, M. B. & Fernandez, D. (2022). A Framework for SDN Forensic Readiness and Cybersecurity Incident Response, Proceedings of the 2022 IEEE Conference on Network Function Virtualization and Software Defined Networks (NFV-SDN), Published by IEEE, pp. 112-116, ISBN 978-1-6654-7334-7, Phoenix, USA. DOI: 10.1109/NFV-SDN56302.2022.9974648, in press.

[2] Tsai, C. & Song, K. (2023). Discussion on Network Security under SDN Architecture, Proceedings of the 2023 26th ACIS International Winter Conference on Software Engineering, Artificial Intelligence, Networking and Parallel/Distributed Computing (SNPD-Winter), Published by IEEE, pp. 53-57., ISBN 979-8-3503-4586-5, Taiyuan, China. DOI: 10.1109/SNPD-Winter57765.2023.10223982, in press.

[3] Liang, H.; Liu, H.; Dang, F.; Yan, L. & Li, D. (2021). Information System Security Protection Based on SDN Technology in Cloud Computing Environment, Proceedings of the 2021 IEEE International Conference on Advances in Electrical Engineering and Computer Applications (AEECA), Published by IEEE, pp. 432-435., ISBN 978-1-6654-3561-1, Dalian, China. DOI: 10.1109/AEECA52519.2021.9574276, in press.

[4] Ahmad, A.; Harjula, E.; Ylianttila, M. & Ahmad, I. (2020). Evaluation of Machine Learning Techniques for Security in SDN, Proceedings of the 2020 IEEE Globecom Workshops (GC Wkshps), Published by IEEE, pp. 1-6., ISBN 978-1-7281-7307-8, Taipei, Taiwan. DOI: 10.1109/GCWkshps50303.2020.9367477, in press.

[5] Hatim, J.; Chaimae, S. & Habiba, C. (2023). SDN/NFV Security Challenges and Proposed Architecture, Proceedings of the 2023 7th IEEE Congress on Information Science and Technology (CiSt), Published by IEEE, pp. 145-149. ISBN 978-1-6654-6133-7, ISSN 2327-1884, Agadir - Essaouira, Morocco. DOI: 10.1109/CiSt56084.2023.10409955, in press.

[6] Tupakula, U.; Karmakar, K. K.; Varadharajan, V. & Collins, B. (2022). Implementation of Techniques for Enhancing Security of Southbound Infrastructure in SDN, Proceedings of the 2022 13th International Conference on Network of the Future (NoF), Published by IEEE, pp. 1-5., ISBN 978-1-6654-7254-8, ISSN 2833-0072, Ghent, Belgium. DOI: 10.1109/NoF55974.2022.9942644, in press.

[7] Ahmed, Z. & Bashir, B. S. A. (2022). Optimized and Secured Utilization of Infrastructure Resources using VMWare Stretched Cluster Multi-Site Solutions, Proceedings of the 2022 International Conference on Computing, Communication, Security and Intelligent Systems (IC3SIS), Published by IEEE, pp. 1-4., ISBN 978-1-6654-6883-1, Kochi, India. DOI: 10.1109/IC3SIS54991.2022.9885490, in press.

[8] Redžepagić, J.; Dakić, V.; Stanešić, J. & Bašić, M. (2023). Analysis of Password Security Policies and Their Implications on Real-Life Security, Proceedings of the 34th DAAAM International Symposium, pp.0077-0081, ISBN 978-3-902734-41-9, ISSN 1726-9679, Vienna, Austria. DOI: 0.2507/34th.daaam.proceedings.011, in press.

[9] Dizdar, K.; Morić, Z., Dakić, V. & Bašić, M. (2023). Denial of Service Attacks Using the Example of Croatian Hosters, Proceedings of the 34th DAAAM International Symposium, pp.0088-0095, ISBN 978-3-902734-41-9, ISSN 1726-9679, Vienna, Austria. DOI: 10.2507/34th.daaam.proceedings.013, in press.

DOI: 10.2507/35th.daaam.proceedings.038

MACHINE LEARNING FOR THREAT DETECTION IN SOCs

Ema Josipovic, Damir Regvart*, Ana Kapulica & Robert Kopal

This Publication has to be referred as: Josipovic, E[ma]; Regvart, D[amir]; Kapulica, A[na] & Kopal, R[obert] (2024). Machine Learning for Threat Detection in SOCs, Proceedings of the 35th DAAAM International Symposium, pp.0279-0285, B. Katalinic (Ed.), Published by DAAAM International, ISBN 978-3-902734-44-0, ISSN 1726-9679, Vienna, Austria

DOI: 10.2507/35th.daaam.proceedings.038

Abstract

This paper explores the integration of Machine Learning (ML) algorithms within Security Operations Centres (SOCs) for enhanced cyber threat detection. It examines the effectiveness of various ML algorithms, such as Random Forest and SVM, in identifying and mitigating cyber threats. The paper delves into the inherent challenges associated with ML in cybersecurity, including vulnerabilities to adversarial attacks and issues of overfitting. Strategies to overcome these challenges are discussed, highlighting the importance of advanced data handling and the necessity of human expertise alongside ML. The paper concludes by emphasizing the transformative potential of ML in cybersecurity, advocating for continuous innovation and ethical considerations in its application.

Keywords: Machine Learning; Cybersecurity; Security Operations Centres; Threat detection; Random Forest

1. Introduction

Security operations centers (SOCs) are teams of IT security specialists responsible for continuously monitoring an organization's entire IT infrastructure. They promptly detect and respond to cybersecurity events, ensuring swift and efficient action [1]. In the absence of immediate threats, SOCs analyze threat data to develop new and improved security solutions, enhancing the organization's overall security posture. SOC duties and activities can be broadly categorized into three areas:

- Preparation, planning, and prevention,
- Monitoring, detection, and response,
- Recovery, refinement, and compliance.

SOCs perform regular maintenance and preparation tasks, such as updating firewalls, applying software patches, and backing up systems. They also develop an incident response plan detailing roles, tasks, and response metrics. Regular tests, including penetration and vulnerability assessments, help SOCs identify potential risks and costs. Staying informed about the latest security threats and solutions ensures business continuity during a cyberattack. In addition, SOCs are responsible for continuous monitoring, identification, and response for networks, servers, system software, applications, computing devices, cloud workloads, and system software.

Security information and event management (SIEM) is the primary technology used for this purpose [2], gathering real-time alerts from network hardware and software, analyzing them, and detecting potential threats. Log management is a crucial monitoring aspect, highlighting anomalies that indicate suspicious activity. When a threat or incident occurs, the SOC mitigates damage, which may involve root cause analysis, root cause investigation, isolating compromised areas, exiting or reversing affected applications, deleting files, and deploying antivirus or anti-malware software.

The SOC is responsible for recovery, improvement, and compliance management after an incident. It mitigates risks, repairs damaged assets and incorporates new intelligence to patch vulnerabilities and update procedures. Additionally, it ensures compliance with data privacy laws such as GDPR, CCPA, PCI DSS, and HIPAA. Post-incident, the SOC ensures incident data is preserved for auditing and evidence and informs users, regulators, law enforcement, and other relevant parties.

The SOC's structure generally includes a SOC Manager, Security Engineers, Security Analysts, and Threat Hunters. The SOC Manager oversees all security operations and reports to the Chief Information Security Officer. Security Engineers design the company's security architecture, while Security Analysts are the first responders to cybersecurity threats or incidents. Threat Hunters specialize in identifying and neutralizing advanced threats that evade automated defenses. Depending on the organization's size and industry, other specialists may be part of the SOC team. Larger organizations may employ a Director of Incident Response to coordinate and manage incident response efforts. Additionally, some SOCs have Forensic Investigators who specialize in recovering data clues from compromised devices during cybersecurity incidents.

2. Overview of Integrating Machine Learning for Threat Detection

Before exploring the use of Machine Learning for threat detection in SOCs, it's essential to define it. Machine Learning, a branch of Computer Science and Artificial Intelligence, employs data and algorithms to mimic human learning processes and improve accuracy over time (IBM, 2024b). Algorithms are trained using statistical techniques to make predictions or classifications and uncover significant information in data mining projects. These insights are then used to inform business and application decisions, ideally positively impacting key growth metrics.

Despite the existing clever marketing messages and some reporting, Machine learning is not a security cure-all. Still, it can substantially support security and IT organizations [3]. While Machine Learning has a long way to go before it can detect threats without human interaction, it can undertake various tasks to improve cybersecurity. Some examples of specific areas where machine learning may be used to simplify and enhance cybersecurity tasks include Phishing and spam filtering, Forensic analysis, Incident response and risk management, penetration testing, and threat modeling and prevention. In the case of Phishing and spam filtering, ML may, by using predefined parameters, assist in accurately classifying internal data, spam, and malicious actions. In Forensic analysis, ML can, by using clustering, assist in obtaining information about the assault strategy and targets. Clustering collects and groups data, making it easier for the team to piece together the elements of the attack. Furthermore, the software learns from the actions taken during various incidents through ML for incident response and risk management. Machine learning can identify actions for incident response and risk mitigation based on patterns and relationships. In the case of penetration testing, programs can generate inputs to test systems for vulnerabilities like a penetration tester searching for flaws. Finally, in Threat modeling and prevention, Machine learning gathers and evaluates data to forecast fraudulent activity. This aids security teams in spotting dangers before they develop into costly and significant data breaches or thefts.

Applying Machine Learning algorithms can streamline and enhance the efficiency of numerous cybersecurity tasks. This technology can significantly reduce workload, decrease human error, and improve decision-making processes. Additionally, these adaptive strategies and tactics can be tailored to fit the specific needs of a business or organization, providing a customized approach to cybersecurity that maximizes effectiveness and aligns with organizational goals. The flexibility and precision of Machine Learning make it a powerful tool in the ongoing effort to protect digital assets and maintain robust security postures.

3. Machine Learning Algorithms for Threat Detection

Threat detection in cybersecurity, including in Security Operations Centres (SOCs), involves using various machine learning algorithms [4]. The choice of which algorithm to use depends on multiple factors, including the specific use case, available data, and the nature of the threats being dealt with. There is also the option of a hybrid approach, which includes combining multiple algorithms, usually done to enhance the overall threat detection capabilities of the Machine Learning model.

The most used algorithms, due to their effectiveness in identifying patterns, anomalies, and potential threats, described by this paper use cases are:

- Supervised Learning Algorithms
- Unsupervised Learning Algorithms
- Reinforcement Learning Algorithms

3.1. *Supervised Learning Algorithms*

Supervised learning algorithms are machine learning algorithms that learn from labeled training data to make predictions or classifications. In supervised learning, each training example consists of an input object (typically a vector) and a desired output value (also called the supervisory signal). The algorithm learns to map inputs to the correct output based on this labeled data, allowing it to predict the production of new, unseen inputs.

Random Forest is an ensemble learning method that can be used for classification tasks in threat detection. It combines multiple decision trees to improve the accuracy and robustness of the model. An example use case could be as follows: Random Forest can analyze web traffic and classify URLs as benign or malicious. By training the model on a labeled dataset of URLs, the algorithm learns patterns indicative of malicious links. In a SOC, this can help identify and block phishing attempts or links leading to malicious sites, enhancing overall web security.

Support Vector Machines (SVM) are practical algorithms for binary and multi-class classification problems. They work well for identifying patterns in data, making them useful for detecting anomalies or malicious activities. An example use case could be as follows: SVMs can be applied to user behavior analytics to identify potential insider threats. By training on historical data, the SVM model learns standard patterns of user activities. The SVM can raise alerts for further investigation when deviations or anomalies occur, such as unusual file access or login behavior.

Naive Bayes is a probabilistic classifier based on Bayes' theorem. It's simple and efficient, making it suitable for real-time threat detection applications. An example use case could be Email Spam Filtering since Naive Bayes is well-suited for email security in SOCs. By analyzing the content and metadata of emails, the algorithm can classify messages as either spam or legitimate. This helps automatically filter out malicious emails and reduce the risk of phishing attacks.

3.2. *Unsupervised Learning Algorithms*

Unsupervised learning algorithms are machine learning algorithms that work with unlabelled data. Unlike supervised learning, which learns from labeled input-output pairs, unsupervised learning identifies patterns and structures in the data without predefined labels. These algorithms are beneficial for discovering hidden structures, clustering similar data points, and reducing data dimensionality.

K-means clustering is a widely used clustering algorithm that identifies patterns and groups similar data points. It can be employed for anomaly detection by identifying clusters that deviate from the norm. An example use case could be Network Intrusion Detection. It can be applied to network traffic data to identify clusters of normal behavior. Any deviations from these clusters may indicate a potential network intrusion. SOC analysts can then investigate these anomalies to detect and respond to security incidents.

Isolation Forest is an ensemble method for anomaly detection. It builds a tree structure to isolate anomalies more efficiently than traditional methods. An example use case could be Anomaly Detection in System Logs. By creating a model on standard log patterns, the algorithm can efficiently isolate and flag abnormal log entries, which may indicate unauthorized access or other security incidents.

Density-Based Spatial Clustering of Applications with Noise (DBSCAN) is a density-based clustering algorithm that can identify clusters of varying shapes and effectively detect outliers, which may represent security threats. One example of its use is Identifying Port Scans. The algorithm can analyze network traffic and identify patterns associated with port scanning activities. Unusual clustering of connections to multiple ports from a single source IP address may suggest a potential reconnaissance attempt, triggering an alert for further investigation.

3.3. *Reinforcement Learning Algorithms*

Reinforcement learning (RL) algorithms are a category of machine learning algorithms that enable an agent to learn how to make decisions by performing actions in an environment to maximize some notion of cumulative reward. Unlike supervised learning, where the model learns from a dataset of labeled examples, reinforcement learning involves learning from actions' consequences rather than direct instruction.

Q-learning is a model-free reinforcement learning algorithm. In cybersecurity, it can train an agent for adaptive firewall rule management. The agent learns to adjust firewall configurations based on evolving threat landscapes, optimizing for security and performance in real-time.

Deep Q Networks (DQN) is an extension of Q-Learning that uses deep neural networks to approximate the Q-function. It can handle more complex and high-dimensional state spaces, making it suitable for cybersecurity applications. An example use case is using DQN for dynamic threat response by training an agent to decide on incident response actions. The model learns optimal strategies for mitigating specific threats, enabling automated and intelligent responses to security incidents.

Multi-Armed Bandits can be used when decisions must be made in uncertain environments. It's relevant for dynamically adapting defense strategies based on evolving threats during incident response. By continuously learning from the effectiveness of different response strategies, the algorithm optimizes resource allocation to maximize the SOC's ability to mitigate and respond to security incidents effectively.

4. Effectiveness of Machine Learning in SOCs - Strengths and Weaknesses

Machine Learning algorithms have demonstrated their proficiency in identifying and hindering cyber threats, capitalizing on their capacity to adapt to changing and emerging patterns of attacks. This section explores the varied ML algorithms and their impact on cyber security, highlighting their adaptability and learning capabilities. In a series of studies, different Machine Learning algorithms were assessed for their effectiveness in cyber-attack detection [5]. Random Forests and Logistic Regression were most efficient in phishing attack detection, with accuracies of 98.64% and 98.59%, respectively. They demonstrated Random Forests [6] superiority in malware detection, achieving a 99.63% accuracy. Random Forests' effectiveness in DDoS and network intrusion detection, respectively. These studies [7][8] collectively underscore the high efficacy of ML algorithms, especially Random Forests shown in Table 1, in combating various cyber threats.

Cyber-Attack Type	Accuracy of Random Forest
Malware Attacks	99,68%
DDoS Attacks	99,48%
Network Intrusion Attacks	99,31%
E-mail SPAM	98,72%

Table 1. Effectiveness of Random Forest in Detecting Different Types of Cyber Attacks

Machine Learning Algorithm:	Advantage:	Disadvantage:
Decision Trees	Easy to interpret and explain, fast training time	Can overfit on training data
Random Forest	Can handle large datasets, less prone to overfitting than decision trees	It can be computationally expensive
Native Bayes	Simple and fast, it requires less training.	Assumes independence between features, which may not be accurate in some cases
Logistical Regression	Easy to interpret and explain, can handle non-linear relationships between features	It may not perform well when features have complex interactions, or the decision boundary is non-linear.
Support Vector Machine (SVM)	Effective for high-dimensional features, can handle non-linear decision boundaries	It can be computationally expensive.
Neural Networks	Can learn complex relationships between features and can handle large datasets.	Requires a lot of training data, prone to overfitting
Clustering	It can detect anomalies in data, which helps identify new types of attacks.	It can be computationally expensive and challenging to interpret.
Reinforcement Learning	Can adapt to changing environments, useful for detection of dynamic attacks	Requires a lot of training data, can be computationally expensive

Table 2. Advantages and Disadvantages of Machine Learning Algorithms for Cyber-Attack Detection and Prevention

As is visible from Table 2, in cyber-attack detection and prevention, various Machine Learning algorithms offer a balance of strengths and weaknesses. Supervised learning algorithms, known for their efficacy in recognizing established threat patterns, are highly accurate due to their reliance on pre-labeled datasets. However, their dependency on prior knowledge makes them less effective against novel threats. On the other hand, unsupervised learning thrives in detecting these unknown threats by identifying anomalies without needing labeled data, although it may yield higher false-positive rates. Moreover, reinforcement learning stands out for its adaptability and continuous learning from the environment, offering a dynamic approach to evolving cyber threats. However, this comes at the cost of increased complexity and computational resources. More specifically, it can be said that Decision Trees are lauded for their quick training and simplicity but are vulnerable to overfitting. Random Forests are robust against overfitting and capable of handling complex data, yet computationally demanding.

Naive Bayes is efficient and requires less data but may falter with interdependent features. Logistic Regression works well with non-linear relationships but struggles with complex interactions. SVMs excel in high-dimensional spaces but are sensitive to kernel functions. Neural Networks can model intricate relationships and manage voluminous data but tend to overfit. Clustering algorithms are adept at anomaly detection but can be complex and resource intensive. It can be concluded that each algorithm's applicability depends on the specific requirements and limitations of the cybersecurity task at hand. Since each machine learning algorithm offers distinct advantages and unique challenges, careful evaluation is needed to determine the most suitable approach for a specific SOC environment and security threat.

5. Challenges in Training Machine Learning Models within SOC

Even though machine learning algorithms have succeeded in identifying and stopping cyberattacks, they have some drawbacks [9]. A few of machine learning's shortcomings in cyber security will be discussed in this paragraph. The susceptibility of machine learning to adversarial attacks is one of its drawbacks. Adversarial attacks are vicious and purposeful attempts to cause disturbances in the input data to trick Machine Learning algorithms. Machine learning algorithms may misclassify data because of adversarial attacks, which could result in successful cyberattacks. The vulnerability of machine learning to overfitting is another drawback. When a machine learning system is trained on a small amount of data, it becomes overfitted and has poor generalization to new data. Overfitting can result in false positives and negatives in the context of cyber security, leading to missed threats and resource loss. Overfitting may happen when machine learning algorithms are used for intrusion detection [10]. Another constraint on machine learning algorithms is the quantity and quality of training data. A machine learning system might not be able to reliably identify or stop cyberattacks if the training data is biased or lacking. Additionally, because the algorithm might need to correctly categorize data that differs from the training set, a lack of diversity in the training data can result in poor generalization. Machine learning algorithms may be computationally costly and need much memory and processing capacity. This may restrict its applicability and scalability, especially for resource-constrained organizations. Finally, the application of ML in cybersecurity raises ethical and privacy issues [11]. It's an enormous task to balance user privacy and security, and it is riddled with moral and legal issues. Ensuring the ethical use of technology requires strict adherence to data governance principles and robust data protection measures to preserve user confidence and comply with legal requirements. Machine learning algorithms have shown considerable promise in detecting and preventing cyberattacks; however, they have limitations. When utilizing machine learning in cybersecurity, it is crucial to consider several factors, including computational complexity, overfitting, the quality and diversity of training data, and susceptibility to adversarial attacks.

6. Strategies to Address the Challenges

A comprehensive strategy is essential to addressing the challenges encountered by Machine Learning in cybersecurity [12]. Effective data management is central to robust model training. This includes employing sophisticated techniques for data acquisition, preprocessing, and feature extraction, enhancing data quality and model performance. Simultaneously, optimizing algorithmic efficiency can be achieved through parallel computing, enabling the scaling of models to handle large and diverse datasets. Continuous model maintenance and adaptation are tackled through transfer learning and ensemble methods, bolstering generalization and resilience against adversarial attacks. Integrating human expertise with Machine Learning ensures a more nuanced approach to threat detection, as humans can provide insights that pure Machine Learning may overlook. Ensuring model interpretability and explainability is equally crucial; transparent models foster trust and facilitate easier management. Upholding legal and ethical standards is imperative, guaranteeing user trust through rigorous data protection and privacy measures. These strategies collectively establish a robust framework for overcoming the inherent limitations of Machine Learning in cybersecurity.

Incorporating Machine Learning into cybersecurity demands significant investment in advanced solutions and expert talent [13]. Equipping experts with AI tools position them better to keep pace with evolving cyber threats. Addressing AI biases is critical; cybersecurity AI must be trained on diverse datasets to ensure fairness and prevent skewed outcomes. Acknowledging vulnerabilities in AI, such as susceptibility to hacker exploitation, is paramount. AI should be viewed as a complementary tool rather than a substitute for human specialists, emphasizing continuous development and establishing legal frameworks to safeguard against AI misuse in cyber activities. This multifaceted approach strengthens AI's role in cybersecurity, balancing innovation with vigilance. Continuous research and development activities are necessary to address these difficulties and improve the robustness, resilience, and efficacy of ML-based cybersecurity systems [14]. Some possible ways to address the challenges include:

- Creating robust machine learning algorithms and resilient structures,
- Improving data gathering and labeling strategies,
- Adding explainability techniques and modifying models in response to evolving risks via ongoing monitoring and learning

7. Real-world Applications and Future Research

Machine learning has many applications for enhancing cybersecurity within SOCs [15]. This paragraph highlights five real-life examples for future research activity.

Firstly, **network traffic analysis** relies on machine learning algorithms to identify and mitigate DDoS attacks, forecast future threats by analyzing traffic volume, and detect botnet activities. For instance, algorithms can swiftly identify spikes in traffic during a DDoS attack, prompting timely action from relevant authorities within the organization. Additionally, these algorithms can pinpoint the attack vector or type of assault, enabling SOC teams to defend against similar cyber threats proactively. Moreover, machine learning techniques in network traffic analysis can detect and counteract bots or botnet activities, thus protecting servers from malicious actors seeking to exploit them.

In **endpoint security**, machine learning complements anti-virus software by detecting and categorizing various types of malware, bolstering defenses against emerging digital threats. Protecting endpoints from viruses and malware is paramount, and machine learning algorithms play a crucial role in uncovering hidden cyber dangers, thereby fortifying the efficacy of firewalls and anti-virus software. Furthermore, machine learning algorithms can classify incoming malicious packages as spyware, ransomware, or malware, empowering SOC teams to prioritize and mitigate potential online threats effectively. ML-assisted endpoint protection can swiftly identify anomalies at endpoints, alert authorities, and expedite threat detection processes.

Machine learning plays a vital role in **enhancing application security** as well. Web Application Firewalls and other application security solutions leverage machine learning to shield servers and systems from cyberattacks targeting the application layer of the OSI model. ML-based solutions, for instance, can be engineered to detect anomalies in HTTP requests and issue alerts in case of an attack. They can also be trained to recognize attack vectors and classify various types of assaults, such as SQL injection and XSS attacks. SOC teams can bolster their cyber protection infrastructure by leveraging insights from these algorithms.

Authentication security also benefits from machine learning. Various authentication security measures, including two-factor authentication, CAPTCHA, and biometric methods like fingerprint and facial recognition, aim to thwart account takeovers and online identity fraud. Facial and fingerprint recognition software employs ML algorithms to provide additional security to computers and mobile devices. Machine learning in authentication security utilizes additional techniques to thwart account takeover attempts better. Adversaries often employ brute force or credential-stuffing techniques to access accounts, potentially leading to organization network breaches. Machine learning methods can identify and thwart such attacks, with trained systems capable of detecting and intercepting the trial-and-error phase of these attacks.

Furthermore, machine learning significantly contributes to **attack surface management** by continuously monitoring organizational vulnerabilities. This support enables SOC teams to uphold a robust security posture. Given the continued expansion of the digital attack surface for significant firms, managing and monitoring an organization's attack surface can be burdensome for SOC teams. Machine learning offers a solution by alleviating the workload on SOC teams and providing error-free monitoring and management capabilities.

8. Conclusion

This paper demonstrates the crucial role of Machine Learning (ML) in bolstering cybersecurity within Security Operations Centres (SOCs). It underscores the benefits of various ML algorithms, such as Random Forest and SVM, in identifying and addressing cyber threats while acknowledging their limitations, including susceptibility to adversarial attacks and overfitting. Furthermore, it proposes comprehensive strategies to overcome these challenges, emphasizing the significance of advanced data management, ongoing model refinement, and the fusion of human expertise. This amalgamation of human and artificial intelligence enhances threat detection and lays the foundation for a more resilient and sophisticated cybersecurity framework. The paper highlights the intersection of ML and human intelligence in SOCs as indicative of the evolving cybersecurity landscape, emphasizing the imperative of continuous innovation and ethical considerations in this dynamic domain.

9. References

[1] https://www.ibm.com/topics/security-operations-center, (2024). IBM: What is a security operations center (SOC)?, Accessed on: 2024-01-08

[2] Suskalo, D., Moric, Z., Redzepagic, J., & Regvart, D. (2023). Comparative Analysis of IBM Qradar and Wazuh for Security Information and Event Management. In DAAAM Proceedings (pp. 0096–0102). DAAAM International Vienna, DOI: 10.2507/34th.daaam.proceedings.014

[3] Nemelka, K. (2023) Ki und ML in der cybersecurity, Security. Available from: https://www.security-insider.de/ki-und-ml-in-der-cybersecurity-a-5a3830cdaf3d26e192de9066e37c4234/ , Accessed: 2024-01-08

[4] Chukhnov, A.P. & Ivanov, Y.S. (2021). "Algorithms for Detecting and Preventing Attacks on Machine Learning Models in Cyber-Security Problems", Journal of Physics: Conference Series, Vol. 2096, Issue 1, p. 012099, DOI 10.1088/1742-6596/2096/1/012099

[5] Saranya, T., Sridevi, S., Deisy, C., Chung, T. D., & Khan, M. K. A. A. (2020). "Performance Analysis of Machine Learning Algorithms in Intrusion Detection System: A Review", In Procedia Computer Science, Vol. 171, pp. 1251–1260, DOI: 10.1016/j.procs.2020.04.133

[6] Geetha, R., & Thilagam, T. (2020). "A Review on the Effectiveness of Machine Learning and Deep Learning Algorithms for Cyber Security", In Archives of Computational Methods in Engineering, Vol. 28, Issue 4, pp. 2861–2879, DOI: 10.1007/s11831-020-09478-2

[7] Apruzzese, G., Colajanni, M., Ferretti, L., Guido, A. & Marchetti, M. (2018). "On the effectiveness of machine and deep learning for cyber security", In 2018 10th International Conference on Cyber Conflict (CyCon), ISBN:978-9-9499-9043-6, ISSN: 2325-5374, DOI: 10.23919/cycon.2018.8405026

[8] Sreekanth, D. et al. (2023) 'An Analysis Of The Effectiveness Of Machine Learning Algorithms In Detecting And Preventing Cyber-Attacks', Analysis of the Effectiveness of Machine Learning Algorithms in Detecting and Preventing Cyber-Attacks [Preprint].

[9] Bharadiya, J. (2023). "Machine Learning in Cybersecurity: Techniques and Challenges", In European Journal of Technology, Vol. 7, Issue 2, pp. 1–14, DOI:10.47672/ejt.1486

[10] Bagaa, M., Taleb, T., Bernabe, J. B., & Skarmeta, A. (2020). "A Machine Learning Security Framework for IoT Systems". In IEEE Access, Vol. 8, pp. 114066–114077, ISSN: 2169-3536, DOI: 10.1109/access.2020.2996214

[11] Sabo, J. (2023) Machine learning and AI in SOC: Enhancing threat detection, LinkedIn. Available at: https://www.linkedin.com/pulse/machine-learning-ai-soc-enhancing-threat-detection-john-sabo-9syqc/, Accessed: 2024-01-05

[12] Intone Networks (2023) Challenges & Solutions for machine learning in Cybersecurity, Intone Networks. Available at: https://intone.com/challenges-and-solutions-in-implementing-machine-learning-for-cybersecurity/, Accessed on: 2024-01-08

[13] Cybernetic Search (2023) Assessing the benefits and challenges of AI in cyber security, Cyber Security and Software Engineering Recruitment. Available at: https://www.cyberneticsearch.com/blog/assessing-the-benefits-and-challenges-of-ai-in-cyber-security/ (Accessed: 05 January 2024).

[14] Bharadiya, J. (2023). "Machine Learning in Cybersecurity: Techniques and Challenges", In European Journal of Technology, Vol. 7, Issue 2, pp. 1–14, DOI: 10.47672/ejt.1486

[15] SOCRadar (2022). "Real-life examples of machine learning in Cybersecurity", In SOCRadar Cyber Intelligence Inc. Available at: https://socradar.io/real-life-examples-of-machine-learning-in-cybersecurity/, Accessed on: 2024-01-05

DOI: 10.2507/35th.daaam.proceedings.039

OPTIMIZING ATHLETE'S COMFORT AND PERFORMANCE: INVESTIGATING LIQUID PERMEABILITY AND DRYING TIME IN SPORTS MATERIALS USING INFRARED THERMOGRAPHY

Antonija Petrov, Goran Cubric & Ivana Salopek Cubric

This Publication has to be referred as: Petrov, A[ntonija]; Cubric, G[oran] & Salopek Cubric, I[vana] (2024). Optimizing Athlete's Comfort and Performance: Investigating Liquid Permeability and Drying Time in Sports Materials using Infrared Thermography, Proceedings of the 35th DAAAM International Symposium, pp.0286-0291, B. Katalinic (Ed.), Published by DAAAM International, ISBN 978-3-902734-44-0, ISSN 1726-9679, Vienna, Austria
DOI: 10.2507/35th.daaam.proceedings.039

Abstract

Determining the fluid permeability of sportswear is crucial for optimizing the comfort and performance of athletes during physical activities. This scientific study aims to investigate the absorption and drying processes of sports materials to gain insights into the dynamics of moisture management in sportswear. Infrared thermography will be used to monitor the drying process over time, providing a non-invasive and efficient method to evaluate material properties. Precise measurements and analysis will help identify materials that absorb liquid efficiently and dry quickly to reduce athlete discomfort during intense training or competition. This research is of paramount importance as it enables the development of sportswear that maximizes comfort and allows athletes to perform at their best without being hindered by moisture in the material.

Keywords: permeability; sportswear; absorption; drying; infrared thermography.

1. Introduction

Regular physical activity is crucial for maintaining good health. In order to achieve a high level of comfort and functionality during various sports activities, athletes today prefer high-performance sportswear [1]. Sportswear can be classified by the level of physical activity into active sportswear and sportswear for leisure [2]. Active sportswear is worn by professional athletes during certain shorter periods of activity. These sports require a high level of physical activity and maximum physical performance, which also leads to heavy sweating. Active sportswear should have optimal thermal and moisture management properties to maintain the thermal balance between the metabolic heat generated by the physical activity and the external environment. Sweat should be wicked away quickly to ensure a dry microclimate next to the skin. This requirement can be met by designing sportswear with low resistance to heat transfer and evaporative heat loss. Sportswear should allow rapid fluid transfer through absorption and have good drying properties to prevent condensation of liquid sweat close to the skin [2].

When the body produces sweat, the feeling of wet clothes is one of the main causes of discomfort. This can be caused by excessive sweat accumulation on the body, as the ability to wick away sweat is limited. Low or no moisture transport causes the body to become wet, resulting in an uncomfortable feeling as the clothing sticks to the wearer's body [3]. Absorption is the transfer of liquid through textiles by capillary action [4]. The absorption properties depend on the type of fibre, the yarn construction, the fabric structure, the density and texture of the yarn, the finishing processes, the viscosity, the relative humidity and the ambient temperature [5], [6]. The capillary channels in the spaces between the fibres and the pore size have a decisive influence on absorption [7].

Sportswear is usually made from polyester, polyamide and combinations with elastane. The literature states that 2% elastane is sufficient to improve the elasticity of the material and maintain its shape. However, in high-performance clothing such as swimsuits and sportswear, the elastane content can increase to 30% [8]. The ability of a material to dry depends on the amount of perspiration of the wearer and environmental factors such as wind speed, relative humidity and temperature [9]. To provide a better environment for athletes, it is important that the materials used in active sportswear quickly wick sweat away from the skin once it has been absorbed [10]. The material dries as the sweat evaporates and the vapour then disperses into the air. The drying process involves two successive operations: wetting and drying [11]. Humidification is the initial process in which the liquid spreads through the fabric by displacing the air interface [11], [12]. Absorption is the capillary force of the fabric, which determines how quickly it can transfer the absorbed sweat to the outer surface. The evaporation process is also crucial, as it helps to release the sweat from the outer surface of the garment to the environment by mass diffusion. The temperature of the sweat in the fabric, the humidity of the air and the speed of the airflow all play a crucial role in how quickly the sweat molecules break down and evaporate [13]. Our previous research has highlighted the importance of the properties of polymeric knitted fabrics in relation to material aging [14] and the transfer of liquid and water vapour through the knitted materials [15]. Furthermore, a study focusing on improving the thermophysiological comfort of athletes through the thermographic evaluation of sportswear materials highlighted the importance of evaluating the absorption and drying processes to optimise the comfort and performance of athletes [16].

Based on the reviewed literature and the outlined requirements for high-performance sportswear, the heat and moisture management properties of different knitted fabrics are evaluated in the experimental part of this study. The focus is on the comparison of polymer materials with and without elastane component, and even more on the comparison of conventional and recycled materials. The reason for this is that the use of recycled materials in sport offers considerable potential for promoting sustainability, achieving economic benefits and fostering innovation. However, the number of studies on this segment has not yet been sufficiently researched, leaving a niche for further research. In the experiment presented in this paper, the absorption and drying rates of artificial sweat on different fabric samples are measured and the data analysed to determine their suitability for high-performance sports activities. The aim is to identify fabric characteristics that ensure optimal comfort by efficiently managing sweat dispersion and evaporation.

2. Materials and methods

2.1. Materials

The materials selected for this study are listed in Table 1. This table gives raw material composition and yarn fineness of the materials tested. The selected materials are different blends of conventional polyester (PES) and elastane (EL) as well as fabrics made from 100% recycled polyester. These differences in composition and structure are analysed to assess their impact on the absorption and drying properties that are essential for high-performance sportswear.

Material ID	Yarn properties	
	Composition	Yarn fineness, tex
P1	100% PES	12
P2	100% PES	10
P3	84% PES + 16% EL	10
P4	87% PES + 13% EL	10
P5	100% PES, recycled yarn	10
P6	100% PES, recycled yarn	10

Legend: PES - polyester, EL - elastane

Table 1. Material description

2.2. Methods

The following properties of the knitted fabrics were analysed: mass per unit area, thickness, stitch density as well as the absorption and drying process of knitted materials intended for sportswear.

The mass per unit area of the knitted fabrics was determined using a Kern ALJ 220-4 analytical balance with an accuracy of 0.00001 g in accordance with the ISO 3801 standard [17]. To measure the mass per unit area, the samples were cut to 100x100 mm. Each sample was weighed five times and the average value of these measurements was calculated. The thickness of the materials was measured using a DM-2000 thickness gauge by Wolf Messtechnik GmbH. During the measurements, a pressure of 1 kPa was applied to an area of 20 cm² of the sample. This test was carried out in accordance with the ISO 5084 standard [18]. A total of 10 measurements were taken at different points on each sample. The thickness measurement points were located in the centre of the sample and along the diagonal (starting from the left corner of the sample). The average thickness was expressed as the mean of these 10 measurements.

The stitch density represents the number of stitches per unit area. It is obtained by the following equation:

$$D_o = D_h \cdot D_v \tag{1}$$

where: D_o - stitch density (stitch/cm^2), D_h - horizontal density, D_v - vertical density.

The horizontal (Dh) and the vertical (Dv) density were determined using a Dino-Lite Edge AM7915MZT digital microscope. Image analysis was performed using the professional software DinoCapture 2.0. The absorption and drying process of the materials were tested as follows: knitted fabric samples with dimensions of 30x30 cm were placed in metal frames (29x29 cm) and elevated above the work surface. Using a pipette, 1 ml of artificial sweat (acidic sweat powder with pH 5.5, prepared according to BS EN ISO 105-E04 [19]) was applied to each sample (Fig. 1). The liquid that passed through the material was collected in plastic cups placed underneath the samples. The amount of liquid that passed through the material and accumulated in the cups was measured using an analytical balance. The difference between the initial amount of liquid (1 ml) and the measured amount in the cup allowed for the calculation of the amount of liquid retained on the surface of the material. In addition, the drying process of the materials was monitored using a TESTO 872 infrared thermographic camera (Fig. 2).

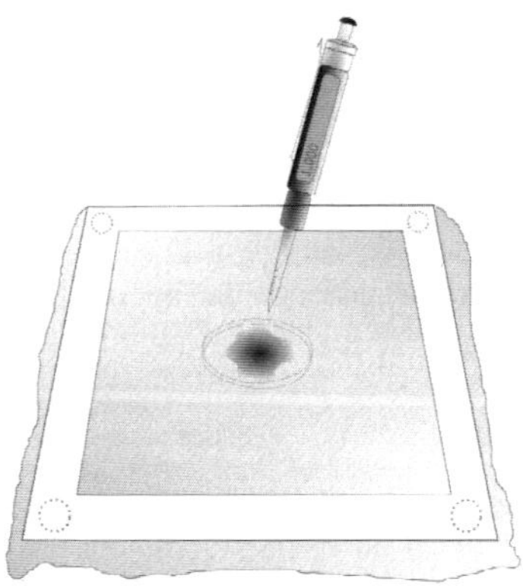

Fig. 1. Application of test liquid on knitted material using a pipette

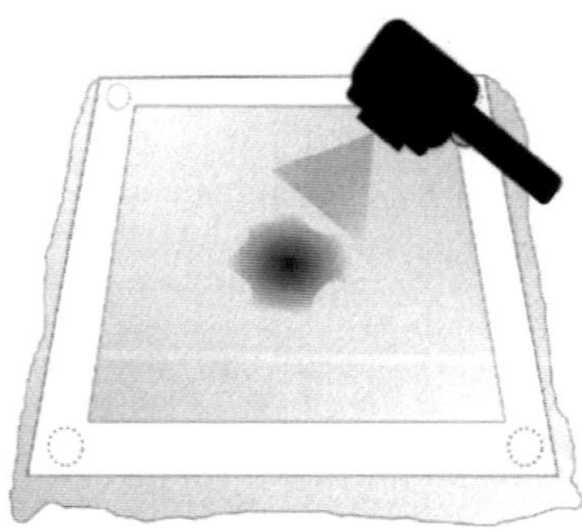

Fig. 2. Drying process monitoring with an infrared thermographic camera

3. Results and discussion

The results of knitted fabric properties testing, including the mass per unit area, thickness and stitch density are shown in Table 2.

Material ID	Fabric properties		
	Mass per unit area, g/m²	Material thickness, mm	Stitch density, stitches/cm²
P1	155	0.618	360
P2	151	0.491	272
P3	180	0.451	888
P4	141	0.351	1175
P5	130	0.451	289
P6	185	0.629	441

Table 2. Results of fabric properties testing

The experimental data show the relationships between the mass per unit area, thickness and stitch density of the knitted materials and their liquid absorption and transmission properties. The material P4 with the highest stitch density of 1175 stitches/cm² absorbs a considerable amount of liquid (0.50752 ml), while 0.49248 ml is transferred into the cup (Fig. 3). This suggests that a higher stitch density can balance both absorption and transfer, providing a denser structure that improves moisture retention and controlled transfer. In contrast, material P5, which is relatively thin (0.451 mm) and lightweight (130 g/m²) and has a low stitch density of 289 stitches/cm², had the highest permeability (0.82459 ml) and the lowest absorption (0.17541 ml). This suggests that a lower stitch density combined with a lighter and thinner fabric allows for faster drying and better liquid permeability, making it suitable for applications where fast moisture management is critical. Material P6, the thickest (0.629 mm) and with the greatest mass per unit area (185 g/m²) of the samples, with a moderate stitch density of 441 stitches/cm², retained a considerable amount of liquid (0.49086 ml) but also allowed a considerable amount of it to pass through (0.50914 ml). This balance suggests that thicker materials with a greater mass per unit area may retain a significant amount of moisture but are also able to allow some of the liquid to pass through, potentially contributing to longer wearer comfort by effectively wicking sweat away.

In summary, the results suggest that fabric structure, particularly stitch density, plays a crucial role in determining the moisture management properties of knitted sportswear. Fabrics with a high stitch density, such as P4, are better at wicking away moisture and keeping the skin dry, which increases comfort during low to moderate activity. On the other hand, fabrics such as P5 with a lower stitch density and lighter mass per unit area, are better suited to high-intensity activities where rapid moisture wicking and evaporation are important. The choice of fabric should therefore be tailored to the specific performance requirements of the sportswear and strike a balance between absorption and permeability to optimise the athlete's comfort and thermoregulation.

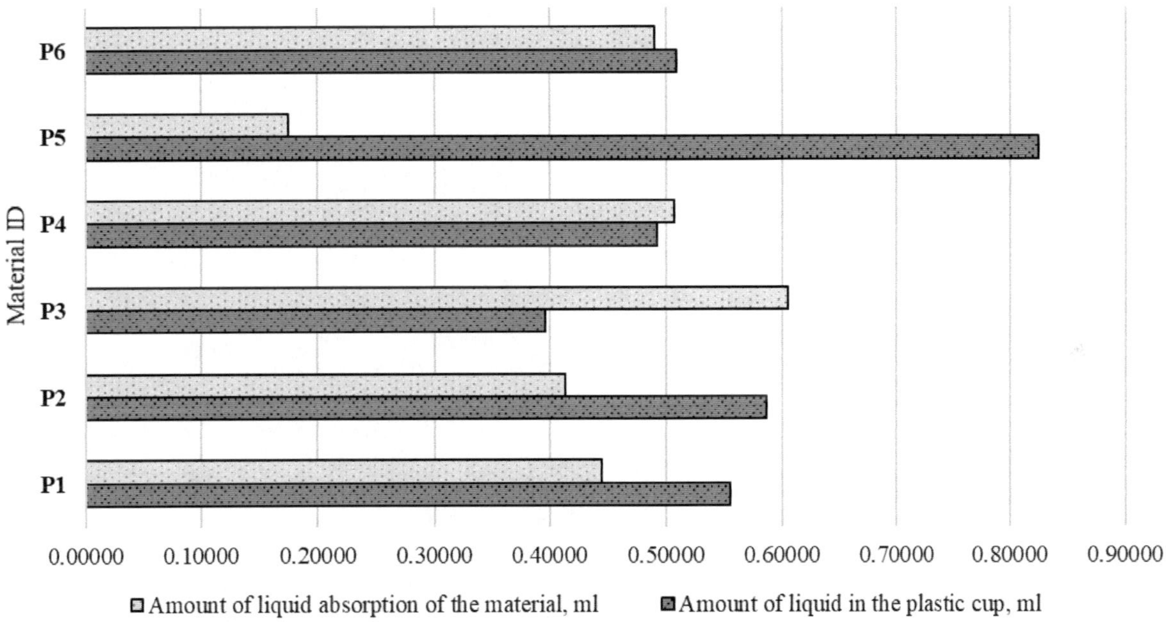

Fig. 3. Distribution of liquid absorption of the material and amount of liquid in the plastic cup

The analysis of the wetting and drying properties is shown in Figure 4. Material P2 is characterised by the fastest wetting time (5.16 seconds) and the highest wetting speed (12.53 mm/s along the y-axis), indicating excellent initial moisture absorption. This material also has a large surface (2608.2 mm²), which indicates that it can wick away moisture quickly. Despite its efficient absorption, P2 has a moderate drying time (1.01 hours), which is related to its relatively low mass (151 g/m²) and moderate thickness (0.491 mm). The stitch density of 272 stitches/cm² enables effective moisture management through a balance between absorption and evaporation. This makes P2 a strong candidate for high performance sportswear where fast moisture management is crucial. On the other hand, material P6 has the slowest wetting time (13.20 seconds) and the lowest wetting speed (2.79 mm/s along the y-axis), indicating less efficient initial moisture absorption. However, it also has the smallest surface (785.8 mm²), which could contribute to faster saturation and slower spread of moisture. P6 also has the longest drying time (1.72 hours), which is due to its higher mass (185 g/m²) and thickness (0.629 mm). These properties suggest that P6 is better suited to cooler climates or less intense activities where prolonged moisture absorption is less important. While material P5 has the highest amount of liquid in the plastic cup (0.82459 mL) and the lowest liquid absorption of the material (0.17541 mL), it also has a slow wetting time (9.87 seconds) and wetting speed (4.66 mm/s along the y-axis).

This material dries the fastest (0.75 hours) and benefits from its relatively low mass (130 g/m²) and moderate thickness (0.451 mm) with a stitch density of 289 stitches/cm². The performance of P5 underlines its suitability for high-intensity activities where quick drying and efficient moisture transport are crucial.

In summary, it can be said that the interaction of mass, thickness and stitch density decisively determines the wetting, absorption and drying behavior of knitted materials. Materials such as P2 (100% PES) with fast wetting and moderate drying time are ideal for active sportswear that requires fast moisture transport and evaporation. Conversely, materials such as P6 with slower wetting and longer drying time are more suitable for conditions where longer moisture absorption is acceptable. P5 (100% recycled PES), which dries quickly and has high permeability, offers excellent performance during intense activities. This underlines the need to select materials based on specific sports and environmental requirements.

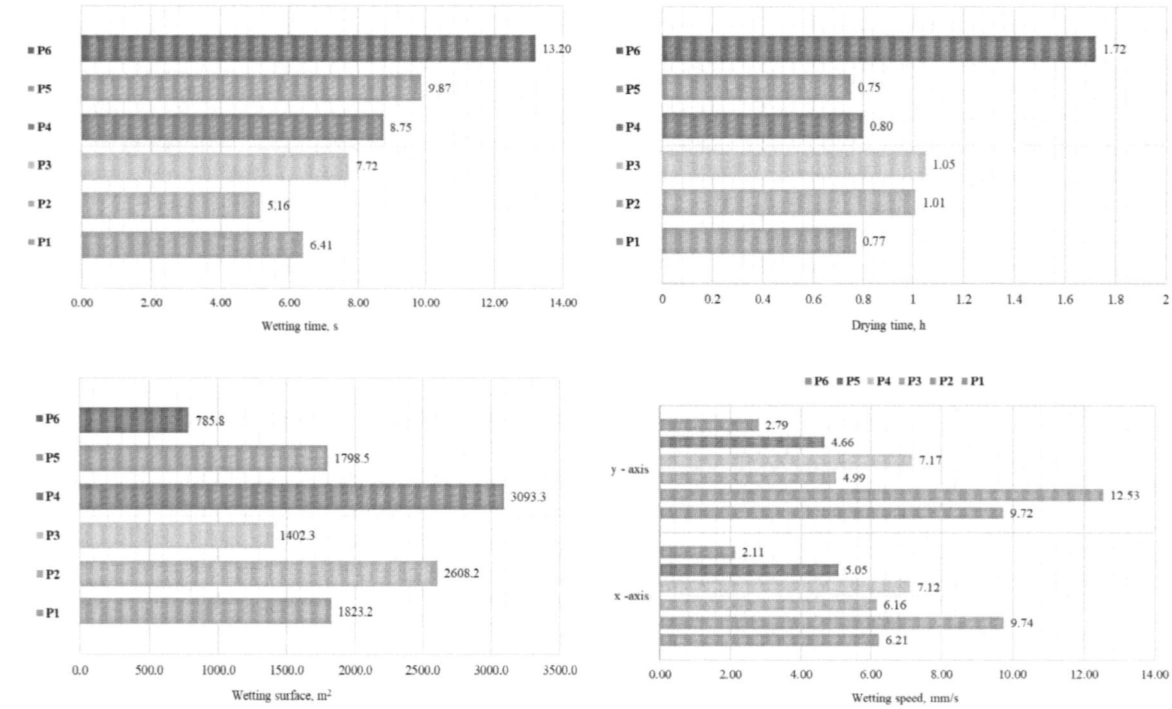

Fig. 4. Results of wetting and drying characteristics of knitted materials - wetting time, drying time, wetting surface, and wetting speed

4. Conclusion

This study addressed the problem of identifying high-performance sportswear materials that optimise the athlete's comfort and performance during physical activity, in particular by improving moisture management. The problem was solved by analysing the absorption and drying processes of different sports materials, using infrared thermography to monitor the drying process over time. This method allowed a precise and non-invasive evaluation of the material properties. The results showed that materials P2 (100% PES) and P5 (100% recycled PES) are the most suitable for high-performance sportswear, as they have an optimal balance between efficient liquid absorption and fast drying, directly reducing athletes' discomfort during intense training or competition.

This study also provided a positive insight into the behaviour of recycled materials, which contrasts with some previous research that suggested otherwise. This emphasises the significant potential for wider application of recycled materials in the sports industry, particularly given the continuing advances in recycling technologies and the growing environmental awareness of consumers and manufacturers. This study had several limitations. These include the number of individual measurements for each sample, which, although sufficient for a preliminary analysis, could limit the generalisability of the results, and possible variations in standard atmospheric conditions during the tests, which could have affected the reliability of the results. In addition, infrared thermography, while accurate, is susceptible to variations in ambient temperature, which could affect the accuracy of the drying rate measurements. For the future, further research should focus on advanced fabric technologies and sustainable materials to continuously improve the functionality of sportswear. This study provides an important basis for the further development of sportswear that effectively wicks away moisture to improve the comfort and performance of athletes.

5. Acknowledgments

This work has been fully supported by Croatian Science Foundation under the project IP-2020-02-5041 Textile Materials for Enhanced Comfort in Sports - TEMPO.

6. References

[1] Guru, K.; Kumar, A. & Kumar, R. (2021). Functional Textile for Active Wear Clothing, In: Textiles for Functional Applications, Kumar, B. (Ed.), IntechOpen.

[2] Jhanji, Y. (2021). Sportswear: Acumen of Raw Materials, Designing, Innovative and Sustainable Concepts, In: Textiles for Functional Applications, Kumar, B. (Ed.), IntechOpen.

[3] Faheem, A.; Khurram Shehzad, A.; Wardah, A.; Bushra, M.; Abher, R.; Sheraz, A.; Farooq, A. & Yasir, N. (2023). Recent Developments in Materials and Manufacturing Techniques Used for Sports Textiles. International Journal of Polymer Science, Vol 2023, 20.

[4] Collier, B. J.; Bide, M. & Tortora, P. G. (2009). Understanding textiles, 7th ed. Upper Saddle River, NJ: Pearson Education.

[5] Nyoni, A. B. & Brook, D. (2010). The effect of cyclic loading on the wicking performance of nylon 6.6 yarns and woven fabrics used for outdoor performance clothing. Textile Research Journal, Vol. 80, No. 8, 720–725.

[6] Hayes, S. G. & Venkatraman, P. (2016). Materials and Technology for Sportswear and Performance Apparel, Taylor & Francis Group, LLC.

[7] Zavec Pavlinic, D.; Balic, J.; & Mekjavic, I.B. (2011). Determination of the Thermoregulatory Specifications for Thermal Manikins. In: DAAAM International Scientific Book 2011, Katalinic, B. (Ed.), 031-044.

[8] Voyce, J.; Dafniotis, P. & Towlson, S. (2005). Elastic textile, In: Textiles in sport, Shishoo, R. (Ed.), Woodhead Publishing Limited and CRC Press LLC©, Cambridge England, p. 204-230.

[9] Chinta, S. & Gujar, P.D. (2013). Significance of moisture management for high performance textile fabrics. International Journal of Innovative Science Engineering and Technology, Vol. 2, 814–819.

[10] Nasrin, S.; Mandal, S.; Islam, M.M.; Petrova, A.; Agnew, R.J. & Boorady, L.M. (2023). Factors Affecting the Sweat-Drying Performance of Active Sportswear—A Review. Textiles, Vol. 3, 319-338.

[11] Kissa, E. (1996). Wetting and wicking. Textile Research Journal, Vol. 66, No. 10, 660–668.

[12] Manshahia, M. & Das, A. (2014). Moisture management of high active sportswear. Fibers Polymers, Vol. 15, No. 6, 1221-1229.

[13] Xu, J.; Xin, B.; Dua, X.; Wang, C.;, Chen, Z.; Zhenga, Y. & Zhou, M. Flexible, portable and heatable non-woven fabric with directional moisture transport functions and ultra-fast evaporation. Rsc Advances, Vol. 10, No. 46, 27512-27522.

[14] Petrov, A.; Salopek Čubrić, I. & Čubrić, G. (2024). Influence of aging on the physical properties of knitted polymeric materials. Polymers, Vol. 16, No. 4, 513.

[15] Petrov, A., Čubrić, G., & Salopek Čubrić, I. (2023). Transfer of liquid and water vapour through knitted materials. HEMIJSKA INDUSTRIJA (Chemical Industry), Vol. 77, No. 3, 211–221.

[16] Salopek Čubrić, I.; Rosić, D. & Petrov, Antonija. (2022). Improvement of thermophysiological comfort of athletes based on thermographic evaluation of sportswear material. Advanced technologies, Vol. 11, No. 2, 55-61.

[17] Textiles — Woven fabrics — Determination of mass per unit length and mass per unit area.

[18] ISO 5084:1996; Textiles—Determination of Thickness of Textiles and Textile Products. International Organization for Standardization: Geneva, Switzerland, 1996.

[19] ISO 105-E04:2013; Textiles. Tests for Colour Fastness. Part E04: Colour Fastness to Perspiration. International Organization for Standardization: Geneva, Switzerland, 2013.

DOI: 10.2507/35th.daaam.proceedings.040

INSIGHTS INTO INDUSTRY 4.0 COMPONENTS AND ACCEPTANCE OF NEW TECHNOLOGIES IN INDUSTRIAL ENTERPRISES

Natália Vraňaková & Zdenka Gyurák Babeľová

This Publication has to be referred as: Vranakova, N[atalia] & Gyurak Babelova, Z[denka] (2024). Insights into Industry 4.0 Components and Acceptance of New Technologies in Industrial Enterprises, Proceedings of the 35th DAAAM International Symposium, pp.0292-0298, B. Katalinic (Ed.), Published by DAAAM International, ISBN 978-3-902734-44-0, ISSN 1726-9679, Vienna, Austria
DOI: 10.2507/35th.daaam.proceedings.040

Abstract

Industry 4.0 is a revolution associated with digitization and automation of processes that are implemented in the company on a daily basis. All types of companies are affected by the mentioned revolution, but industrial enterprises are currently facing it to the highest extent. Among the most important components associated with Industry 4.0 are autonomous robots, big data, cloud solutions, simulations and virtual reality, internet of things, system integration, cybersecurity and cyberphysical systems. These components are developed and implemented gradually, while the very prioritization of individual components in industrial enterprises is influenced by the current situation in the enterprise, the focus of the enterprise, the need for digitization of selected processes, and the pandemic situation, which accelerated the implementation of selected components, was no less important. The main aim of the paper is to present the results of research focused on the components of Industry 4.0, their introduction in industrial enterprises and the presentation of results taking into account the perception of employees regarding the introduction of new technologies in industrial enterprises. The research was carried out through a questionnaire, which contains answers from 337 representatives of industrial enterprises in Slovakia.

Keywords: automation; components of Industry 4.0; digitization; Industry 4.0; industrial enterprises.

1. Introduction and theoretical framework

The fourth industrial revolution (Industry 4.0) is a forced and inevitable transformation process that is gradually reaching industries and global supply chains. One of its missions is the efficiency of operation management without disruptive effects on manufacturing companies and supply chains [1]. Choosing the right strategy is an important part of the plan that will help ensure digital transformation. However, choosing the right and appropriate strategy is a complex and laborious process, as it is necessary to evaluate all the factors and ambiguities that arise from Industry 4.0 [2]. The digital transformation itself helps to increase productivity, reduce costs, increase efficiency and competitive advantage. Due to the mentioned huge advantages, digitalization is adopted by companies as a strategic tool to gain a competitive advantage. Among other things, the outbreak of the pandemic accelerated digitization itself, as restrictions forced organizations to continue to operate without problems [3].

In addition to the potential benefits, the transition from people to technology can also present risks for the company and for employees. Simple tasks can be replaced by machines, which can mean the loss of jobs for certain groups of employees and for personnel without qualifications [4]. Another concern of employees is that Industry 4.0 technologies may increase the degree of supervision over their work [5], eventually, it will be necessary for them to increase their qualifications, and thus there is an assumption that says that Industry 4.0 requires employees with specific qualifications [6].

Industry 4.0 is not only an industrial revolution, but also a challenge in terms of the integration of people, data, services and objects. It requires the interaction of many components, with key aspects including devices, connectivity, properly managed services and appropriate data [7]. Technologies such as cyber-physical systems, the Internet of Things, big data [8], and cloud solutions are considered to be the next generation of the industrial revolution [9]. Autonomous robots are also important [10], virtual reality and simulations [11]. With the right combination and involvement of the right components of Industry 4.0, depending on the type of company and current conditions, it is possible to ensure increased production, production per capita and productivity [8].

2. Materials and methods

The main aim of the paper is to present the results of research focused on the components of Industry 4.0, their implementation in industrial enterprises and the presentation of results taking into account the perception of employees regarding the introduction of new technologies in industrial enterprises. In order to fulfil the main aim of the paper, three research questions were determined. Based on the above-mentioned framework, we decided to focus on the employees' point of view, while it is necessary to differentiate their point of view on the perception of the introduction of new technologies and, on the other hand, their point of view on the importance of Industry 4.0 itself. Therefore, the following research questions were set:

Research question 1: How do employees of industrial enterprises perceive the introduction of new technologies?
Research question 2: How do employees of industrial enterprises perceive the importance of Industry 4.0?
Research question 3: What components of Industry 4.0 are introduced in industrial enterprises?

For the purposes of data collection, a questionnaire was constructed and distributed online. A total of 337 responses were received from representatives of industrial enterprises in Slovakia. For the purpose of a clearer presentation of the results, the companies were divided according to the industry sector in which they operate into the following categories: automotive sector; engineering, metalworking and metallurgy and other production; electrical engineering and energy; chemical industry and plastics; construction industry; food industry, agriculture and forestry; projecting and engineering; transport and logistics; information technology, communication. A basic overview of the research sample can be seen in Table 1.

Sector of industry	**Number of employees in company**			**Respondent's gender**		**Sum**
	1 to 49	**50 to 249**	**249 and more**	**male**	**female**	
Automotive sector	1	22	111	76	58	134
Engineering, metalworking and metallurgy and other production	16	11	61	67	21	88
Electrical engineering and energy	1	9	23	28	5	33
Chemical industry and plastics	2	10	16	17	11	28
Construction industry	8	2	6	6	10	16
Food industry, agriculture and forestry	2	7	6	7	8	15
Projecting and engineering	4	1	5	5	5	10
Transport and logistics	2	2	4	7	1	8
Information technology, telecommunications	4	0	1	5	0	5
Sum	40	64	233	218	119	337

Table 1. Overview of research sample (own elaboration, 2024)

In Table 1, it can be seen that the automotive sector has the largest representation with the number of employees 249 and more. As for the gender of the respondents, almost 65% (218) are men. The collected data were evaluated using descriptive statistics, in the form of tables and figures with relative frequencies.

3. Results

In the third part of the article, the research questions established in the first chapter of the article will be evaluated; specific results will be presented.

3.1. Perception of the introduction of new technologies by employees of industrial enterprises

The aim of the first research question is to evaluate how employees perceive the introduction of new technologies in the company. The respondents had the opportunity to mark one answer, while they had the following options to choose from: as a necessity, as work facilitation, I do not observe it in our company, as a burden. The evaluation in relative frequencies can be seen in the following Table 2.

Sector of industry	**as a necessity [%]**	**as a burden [%]**	**as work facilitation [%]**	**I do not observe this in our company [%]**	**Sum [%]**
Automotive sector	38.06	5.22	51.49	5.22	100
Engineering, metalworking and metallurgy and other production	39.77	5.68	42.05	12.50	100
Electrical engineering and energy	60.61	9.09	24.24	6.06	100
Chemical industry and plastics	46.43	3.57	46.43	3.57	100
Construction industry	18.75	6.25	68.75	6.25	100
Food industry, agriculture and forestry	40.00	6.67	46.67	6.67	100
Projecting and engineering	40.00	0.00	40.00	20.00	100
Transport and logistics	25.00	12.50	37.50	25.00	100
Information technology, telecommunications	0.00	0.00	100.00	0.00	100

Table 2. Perception of the introduction of new technologies by employees of industrial enterprises based on the sector (own elaboration, 2024)

Based on the data in Table 2, it can be seen that within the automotive sector, more than half of the respondents consider new technologies to facilitate work and almost 70% in the construction industry sector. Technologies are perceived as a necessity to the highest extent in the electrical engineering and energy sector. A total of 12.5% of respondents from the transport and logistics sector consider technology as a burden. Overall, it is possible to positively assess the fact that respondents in almost all sectors perceive new technologies positively (as making work easier). The following Figure 1 contains an overall assessment of the first research question.

Research question 1: How do employees of industrial enterprises perceive the introduction of new technologies?

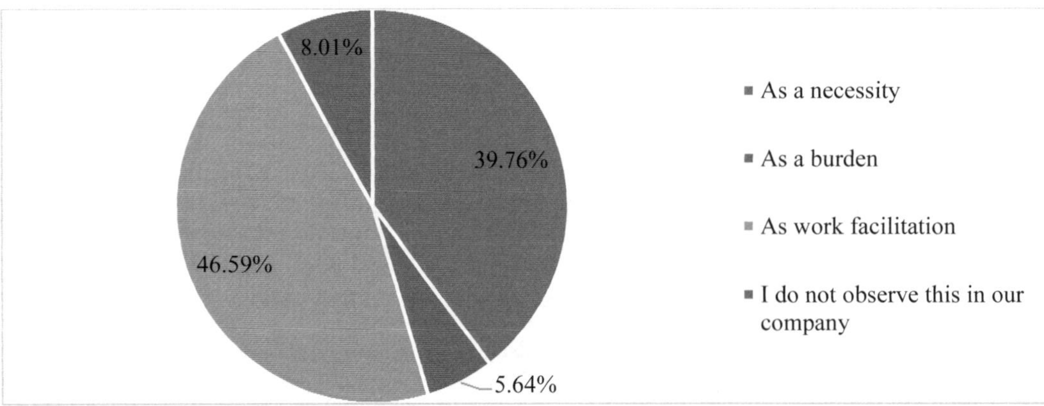

Fig. 1. Evaluation how do employees of industrial enterprises perceive the introduction of new technologies (own elaboration, 2024)

According to the data from Figure 1, it can be argued that, regardless of the sector, almost half of the respondents consider technology to facilitate work, followed by respondents who consider technology to be a necessity. On the other hand, more than 5% of respondents consider technology as a burden and 8% of respondents do not observe the introduction of new technologies in the company.

3.2. Perception of the importance of Industry 4.0 by employees in industrial enterprises

The second research question deals with employees' opinion on Industry 4.0. Based on the theoretical starting points, the advantages and risks resulting from the industrial revolution are obvious, but this research question aims to identify the opinion of the employees. Similar to the previous case, employees had the opportunity to choose one of four answers, while the actual evaluation in relative frequencies by sector can be seen in the following Table 3.

Sector of industry	**very important for the future of the company [%]**	**increasing importance due to the pandemic [%]**	**I do not know [%]**	**currently low priority [%]**	**Sum [%]**
Automotive sector	55.97	7.46	20.15	16.42	100
Engineering, metalworking and metallurgy and other production	53.41	9.09	20.45	17.05	100
Electrical engineering and energy	39.39	3.03	30.30	27.27	100
Chemical industry and plastics	46.43	3.57	35.71	14.29	100
Construction industry	43.75	6.25	43.75	6.25	100
Food industry, agriculture and forestry	66.67	0.00	20.00	13.33	100
Projecting and engineering	40.00	10.00	50.00	0.00	100
Transport and logistics	75.00	0.00	12.50	12.50	100
Information technology, telecommunications	20.00	20.00	40.00	20.00	100

Table 3. Perceptions of the importance of Industry 4.0 by employees of industrial enterprises based on the sector (own elaboration, 2024)

The data in Table 3 show that, to the greatest extent, employees consider Industry 4.0 necessary for the successful future of the company. Within the electrical engineering and energy sector, almost 28% of employees indicated that Industry 4.0 currently has a low priority. Respondents also reflected on the recent pandemic situation, for example in the information technology and telecommunications sector, a total of 20% of respondents indicated increasing importance since the pandemic itself. Similar to the previous case, the aggregated results for all sectors can be seen in Figure 2.

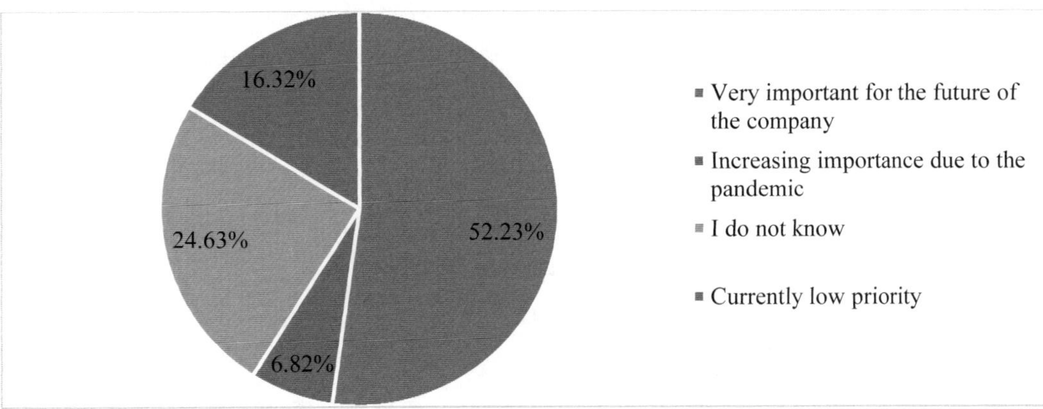

Fig. 2. Evaluation how do employees of industrial enterprises perceive the importance of Industry 4.0 (own elaboration, 2024)

Figure 2 shows that more than half of the respondents perceive Industry 4.0 as important for the future of the company, which can be considered a positive finding and it can be concluded that the employees themselves are aware that if the company is to prosper, the path of digitization and automation is necessary. Almost a quarter of respondents could not comment on the given area, and more than 16% of respondents think that Industry 4.0 is currently of low importance in the company.

3.3. Components of Industry 4.0 introduced in industrial enterprises in Slovakia

Based on the established third research question, it is possible to evaluate the current implementation of the components of Industry 4.0 in industrial enterprises. In the theoretical framework, specific components of the industrial revolution were defined, while the aim is to identify which of them are used to the highest or lowest extent. Respondents had the opportunity to select multiple answers and their relative frequencies can be seen in the Table 4.

Sector of industry	autonomous robots [%]	big data [%]	cloud based solutions [%]	simulation and virtual reality [%]	internet of things (IoT) [%]	system integration [%]	cybersecurity [%]	cyber-physical systems [%]	I do not know [%]
Automotive sector	21.02	9.94	14.20	7.39	4.26	9.66	11.93	13.35	8.24
Engineering, metalworking and metallurgy and other production	13.00	7.50	16.00	10.50	7.00	8.50	12.50	7.00	18.00
Electrical engineering and energy	15.66	18.07	20.48	4.82	3.61	9.64	13.25	9.64	4.82
Chemical industry and plastics	7.79	16.88	19.48	6.49	3.90	11.69	18.18	6.49	9.09
Construction industry	3.23	6.45	25.81	3.23	6.45	12.90	16.13	6.45	19.35
Food industry, agriculture and forestry	5.88	5.88	20.59	2.94	5.88	8.82	20.59	11.76	17.65
Projecting and engineering	4.76	9.52	19.05	4.76	19.05	9.52	9.52	4.76	19.05
Transport and logistics	25.00	12.50	12.50	6.25	6.25	18.75	6.25	0.00	12.50
Information technology, telecommunications	0.00	20.00	40.00	0.00	0.00	0.00	0.00	0.00	40.00

Table 4. Components of Industry 4.0 introduced in industrial enterprises in Slovakia based on the sector (own elaboration, 2024)

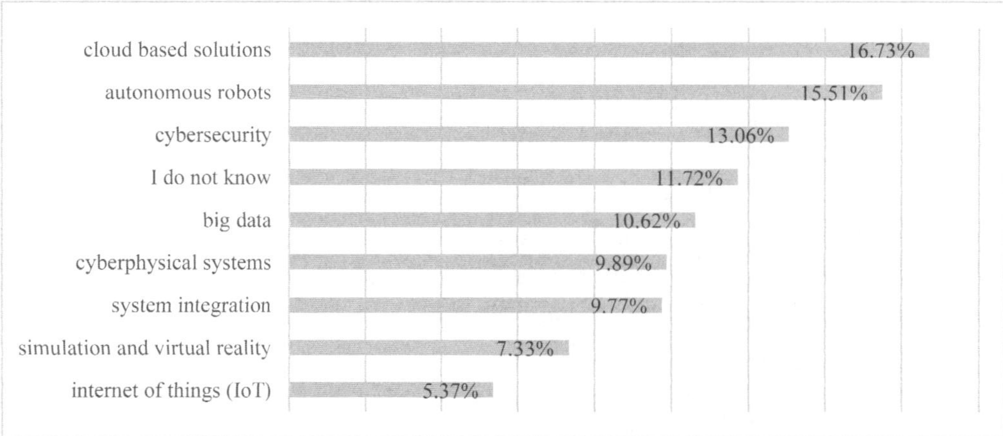

Fig. 3. Evaluation how do employees of industrial enterprises perceive the implementation of components of Industry 4.0 in industrial enterprises in Slovakia (own elaboration, 2024)

The relative frequencies in Table 4 show that autonomous robots are the most used element within the automotive and transport and logistics sectors. Within other sectors of the industry, cloud solutions are the most used element. Cloud solutions are generally considered to be an efficient and less expensive element. Internet of things, simulations and virtual reality are among the least used elements. For the construction industry, autonomous robots are the least used element. Aggregate results for all industry sectors are shown in Figure 3.

In summary, according to Figure 3, it can be concluded that Cloud solutions, Autonomous robots and Cybersecurity are the most used elements of Industry 4.0 by industrial enterprises. Among the least used are Internet of things, Simulations and virtual reality, Cyberphysical systems and System integration. Almost 12% of respondents could not answer the question.

4. Discussion

Digitization in the Industry 4.0 environment has become essential for improving the production process and related services. The increasing demand for sustainability due to external pressures such as strict government regulations and due to internal factors such as the need to reduce costs or the limited availability of resources is forcing manufacturing industries to adopt new approaches and implement new initiatives [12].

The challenge is the universal transformation and support of Industry 4.0 culture in small, medium and large enterprises. Although significant progress has been made in various industries such as the automotive sector, there will be a need to increase the pressure to raise awareness of the long-term benefits through consultancy, training, audit assessments and possibly competence development programmes. Industry-relevant transformation and learning need special attention to build long-term skills and capabilities [13]. Rapid technology modernization, adoption of new technologies was common during the pandemic, when remote work created operational uncertainty, often caused by a lack of employee skills [14]. For this reason, further research will also be directed to the education of employees in the conditions of Industry 4.0. However, since education is an important personnel process, further research will also focus on other personnel processes (such as employee evaluation, employee motivation, recruitment and selection of employees, etc.) and the need for their adaptation to the new conditions resulting from ongoing digitization and automation.

As part of Industry 4.0 technologies, an increase in productivity, better and more customized products and a better working environment are expected. However, one of the problems is employment, as it is assumed that the rate of automation will increase exponentially until it reaches its peak. One reason is that robots are becoming cheaper and more flexible. Overall, the impact of Industry 4.0 on employees is still a relatively unexplored topic and there are countries and industries where such research has not been conducted. The available publications do not provide a clear assessment of the impact of Industry 4.0 on employees and therefore further in-depth research is needed [15]. Our research has shown that employees have a relatively positive perception of the benefits of Industry 4.0 technologies, but their opinion and a deeper view of the barriers associated with digitalization need to be further explored.

Industry 4.0 contains various components that are gradually being developed and implemented. Individual sectors of the industry are characterized by the fact that they use certain elements to a greater extent [16]. Therefore, the right combination of Industry 4.0 elements can be crucial for industrial enterprises in order to increase productivity and profitability.

Among the limitations of the research, it is possible to include the fact that the research was carried out exclusively within enterprises in Slovakia. Employees of different job categories took part in the research, and therefore it would be useful to compare the results also within different job positions and thus to identify differences in the perception of importance between employees at different levels of management. On the basis of theoretical premises, Industry 4.0 affected all sectors and industries, and therefore it would be appropriate to expand the sample of respondents to include companies providing services, or public or state institutions.

5. Acknowledgments

This work was supported by the call for doctoral students and young researchers of Slovak University of Technology in Bratislava to start a research career (Grant 23-06-05-B).

6. References

[1] Choi, T.-M.; Kumar, S.; Yue, X. & Chan, H.-L. (2022). Disruptive Technologies and Operations Management in the Industry 4.0 Era and Beyond. Production and Operations Management 31, 9–31

[2] Görçün, Ö. F.; Mishra, A. R.; Aytekin, A.; Simic, V. & Korucuk, S. (2024). Evaluation of Industry 4.0 strategies for digital transformation in the automotive manufacturing industry using an integrated fuzzy decision-making model. Journal of Manufacturing Systems 74, 922–948

[3] Nadeem, K.; Wong, S. I.; Za, S. & Venditti, M. (2024). Digital transformation and industry 4.0 employees: Empirical evidence from top digital nations. Technology in Society 76, 102434

[4] Schroeder, S.; Lehberger, M. & Sparke, K. (2024). Threat or opportunity? - Managers' and employees' perception of automation and digitalization in the horticultural sector. Procedia Computer Science 232, 564–573

[5] Moeuf, A. et al. (2020).Identification of critical success factors, risks and opportunities of Industry 4.0 in SMEs. International Journal of Production Research 58, 1384–1400
[6] Cugno, M.; Castagnoli, R. & Büchi, G. (2021). Openness to Industry 4.0 and performance: The impact of barriers and incentives. Technological Forecasting and Social Change 168, 120756
[7] Zhang, C.; Chen, Y.; Chen, H. & Chong, D. (2021). Industry 4.0 and its Implementation: a Review. Inf Syst Front
[8] Calış Duman, M. & Akdemir, B. (2021).A study to determine the effects of industry 4.0 technology components on organizational performance. Technological Forecasting and Social Change 167, 120615
[9] Aryskin, A.; Grigoriev, A.; Petrakov, M.; Pryanichnikov, V.; Travushkina, A. & Travushkin, A. (2021). Industry 4 0 - Automation in the Conditions of Reconfiguration of Production Lines, Proceedings of the 32nd DAAAM International Symposium, pp.0660-0665, B. Katalinic (Ed.), Published by DAAAM International, ISBN 978-3-902734-33-4, ISSN 1726-9679, Vienna, Austria
[10] Sidiropoulos, A.; Sidiropoulos, V.; Bechtsis, D. & Vlachos, D. (2021). An Industry 4.0 Tool to Enhance Human-Robot Collaboration, Proceedings of the 32nd DAAAM International Symposium, pp.0617-0622, B. Katalinic (Ed.), Published by DAAAM International, ISBN 9783-902734-33-4, ISSN 1726-9679, Vienna, Austria
[11] Kolla, S.; Minufekr, M. & Plapper, P. (2019). Deriving essential components of lean and industry 4.0 assessment model for manufacturing SMEs. Procedia CIRP 81, 753–758
[12] Iyer, S. V., Sangwan, K. S., & Dhiraj. (2024). Development of an Industrial Symbiosis Framework through Digitalization in the Context of Industry 4.0. Procedia CIRP 122, 515–520
[13] Javaid, M., Haleem, A., Singh, R. P. & Sinha, A. K. (2024). Digital economy to improve the culture of industry 4.0: A study on features, implementation and challenges. Green Technologies and Sustainability 2, 100083
[14] Nakandala, D. Impact of Industry 4.0 Technologies on Operations Resilience: Adverse Effects of COVID-19 as a Moderator. Procedia Computer Science 232, 3258–3267 (2024).
[15] Ejsmont, K. (2021). The Impact of Industry 4.0 on Employees—Insights from Australia. Sustainability 13, 3095
[16] Oliveira, J., Nunes, M. & Afonso, P. (2019). New Product Development in the Context of Industry 4.0: Insights from the Automotive Components Industry. in Industrial Engineering and Operations Management II (eds. Reis, J., Pinelas, S. & Melão, N.) 83–94 (Springer International Publishing, Cham, 2019). doi:10.1007/978-3-030-14973-4_8.

DOI: 10.2507/35th.daaam.proceedings.041

SERVERLESS ARCHITECTURE AND SECURITY

Marko Harambasa, Karlo Josic & Matej Basic*

This Publication has to be referred as: Harambasa, M[arko]; Josic, K[arlo] & Basic, M[atej] (2024). Serverless Architecture and Security, Proceedings of the 35th DAAAM International Symposium, pp.0299-0305, B. Katalinic (Ed.), Published by DAAAM International, ISBN 978-3-902734-44-0, ISSN 1726-9679, Vienna, Austria
DOI: 10.2507/35th.daaam.proceedings.041

Abstract

This paper explores the serverless architecture, detailing its key components, benefits, and security concerns. It contrasts two main types: Function as a Service (FaaS) and Backend as a Service (BaaS), highlighting how they allow for building applications without the need for managing underlying servers. The paper continues into serverless architecture's operational and financial benefits, such as scalability, reduced maintenance, and cost efficiency. Security challenges unique to serverless computing are examined, stressing the importance of adopting robust security measures like comprehensive monitoring and logging to mitigate risks. The paper discusses the inherent security benefits of serverless computing, including minimised attack surfaces and automated updates. It outlines best practices for ensuring secure serverless environments, such as secure coding, stringent access controls, data encryption, and continuous monitoring.

Keywords: Serverless; Function as a Service (FaaS); Backend as a Service (BaaS), network security.

1. Introduction

In the last few years, serverless architecture has become the revolutionary approach in cloud computing, providing a way to build and deploy applications and services without the need to know and configure the underlying infrastructure. The heart of the serverless architecture is deploying the code in response to events, which means the application is running only when needed, eliminating the need for traditional server management. Two key concepts are Function as a Service (FaaS) and Backend as a Service (BaaS). FaaS focuses on executing code in response to some action or events, while BaaS provides automated backend services [1].

These two approaches speed up the development life cycle, providing quicker feature release and application updates. The serverless architecture can be used with today's three most extensive serverless technologies: AWS Lambda, Azure Functions, and Google Cloud Functions. Those three serverless technologies are also called functions since the serverless architecture combines multiple small functions. All other resources on the internet that we are using have some pros and cons, some of which are serverless. This paper will cover all the excellent and tricky parts of serverless architecture, such as scalability and cost-efficiency, and critically examine the security landscape unique to serverless environments [2]. Serverless architecture inherently enhances development agility by abstracting the complexity of the underlying infrastructure, allowing developers to focus more on coding than operational concerns. However, it also introduces challenges in monitoring and debugging because traditional tools are often not equipped to handle the ephemeral nature of serverless components.

The paper aims to provide a comprehensive examination of serverless architecture, focusing on its advantages and challenges. In the paper, we will present the benefits of serverless architecture, including its scalability, cost-efficiency, and ability to speed up the development life cycle and compare the most extensive serverless technologies: AWS Lambda, Azure Functions, and Google Cloud Functions.

2. Core Components of Serverless Architecture

Serverless architecture is a cloud computing model where the cloud service provider dynamically manages the allocation of machine resources. In a serverless architecture, developers focus on writing code without worrying about the underlying infrastructure, such as servers or virtual machines [3]. Figure 1 shows the Serverless Architecture diagram from a high-level design view.

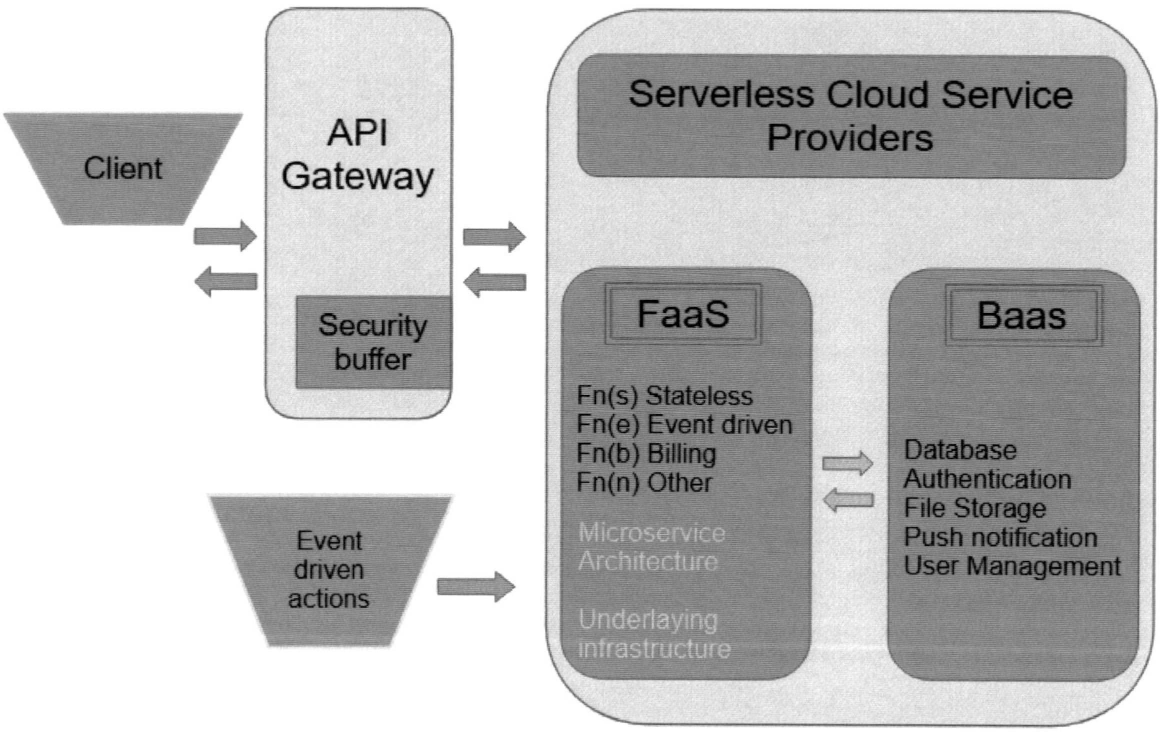

Fig. 1. Diagram of Serverless Architecture

2.1. Function as a Service (FaaS)

"FaaS" is the heart of the serverless architecture, allowing developers to write and deploy code executed in response to events without the complexities of managing the underlying infrastructure. Focusing on the individual functions and excludes servers entirely. There are a few steps to how "FaaS" works:

- Event-Driven Execute: functions are invoked by specific events, for example, HTTP requests, file uploads, or database operations
- Stateless Functions: each function call is treated as an independent event, with no shared state between executions
- Scalability and Auto-Scaling: as the demand or workload increases, more instances of the function are automatically created to handle the load
- Micro billing: billing is based on the actual usage, typically calculated as a combination of the execution time and the number of times the function is triggered.

2.2. Backend as a Service (Baas)

Backend as a Service (BaaS) is a cloud computing model that provides backend services and functionalities to developers without requiring them to manage the underlying infrastructure. In the context of serverless architecture, BaaS offers pre-built backend services, such as databases, authentication, file storage, push notifications, and user management, which developers can easily integrate into their applications.

The roles the "BaaS" involved in the serverless architecture:

- Managed Backend Services: The cloud provider manages several backend services BaaS offers, including database administration, user authentication, cloud storage, and server-side logic
- Emphasis on Frontend Development: Since the service provider handles the backend of apps when using BaaS, developers may concentrate more on creating the front end of their projects
- Integration with FaaS: BaaS and FaaS are frequently combined to provide a complete serverless architecture, with BaaS managing standard backend operations and FaaS handling business logic in response to events.

2.3. Key Players and Technologies

Several major companies dominate the serverless market and provide reliable FaaS and BaaS solutions:

- One of the first offerings is AWS Lambda (Amazon Web Services), which enables running code without the need to provision or manage servers. Its seamless integration with other AWS services enables a serverless architecture
- Microsoft Azure's Azure Functions: Provides event-driven serverless computing that seamlessly integrates into the more extensive Azure network of services
- Google Cloud Functions: A fully managed environment within the Google Cloud Platform runs code in response to events. It is a scalable, pay-as-you-go function as a service solution [4].

3. Benefits of Serverless Architecture

Serverless architecture presents many advantages, such as minimised operational burdens, enhanced cost-effectiveness, scalability, and expedited time-to-market. It is a compelling option for numerous organisations developing contemporary and scalable applications.

3.1. Cost Efficiency in Serverless Architecture

Cost efficiency is the most valuable thing when implementing serverless architecture. That stood out from the unique billing model and reduced operational costs. It allows us to optimise the price of the architecture at the minimum level. Nowadays, there are different types of cost models for serverless architecture:

- Pay-Per-Use Billing Model: Unlike traditional architectures, where resources are paid for regardless of use, serverless computing follows a pay-per-use model. Costs are incurred based on the number of executions and the duration of the code execution, down to the millisecond. This granular billing approach means no charge if a function is not running.
- Reduced Infrastructure Costs: Since the cloud provider manages the infrastructure, organisations save on the costs associated with provisioning, maintaining, and scaling physical servers or virtual machines
- Decrease in Over-Provisioning: Serverless computing eliminates the need for guesswork in capacity planning [5].

3.2. Scalability and Performance

After cost considerations, scalability and performance are the primary focal points in an application's development. The application must function consistently, whether accessed by a single user or thousands. In serverless architecture, there is a couple of stuff that is different from traditional architecture, such as:

- Automatic Scalability: Serverless functions automatically scale depending on the workload. This means that during periods of high demand, the architecture can scale up seamlessly to handle the increase in requests and scale down during low-demand periods, ensuring efficient resource utilisation.
- Performance Optimization: The underlying serverless computing infrastructure is optimised for performance. Cloud providers continually upgrade and maintain their systems, ensuring serverless functions run on high-performance computing resources.
- Instantaneous Scaling: The ability to instantly scale up or down in response to incoming requests ensures that serverless applications can maintain performance levels regardless of the number of requests they receive.

3.3. Maintenance and Operational Efficiency

Maintenance and Operations are necessary for ongoing server upkeep and slashing expenses linked with hosting applications. This technology empowers businesses to trim server maintenance costs, enhancing efficiency and scalability. While serverless computing eliminates hardware expenses tied to conventional web servers, there might still be some operational overhead related to overseeing serverless applications, such as configuring logging and monitoring metrics from various origins. Over recent years, serverless computing has become a preferred solution for bespoke software developers. It reduces maintenance and improves operation efficiency in server ways, such as:

- Reduced Maintenance Overhead: With the server management and operational responsibilities shifted to the cloud provider, developers and IT teams are freed from routine maintenance tasks such as patching, updating, and managing server health
- Focus on Core Product Development: Thanks to this infrastructure management offloading, organisations may refocus their attention and resources on core product development and innovation instead of being mired in operational issues.

4. Security in Serverless Architecture

Since serverless users work with cloud providers, they shift their security responsibility to them. In deploying serverless applications, we keep control over most of the stack to our cloud provider, and they provide services such as key management. No longer own OS hardening, admin rights, SSH, and segmentation. Cloud providers are reliable in keeping their parts of the stack patched and secured, so giving them a more significant portion of the stack certainly improves things. Additionally, the temporary, stateless nature of serverless computing makes the attacker's liver harder. Serverless functions like AWS Lambda run for a few seconds and then die, and containers get recycled. The faster serverless function comes and goes, the less memory is required, and having no memory reduces the risk of long-term attacks. The fact that serverless applications are structured as a more significant number of small functions in the cloud provides a fantastic opportunity for security. Application security tools often go to incredible lengths to analyse and instrument our packaged application to observe or filter the internal flow of applications. Moving to smaller microservices enables more fine-grained IAMs. The opportunity to apply security policies to each of those small things can significantly reduce the attack surface. In serverless architecture, security assumes a different dimension than traditional server-based setups. While the cloud provider manages the infrastructure's security, application-level security, particularly the code and data, essentially remains the developer's responsibility.

Security in serverless architecture encompasses various crucial topics to protect data, applications, and resources. Some critical issues include:

- **Shift in Security Responsibility**: The serverless model shifts specific security responsibilities to the cloud service provider. However, this means only some security concerns are outsourced. Issues like application vulnerabilities, data encryption, and access controls still need vigilant management.
- **Unique Security Context:** Serverless functions are stateless and ephemeral, meaning they do not traditionally persist. Their transient nature requires a different approach to monitoring and securing them than persistent server environments [6].

4.1. Common Security Challenges

In serverless architecture, several common security challenges must be addressed to ensure the robust protection of applications and data. Here are some of these challenges:

- **Third-party Dependencies:** Third-party libraries and APIs are frequently used by serverless applications. Each external component introduces potential vulnerabilities. It is essential to make sure these dependencies are current and safe
- **Insecure Serverless Deployment Configurations:** Security vulnerabilities in serverless apps might arise from deployment errors. These include failing to establish appropriate authentication and permission procedures, having too liberal access rights, and not keeping enough logs.
- **Vulnerabilities in Application Code:** Since serverless computing strongly emphasises function execution, security breaches may result from any weakness in the code, including injection errors or incorrect exception handling
- **Limited Monitoring and Visibility:** Standard monitoring solutions might also not work in a serverless system. The monitoring services the cloud provider provides where the serverless is hosted can enable the logging application.
- **Risks of Denial of Service (DoS):** Even if serverless architecture can handle heavy loads, DoS attacks can still occur.

These attacks could cause many function executions, resulting in significant costs and possible service degradation.

4.2. Serverless Security Threats

While attackers may still leverage similar motivations, they will employ different tactics when targeting serverless applications, as these applications are architecturally different from those we're used to. Below are the specific security risks inherent to this novel application architecture.

4.2.1. The Threat of Over-Privileged Functions

With serverless applications, applying permissions to specific functions and ensuring those privileges are limited to the bare minimum is possible. It will be able to lessen the impact of any strike and drastically reduce the attack surface by doing this. Regretfully, a recent Check Point study revealed [7] that few developers utilise this possibility.

According to analysis, 16% of serverless application tasks are deemed "serious," and 98% of functions are in danger. Additionally, to increase the security of the function and the application, most of these functions could have their permissions granted less than necessary. Check Point rates the risk associated with each function it analyses. That is based on the identified posture flaws, considering the type of weakness and the environment in which it manifests. Tens of thousands of functions in live applications were scanned, and the results showed that most serverless apps are just not implemented as securely as they should be to reduce risks, something that OWASP (The Open Worldwide Application Security Project) controls mention more than a few times in their top 10 list. Check Point found that unnecessary permissions are the main security posture problems; the remaining issues are insecure code and setups.

4.2.2. The Groundhog Day Attack

Attackers find it harder to stay in applications over time since serverless functions are transient and fleeting. Attackers will still carry out attacks; their tactics will vary, making their lives more difficult. Because serverless functions have a short lifespan, serverless security risks could evolve. An attacker might create a far shorter attack that only takes a little information, such as a few credit card numbers. What is known as the "Groundhog Day" attack is a single attack round that keeps repeating itself.

4.2.3. Poisoning the Well

Despite the short lifespan of cloud-native resources, attackers can still establish long-term application persistence. They achieve this by exploiting a technique known as "Poisoning the Well," which circumvents the transient nature of serverless apps. Cloud-native applications often comprise multiple modules and libraries with code sourced from various third-party providers. Attackers aim to insert malicious code into these applications. Subsequently, the malicious code can communicate with its source, receive instructions, and wreak havoc after poisoning the well [8].

5. Security Best Practices for Serverless Architecture

Like any other technology, serverless architecture is not immune to vulnerabilities. Therefore, it's crucial to minimise risks as much as possible. Implementing best practices for serverless architecture, also recommended by cloud providers, is essential in achieving this goal.

5.1. Use API Gateways as Security Buffers

The most common protocol for communication between the client and server is HTTP protocol. The function shouldn't be exposed directly to the front end; it must be wrapped using the API gateway service. Using API HTTPS endpoint gateways to divide data from functions is one technique to stop event-data injection in serverless apps. An API gateway will serve as a security buffer when data is collected from various sources, separating app users on the client side from serverless functions on the backend. Implementing a reverse proxy to provide several security checks lowers the attack surface. Built-in security measures, such as data encryption and key management provided by the provider, can be used by HTTP endpoints. These measures help to protect sensitive data, environment variables, and stored data.

5.2. Data Separation and Secure Configurations

Implementing preventive measures such as code scanning, isolating commands and queries, identifying any exposed secret keys or unlinked triggers, and configuring them according to the recommended practices by the CSP's serverless application are vital strategies to thwart DoS attacks. Additionally, to mitigate the risk of DoS attacks disrupting execution calls, setting function timeouts as low as possible is advisable.

5.3. Dealing with Insecure Authentication

Implementing various specialised access control and authentication services is crucial to minimise the likelihood of authentication failures. Leveraging the CSP's access control solutions, such as OAuth, OIDC, SAML, OpenID Connect, and multi-factor authentication (MFA), can enhance the complexity of authentication, making it more difficult to compromise. Additionally, imposing unique password complexity requirements, including character type and length constraints, impedes hackers' attempts to crack passwords.

5.4. Sufficient Serverless Monitoring and Logging

Monitoring and logging provide users and operators with a comprehensive view of tracking the behaviour of our functions and workflow. Implementing monitoring and logging is essential to gain insights into every serverless application operation.

Relying solely on the CSP's logging and monitoring capabilities is inadequate as they do not encompass the application layer. It's also imperative to consider what data is being logged, ensuring that sensitive information that should remain undisclosed is not logged, as it could serve as an entry point for attackers.

5.5. Minimize Privileges

They diverge the functions from each other and limit their rules by provisioning IAM roles based on their rights. Minimising the privileges in independent functions is crucial. Another way to do this would be to make sure the code executes with the fewest permissions necessary to act.

5.6. Separate Application Development Environments

One of the most significant development practices is maintaining continuous development, integration, and deployment (CI/CD) by keeping the different environments apart from staging, development, and production [9]. This guarantees that appropriate vulnerability management is prioritised at every development stage before pushing that code version forward. Additionally, it ensures that the program is continuously tested and improved by prioritising patches, protecting updates, and detecting vulnerabilities, all of which help developers stay one step ahead of attackers.

6. Test case example

Netflix, a streaming service with 125 million customers, delivers 10 billion hours of video content each quarter. This feat is accomplished mainly through its infrastructure, which is built predominantly on Amazon Web Services (AWS). With hundreds of thousands of files and petabytes of data, scaling the application becomes imperative. The service reaches millions of users across 55 countries, providing regularly updated content. Over seven years, Netflix underwent a significant strategic transition, fully migrating to AWS cloud infrastructure. Leveraging serverless technology, particularly AWS Lambda, has become integral to many aspects of its operations. For instance, Lambda functions are vital in processing the vast number of video clips submitted daily, dividing them into segments, and encoding them into the various streams required for Netflix's functionality. Additionally, Lambdas are employed in backend systems for tasks such as verifying file integrity, creating data backups, and resolving issues with the backup process, given the substantial number of daily file modifications. Lambda is also utilised for security purposes, ensuring that all operations, including instance launches and terminations, adhere to established standards. It actively detects unauthorised access attempts and notifies users accordingly. Furthermore, Netflix utilises monitoring tools provided by AWS to track application traffic, enabling the detection of suspicious behaviour.

By embracing serverless architecture, Netflix effectively delegates server management responsibilities to the provider, ensuring that Lambda manages server deployment, compliance, and configuration. This approach guarantees efficient provisioning processes and rapid adaptation to evolving business requirements. Netflix has demonstrated ingenuity in harnessing serverless technologies, particularly AWS Lambda, to manage extensive data volumes, uphold security and compliance standards, and optimise operational efficiency within its cloud-based streaming service [10].

7. Future research

Serverless computing presents several promising scientific research areas, especially security and architecture. One vital research direction addresses the security challenges unique to serverless models, which heavily depend on external services and infrastructure. Studies could investigate how to adapt or redesign conventional security frameworks to suit better the serverless environment, where the short-lived nature of functions makes traditional security monitoring and incident handling methods less effective. This research might include developing new strategies for detecting and addressing security threats tailored to the fleeting characteristics of serverless components.

Another critical area of research focuses on improving resource management and performance within serverless architectures. The on-demand resource provisioning central to serverless computing introduces issues regarding efficient resource use and the reduction of latency during function initialisation, known as cold starts. Future studies might look into creating algorithms and architectural improvements to predict better and manage resource needs, possibly using artificial intelligence to forecast application demands dynamically. Furthermore, there is potential for exploring how serverless computing could be effectively merged with cutting-edge technologies like edge computing. This exploration could enhance how computational tasks are managed and performed across hybrid architectures, significantly benefiting applications that require immediate data processing, such as those involving the Internet of Things (IoT).

8. Conclusion

Flexibility and efficiency are significant benefits that serverless architecture provides, and application to application can handle enormous amounts of users simultaneously. Also, it speeds up the development lifecycle, providing quicker feature releases and bug fixes.

Moving risks to the cloud provided and having available cloud-supplied support is a significant improvement compared to traditional application management, where we are responsible for the server and the infrastructure. By controlling permissions on Lambdas to specify which services in the cloud can call them and which methods within those services can access them, we gain complete flexibility in managing access to Lambdas. Implementing effective data encryption, secure coding techniques, and robust identification and access controls is imperative. Monitoring and logging play crucial roles in ensuring the security of a serverless system. The scalability and reduced operating costs offered by serverless applications are compelling reasons many organisations adopt this approach. Serverless technology represents the future of development and application execution, serving as an entry point into the microservices architecture, a pivotal architectural paradigm in today's and future technological landscape.

By offloading infrastructure management to the cloud provider, serverless architecture frees up valuable time and resources for organisations to focus on innovation and core business objectives. Developers can concentrate on building innovative features and applications, rather than spending time on infrastructure maintenance tasks. This focus on innovation can drive competitive advantage and differentiation in the marketplace.

9. References

[1] A. Sabbioni, C. Mazzocca, M. Colajanni, R. Montanari & A. Corradi (2022). A Fully Decentralized Architecture for Access Control Verification in Serverless Environments, Proceedings of 2022 IEEE Symposium on Computers and Communications (ISCC), ISBN:978-1-6654-9792-3, ISSN: 2642-7389,pp. 1-6, DOI: 10.1109/iscc55528.2022.9912764

[2] https://www.cloudflare.com/en-gb/learning/serverless/what-is-serverless/ (2013), Cloudflare, Accessed on: 2023-11-16

[3] A. Koschel, S. Klassen, K. Jdiya, M. Schaaf & I. Astrova (2021). Cloud Computing: Serverless, Proceedings of 2021 12th International Conference on Information, Intelligence, Systems & Applications (IISA), ISBN:978-1-6654-0032-9, pp. 1-7 DOI: 10.1109/iisa52424.2021.9555534

[4] https://www.hitechnectar.com/blogs/baas-vs-faas-explaining-the-two-serverless-architectures/ (2023). HiTechNectar Trending IT Analysis & Technology News, Accessed on: 2023-11-16

[5] https://www.educative.io/answers/cost-scalability-and-performance-in-a-serverless-architecture (2023). Educative: Interactive Courses for Software Developers, Accessed on: 2023-11-16

[6] A. Serckumecka, I. Medeiros, and A. Bessani (2019). Low-Cost Serverless SIEM in the Cloud, Proceedings of 38th Symposium on Reliable Distributed Systems (SRDS), ISBN:978-1-7281-4222-7, ISSN: 2575-8462, pp. 381-3811, DOI: 10.1109/srds47363.2019.00057

[7] https://www.checkpoint.com/cyber-hub/cloud-security/what-is-serverless-security/ (2023). Cyber Security Solutions, Check Point Software, Accessed on: 2023-11-16

[8] Moric, Z.; Redzepagic, J. & Gatti, F. (2021). Enterprise Tools for Data Forensics. Annals of DAAAM & Proceedings, 10(2), Published by DAAAM International, ISBN 978-3-902734-33-4, ISSN 1726-9679, Vienna, Austria, DOI: 10.2507/32nd.daaam.proceedings.014

[9] Dakić V.; Redžepagić J.; Bašić M. (2022). CI/CD Toolset Security, 33rd DAAAM International Symposium on Intelligent Manufacturing and Automation, Published by DAAAM International, ISBN 978-3-902734-36-5, ISSN 1726-9679, Vienna, Austria, DOI: 10.2507/33rd.daaam.proceedings.022

[10] https://dashbird.io/blog/serverless-case-study-netflix/ (2023). Dashbird blog, Accessed on: 2023-12-28

DOI: 10.2507/35th.daaam.proceedings.042

WEB APPLICATION FIREWALLS (WAFS) IN PROTECTING SOFTWARE

Ivana Bilic, Karlo Josic*, Drazen Pranic & Sanja Ribaric

This Publication has to be referred as: Bilic, I[vana]; Josic, K[arlo]; Pranic, D[razen] & Ribaric, S[anja] (2024). Web Application Firewalls (WAFs) in Protecting Software, Proceedings of the 35th DAAAM International Symposium, pp.0306-0311, B. Katalinic (Ed.), Published by DAAAM International, ISBN 978-3-902734-44-0, ISSN 1726-9679, Vienna, Austria
DOI: 10.2507/35th.daaam.proceedings.042

Abstract

Web Application Firewalls (WAFs) are vital in safeguarding web applications by inspecting and managing HTTP traffic between the application and the Internet, adhering to the principles of the CIA (Confidentiality, Integrity, Availability) triangle. This paper explores the function of WAFs in enhancing security through strict traffic filtering based on pre-defined rules. It also highlights the importance of balancing application security and performance to prevent potential slowdowns due to extensive security checks. The paper's topic extends to the likely consequences of imprecise rules, which could block legitimate requests, reducing user satisfaction and possible loss of clientele.

Keywords: Web Application Firewall (WAF); security policies; OWASP; traffic filtering; cybersecurity.

1. Introduction

The globalisation of the market and technological advancements have encouraged the adaptation of software development to enable web applications to scale a large volume of requests. The transition from monolithic applications to microservices, which communicate through diverse networks, has significantly evolved. To safeguard data and users, these microservices must adhere to the principles of confidentiality, integrity, and availability (CIA). Achieving this requires transforming and scaling up security measures and application development. Given the alarming rate of nearly 4000 cyberattacks daily worldwide, the frequency of such incidents serves as a crucial metric, offering insights into the extent and prevalence of cyber threats organisations face [1]. Continuous awareness among software developers and security experts is essential, as they must remain on track with emerging vulnerabilities and consistently assess their assets to integrate the most effective tools for protection.

The Open Web Application Security Project (OWASP), the organisation that annually publishes the "OWASP Top 10," recommends [2] the adoption of web security firewalls as one of the best practices to mitigate security risks. Integration of Web Application Firewalls involves configuring them as reverse proxies, where they intercept each incoming request and assess it against a predefined set of rules to determine the presence of malicious content. While this enhances the system's security, it does come at the expense of potentially slowing down the application response.

Also, admins of Web Application Firewall's rules need to be careful. If the rules aren't precise, the system might classify false negatives, leading to losing customers.

2. Security challenges in software protection

Applications and APIs have become increasingly critical to business success [3]. Employees, partners, providers, customers, and other users rely on various applications to communicate, collaborate, and transact business. APIs have exploded recently as organisations power mobile applications, the Internet of Things (IoT), internal applications, partner applications, and cloud-based customer services. The more organisations rely on applications and APIs, the more attractive these digital assets are to attackers. Hackers today use automated bots to crawl websites at random, looking for vulnerabilities in applications they can use to access a database, load malicious files onto a web server, or take down a server with an overwhelming amount of traffic [4].

To effectively mitigate risks, it is essential to begin by identifying the assets that require safeguarding. This entails a comprehensive recognition and classification of the organisation's critical components. Following asset identification, the next crucial step involves pinpointing potential risks that could compromise the security and functionality of these assets. The MATA model then comes into play, providing a structured approach to risk management. This model involves Mitigating risks by implementing preventative measures, accepting certain risks, transferring risk responsibility through mechanisms like insurance or outsourcing, and avoiding risks by eliminating or redesigning processes. In tandem with this risk management model, the Open Web Application Security Project (OWASP) is a valuable resource, particularly in web applications.

A Security Information and Event Management (SIEM) system is instrumental in addressing security challenges within software protection. SIEM enables real-time detection of suspicious behaviour and unauthorised access attempts by continuously monitoring events and activities across the software infrastructure, mitigating potential security threats. SIEM provides insights into vulnerabilities, unauthorised changes, and other security issues by aggregating and correlating log data from various sources within the software environment. Leveraging advanced analytics and threat intelligence, SIEM identifies patterns indicative of malware infections, insider attacks, and other threats, triggering automated responses or alerting security personnel for further investigation and remediation. Additionally, SIEM facilitates compliance monitoring by ensuring adherence to security policies, standards, and regulations and supports incident investigation and forensic analysis in case of a security breach involving software assets. Through integration with other security controls and technologies, such as intrusion detection/prevention systems and antivirus solutions, SIEM offers a centralised approach to security management, enhancing overall protection against emerging threats and bolstering software security measures [5].

One of the most famous projects is OWASP Top 10, a list of the top 10 security risks web applications face, which uses the SIEM tools approach. The OWASP Top 10 Security Risks for 2023 are:

- Injection Attacks
- Broken Authentication
- Sensitive Data Exposure
- XML External Entities (XXE)
- Broken Access Control
- Security Misconfigurations
- Cross-Site Scripting (XSS)
- Insecure Deserialization
- Using Components with Known Vulnerabilities
- Insufficient Logging and Monitoring logging and monitoring [6].

Creating a secure environment for web applications is paramount, whether the application is hosted on-premises or in the cloud. Various tools play a crucial role in adding an extra layer of protection to mitigate potential threats and vulnerabilities.

3. Concept of Web Application Firewall (WAF)

A Web Application Firewall helps protect web applications by filtering and monitoring HTTP traffic between them and the Internet. It operates at layer 7 (the OSI model) and protects web applications from attacks such as forgery, cross-site scripting (XSS), file inclusion, and SQL injection [7]. Deploying a Web Application Firewall before a Web application creates a shield between it and the Internet. A web application firewall is a reverse proxy that acts as a single entry point for external systems to access resources on private subnets. Web Application Firewalls can run as a network appliance, server plugin, or cloud service. Advanced Web Application Firewalls allow analysts, security experts, and other users to define advanced rules with granular customisation options [8]. WAF is equipped with robust security features such as access control, content filtering, and virtual patching, helping to minimise security vulnerabilities and ensure compliance with industry standards and best practices.

Organisations can customise security policies and rule sets within WAF configurations to meet specific security requirements, effectively combating emerging threats and adapting to evolving attack strategies. Additionally, WAFs often include logging and reporting functions, empowering security teams to monitor and analyse web traffic, identify security breaches, and conduct thorough investigations as needed. WAF plays a crucial role in proactively preventing threats, managing access control with precision, and providing comprehensive visibility into web application traffic, thereby significantly enhancing the security posture of web applications and protecting sensitive data and critical assets from cyber threats.

3.1. Web Application Firewall Rules

Web application firewall rules define how to inspect HTTP/HTTPS web traffic (requests) to an application, where and what parameters and conditions to look for in the request, and what action the WAF should take when a request matches those definitions. WAF comes with predefined rules. Vendors often include a database with known signatures of malicious software and IP addresses. Advanced Web Application Firewalls allow analysts, security experts, and other users to define advanced rules with granular customisation options to improve their applications' security and performance and reduce network load on application servers.

Some examples of custom rules in WAF:

- Block access to specific sections of a website based on IP address or Headers,
- Prevent search engine bots from accessing a website,
- Redirect traffic to a maintenance page,
- Add header to specific requests,
- Change the value of a header to another value.

Conditions, actions, priorities, and descriptions can be defined for each rule. The conditions describe the request; they define desired or unwanted requests. The action is triggered when the request meets the condition. Priorities exist to initiate the correct action if a request meets multiple conditions.

3.2. WAFs deployment types

Software-based WAFs run as any other service in the system. They can run in a Docker container, which will intercept user requests. Software-based WAFs are easy to maintain, upgrade, and scale, especially in Kubernetes/OpenShift platforms. Containers/pods can be configured to scale based on traffic, which can prevent latency. The load balancer in OpenShift will automatically create another pod based on the amount of memory used or processing time. Cloud-hosted WAFs are WAFs that are hosted and managed by a third-party provider. They inspect incoming traffic to a web application and block any traffic that does not meet the configured security rules. Cloud-hosted WAFs are typically deployed as a service, with the WAF provider managing the hardware and software infrastructure required to run the WAF.

A hardware-based WAF (commonly referred to as network-based WAF) is installed locally on a network. These are often the most expensive WAFs, requiring maintenance and storage space. Their primary purpose is to minimise latency. Hardware-based WAFs are commonly leveraged by large organisations with the budget and headcount to manage on-premises appliance and IT infrastructure [9].

3.3. Pros and cons

A Web Application Firewall (WAF) acts as a security layer for your application by scanning and filtering all incoming traffic. It effectively manages requests by applying rules to allow or block specific content. Additionally, a WAF is a preventive measure against various threats, including DDoS attacks, injection attacks, and cross-site scripting (XSS).

Using a WAF also has some disadvantages. First, a WAF can introduce latency and overhead to the HTTP traffic, as it must process and analyse each request and response. It can also affect the functionality and usability of the web application, as it may block some legitimate traffic or trigger some false positives. Second, a WAF can be bypassed or evaded by sophisticated attacks, such as obfuscation, encryption, fragmentation, and zero-day attacks. It can also be vulnerable to attacks like brute force, flooding, or poisoning. Third, a WAF can be costly and complex to deploy and maintain, requiring specialised hardware, software, or services. It also requires constant monitoring, tuning, and updating to keep up with the web application changes and security requirements.

4. Web Application Firewalls Case study

Table 1 examines the top 10 implementations of WAFs for case study research. This was based on the diverse array of solutions tailored to meet the varied security needs of organisations across different industries and sizes. From industry giants like Akamai and AWS to specialised providers such as Barracuda Networks and Sucuri, each WAF solution brings unique features and strengths.

Solution	Company size	Options	Integration with
Akamai	all (with limitations)	Bot mitigation API security Layer 7 DDoS protection DevOps integration	Existing systems
AWS	medium to large	Protection from common threats like web exploits and bots Monitoring, filtering, and rate-limiting capabilities	AWS services and infrastructure
Barracuda networks	medium to large	Reliable traffic inspection Filtering of network traffic Outbound network scanning	Existing security infrastructure
Cloudflare	all (with limitations)	Integration with CDN and global network Instant access to Cloudflare's global network	Cloudflare's CDN and global network
F5	small	Spam protection Implementation of SMTP and FTP protocol checks	Existing systems and protocols
Fastly	medium	Next-gen WAF with adaptive protection against advanced threats DDoS protection and TLS encryption	Existing infrastructure and applications
Fortinet	medium to large	Machine learning for accurate protection against DDoS attacks and common threats Comprehensive visibility and support for intelligent threat identification	Machine learning
Imperva	all (with limitations)	Application security across all cloud environments (SaaS, PaaS, IaaS) Fast setup and automated protection	Existing security stacks and workflows
NetScaler	medium to large	Bot mitigation API protection Advanced security features Integration with existing security stacks	Existing security infrastructure and protocols
Sucuri	small to medium	Robust protection against threats SSL certificate creation High-performance caching Access to the Sucuri WAF network	Various environments and platforms

Table 1. Case study of WAF implementations

Businesses have many options, whether robust protection against common threats, seamless integration with existing systems, or specialised functionalities like bot mitigation and API security. As cybersecurity threats continue to evolve, the importance of implementing effective WAF solutions cannot be overstated. By carefully evaluating their requirements and selecting the most suitable WAF provider, organisations can bolster their defences, safeguard their web applications, and protect sensitive data from cyber threats in an increasingly digital world.

5. Future Security challenges and research

If "Zero Day Charlie" were to exploit vulnerabilities in an application that traditional Web Application Firewalls (WAFs) were unaware of, it could lead to a zero-day attack. In this scenario, Zero Day Charlie would capitalise on security flaws or weaknesses within the application that developers or the security community have not yet identified or patched. Since traditional WAFs rely on established patterns and signatures for threat detection, they likely prove ineffective against zero-day exploits. The ramifications of a zero-day attack are severe, granting attackers the ability to infiltrate systems and exfiltrate sensitive data without detection. Organisations may need robust defences to detect and respond to zero-day attacks, such as sophisticated threat detection mechanisms and real-time monitoring. Furthermore, zero-day attacks present a significant challenge for security teams, necessitating rapid response and mitigation strategies to mitigate the impact on affected systems and prevent further exploitation. Well-known examples are NotPetya and Stuxnet, which have multiple zero-day vulnerabilities in their code [10].

To mitigate the risk of zero-day attacks, organisations must adopt a multi-layered security approach that includes proactive threat intelligence, comprehensive vulnerability management, and advanced security solutions capable of detecting abnormal behaviour and unidentified threats. Additionally, staying informed about emerging security threats and promptly updating systems and software are crucial to reducing vulnerability to zero-day exploits. By implementing these measures, organisations can bolster their defences and minimise the potential impact of zero-day attacks on their systems and data. A next-generation firewall is designed to address advanced security threats at the application level through intelligent, context-aware security features. Next-generation firewall specifications vary by provider, but they generally include some combination of the following features:

- Deep-packet inspection, which inspects the data contained in packets on layer 7 of the OSI model.
- Intrusion Prevention System (IPS) monitors the network for malicious activity and blocks it where it occurs. Monitoring can be conducted based on malware signatures, or the request can be processed in sandboxes. Sandbox environments are isolated environments that exist to test potential malicious programs.
- External threat intelligence, or communication with a threat intelligence network to ensure that threat information is current and help identify bad actors.
- High performance, which can handle and scale large amounts of network traffic.

Firewall as a Service (FWaaS) represents a cloud-based approach to firewall implementation, offering enhanced scalability and streamlined maintenance procedures. In this model, the firewall software's responsibility rests with the service provider, alleviating enterprises from the burden of routine tasks such as patch management, upgrades, and resource allocation [11]. With FWaaS, organisations benefit from the inherent scalability of cloud services, as resources dynamically scale up or down based on processing demands. This ensures optimal performance and availability without manual intervention from internal IT teams.

Moreover, FWaaS provides flexibility and agility that traditional on-premises firewall solutions may struggle to match. By offloading firewall management to a specialised provider, organisations can redirect their internal resources toward strategic initiatives rather than routine maintenance tasks. Additionally, FWaaS offers the advantage of centralised management and monitoring, allowing for comprehensive oversight of firewall policies and security posture across distributed environments. This centralised approach streamlines security operations, enhances visibility, and facilitates rapid response to emerging threats, ultimately bolstering the overall resilience of the organisation's network infrastructure.

6. Conclusion

The dynamic nature of the internet and the continuous evolution of technology underscore the critical importance of robust security measures for web applications. The globalisation of markets and the shift from monolithic applications to microservices have compelled organisations to embrace agile software development practices and adopt proactive risk mitigation strategies. OWASP's annual Top 10 highlights the pivotal role of Web Application Firewalls (WAFs) in defending against prevalent cyber threats. While WAFs serve as effective reverse proxies, offering an additional layer of security by scrutinising incoming requests, striking a delicate balance between security and application response speed is imperative. Administrators tasked with defining WAF rules must navigate this balance cautiously, mindful of the potential for false negatives that could inadvertently impact legitimate users.

As cyberattacks become increasingly frequent and sophisticated, organisations must adopt proactive measures to safeguard their data and users effectively. This entails regularly assessing vulnerabilities, implementing robust security tools, and fostering ongoing collaboration between security experts and software developers [12]. By promoting a culture of cooperation and vigilance, organisations can develop resilient systems capable of adapting to the ever-changing cybersecurity landscape, bolstering their defences against emerging threats, and ensuring the integrity and security of their web applications.

Future research in web application security could encompass several areas to address emerging challenges and technological advancements. One avenue for exploration lies in integrating machine learning and artificial intelligence techniques into Web Application Firewalls (WAFs), aiming to enhance threat detection and mitigation capabilities. Additionally, research could focus on developing innovative approaches for early detection and mitigation of zero-day threats, leveraging anomaly detection and threat intelligence to bolster security measures.

Another promising area for investigation is the security implications of serverless architectures, particularly in securing serverless functions and microservices. Research in this domain could explore new strategies and technologies to mitigate potential vulnerabilities and ensure robust protection against attacks in serverless environments. Furthermore, exploring privacy-preserving techniques for web applications, considering the increasing concerns around data privacy and evolving privacy regulations, could lead to the development of more secure and privacy-centric web applications. By delving into these research directions, the field of web application security can advance towards more effective and resilient solutions to safeguard against evolving cyber threats and ensure the integrity and security of web applications and their data.

7. References

[1] https://www.getastra.com/blog/security-audit/how-many-cyber-attacks-per-day, (2024). How Many Cyber Attacks Per Day: The Latest Stats and Impacts in 2024, Accessed on: 2024-1-16

[2] https://owasp.org/, (2024). Open Worldwide Application Security Project (OWASP), Accessed on: 2024-1-18

[3] Dakic, V., Jakobovic, K., & Zgrablic, L. (2022). Linux Security in Physical, Virtual, and Cloud Environments. In DAAAM Proceedings (pp. 0151–0160). DAAAM International Vienna, DOI: 10.2507/33rd.daaam.proceedings.021

[4] Islam, S. A., MohanKumar, M., & Jannat, U. K. (2023). Exploring the Effectiveness of Web Application Firewalls Against Diverse Attack Vectors. In 2023 7th International Conference on Electronics, Communication and Aerospace Technology (ICECA). 2023 7th International Conference on Electronics, Communication and Aerospace Technology (ICECA), DOI: 10.1109/iceca58529.2023.10395379

[5] Suskalo, D., Moric, Z., Redzepagic, J., & Regvart, D. (2023). Comparative Analysis of IBM Qradar and Wazuh for Security Information and Event Management. In DAAAM Proceedings (pp. 0096–0102). DAAAM International Vienna, DOI: 10.2507/34th.daaam.proceedings.014

[6] https://www.cloudflare.com/learning/ddos/glossary/web-application-firewall-waf/ (2024), Cloudflare WAF Accessed on: 2024-1-16

[7] Chen, H.-C., Nshimiyimana, A., Damarjati, C., & Chang, P.-H. (2021). Detection and Prevention of Cross-site Scripting Attack with Combined Approaches, 2021 International Conference on Electronics, Information, and Communication (ICEIC), DOI:10.1109/iceic51217.2021.9369796

[8] https://expertinsights.com/insights/the-top-web-application-firewalls/ (2024). Expert InSights, Accessed on: 2024-1-18

[9] Ghanbari, Z., Rahmani, Y., Ghaffarian, H., & Ahmadzadegan, M. H. (2015). Comparative approach to web application firewalls. In Proceedings of 2nd International Conference on Knowledge-Based Engineering and Innovation (KBEI), pp. 808-812, ISBN:978-1-4673-6506-2, DOI: 10.1109/kbei.2015.7436148

[10] Kaminska, M., Broeders, D., & Cristiano, F. (2021). Limiting Viral Spread: Automated Cyber Operations and the Principles of Distinction and Discrimination in the Grey Zone, In Proceedings of 13th International Conference on Cyber Conflict (CyCon 2021), pp. 59-72, ISBN:978-9916-9565-5-7, ISSN: 2325-5374, DOI: 10.23919/cycon51939.2021.9468290

[11] https://www.cloudflare.com/learning/security/what-is-next-generation-firewall-ngfw/ (2024), Cloudflare NGFW, Accessed on: 2024-1-26

[12] Liang, J., & Kim, Y. (2022). Evolution of Firewalls: Toward Securer Network Using Next Generation Firewall, Proceedings of 12th Annual Computing and Communication Workshop and Conference (CCWC), pp. 0752-0759, ISBN:978-1-6654-8303-2, DOI: 10.1109/ccwc54503.2022.9720435

DOI: 10.2507/35th.daaam.proceedings.043

LEARNING WITH SMALL DATA: A NOVEL FRAMEWORK FOR BLUEBERRY ROOT COLLAR DETECTION

A G M Zaman[1], Nesma Mahmoud[2], Indrek Virro[1], Olga Liivapuu[1], Tormi Lillerand[1], Kallol Roy[2] & Jüri Olt[1]

[1]Estonian University of Life Sciences and [2]University of Tartu, Tartu, Estonia

This Publication has to be referred as: Zaman, A[] G[M]; Mahmoud, N[esma]; Virro, I[ndrek]; Liivapuu, O[lga]; Lillerand, T[ormi]; Roy, K[allol] & Olt, J[uri] (2024). Learning with Small Data: A Novel Framework for Blueberry Root Collar Detection, Proceedings of the 35th DAAAM International Symposium, pp.0312-0321, B. Katalinic (Ed.), Published by DAAAM International, ISBN 978-3-902734-44-0, ISSN 1726-9679, Vienna, Austria
DOI: 10.2507/35th.daaam.proceedings.043

Abstract

The integration of artificial intelligence and data science in agriculture is significantly transforming traditional human knowledge-based agricultural practices into a data-driven decision-making domain, known as precision agriculture. This study focuses on enhancing precision farming by using computer vision models to detect root collar of blueberry plants. Accurate detection of the root collar is crucial for optimizing agricultural practices such as fertilizer application, pesticide usage, and water management. Integrating computer vision model with agrobotic promises to maximize food production, reduce operational costs, minimize environmental impact, and ensure food quality and sustainability. We collected 118 images of blueberry plants from a plantation in Vehendi village, Tartu, Estonia, using an Insta360 One X2 camera mounted 1.6 meters above the plant rows. We developed and trained a machine learning model to detect blueberry root collar in a diverse environment. A pre-trained YOLOv5 model is trained and tested for blueberry root collar detection. The model's performance was evaluated, achieving 74.7% precision, 78.3% recall, and a 76.2% mAP50. Our results demonstrate the model's generalization capability with limited numbers of training samples. Future work will focus on expanding the dataset and integrating the model end-to-end with agrobotic systems to further advance precision farming practices.

Keywords: precision farming; YOLO; agrobotic; computer vision; machine learning.

1. Introduction

The world's population reached 8 billion in November 2022, raises its implications on lives, rights, health, and future generations [1]. Modern technologies have enabled human society to meet the increasing demand for food production. However, numerous challenges persist in ensuring food security, including plant diseases, climate change, extreme weather events, conflict and instability, water scarcity, land degradation, land grabbing, and many more.

To address the crucial need for food safety and also meeting the present challenges in agricultural sustainability, integration of advanced technologies such as machine learning becomes central to smart farming [2] and precision agriculture [3]. Precision farming practices optimize productivity, reduce environmental impact, ensures food security, and promote sustainability. These practices have the potential to significantly increase global food production while maintaining availability, accessibility, and nutritional quality worldwide. Moreover, sustainable farming practices can safeguard natural ecosystems, benefiting the future generations [4], [5]. The combination of machine vision, big data, IOT, and robotics has emerged as a promising set of tools for smart farming and precision agriculture. In precision farming, object detection plays a crucial role in different automated applications, such as, root collar detection, leaf detection, fruit counting, weed identification, phenotyping, and many more, by applying the computer vision and machine learning techniques. The collaboration of these technologies is a key enabler for precision farming, with the potential to transform the entire agricultural lifecycle from planting to harvesting. To achieve such advanced automation, a high-accuracy object detection model is a primary requirement. In this study, we applied the transfer learning technique using the pre-trained YOLOv5 [6], small variant model to train and test the detection of the root collar of low bush blueberry (*Vaccinium angustifolium*) plants. Accurate detection of the plant's root collar can enhance fertilization use, optimize irrigation, and enable targeted pesticides application, ultimately promoting sustainable agriculture and increase crop yield, and lowering production costs.

This study makes several key contributions to precision agriculture. Firstly, a specialized platform was designed and developed for collecting high-quality images of blueberry plants, which enables precise data acquisition. Secondly, a dataset of high-quality blueberry images was assembled, which is crucial for training and evaluating detection models. Finally, the effectiveness of a pre-trained YOLOv5 model for detecting root collars was demonstrated, offering valuable insights into its applicability in precision farming.

2. Literature review

Machine vision is pivotal in precision agriculture, enhancing efficiency through automated analysis of plant health, fruit count, weed detection, crop monitoring, disease identification, and yield estimation. This technological advancement parallels the evolutionary significance of vision, as zoologist Andrew Parker links to the Cambrian explosion [7], [8]. Precision agriculture, initially conceptualized in the 1980s by Dr. Pierre C. Robert, is often credited as the cornerstone of modern precision agriculture. It is defined as an information revolution driven by new technologies, leading to a more precise farm management system. Later, it evolved to be defined as "the application of modern information technologies to provide, process and analyze multisource data of high spatial and temporal resolution for decision making and operations in the management of crop production" [9]. Precision agricultural applications and experiments can be classified into four distinct groups, such as, crop management, soil management, water management and livestock management [10]. Its main goal is to promote sustainable, efficient, and resilient farming practices that address the challenges of growing global population, environmental concerns, and changing climate conditions, while optimizing resource use and ensuring food security. Recent, advancements in artificial intelligence and computer vision have significantly heightened research interest in this field.

In the field of object detection, numerous experiments have been conducted such as, plant diseased detection, yield estimation, root collar detection, autonomous robotic harvesting guidance, vegetation index calculation and many more. Plant disease detection from images is a highly explored area and can be categorized into three main categories, i.e., diseases detection, measurement of disease severity, and classification [11]. Among different image processing techniques convolutional neural network (CNN) has achieved impressive results in the field of image classification and detection. CNN has shown remarkable performance in image classification, successfully identifies 13 distinct types of plant diseases with accuracy ranging from 91% to 98% in various individual class tests [12]. In another study, disease detection from cucumber leaves employed a classical method involving three key stages i.e., segmenting diseased leaf images using the K-means clustering algorithm, extracting shape and color features from data and classifying diseased leaf images through sparse representation. This approach effectively recognized seven major cucumber diseases, with an overall recognition rate of 85.7%, surpassing the performance of alternative methods [13].

A novel neural network model integrated with semantic graphics has been proposed for detecting the crop lines and weeds in paddy fields, where the detected line serves as a guiding path for autonomous weeding robot aimed to enhance agricultural efficiency [14]. Another study employed a deep learning-based two-step method to identify weeds. This method used a CenterNet model trained to detect vegetables, achieving impressive performance with a precision 95.6%, recall 95.0%, and an F1 score of 0.953. Any remaining green objects in the image were subsequently classified as weeds [15]. Additionally, the study [16] shows that an automated system for targeted spraying in orchards using RGB imaging demonstrated a 23% reduction in pesticide use without compromising spray coverage by utilizing machine vision to adjust the pesticide spray based on the detected shape of the apple tree canopy.

A real-time computer vision approach was proposed for locating crop stems, with applications in mechanical hoeing within precision farming. This method integrates an object-detecting neural network to identify stems in RGB images, followed by the utilization of an aggregation algorithm that enhances detection by considering temporal data from consecutive frames. Testing was conducted on maize and bean crops in two configurations demonstrated promising results, achieving F1-score of 94.74% and 93.82%, respectively, for detecting maize and bean stems, with a location accuracy of 0.7 cm and 0.5 cm [17].

In this study, we focus on blueberry root collar detection, an increasingly important crop in Europe and North America [18]. Due to high production costs and a labour shortage, there is a rising need for the successful deployment of agrobotic platforms [19]. Accurate root collar detection could enable the precise application of fertilizers, pesticides, and watering to plants in an informed manner leading to efficient farming practices. Additionally, developing detection models for blueberry root collar holds the potential for generalization to other bush-structured fruits such as raspberries and blackberries, thereby broadening the application of autonomous robots in precision farming.

3. Materials and methods

In this study, we employ advanced object detection techniques that have evolved considerably over the past two decades. Initially dependent on handcrafted feature detection methods, the field underwent a significant transformation in 2012 with the introduction of convolutional neural networks (CNNs) [20]. Modern object detectors are categorized into two-stage and one-stage approaches [21]. For our study, we utilize YOLO (You Only Look Once) [22], a single-stage object detector. YOLO simultaneously performs feature extraction and region of interest pooling by framing detection as a regression problem. It predicts bounding boxes and class probabilities directly from a grid, which provides enhanced processing speed and competitive accuracy, making it particularly suited for detecting blueberry root collars. Building on these advantages, our future goal is to deploy this machine vision model within an agricultural robot to automate precision farming tasks. This integration will also address constraints such as computational resources, power consumption, and environmental uncertainties, enabling real-time detection of blueberry root collars in a field environment. The agrobotic system, which is currently under development is shown in figure 1 below.

Fig. 1. Agricultural robot for precision farming applications

The figure 2 below shows the full process of the blueberry root collar detection experiment. This process has four fundamental stages: data acquisition, data pre-processing and labelling, YOLOv5 architecture and training, and results and discussion.

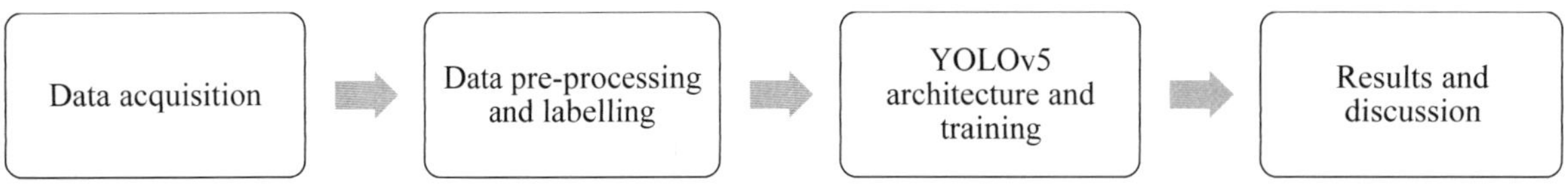

Fig. 2. End-to-end process for YOLOv5 based blueberry root collar detection

3.1. Data acquisition

Agricultural data is often not readily available in open sources due to high costs, the need for specialized equipment, extensive manual effort, and the significant time required for collection. In this initial phase, we focused on capturing images of individual low bush blueberry plants from above, with an average height of 1.5 meters. The data was collected from a plantation field in Vehendi village, Elva Municipality, Tartu County (GPS 58.20, 26.13), Estonia. The Insta360 One X2 camera was used for image acquisition, mounted on a mobile platform that navigates through the blueberry field rows and capture continuous frames of the plants. The design of the data acquisition platform, shown in figure 3(a), was specifically developed for this task. It features an adjustable arm reaching a height of 1.6 meters and ensures stability and mobility in the field. Figure 3(b) illustrates the on-field data collection process, highlighting the camera's positioning and the captured image frame in a red bounding box. This setup allowed for efficient and consistent data collection across the field. The collected blueberry dataset was relatively small, consisting of 118 blueberry images in total.

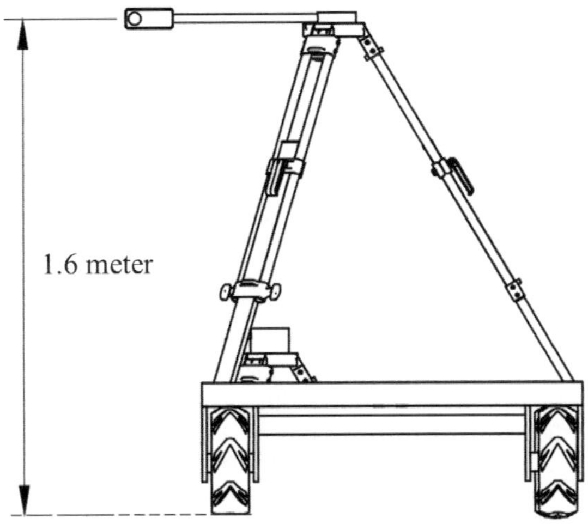

a) Data acquisition platform

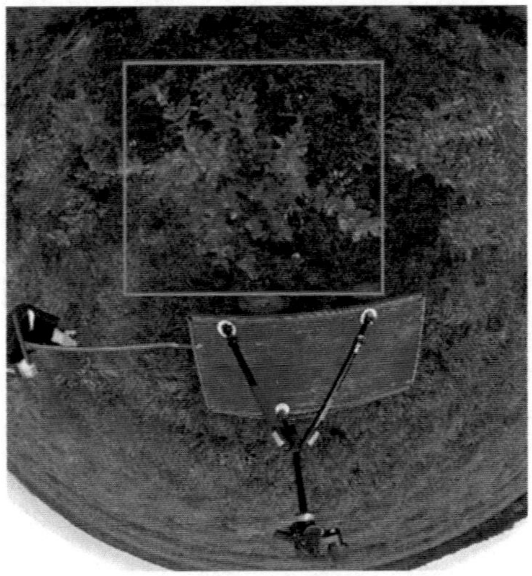

b) On-field data collection process

Fig. 3. Blueberry data collection platform and process

3.2. Data pre-processing and labelling

Specific frames of blueberry images were extracted using the Insta360 Studio software. The original images have a resolution of 1440 by 1440 pixels and a density of 24 dpi. Accurate annotation of bounding boxes and assignment of class labels are essential for effective supervised learning. To facilitate this process, we used Roboflow software [23] which allows for direct annotation within its interface. Each bounding box, i.e., the yellow rectangle outlines a distinct spatial region within the image and is associated with a class label - in this study, "blueberry root collar" as shown in figure 4(a). Figure 4(b) depicts the annotation process, detailing how the class label, the coordinates (x, y), and dimensions (width, height) of the bounding box are specified in normalized units relative to the image dimensions. This structured data format enables the model to learn to detect and classify objects with precision during both training and inference stages.

a) Bounding box (yellow) for root collar

Root collar coordinates details:

0	Blueberry root collar (class label)
0.49	Normalized centre X-coordinate with respect to image width
0.69	Normalized centre Y-coordinate with respect to image height
0.24	Normalized width of bounding box with respect to image width
0.18	Normalized height of bounding box with respect to image height

b) YOLO-specific bounding box coordinates

Fig. 4. Bounding box and annotations for blueberry root collar

3.3. YOLOv5 architecture and training

In this study we present a baseline method for blueberry root collar detection utilizing the YOLOv5 [6] model. This model has consistently demonstrated high detection accuracy and rapid inference speed across variety of applications [24], [25], [26]. The YOLOv5 model is available in different configurations: nano (n), small (s), medium (m), large(l), and extra-large (x). Though larger models with a greater number of parameters generally exhibit increased accuracy, the trade-off between accuracy and inference speed makes lighter models more suitable for real-time, edge-based applications due to their lower computational complexity and reduced resource requirements. The architecture of YOLOv5 consists primarily of three main components: the backbone, neck, and head. The figure 5 below illustrates how these components work together to process input images and predict bounding boxes [6].

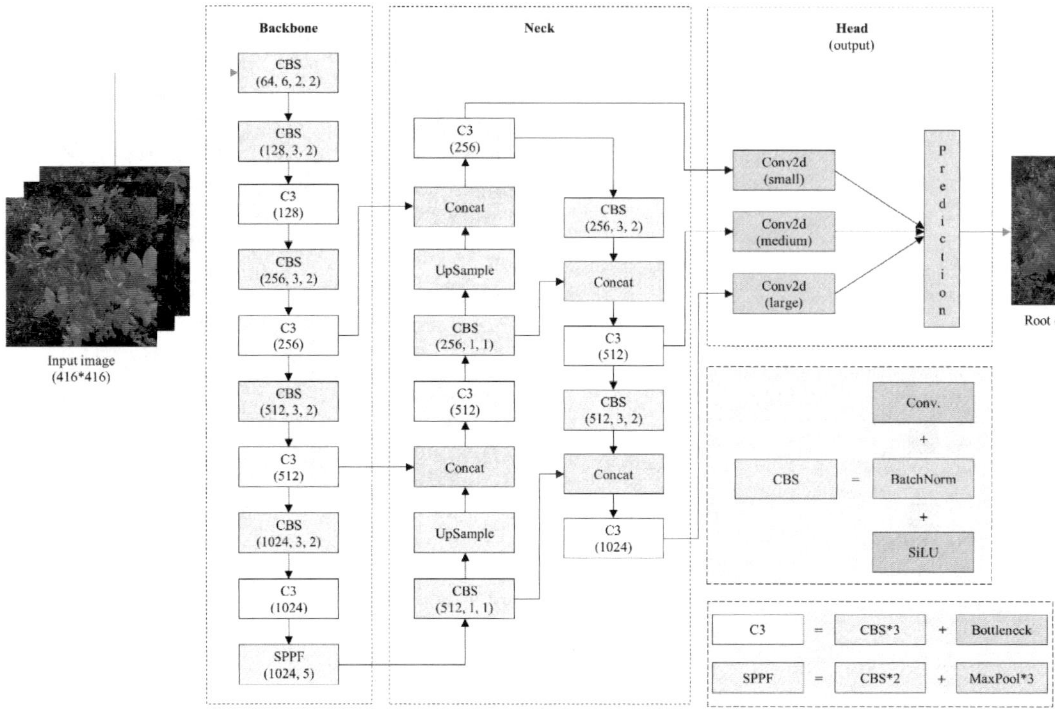

Fig. 5. Simplified YOLOv5 network architecture

The model processes the input image through a comprehensive end-to-end pipeline, where the model extracts relevant features and regresses the coordinates of bounding boxes, alongside calculating the confidence scores for each class. This approach allows YOLOv5 to efficiently detect objects by simultaneously considering spatial and class information, resulting in a fast and accurate detection process suitable for real-time applications. Throughout the network, the bounding boxes are refined, ensuring precise localization and classification of objects in diverse and complex scenes.

The backbone of YOLOv5 employs CSPDarknet53 [27], which enhances feature extraction while minimizing computational complexity. The backbone consists of multiple CBS (Conv. + BatchNorm + SiLU) modules, C3 modules, and an SPPF (Spatial Pyramid Pooling-Fast) module. The CBS layers initiate feature extraction with convolution operations, batch normalization, and the SiLU (Sigmoid Linear Unit) activation function, refining the learning process. For instance, an input blueberry image resized to 416x416 pixels passes through the first CBS layer configured with 64 filters, a kernel size of 6, stride of 2, and padding of 2, resulting in a 208x208x64 feature map. This step not only extracts essential features but also normalizes the outputs through batch normalization, ensuring a more stable and consistent range of activations. The SiLU activation function introduces non-linearity, enabling the network to capture complex patterns. These combined processes refine the feature representation, enhancing the model's capability to distinguish between subtle details and variations in the data, which is crucial for accurate detection in challenging conditions.

The backbone also includes C3 layers, composed of three consecutive CBS layers followed by a bottleneck layer, which capture spatial hierarchies and contextual information. The final feature maps undergo processing in the SPPF layer, which differs from the original SPPNet implementation [28]. In YOLOv5, the SPPF layer does not output a fixed-size vector but instead maintains spatial dimensions that depend on the input size, resulting in a fixed number of channels. The SPPF layers aggregate context at multiple scales, enhancing the receptive field and improve detection of objects at various levels of abstraction.

The neck component receives various features from the backbone ranging from low to high level, and applies fusion techniques, that combine principles from FPN (Feature Pyramid Network [29]) and PANet (Path Aggregation Networks) [30]. This architecture utilizes UpSample layers and Concat operations to construct a multi-scale feature pyramid, crucial for detecting small objects and preserving fine details. The UpSample layers enhance the spatial resolution, while the Concat layers merge these up-sampled features with lower-level maps, enabling a comprehensive representation across object sizes. The integration of PANet principles further refines and aggregates features through CBS and C3 modules, enhancing the model's localization and detection accuracy. The head section is responsible for the final predictions, which includes object classes, objectness scores, and bounding box coordinates. It utilizes anchor boxes, which are pre-defined shapes used to assist in predicting object locations and sizes, facilitating detection across different scales - small, medium, and large. The head processes inputs from three distinct feature maps obtained from the neck, each corresponding to a different level of down-sampling.

This approach enables multi-scale predictions, allowing the model to effectively identify objects of various sizes within an image, from small details to larger objects, thereby enhancing overall accuracy and robustness of the detection system. By integrating feature maps of different resolutions, the head section refines and synthesizes both spatial and semantic information, ensuring that fine-grained details and broader contextual context are captured. This detailed, layered prediction strategy is crucial for accurate detection across complex and varied images.

In our experiments for blueberry root collar detection, the pre-trained YOLOv5 's' variant was utilized which was initially pre-trained on the COCO dataset [31]. The input image size for the model was set to 416x416 pixels. Various augmentation techniques were applied to the training images to enhance the diversity and robustness of the dataset. These common augmentations include random scaling, cropping, flipping, colour adjustments, and affine transformations. Such techniques help the model generalize better by simulating different variations and distortions that might occur in real-world scenarios. This process enhances model performance on unseen data and prevents the model from overfitting to the training data.

We have trained the model using stochastic gradient descent (SGD) optimizer, with a learning rate of 0.01, a batch size of 16, a weight decay of 0.0005, and a momentum value set at 0.937 to ensure efficient convergence. The loss function is defined as the sum of three losses (i) bounding box regression loss (ii) classification loss, and (iii) objectness loss, as shown in equation (1). The overall loss is a weighted sum of these individual losses, reflecting their relative contributions to the training objective.

$$loss = loss_{box} + loss_{cls} + loss_{obj} \tag{1}$$

The bounding box regression loss function measures how well the predicted bounding box matches the ground truth bounding box. This loss is defined in equation (2) below. Where λ_{coord} is a hyperparameter that regulates the relative contribution of bounding box regression loss compared to other loss factors, s^2 represents the number of grid cells into which the image is divided. The indicator $I_{obj}^{(i,j)}$ is set to 1 if the jth bounding box in the ith grid cell is responsible for the prediction. The term $b_j(2 - w_i h_i)$ acts as a scaling factor, emphasizing the significance of different bounding box sizes, smaller bounding boxes are given higher influence, while larger bounding boxes receive relatively lower influence. The relative coordinates and dimensions of the ground truth bounding box are denoted by x_i, y_i, w_i, h_i where x_i and y_i represent the relative coordinates of the centre of the root collar bounding box, and w_i and h_i represents its relative width and height, respectively. Similarly, $\hat{x}_i, \hat{y}_i, \widehat{w}_i, \widehat{h}_i$ denote the coordinates and dimensions of the predicted bounding box, where $\hat{x}_i$ and $\hat{y}_i$ are the predicted relative coordinates of the centre, and $\widehat{w}_i$ and $\widehat{h}_i$ are the predicted relative width and height.

$$loss_{box} = \lambda_{coord} \sum_{i=0}^{s^2} \sum_{j=0}^{B} I_{obj}^{(i,j)} b_j (2 - w_i h_i) \left[(x_i - \hat{x}_i)^2 + (y_i - \hat{y}_i)^2 + (w_i - \widehat{w}_i)^2 + \left(h_i - \widehat{h}_i\right)^2 \right] \tag{2}$$

The classification loss function, as shown in equation (3), quantifies the accuracy of predicted class probabilities against the ground truth class. Here, λ_{cls} is a loss weighting factor for the classification loss. The terms, $p_i(c)$ and $\widehat{p}_i(c)$ denote the ground truth and predicted probabilities for class c in the ith grid cell, respectively. This formulation ensures that the model's prediction aligns closely with the actual class distribution, thereby enhancing classification accuracy.

$$loss_{cls} = \lambda_{cls} \sum_{i=0}^{s^2} \sum_{j=0}^{B} I_{obj}^{(i,j)} \sum_{c \in classes} p_i(c) \, log\left(\widehat{p}_i(c)\right) \tag{3}$$

The objectness loss function, as represented in equation (4), evaluates the accuracy of the predicted bounding box's confidence score in relation to the ground truth. The parameters, λ_{no_obj}, and λ_{obj} act as weighting factors for the loss in scenarios where no object is present and where an object is present, respectively. The terms c_i and $\hat{c}_i$ are the ground truth and predicted confidence score for the ith grid cell, respectively. This loss function ensures that the model effectively distinguishes between object and non-object regions, thereby enhancing its detection performance.

$$loss_{obj} = \lambda_{no\_obj} \sum_{i=0}^{s^2} \sum_{j=0}^{B} I_{no\_obj}^{(i,j)} (c_i - \hat{c}_i)^2 + \lambda_{obj} \sum_{i=0}^{s^2} \sum_{j=0}^{B} I_{obj}^{(i,j)} (c_i - \hat{c}_i)^2 \tag{4}$$

The YOLOv5 model significantly enhances object detection accuracy through its advanced use of anchor boxes and Intersection over Union (IoU). It automatically generates anchor boxes by applying K-means clustering to the bounding box dimensions in the training dataset, and then refines these boxes using genetic algorithm to optimize their shapes and sizes for better matching of the objects being detected. By assigning these dynamically generated anchors based on the highest IoU and utilizing multiple anchors per grid cell effectively identifies and localizes small or closely spaced objects.

Its multi-scale prediction approach and non-maximum suppression (NMS) further refine detection results, ensuring that overlapping detections are minimized and only the most confident predictions are retained. In our blueberry root collars detection it shows high accuracy, even in smaller image sizes such as 118 pixels.

4. Results and discussion

The model training was conducted on a system running Windows 10 Pro with Python 3.12. The hardware configuration included an Intel® Core™ i5-7400 CPU @ 3.00 GHz with 4 cores, 16 GB of RAM, and the training was performed exclusively on the CPU. The model was trained and validated over 150 epochs using the datasets. The training curves are represented by the box loss and object loss curves for both training and validation dataset, as shown in figure 6. The curves for box loss and object loss reflect the model's convergence and performance improvement throughout the training process. Specifically, the reduction in validation loss values indicates effective adaptation and learning, demonstrating the model's ability to generalize and optimize performance over successive epochs.

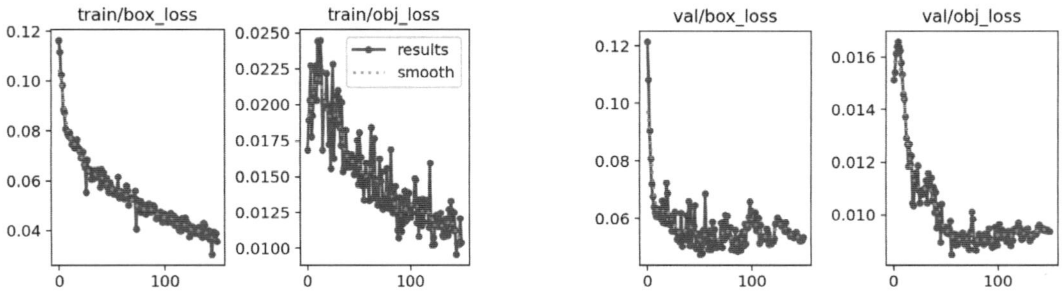

a) Model training loss b) Model validation loss

Fig. 6. Convergence of model training and validation loss

The performance of the model is assessed using precision (P), recall (R), and average precision (AP) metrics, which are defined in equations (5), (6), and (7), respectively.

$$P = \frac{TP}{(TP + FP)} \tag{5}$$

$$R = \frac{TP}{(TP + FN)} \tag{6}$$

$$AP = \int_0^1 P(R)dR \tag{7}$$

Precision (P) measures the proportion of true positive (TP) detections among all positive detections, which includes both true positives (TP) and false positives (FP). Recall (R), on the other hand, measures the proportion of true positive (TP) detections among all actual positive examples, which includes both true positives (TP) and false negatives (FN). Given that our evaluation involves a single class, the average precision is synonymous with the mean average precision (mAP) computed at a 50% confidence threshold. This metric represents the area under the precision-recall curve, which is derived by integrating the precision with respect to recall. Figure 7 presents the test results of the model's performance metrics including precision, recall, mAP.

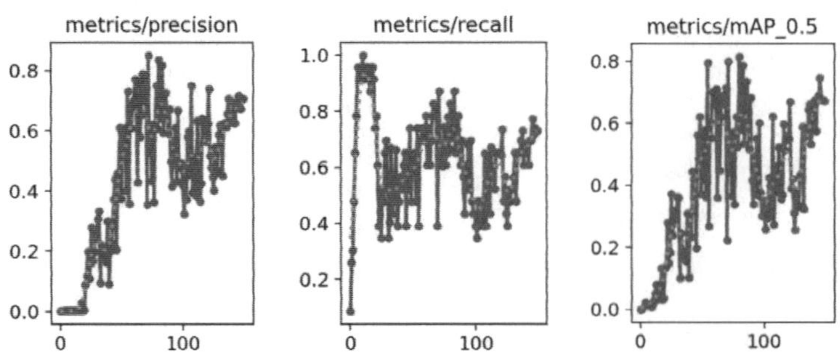

Fig. 7. Evaluation metrics: precision, recall, and mAP@50 results

The precision of the model is 74.7%, indicating that 74.7% of the predicted positive detections are true positives. The recall is 78.3%, reflecting that 78.3% of actual positive examples are correctly detected. Additionally, the mAP@50 is 76.2%, shows the model's effectiveness in detecting root collars with a high level of accuracy at a 50% IoU threshold. These results collectively show the model's strong performance in object detection. Figure 8 shows three predictions of blueberry root collar detection. The first two images show correct predictions, with the red bounding boxes representing the model's predictions with confidence score, and the yellow boxes denoting the ground truth. In contrast the third image (right most) shows a failed detection, where the predicted bounding box does not align with the actual root collar, resulting in a low IoU score. These examples show the challenges of the detection task, highlighting the difficulty due to the high similarity of features within the image context.

Fig. 8. Blueberry root collar detection: correct and misaligned predictions

Our results demonstrate the potential of automated machine vision in scenarios with scarce data, particularly in agricultural applications where large datasets are often unavailable, making model training challenging. Despite the limited size of our dataset, the model exhibited significant convergence and robust performance, effectively detecting blueberry root collars. This highlights the model's potential for practical applications in precision farming, suggesting it could be a valuable tool for enhancing agricultural practices and decision-making.

5. Conclusion

In this study, advanced machine learning techniques were utilized to detect blueberry root collar positions using a pre-trained YOLOv5s model. Initially, a data acquisition platform was designed and developed to gather images from blueberry plantations in Tartu, Estonia. The dataset underwent pre-processing for supervised learning, where the root collar bounding box coordinates were prepared in a specified format. The model was systematically trained, validated, and tested on this dataset. Our results indicate that the model effectively learned from the dataset despite its small size and the complex learning environment, which included similar features from other plants such as grass and weeds. The model achieved a precision of 74.7%, a recall of 78.3%, and a mAP@50 of 76.2%, marking a significant advancement in the field of precision farming.

Future work will focus on expanding the dataset and integrating the model into an agrobotic platform for real-time field operations. This integration aims to optimize agricultural practices by accurately identifying the root collar positions to facilitate precise applications of fertilizers, pesticides, and watering.

6. Acknowledgments

This research was supported by development fund PM210001TIBT from the Estonian University of Life Sciences and proof-of-concept grant EAG304 from the Estonian Research Council.

7. References

[1] The UNFPA State of World Population 2023 report. (2023). Population bomb, bust - or boon? New UNFPA report debunks 8 myths about a world of 8 billion, The UNFPA State of World Population 2023 Report, Available from: https://www.unfpa.org/news/population-bomb-bust-or-boon-new-unfpa-report-debunks-8-myths-about-world-8-billion Accessed: 2024-09-06

[2] Tyagi, A. C. (2016). Towards a Second Green Revolution, Irrigation and Drainage, Vol. 65, No. 4, pp. 388-389, ISSN 1531-0361, DOI: 10.1002/IRD.2076

[3] Zhang, N.; Wang, M. & Wang, N. (2002). Precision agriculture a worldwide overview, Computers and Electronics in Agriculture, Vol. 36, No. 2-3, pp. 113-132, ISSN 1872-7107, DOI: 10.1016/S0168-1699(02)00096-0

[4] Gebbers, R. & Adamchuk, V. I. (2010). Precision agriculture and food security, Science, Vol. 327, No. 5967, pp. 828-831, ISSN 0036-8075, DOI: 10.1126/science.1183899

[5] Vekic, A.; Borocki, J.; Stankovski, S. & Ostojic, G. (2017). Development of innovation in field of precision agriculture, Proceedings of the 28th DAAAM International Symposium, pp.0787-0795, B. Katalinic (Ed.), Published by DAAAM International, ISBN 978-3-902734-11-2, ISSN 1726-9679, Vienna, Austria, DOI: 10.2507/28th.daaam.proceedings.111

[6] Jocher, G. (2020). Ultralytics YOLOv5, DOI: 10.5281/zenodo.3908559, Available from: https://github.com/ultralytics/yolov5 Accessed: 2024-09-06

[7] Douglas, E. & Valentine, J. W. (2013). The Cambrian explosion: the construction of animal biodiversity, Roberts and Company Publishers Inc., ISBN 1936221039, Greenwood Village, CO, USA

[8] Parker, A. (2003). In the blink of an eye: how vision sparked the big bang of evolution, Basic Books, ISBN 0465054382, New York, USA

[9] National Research Council & others (1997). Precision Agriculture in the 21st Century: Geospatial and Information Technologies in CROP Management, National Academy Press, ISBN 978-0-309-05893-3, DOI: 10.17226/5491, Washington, D.C., USA

[10] Liakos, K.; Busato, P.; Moshou, D.; Pearson, S. & Bochtis, D. (2018). Machine Learning in Agriculture: A Review, Sensors, Vol. 18, No. 8, pp. 2674, ISSN 1424-8220, DOI: 10.3390/s18082674

[11] Garcia, J. & Barbedo, A. (2013). Digital image processing techniques for detecting, quantifying and classifying plant diseases, SpringerPlus, Vol. 2, No. 1, pp. 660-672

[12] Sladojevic, S.; Arsenovic, M.; Anderla, A.; Culibrk, D. & Stefanovic, D. (2016). Deep Neural Networks Based Recognition of Plant Diseases by Leaf Image Classification, Computational Intelligence and Neuroscience, Vol. 2016, No. 1, ISSN 16875273, DOI: 10.1155/2016/3289801

[13] Zhang, S.; Wu, X.; You, Z. & Zhang, L. (2017). Leaf image based cucumber disease recognition using sparse representation classification, Computers and Electronics in Agriculture, Vol. 134, pp. 135-141, ISSN 0168-1699, DOI: 10.1016/j.compag.2017.01.014

[14] Adhikari, S. P.; Yang, H. & Kim, H. (2019). Learning semantic graphics using convolutional encoder–decoder network for autonomous weeding in paddy, Frontiers in Plant Science, Vol. 10, pp. 1404, ISSN 1664-462X, DOI: 10.3389/fpls.2019.01404

[15] Jin, X.; Che, J. & Chen, Y. (2021). Weed identification using deep learning and image processing in vegetable plantation, IEEE Access, Vol. 9, pp. 10940-10950, ISSN 2169-3536, DOI: 10.1109/ACCESS.2021.3050296

[16] Stajnko, D.; Vindis, P. & Mursec, B. (2011). Automated system for targeted spraying in orchards by using RGB imaging, Chapter 23 in DAAAM International Scientific Book 2011, pp. 283-300, B. Katalinic (Ed.), Published by DAAAM International, ISBN 978-3-901509-84-1, ISSN 1726-9687, Vienna, Austria, DOI: 11.2507/daaam.scibook.2011.23

[17] Lac, L.; Da Costa, J.-P.; Donias, M.; Keresztes, B. & Bardet, A. (2022). Crop stem detection and tracking for precision hoeing using deep learning, Computers and Electronics in Agriculture, Vol. 192, pp. 106606, ISSN 01681699, DOI: 10.1016/j.compag.2021.106606

[18] Medeiros, J. G. S.; Biasi, L. A.; Bona, C. M. D. & Cuquel, F. L. (2018). Phenology, production and quality of blueberry produced in humid subtropical climate, Revista Brasileira de Fruticultura, Vol. 40, No. 3, pp. e- -520, ISSN 1806-9967, DOI: 10.1590/0100-29452018520

[19] Stefanović, D.; Antić, A.; Otlokan, M.; Ivošević, B.; Marko, O.; Crnojević, V. & Panić, M. (2022). Blueberry row detection based on uav images for inferring the allowed ugv path in the field, Proceedings of the fifth Iberian Robotics Conference, pp. 401-411, Published by Springer Cham, ISBN 978-3-031-21062-4, Zaragoza, Spain, DOI: 10.1007/978-3-031-21062-4_33

[20] Krizhevsky, A.; Sutskever, I. & Hinton, G. E. (2012). Imagenet classification with deep convolutional neural networks, Proceedings of twenty fifth Advances in Neural Information Processing Systems, ISBN 9781627480031, pp. 1097-1105

[21] Zou, Z.; Chen, K.; Shi, Z.; Guo, Y. & Ye, J. (2023). Object detection in 20 years: A survey, Proceedings of the IEEE, Vol. 111, No. 3, pp. 257-276, ISSN 0018-9219, DOI: 10.1109/JPROC.2023.3238524

[22] Redmon, J.; Divvala, S.; Girshick, R. & Farhadi, A. (2016). You only look once: Unified, real-time object detection, Proceedings of the IEEE Conference on Computer Vision and Pattern Recognition (CVPR), ISBN 978-1-4673-8851-1, pp. 779-788, IEEE, DOI: 10.1109/CVPR.2016.91

[23] Dwyer, B.; Nelson, J.; Hansen, T. & et al. (2024). Roboflow (Version 1.0), Available from: https://roboflow.com Accessed: 2024-09-06

[24] Ge, Z.; Liu, S.; Wang, F.; Li, Z. & Sun, J. (2021). Yolox: Exceeding yolo series in 2021, arXiv preprint arXiv:2107.08430

[25] Qi, D.; Tan, W.; Yao, Q. & Liu, J. (2022). YOLO5Face: Why reinventing a face detector, Proceedings of European Conference on Computer Vision, ISBN 978-3-031-25072-9, pp. 228-244, Tel Aviv, Israel, DOI: 10.1007/978-3-031-25072-9_15

[26] Wu, W.; Chang, H.; Zheng, Y.; Li, Z.; Chen, Z. & Zhang, Z. (2022). Contrastive learning-based robust object detection under smoky conditions, Proceedings of the IEEE/CVF Conference on Computer Vision and Pattern

Recognition, ISBN 978-1-6654-8739-9, pp. 4295-4301, New Orleans, LA, USA, DOI: 10.1109/CVPRW56347.2022.00475

[27] Wang, C.-Y.; Liao, H.-Y. M.; Wu, Y.-H.; Chen, P.-Y.; Hsieh, J.-W. & Yeh, I.-H. (2020). CSPNet: A new backbone that can enhance learning capability of CNN, Proceedings of the IEEE/CVF Conference on Computer Vision and Pattern Recognition Workshops, ISBN 978-1-7281-9360-1, pp. 390-391, Seattle, WA, USA, DOI: 10.1109/CVPRW50498.2020.00203

[28] He, K.; Zhang, X.; Ren, S & Sun, J. (2015). Spatial pyramid pooling in deep convolutional networks for visual recognition, IEEE Transactions on Pattern Analysis and Machine Intelligence, Vol. 37, No. 9, pp. 1904-1916, ISSN 0162-8828, DOI: 10.1109/TPAMI.2015.2389824

[29] Lin, T.-Y.; Dollár, P.; Girshick, R.; He, K.; Hariharan, B. & Belongie, S. (2017). Feature pyramid networks for object detection, Proceedings of the IEEE Conference on Computer Vision and Pattern Recognition, ISBN 978-1-5386-0457-1, pp. 936-944, IEEE, Honolulu, HI, USA, DOI: 10.1109/CVPR.2017.106

[30] Liu, S.; Qi, L.; Qin, H.; Shi, J. & Jia, J. (2018). Path aggregation network for instance segmentation, Proceedings of the IEEE Conference on Computer Vision and Pattern Recognition, ISBN 978-1-5386-6420-9, pp. 8759-8768, Salt Lake City, UT, USA, DOI: 10.1109/CVPR.2018.00913

[31] Lin, TY. & et al. (2014). Microsoft COCO: Common Objects in Context, Proceedings of Computer Vision - ECCV 2014, ISBN 978-3-319-10602-1, pp. 740-755, Springer, Zurich, Switzerland, DOI: 10.1007/978-3-319-10602-1_48

DOI: 10.2507/35th.daaam.proceedings.044

A MODEL FOR TRAINING OPTIMISATION IN A HUMAN-ROBOT COLLABORATIVE 3D PRINTING POST PROCESS

Sergio Scibilia & Giuseppe Casalino

This Publication has to be referred as: Scibilia, S[ergio] & Casalino, G[iuseppe] (2024). A Model for Training Optimisation in a Human-Robot Collaborative 3D Printing Post Process, Proceedings of the 35th DAAAM International Symposium, pp.0322-0331, B. Katalinic (Ed.), Published by DAAAM International, ISBN 978-3-902734-44-0, ISSN 1726-9679, Vienna, Austria
DOI: 10.2507/35th.daaam.proceedings.044

Abstract

This study introduces a new downtime optimisation criterion in the production line of a dynamic additive manufacturing system applied to 3D printing. The focus is on post-processing activities, where an anthropomorphic robot with 6 degrees of freedom is employed. In Human-Robot Collaboration (HRC) and Human-Robot Interaction (HRI), data from a sensor system are analysed to reduce the training time of the cobot in the post-processing activities of new parts. This study proposes a theoretical criterion for planning a new cobot training cycle, which no longer follows the chronological evolution of the production cycle, but an ordered system of instruction packages based on the mono-modal command type. Each instruction is assigned to a mono-modal command based on the shortest time required to achieve optimal cobot training and then grouped into similar instruction packages. These packets identify the scheduling of a new training process in real time, and not offline, when the production of a new finished part is required.

Keywords: Human robot training; Physical human-robot interaction; Training collaborative environment; Industrial robot programming.

1. Introduction

Advances in collaboration between collaborative robots (cobots) and humans represent a significant breakthrough in the manufacturing sector. Cobots integrate human and robotic skills, combining precision and productivity with the added value of human intervention. Compared to conventional robots and automation technologies such as Computer Numerical Control (CNC) machines, cobots offer improved safety, simplified programming and flexibility without the need for isolated work cells, enabling rapid deployment. While conventional robots excel at repetitive tasks, cobots are designed to interact directly with operators, promoting creative solutions and problem-solving skills that are fundamental for customised and flexible production. These characteristics make cobots a technology that can improve the effectiveness of human activities in various production sectors. [1],[2] As cobot-based solutions balance the benefits of advanced automation with adaptability and human decision-making, targeted design strategies are crucial. [3] These must consider systems efficiency and the human-machine interface, ensuring smooth and safe collaboration. It is essential to create working environments in which the unique capabilities of humans and cobots integrate harmoniously, maximising productivity, innovation and operational safety.

With regard to safety in collaborative systems between small industrial robots and human operators, it is important to mention two main problems: the control of the human safety monitoring system and the downtime of the industrial robot during the activation of the safety system. [4] However, the automation of many industrial tasks requires significant effort due to the complexity of technical systems and lack of expertise. [5] A further significant obstacle to automation is the lack of interoperability standards between technology providers, which complicates the process and increases costs. Maintenance engineers must manage increasingly complex automated production systems characterised by high computerisation, high-quality data acquisition in real-world environments, storage and fusion of information from different sensors and devices, and optimisation of data flow to ensure continuous interoperability in automated production. [6],[7].[8].[9],[10] This lack of standards generates intricate interdependencies between process parameters and system components, making it difficult to assess the effects of changes or replacements.

In this context, human-robot collaboration focuses on key elements such as safety, interaction dynamics in cooperative contexts and the degree of autonomy and adaptability of cobots. [11] Robots, equipped with intelligent sensors, are designed to operate safely in close contact with humans. Cobots are able to adapt to human actions and learn new processes intuitively, without requiring complex programming. Designed to interact directly with humans and other robots, they share the workload by considering the skills and strengths of each team member. [12] In addition, they develop advanced functions such as visual perception, action recognition, prediction of intentions and safe planning in real time, avoiding collisions through specific rules. [13]

The goal is to train robots to perform tasks safely and effectively, balancing human intervention and automation to improve industrial efficiency. However, many robots are controlled by rigid codes that hinder smooth collaboration with human operators. [6] Control requires robust policies for safe and efficient operation, but their creation is often complex and costly. It is therefore essential to develop modal or multimodal methods of communication and control, such as. Speech processing, gesture recognition and haptic interaction, to enable robots to dynamically adapt pre-planned tasks and improve interaction. [14],[15],[16],[17] These approaches allow users to quickly train agents without advanced programming skills, facilitating smoother collaboration by reducing training time. In addition, the operator needs information from the system to orientate himself in his work. [18]

Programming industrial robots is a challenge, requiring specialised knowledge. Offline programming tools based on simulations can simplify the process, but human intervention is often necessary to optimise the programmes. [19] A crucial aspect is the integration of cobots with sensors for real-time feedback, facilitating collaboration in hybrid environments, speeding up training and minimising downtime. [20] This article focuses on the human-robot training cycle, not production. In this context, it is essential to explore approaches that simplify the development of control and communication policies, increasing the operational flexibility of cobots and reducing human errors. Machine learning algorithms and automatic optimisation techniques stand out for their ability to generate dynamic and adaptive solutions, facilitating programming and decreasing the need for manual intervention, thus improving the efficiency and reliability of robotic operations during training. [13],[21],[22],[23],[24],[25],[26]

Considering the different abilities and preferences of the operators, the allocation of tasks must take them into account, based on characteristics such as attention, experience and reliability. It is crucial that the robot's mental model encodes not only the sequence of necessary actions, but also expectations of operator preferences.[21],[27]. This approach allows cobots to adapt to user techniques and preferences, making collaboration in production tasks more efficient, especially during real-time adjustments. [28],[29] The main goal is for the robot to autonomously perform tasks and learn user preferences, improving the acceptance and quality of human-robot interaction. [30]

The demand for rapid reconfiguration of production systems is growing due to frequent changes in operational requirements. Systems must adapt quickly, optimising processes and reducing reconfiguration time and costs. This implies a decrease in training or re-training time, improving learning performance through knowledge transfer between human and robot, and vice versa, as well as between robot and robot in multi-agent systems. In this context, the training programme follows the chronological order of production with multimodal interactions. Crucially, machine learning algorithms are crucial for optimising training cycles by analysing large amounts of data from multiple devices. Although machine learning and data management are not the subject of this project, some articles are cited for their contribution to the literature. [7] Recent advances in machine learning make it possible to automate multimodal and sequential operations, but they have significant disadvantages:

- Higher margin of error due to multiple devices.
- Need for integrated configuration and ideal working environments.
- High computational costs and large data processing. [31]
- Training of operators in multiple control methods.

In dynamic production systems, these problems can cause prolonged downtime during training processes.

1.1 Scope of the paper

This research project focuses on a dynamic system in which continuous adaptation and learning are essential to manage variations in production requirements. The paper addresses the challenge of optimising real-time symbiotic training time between a cobot and a human operator in a Human-Robot Collaboration (HRC) system during post-processing in 3D printing.

The interaction is based on multiple devices and sensors, favouring modal communication (Human Robot Collaboration, HRI) instead of multimodal, without necessarily following the chronological order of production activities. In a dynamic industrial environment with frequent new parts to be produced, the aim is to optimise cobot and operator training time as well as system set-up time. [32] When the system is up and running, the operator and cobot know their tasks; however, a change in production generates uncertainty. This study proposes a rapid reconfiguration, training and reprogramming approach to reduce downtime, with a long-term positive impact through the continuous acquisition of data and instructions. The use of machine learning algorithms will help minimise economic losses due to downtime associated with new training cycles.

In the reconfiguration phase, it is crucial to understand how the symbiotic communication between man and robot takes place to ensure fast and efficient training. Is it more advantageous to follow a sequential approach that respects the production cycle or to use intelligent algorithms that memorise instructions? Is multi-modal training more effective than a system that breaks down operations into simple single-mode instructions? An alternative approach is to break down each activity into small instructions, identifying the fastest modal command method to achieve the optimal instruction. These instructions can be grouped into packages with homogeneous interaction modes, implementing criteria to optimise the long-term training cycle, considering the system's ability to learn autonomously. Let us put the project into context.

2. Working context

In order to test the logic of the research project, a 3D printing process of two pieces (hollow circular cylinders) on the same bed and printing process is considered. The two printed parts will then be joined and painted, resulting in a single hollow cylinder. The project therefore concerns the automation of a production cell dedicated to the machining of a part obtained from two parts produced by 3D printing, using the FFF (Fused Filament Fabrication) or FDM (Fused Deposition Modelling) technique with PLA (Polylactic Acid) thermoplastic polymer. The focus of the study is on post-processing activities, which include removing the support material, joining the parts by gluing, sanding, painting and polishing, with the aim of achieving a high-quality final part. Despite advances in 3D printing, the parts produced have surface defects that compromise their functionality and aesthetics. Post-processing operations are therefore essential to ensure the quality and reliability of the final product. [33],[34]

2.1 Production Cell

The production cell is equipped with an ecosystem of command, control and sensor devices, including a Programmable Logic Controller (PLC) and two Graphical User Interfaces (GUIs). It consists of two workstations, delimited by security fences: the Printing Station and the Post-Processing Station. [35] The two stations are connected by an Automated Guided Vehicles (AGV). The robot base can be positioned in three positions: two in the Print Station (Pos. 1 for reel replacement and Pos. 2 for loading the print bed) and one in the Post-Processing Station (Pos. 3).

Print Station (Pos. 1 and Pos. 2). The printing station consists of the 3D printer equipped with ventilation, shelving for new filament spools and print beds, and a dedicated storage tray. Post-Processing Station (Pos. 3). The robot keeps the base in a fixed position (x3, y3, z3) except when it has to move to Positions 1 and 2. It supports and/or performs all post-processing activities on a circular workbench divided into three sections according to tasks: A: Removing and gluing, B: Sanding and polishing, C: Painting. The robot arm has 360° manoeuvrability around the Z axis of the base. The unused space of the bench is intended for the AGV system and the storage of the finished part, which will be transported out of the cell by a conveyor belt. There are two tool stands on the workbench, one for the effector and one for the operator. [28],[36] The collaborative manipulator is an anthropomorphic robot with 6 degrees of freedom, a payload of 3 kg and an outreach of 0.58 metres. It features a built-in force/torque sensor and a dual quick-change system that does not compromise reach and extension capability even when two effectors are mounted simultaneously, allowing for automated changeover. [37] The quick-change enables two operations:

- Replacement Changeover: depositing one effector and replacing it with another.
- Rotary change: selection of the effector from the two available on the robot flange.

The system is equipped with seven connectors, such as grippers and tools, to guarantee all tasks: grippers to load and unload filament reels and print beds, grippers to move and place workpieces, a glue gun, a spindle with rotating heads for sanding and polishing, and a paint gun. [38] The proposed layout aims to ensure quick and easy set-up or reconfiguration of the HRC workstation, in line with the goal of increasing production speed, reducing cycle time and containing costs.

2.2 Human-Robot Interactive Ecosystem

Our system is a generalised unified model for active and passive resources, including humans, robots, printers and structures, as well as a system of sensing devices. [39] The latter captures arm, body, hand and finger movements, as well as sonic detections during training, providing valuable data for machine learning and enabling cobots to self-train for future tasks. [13] The project's HRI ecosystem hardware, crucial for sensing, visualisation, command and control, enables eleven (11) types of interaction as is illustrated below, enabling real-time training between human and cobot. [22],[40],[41].

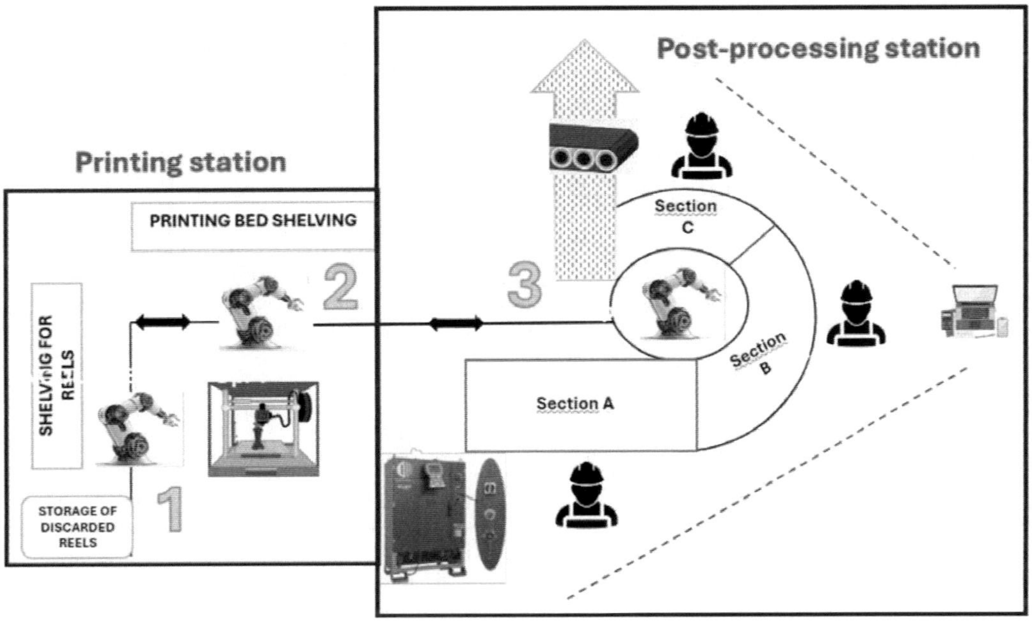

Fig. 1. Layout production cell

The interaction modes required for training and related devices are classified according to the degree of virtualisation:

Real:

- Leap Motion optical sensor for hand gestures.
- Technological gloves for tactile and precision interaction.
- Visual instructions via joypad, touch pendant and monitor/tablet.
- Voice recognition with Audio Science Review (ASR) microphones mounted on head and positioned in the cell.
- Manual guidance with direct robot control, force/torque sensing for precision and safety.
- Manual guidance without direct robot contacts via joypad or touch pendant.
- Real-time instruction via interactive video tutorials on monitor/tablet.
- Man-to-man training with direct learning.

Semi-Real:

- Monitor/tablet with touch controls and 2D vision for Augmented Reality (AR).
- 3D HDM (Head-Mounted Display) helmets for virtual controls in augmented reality (AR).

Virtual:

- RV helmets in 3D for virtual commands through virtual reality (VR) viewers.

A unified communication language has been implemented, enabling the integration and use of all components and services within the device architecture, based on a three-tier Service Oriented Architecture (SOA). As a Reconfigurable Manufacturing Systems (RMS), the system respects fundamental characteristics such as modularity, integrability, customisation, convertibility, scalability and diagnostics. [24],[32],[42],[43],[44] This architecture is designed to be service-oriented, ensuring compliance and interoperability between autonomous systems. By emulating our system through a virtual Digital Twin, we have identified a structure organised on three distinct levels, which are subject to constraints and operating parameters, and which communicate bilaterally and share common information. [45],[46],[47]

1.System and technological/production conditions:

- System: production times and volumes, progress, downtime reduction, technological requirements, training, working conditions and safety regulations, etc.
- Product: PLA material, weight, dimensions, cooling, handling, etc.
- Economic efficiency: resource costs, energy, hourly costs, installation, filament and technologies, etc.

2.Tasks:

- Human performance: hazards, workload, ergonomics, fatigue, physical characteristics and experience, etc.
- Robot performance: speed, reach, load, safety, graspability, etc.
- Human-robot team: control, trust, monitoring and feedback on robot performance, etc.

3.Human-robot communication:

This layer includes a two-way Human-Robot/System channel, using interface devices and sensors to transmit data and targets between the operator and the system.

This allows visualisation of pending and ongoing tasks, robot trajectories and notifications of events and results of operations. Communication devices, such as augmented reality and tablet interfaces, HD (High Definition) monitors and VR or HMD RA helmets, are adaptable to the operator's preferences through the knowledge base ontology. The HRI ecosystem will be constrained by specific metrics and constraints related to device combination (multimodal) and modal level (unique) functionality associated with each device.

- A combinatorial/multimodal level: reliability, robustness and cognitive load.
- A modal level (single device): extent of use, flexibility, wearability, connectivity and durability.

Durability: time required to interpret and execute the instruction, estimated numerically. This value is derived from the time identified for the human or robot to complete the optimal instruction in the training process. We can identify technologies with specific applicability for instructions/operations with shorter response times than others. For example, the use of a joypad will take longer than manual guidance with robot arm contact in a positioning instruction. [16],[28],[48]

3. Approach

Analysing the entire production process examined in this study, eight main activities were identified: three in the printing station and five in the post-processing station. [35],[49],[50] To simplify the conceptual analysis, we will focus on four post-processing activities, ordered chronologically: removal, gluing, sanding, and coating. The study explores different scenarios of human-robot interaction, as shown in Tables 1 and 2, highlighting specific characteristics, contributions, and tasks for each activity. It should be emphasized that the interaction in the backing material removal and gluing activities is collaborative human-robot, avoiding a fully automated system, which is not suitable for the demonstration purposes of this study.

Location and Workstation	Activities	Level of interaction	Human task	Cobot task	Joint effort
Pos.3_Post-processing	Substrate Removal	Assistance. Man - Robot Collaboration	Man manually removes material	Robot grabs and holds workpieces at a fixed position and manages their weight	Yes
Pos.3_Post-processing	Past Processing	Assistance. Man - Robot collaboration	Man distributes glue and glues the workpiece via Robot effector	Robot maintains glue gun with orientation and position of workpiece controlled by man	Yes
Pos.3_Post-processing	Levelling	Robot programmed and man can intervene	Man supervises activity from close range	Robot smoothes surface, moving parallel to it automatically	Man supervises
Pos.3_Post-processing	Painting	Fully programmed robot	None	Robot sprays the surface automatically	No

Table 1. Human-robot collaborative interactions and scenarios, contributions and characteristics 1

Location and Workstation	Activities	Physical contact	Mode of operation Robot	Type of workspace	Working tools and instruments	Safety conditions
Pos.3_Post-processing	Substrate Removal	Required	Required	Adaptive	Cutter, sandpaper, workpiece gripper	Robot is moved manually
Pos.3_Post-processing	Past Processing	Required	Adaptive	Shared	Glue gun, workpiece gripper	Robot is moved manually
Pos.3_Post-processing	Levelling	Allowed	Maximum speed / Full stop on approach Man	Shared with side barriers	Sanding spindle, workpiece gripper	Robot stops when Man enters work section
Pos.3_Post-processing	Painting	Excluding	Maximum speed / Full stop on approach Man	Not shared	Sprayer Spray, workpiece clamp	Robot stops when Man enters work section

Table 2. Human-robot collaborative interactions and scenarios, contributions and characteristics 2

The total of these four activities requires a total of thirty-four (34) operations. If we decompose these operations into instructions, we obtain fifty-seven (57) instructions. [51] Of the fifty-seven instructions, many coincide, leading to the identification of seventeen (17) basic instructions distinct from each other in the four observed post-processing activities. At this point, we pose the question as to which method of communication, using a single mode for each instruction, allows the optimal level of execution during training to be achieved in the shortest possible time. To answer this conceptually, we imagine various interactions were tested in the laboratory, assigning numerical values to each communication mode associated with a single instruction. The interaction with the shortest duration was then chosen, always guaranteeing optimal execution. The main criterion is to use a single communication mode for each instruction, applying a single device for each basic instruction. Therefore, each of the 17 instructions will be assigned a specific type of human-robot, robot-human or human-robot interaction (robot-robot, if it were a multi-agent system). Similar instructions with the same interaction mode can be grouped into packages, classified according to the single-mode command mode with the shortest execution time in the training process. In the following table, the result obtained considering the four selected post-processing activities is summarised.

N. OF BASIC INSTRUCTIONS	SIMILAR INSTRUCTION PACKET	N° OF INSTRUCTIONS / PACKAGE	MONO MODALITY CONTROL MODE
1	TILE piece 1	4	Gestures with Sensors
2	TILE piece 2	3	HDM for RA
3	COLLOC piece 1	3	Gestures with Sensors
4	COLLOC piece 2	3	HDM for RA
5	MOOVE Flange piece 1	11	Manual guide with force/torque
6	MOOVE Flange piece 2	10	Glove Handling
7	STOP flange position	2	Visual instruction
8	Man Separates with Nylon piece 1	2	Man - Man
9	Man Removes support piece 1	2	Man - Man
10	Man clears table	1	Man - Man
11	New effector assembly	1	Visual instruction
12	Replacement effector	4	Visual instruction
13	Rotational effector change	7	Vocal Instruction
14	Man distributes glue with guidance	1	Manual guide with force/torque
15	Man glues 1b onto 1a with guide	1	Manual guide with force/torque
16	Robot smoothes workpiece	1	Virtual simulation VR helmets
17	Robot sprays paint onto workpiece	1	Manual guide with force/torque

Table 3. Packing similar instructions according to a single mode of interaction

In this table, piece 1 refers to one of the two cylindrical pieces 1a and 1b obtained simultaneously by the printing process 3, and piece 2 to the part obtained after the gluing activity of the two pieces 1. Considering this methodology, we do not follow a human-robot training in a chronological order related to production time; instead, we contemplate a training cycle aimed at optimising and distributing a set of similar instruction packages requiring the same type of interaction. The Figure 2 shows the flowchart of the logical steps of this proposed study, representing the generic planning of the training cycle.

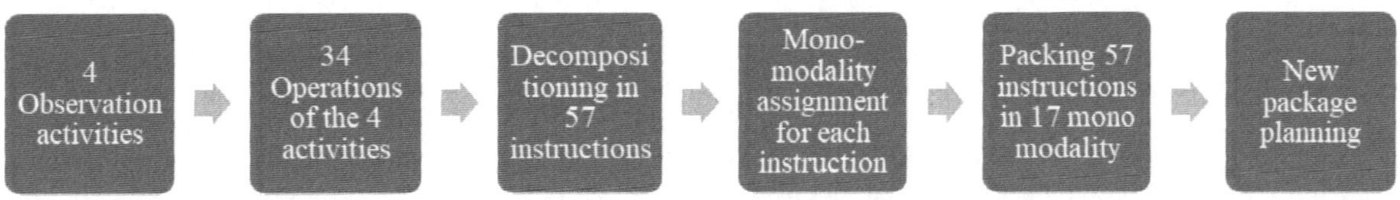

Fig. 2. Generic planning of the training cycle.

During the training cycle, all obtained optimal instructions related to human-robot interaction are stored and subsequently reassigned to the original activities and operations, following the planned chronology of the production cycle. This process, guaranteed by an intelligent decision-making algorithm, contributes to task planning between man and cobot in the production phase. [52] In other words, all activities already broken down into individual instructions are sent to a planning algorithm, which returns the result to the activity management system once the instruction is completed. If the instruction is executed correctly during training, it proceeds to the next one; in the event of an error, it is repeated until the instruction has been executed perfectly. [29]

4. Discussion

Due to the vastness of the content, it was necessary to identify a structure of assumptions that would keep the concept of this proposed approach consistent and avoid dispersion of information. In this section, we present these assumptions and system limitations in a clear and concise manner. A significant limitation was the difficulty in collecting information from the companies involved. The survey found that cobots are mainly used for simple tasks and long cycles, with limited flexibility and reusability. Training, still focused on traditional robotics, lacks standardization and requires specific skills for each cobot model. Although the interfaces are intuitive, advanced programming requires the support of specialized engineers. Therefore, training focused exclusively on cobots is needed. These are chosen for their low cost, space savings, and safety, but their potential remains underutilized.

Another limitation was and the lack of a laboratory to perform the necessary experiments, making it impossible to validate through practical tests the "Duration" parameter of the instruction to achieve optimal execution. However, the approach presented in this study remains conceptually sound, as the numerical value of the "Duration" parameter can still be estimated through specific tests. Moreover, the exact accuracy of this numerical value is not essential for understanding the content of the article. In this study therefore, the only analysis metric considered is "Duration". Although other metrics may influence the choice of the most appropriate mode of interaction in real-world scenarios, we assume that all these variables and constraints have already been analysed and evaluated in the three levels of our digital twin model (System, Tasks and Communication). Therefore, the decision criterion is based solely on the "Duration" constraint/metric.

The project focuses on proximal, online and real-time communication technologies, excluding default scheduling systems in offline or remote virtual environments. The system analysed involves the optimal implementation of devices for human-system interaction, adopting a multi-view configuration that ensures robustness against occlusions, blurring and lighting variations, accurately detecting both visual and acoustic attributes. Each device operates without interference, ensuring optimal execution of instructions, operations and activities. At the level of system structure and layout, analyses regarding compliance with safety standards or the technologies needed to ensure them are not considered, as they are beyond the scope of this study. In addition, human-human interactions and self-training are considered, in addition to human-cobot/system interactions, as the latter may be among the most effective in terms of quality and instruction execution time. This choice aims to make the study scenario more generic and representative of reality.

5. Conclusion

Through the analysis of this new training criterion, two fundamental conclusions emerge.

1. Training of instruction packages according to mono-modality.

- The human operator, by repeating very similar instructions through the 'Learning by Demonstration' method, would quickly attain sufficient mastery to execute all optimal instructions in a single session. This applies not only to the four post-processing tasks in our case study but could extend to the entire production cycle.
- As this is a mono-mode rather than multi-mode instruction block, the system is able to process a reduced data flow. This approach decreases the number of errors, reduces delays caused by latencies and interface updates, and lowers computational costs. Looking into the future, and considering the dynamic nature of the system, we are processing many similar data and commands, allowing the system to self-learn more quickly and easily, facilitating optimised instruction execution and reducing deviations, in accordance with a Gauss curve.
- The mono-mode characteristic of instruction packages allows for continuity of training without the need to reconfigure the system until switching to another package classified with a different mode of interaction. This has a positive impact on the training cycle of the entire production process.

2. Planning instruction packages and optimising the training cycle.

Once the instruction packages characterised by mono-modality have been identified, it is necessary to define the execution sequence of these packages. We can consider various parameters and constraints, such as:

- Technological flexibility of the device-mode. It is possible to use the device for other instructions, reducing the hardware setup time. For example, sequencing manual force/torque guidance to execute the instructions 'MOVE Flange piece 1' and 'Man distributes glue with guidance" could facilitate technological reconfiguration.
- Physical device proximity and wearability. If the sensor for gesture recognition is mounted on the robot base, and the microphone for speech recognition is also placed on the robot base, the training packages could be executed in succession to optimise the process.
- Margin of error: Although RV helmets may offer shorter execution times, they have a larger margin of error. Therefore, the operator could adopt a training logic that involves "first the complex activities and then the simpler ones", ordering the packages from the most difficult to the simplest.

In addition, the adoption of other parameters could be considered, one of them being machine learning algorithms that, based on historical data from previous production and training orders, are able to recognise the optimal sequence for training packages or individual instructions.

5. Future works

The method presented is an important step toward integrating human and robot operators in shared workspaces, with the goal of assigning tasks dynamically. However, it has not yet been tested experimentally, so aspects remain to be further investigated. Artificial intelligence, particularly potential machine learning algorithms (Supervised Learning, Unsupervised Learning, Semi-supervised Learning, Reinforcement Learning, and Deep Learning) has not been considered. Future studies could explore which algorithm is best suited for each instructional package. For example, Supervised Learning is ideal for simple instructions, while Reinforcement Learning is better suited for complex instructions, especially an already trained system. In the long run, this approach may accelerate the self-learning process.

Another aspect to consider concerns the possible adaptation of the algorithms as the system's learning progresses. The planning algorithm, at a higher level, could optimize the sequence of instruction packets and, in the future, fully automate planning, reducing the need for human supervision. As technology evolves, it is expected that human-robot interaction tools will become more performant, and machine learning will play a crucial role in managing and optimizing data streams, fostering continuous system improvement.

6. Declaration of conflicts of interest

The authors declare that they have no known financial conflicts of interest or personal relationships that could have influenced the work reported in this article.

7. Acknowledgments

We would like to express our sincere thanks for the financial support received from Philip Morris Italy and Birex Competence Center, which provided crucial initial funding for the development of this first conceptual approach, as well as for study and research activities.

8. References

[1] Michaelis, Joseph & Siebert-Evenstone, Amanda & Shaffer, David & Mutlu, Bilge. (2020). Collaborative or Simply Uncaged? Understanding Human-Cobot Interactions in Automation. 1-12. 10.1145/3313831.3376547.

[2] Mohd Javaid, Abid Haleem, Ravi Pratap Singh, Shanay Rab, Rajiv Suman. (2022). Significant applications of Cobots in the field of manufacturing, Cognitive Robotics, Volume 2, Pages 222-233, ISSN 2667-2413, https://doi.org/10.1016/j.cogr.2022.10.001.

[3] Djuric, Ana & Urbanic, Ruth Jill & Rickli, Jeremy. (2016). A Framework for Collaborative Robot (CoBot) Integration in Advanced Manufacturing Systems. SAE International Journal of Materials and Manufacturing. 9. 10.4271/2016-01-0337.

[4] Kuts, Vladimir & Šarkans, Martinš & Otto, Tauno & Tahemaa, Toivo. (2017). Collaborative Work Between Human And Industrial Robot In Manufacturing By Advanced Safety Monitoring System. 10.2507/28th.daaam.proceedings.138.

[5] Eike Schäffer, Matthias Bartelt, Tobias Pownuk, Jan-Peter Schulz, Bernd Kuhlenkötter, Jörg Franke. (2018). Configurators as the basis for the transfer of knowledge and standardized communication in the context of robotics,Procedia CIRP, Volume 72, Pages 310-315, ISSN 2212-8271, https://doi.org/10.1016/j.procir.2018.03.190.

[6] Debasmita Mukherjee, Kashish Gupta, Li Hsin Chang, Homayoun Najjaran. (2022) .A Survey of Robot Learning Strategies for Human-Robot Collaboration in Industrial Settings, Robotics and Computer-Integrated Manufacturing, Volume 73, 102231, ISSN 0736-5845, https://doi.org/10.1016/j.rcim.2021.102231.

[7] Benjamin Maschler, Timo Müller, Andreas Löcklin, Michael Weyrich. (2022). Transfer Learning as an Enhancement for Reconfiguration Management of Cyber-Physical Production Systems, Procedia CIRP, Volume 112, Pages 220-225, ISSN 2212-8271, https://doi.org/10.1016/j.procir.2022.09.095.

[8] Angelos Argyrou, Christos Giannoulis, Andreas Sardelis, Panagiotis Karagiannis, George Michalos, Sotiris Makris. (2018). A data fusion system for controlling the execution status in human-robot collaborative cells, Procedia CIRP, Volume 76, Pages 193-198, ISSN 2212-8271, https://doi.org/10.1016/j.procir.2018.01.012.

[9] Ramasubramanian, A.K., Mathew, R., Kelly, M., Hargaden, V., & Papakostas, N. (2022). Digital Twin for Human–Robot Collaboration in Manufacturing: Review and Outlook. Applied Sciences, 12(10), 4811; https://doi.org/10.3390/app12104811

[10] Cohen, Y., Naseraldin, H., Chaudhuri, A. et al. (2019). Assembly systems in Industry 4.0 era: a road map to understand Assembly 4.0. Int J Adv Manuf Technol 105, 4037–4054 https://doi.org/10.1007/s00170-019-04203-1

[11] Shirine El Zaatari, Mohamed Marei, Weidong Li, and Zahid Usman. (2019). Cobot programming for collaborative industrial tasks: An overview. Robot. Auton. Syst. 116, C (Jun 2019), 162–180. https://doi.org/10.1016/j.robot.2019.03.003

[12] Malik, Ali Ahmad & Bilberg, Arne. (2017). Framework to Implement Collaborative Robots In Manual Assembly: A Lean Automation Approach. 1151-1160. 10.2507/28th.daaam.proceedings.160.

[13] Reinhardt, Dagmar & Haeusler, M. Hank & London, Kerry & Loke, Lian & Feng, Yingbin & Barata, Eduardo & Firth, Charlotte & Dunn, Kate & Khean, Nariddh & Fabbri, Alessandra & Wozniak-O'Connor, Dylan & Masuda,

Lynn. (2020). CoBuilt 4.0: Investigating the potential of collaborative robotics for subject matter experts. International Journal of Architectural Computing. 18. 147807712094874. 10.1177/1478077120948742.

[14] L. Wang, R. Gao, J. Váncza, J. Krüger, X.V. Wang, S. Makris, G. Chryssolouris. (2019). Symbiotic human-robot collaborative assembly, CIRP Annals, Volume 68, Issue 2, Pages 701-726, ISSN 0007-8506, https://doi.org/10.1016/j.cirp.2019.05.002.

[15] Kardos, Csaba & Kemény, Zsolt & Kovacs, Andras & Pataki, Balázs & Váncza, J.. (2018). Context-dependent multimodal communication in human-robot collaboration. Procedia CIRP. 72. 15-20. 10.1016/j.procir.2018.03.027.

[16] Patrik Gustavsson, Magnus Holm, Anna Syberfeldt, Lihui Wang. (2018). A Human-robot collaboration – towards new metrics for selection of communication technologies, Procedia CIRP, Volume 72, Pages 123-128, ISSN 2212-8271, https://doi.org/10.1016/j.procir.2018.03.156.

[17] Hongyi Liu, Tongtong Fang, Tianyu Zhou, Yuquan Wang, Lihui Wang. (2018). Deep Learning-based Multimodal Control Interface for Human-Robot Collaboration, Procedia CIRP, Volume 72, Pages 3-8, ISSN 2212-8271, https://doi.org/10.1016/j.procir.2018.03.224.

[18] Gustavsson, Patrik & Holm, Magnus & Syberfeldt, Anna & Wang, Lihui. (2018). Human-robot collaboration – towards new metrics for selection of communication technologies. Procedia CIRP. 72. 10.1016/j.procir.2018.03.156.

[19] Alejandro Magaña Flores, Philipp Bauer, Gunther Reinhart. (2019). Concept of a learning knowledge-based system for programming industrial robots, Procedia CIRP, Volume 79, Pages 626-631, ISSN 2212-8271, https://doi.org/10.1016/j.procir.2019.02.076.

[20] Nikolaos Nikolakis, Kostantinos Sipsas, Panagiota Tsarouchi, Sotirios Makris. (2018). On a shared human-robot task scheduling and online re-scheduling, Procedia CIRP, Volume 78, Pages 237-242, ISSN 2212-8271, https://doi.org/10.1016/j.procir.2018.09.055.

[21] Nikolaidis S, Lasota P, Ramakrishnan R, Shah J. (2015). Improved human–robot team performance through cross-training, an approach inspired by human team training practices. The International Journal of Robotics Research. 2015;34(14):1711-1730. doi:10.1177/0278364915609673

[22] Valeria Villani, Fabio Pini, Francesco Leali, Cristian Secchi. (2018). Survey on human–robot collaboration in industrial settings: Safety, intuitive interfaces and applications, Mechatronics, Volume 55, Pages 248-266,ISSN 0957-4158, https://doi.org/10.1016/j.mechatronics.2018.02.009.

[23] Francesco Semeraro, Alexander Griffiths, Angelo Cangelosi. (2023). Human–robot collaboration and machine learning: A systematic review of recent research, Robotics and Computer-Integrated Manufacturing, Volume 79, 102432, ISSN 0736-5845, https://doi.org/10.1016/j.rcim.2022.102432.

[24] W. Bradley Knox and Peter Stone. (2009). Interactively shaping agents via human reinforcement: the TAMER framework. In Proceedings of the fifth international conference on Knowledge capture (K-CAP '09). Association for Computing Machinery, New York, NY, USA, 9–16. https://doi.org/10.1145/1597735.1597738

[25] C. Wang, X.P. Tan, S.B. Tor, C.S. Lim. (2020). Machine learning in additive manufacturing: State-of-the-art and perspectives, Additive Manufacturing, Volume 36,101538, ISSN 2214-8604, https://doi.org/10.1016/j.addma.2020.101538.

[26] Rong Zhang, Qibing Lv, Jie Li, Jinsong Bao, Tianyuan Liu, Shimin Liu. (2022). A reinforcement learning method for human-robot collaboration in assembly tasks, Robotics and Computer-Integrated Manufacturing, Volume 73, 102227, ISSN 0736-5845, https://doi.org/10.1016/j.rcim.2021.102227.

[27] L. Johannsmeier and S. Haddadin. (2017). A Hierarchical Human-Robot Interaction-Planning Framework for Task Allocation in Collaborative Industrial Assembly Processes, in IEEE Robotics and Automation Letters, vol. 2, no. 1, pp. 41-48, doi: 10.1109/LRA.2016.2535907.

[28] Ciccarelli, M., Forlini, M., Papetti, A. et al. (2024). Advancing human–robot collaboration in handcrafted manufacturing: cobot-assisted polishing design boosted by virtual reality and human-in-the-loop. Int J Adv Manuf Technol 132, 4489–4504. https://doi.org/10.1007/s00170-024-13639-z

[29] Umbrico, A.; Orlandini, A.; Cesta, A.; Faroni, M.; Beschi, M.; Pedrocchi, N.; Scala, A.; Tavormina, P.; Koukas, S.; Zalonis, A.; et al. (2022). Design of Advanced Human–Robot Collaborative Cells for Personalized Human–Robot Collaborations. Appl. Sci. 12, 6839. https://doi.org/10.3390/app12146839

[30] Munzer, T., Toussaint, M. & Lopes, M. (2018). Efficient behavior learning in human–robot collaboration. Auton Robot 42, 1103–1115. https://doi.org/10.1007/s10514-017-9674-5

[31] Nikolaos Nikolakis, Konstantinos Sipsas, Sotiris Makris. (2018). A cyber-physical context-aware system for coordinating human-robot collaboration, Procedia CIRP, Volume 72, Pages 27-32, ISSN 2212-8271, https://doi.org/10.1016/j.procir.2018.03.033.

[32] Qizhang Zhu, Sihan Huang, Guoxin Wang, Shokraneh K. Moghaddam, Yuqian Lu, Yan Yan. (2022). Dynamic reconfiguration optimization of intelligent manufacturing system with human-robot collaboration based on digital twin, Journal of Manufacturing Systems, Volume 65, Pages 330-338, ISSN 0278-6125, https://doi.org/10.1016/j.jmsy.2022.09.021.

[33] Kristiawan, Ruben Bayu, Imaduddin, Fitrian, Ariawan, Dody, Ubaidillah, and Arifin, Zainal. (2021). A review on the fused deposition modeling (FDM) 3D printing: Filament processing, materials, and printing parameters" Open Engineering, vol. 11, no. 1, pp. 639-649. https://doi.org/10.1515/eng-2021-0063

[34] Wei Gao, Yunbo Zhang, Devarajan Ramanujan, Karthik Ramani, Yong Chen, Christopher B. Williams, Charlie C.L. Wang, Yung C. Shin, Song Zhang, Pablo D. Zavattieri, (2015). The status, challenges, and future of additive manufacturing in engineering, Computer-Aided Design, Volume 69, Pages 65-89, ISSN 0010-4485, https://doi.org/10.1016/j.cad.2015.04.001.

[35] Iina Aaltonen, Timo Salmi, Ilari Marstio. (2018). Refining levels of collaboration to support the design and evaluation of human-robot interaction in the manufacturing industry, Procedia CIRP, Volume 72, Pages 93-98,ISSN 2212-8271, https://doi.org/10.1016/j.procir.2018.03.214.

[36] Fan Mo, Hamood Ur Rehman, Fabio Marco Monetti, Jack C. Chaplin, David Sanderson, Atanas Popov, Antonio Maffei, Svetan Ratchev, A framework for manufacturing system reconfiguration and optimisation utilising digital twins and modular artificial intelligence. (2023). A framework for manufacturing system reconfiguration and optimisation utilising digital twins and modular artificial intelligence, Robotics and Computer-Integrated Manufacturing, Volume 82, 02524, ISSN 0736-5845, https://doi.org/10.1016/j.rcim.2022.102524.

[37] Manuel Beschi, Marco Faroni, Cosmin Copot, Nicola Pedrocchi. (2020). How motion planning affects human factors in human-robot collaboration, IFAC-PapersOnLine, Volume 53, Issue 5, Pages 744-749, ISSN 2405-8963, https://doi.org/10.1016/j.ifacol.2021.04.167.

[38] Bajrami, Albin & Palpacelli, Matteo. (2023). A Flexible Framework for Robotic Post-Processing of 3D Printed Components. 10.1115/DETC2023-109746.

[39] Panagiota Tsarouchi, Jason Spiliotopoulos, George Michalos, Spyros Koukas, Athanasios Athanasatos, Sotiris Makris, George Chryssolouris. (2016). A Decision Making Framework for Human Robot Collaborative Workplace Generation, Procedia CIRP, Volume 44, Pages 228-232, ISSN 2212-8271, https://doi.org/10.1016/j.procir.2016.02.103.

[40] Jonas Wassermann, Axel Vick, Jörg Krüger. (2018). Intuitive robot programming through environment perception, augmented reality simulation and automated program verification, Procedia CIRP, Volume 76, Pages 161-166, ISSN 2212-8271, https://doi.org/10.1016/j.procir.2018.01.036.

[41] Bauer, Andrea & Wollherr, Dirk & Buss, Martin. (2008). Human-Robot Collaboration: a Survey.. I. J. Humanoid Robotics. 5. 47-66. 10.1142/S0219843608001303.

[42] Jeff Morgan, Mark Halton, Yuansong Qiao, John G. Breslin. (2021). Industry 4.0 smart reconfigurable manufacturing machines, Journal of Manufacturing Systems, Volume 59, Pages 481-506, ISSN 0278-6125, https://doi.org/10.1016/j.jmsy.2021.03.001.

[43] Colombo, AW., Karnouskos, S., Mendes, JM. (2010). Factory of the Future: A Service-oriented System of Modular, Dynamic Reconfigurable and Collaborative Systems. In: Benyoucef, L., Grabot, B. (eds) Artificial Intelligence Techniques for Networked Manufacturing Enterprises Management. Springer Series in Advanced Manufacturing. Springer, London. https://doi.org/10.1007/978-1-84996-119-6_15

[44] Yelles-Chaouche, A. R., Gurevsky, E., Brahimi, N., & Dolgui, A. (2020). Reconfigurable manufacturing systems from an optimisation perspective: a focused review of literature. International Journal of Production Research, 59(21), 6400–6418. https://doi.org/10.1080/00207543.2020.1813913

[45] Dimitris Mourtzis, Thodoris Togias, John Angelopoulos, Panos Stavropoulos. (2021). A Digital Twin architecture for monitoring and optimization of Fused Deposition Modeling processes, Procedia CIRP, Volume 103, Pages 97-102, ISSN 2212-8271, https://doi.org/10.1016/j.procir.2021.10.015.

[46] W. Xiang, K. Yu, F. Han, L. Fang, D. He and Q. -L. Han. (2024). Advanced Manufacturing in Industry 5.0: A Survey of Key Enabling Technologies and Future Trends, in IEEE Transactions on Industrial Informatics, vol. 20, no. 2, pp. 1055-1068, Feb. doi: 10.1109/TII.2023.3274224.

[47] Kousi, N.; Gkournelos, C.; Aivaliotis, S.; Lotsaris, K.; Bavelos, A.C.; Baris, P.; Michalos, G.; Makris, S. (2021). Digital Twin for Designing and Reconfiguring Human–Robot Collaborative Assembly Lines. Appl. Sci. 11, 4620. https://doi.org/10.3390/app11104620

[48] Yue Yin, Pai Zheng, Chengxi Li, Lihui Wang. (2023). A state-of-the-art survey on Augmented Reality-assisted Digital Twin for futuristic human-centric industry transformation, Robotics and Computer-Integrated Manufacturing, Volume 81, 102515, ISSN 0736-5845, https://doi.org/10.1016/j.rcim.2022.102515.

[49] Castro, A.; Silva, F.; Santos, V. (2021). Trends of Human-Robot Collaboration in Industry Contexts: Handover, Learning, and Metrics. Sensors, 21, 4113. https://doi.org/10.3390/s21124113

[50] Schmidbauer, C. (2022). Adaptive task sharing between humans and cobots in assembly processes [Dissertation, Technische Universität Wien]. repositUm. https://doi.org/10.34726/hss.2022.81342

[51] Nikolaos Nikolakis, Niki Kousi, George Michalos, Sotiris Makris. (2018). Dynamic scheduling of shared human-robot manufacturing operations, Procedia CIRP, Volume 72, Pages 9-14, ISSN 2212-8271, https://doi.org/10.1016/j.procir.2018.04.007.

[52] Tsarouchi, P., Matthaiakis, A. S., Makris, S., & Chryssolouris, G. (2016). On a human-robot collaboration in an assembly cell. International Journal of Computer Integrated Manufacturing, 30(6), 580–589. https://doi.org/10.1080/0951192X.2016.1187297

DOI: 10.2507/35th.daaam.proceedings.045

REMOTE TRANSMISSION OF INFORMATION DURING THE OPERATION OF ROBOTS FANUC BY USING FANUC ROBOGUIDE

Boryan Vladimirov & Stiliyan Nikolov

This Publication has to be referred as: Vladimirov, B[oryan] & Nikolov, S[tiliyan] (2024). Remote Transmission of Information During the Operation of Robots FANUC by using FANUC ROBOGUIDE, Proceedings of the 35th DAAAM International Symposium, pp.0332-0339, B. Katalinic (Ed.), Published by DAAAM International, ISBN 978-3-902734-44-0, ISSN 1726-9679, Vienna, Austria
DOI: 10.2507/35th.daaam.proceedings.045

Abstract

Many technical solutions are currently being used to achieve remote programming and management of production equipment. This document reviews the communication protocols that could support the remote operation/programming/ of industrial robots FANUC, through the use of a ROBOGUIDE system. Defined here are the major steps necessary to establish a link between the controller of a real industrial robot and the PC on which a FANUC ROBOGUIDE system has been installed, including the opportunity of providing communication between them. A server and a customer have been developed for the remote transmission of data to industrial robot FANUC LR Mate 200iD7L.

Keywords: Fanuc Roboguide; Ethernet; Profibus; Profinet; Client-Server System.

1. Introduction

Industrial robots are increasingly used in the modern industry. They are the basis for bringing the production equipment in line with the requirements of Industry 5.0 [1]. As a result of that, the range (envelope) of tasks performed by them has increased, which has impacted the complexity of the programs the industrial robots work on. At the same time, the possibilities for remote transmission of the information necessary for the operation of industrial robots are also being developed. This has created prerequisites for the widespread introduction in the engineering practice of specialized computer systems for the programming of industrial robots. These are the CARC (Computer Aided Robot System) systems, also known as the OLPE (Off-line Programming Environments). CARC systems take advantage of the recent successes in the field of computing technology and engineering computer graphics [2]. By employing 3D models of industrial robots and other production equipment, these systems allow for the development of robotic production schemes, the simulation and analysis of their operation, and the generation of management programs for the industrial robots included in them. Nowadays, all leaders in the manufacturing of industrial robots offer their users CARC programming systems for the industrial robots sold by them.

One of the problems in using CARC systems is related to the data transfer between the system and the robot controller. This document examines the possibilities which FANUC ROBOGUIDE provides to the users for remote transmission of information during the operation of industrial robots FANUC. Section 2 discusses the characteristics of the main communication protocols and their application when working with ROBOGUIDE. Section 3 describes the basic steps required to ensure communication between FANUC ROBOGUIDE and the real robot controller. The developed server and client necessary for the implementation of the data transfer in the operation of FANUC robots with using FANUC ROBOGUIDE are presented.

2. Communication protocols used in FANUC ROBOGUIDE

A lot of technical solutions are being used these days for the remote programming and management of industrial equipment that performs a specific manufacturing program. When operating industrial robots, two interconnected groups of tasks can be distinguished: the generation of control programs and the transfer of data between the robot controller and the platform on which the remote programming and control software is installed. To solve the tasks of the first group, when working with FANUC industrial robots, the CARC system ROBOGUIDE [7] is used. ROBOGUIDE can be used to: design robotic systems; generate control programs for FANUC industrial robots; create multiple 3D simulations and more.

In order to complete the tasks included in the second group, the controllers of industrial robots FANUC employ communication protocols through which a certain type of information is being transmitted. For each communication protocol, FANUC has assigned a software option and an application used by the industrial robot. Most developments are currently focused on the design of robotic systems and the generation of programs aimed at controlling the industrial robots included in them and any additional equipment [3], [4], [5], [6]. As a result, tasks related to data transfer between individual components of robotic systems and remote-control systems remain in the background. The communication protocols through which FANUC industrial robots can be remotely operated using the ROBOGUIDE CARC system are discussed here.

2.1. Ethernet

Ethernet/IP (Ethernet/Industrial Protocol) [8] is a communication system suitable for industrial environments. Ethernet/IP enables the industrial devices to exchange information that is critical in terms of both time and their applications. These devices include simple I/O (Input/Output) equipment such as sensors, as well as complex machinery such as industrial robots and PLC (Programmable Logic Controller). Ethernet/IP uses standard Ethernet and TCP/IP (Transmission Control Protocol/Internet Protocol) technology to transport CIP (Common Industrial Protocol) communication packages. The result is a common application layer on top of the open and very popular Ethernet and TCP/IP protocols. Ethernet/IP provides a producer/consumer model for the exchange of time-critical control data. The producer/consumer model allows for the application information to be exchanged between a sending device (the producer) and many receiving devices (the consumers), without the need of sending data multiple times to a number of destinations. Ethernet/IP employs standard IEEE 802.3 technology. It is recommended to use switch technology with 100 Mbps bandwidth and full duplex so as to ensure deterministic performance.

In a free configuration of robots FANUC, Ethernet can only be used as a network scanner. Furthermore, when working with EDA (Enhanced Data Access), Ethernet/IP allows the users to configure data that will be sent to a PLC. This provides an easy method of setting up and communicating production data that is not normally mapped as normal I/O. Default data structures selected by the user from PLC logic can be directly imported into a PLC. Names of variables are imported in the form of comments defined by the robot into the PLC data structures without re-entering the comments in the PLC. Available data types include: Registers (R), String Registers (SR), Positional Registers (PR) and System Variables.

2.2. PROFINET

PROFINET (Process Field Network) [9] is an industry technical standard for data communication over industrial Ethernet, designated for data collection from and control of equipment in industrial systems. It is suitable for data delivery under short time constraints. The robot controller can communicate in real time with other PROFINET devices such as PLC or peripherals through PROFINET in real time. There are I/O controllers that regulate the network and I/O devices which are connected to an I/O controller. The robot controller functions as an I/O device for communication with PLC. On the other hand, the robot controller operates as an I/O controller that communicates with peripherals.

The hardware configuration of PROFINET in FANUC industrial robots is accomplished by using two hardware devices, a PCB device (FANUC hardware [14]) and a PROFINET PCB BOARD (Siemens hardware). It is possible to use SKI (SIMATIC Kinematics Integrator). This is a software solution for controlling a robot (kinematics) with Siemens S7-1500T industrial controllers. It consists of a function block to execute (added library for the corresponding robot) a certain motion program, including robot, additional axes and inputs/outputs, such as conveyor, gripper, etc. The application can optionally be extended with several HMI screens that allow the user to operate, program, commission and diagnose the robot program. If necessary, the functionality of SKI can be extended by so-called technological adds, such as palletizing, product register or G-code processing.

Simatic Robot Integrator works with Siemens S7-1500T technology controllers under TIAv18. In the future, it is planned to be able to be integrated into the entire range of Siemens S7-1500 and S7-1200 series controllers and for all versions of the TIA Portal [12].

2.3. PROFIBUS

PROFIBUS (Process Field Bus) [10] enables the devices of various manufacturers to communicate between them without any special interface settings. Two different types of PROFIBUS are used nowadays:

- PROFIBUS DP – it is used to manage sensors and actuators through a centralized controller in manufacturing automation applications.
- PROFIBUS DA – it is used to monitor measuring equipment through a process control system in process automation applications.

Industrial robots FANUC use the interface option PROFIBUS DP implemented on two interface boards: PROFIBUS

DP Master and PROFIBUS Slave. These Master/Slave functions can be connected to different networks. In one of the networks, it is connected to, the robot operates (main) as a master for the exchange of I/O data with the peripheral equipment. In the other network, the robot works as a Slave (subordinate) device for the exchange of I/O data with PLC and other equipment. A single Master device can communicate with multiple Slave devices.

2.4. MODBUS

MODBUS [11] supports the I/O exchange between activated MODBUS TCP devices over Ethernet. The TCP/IP used is based on the customer-server model. The industrial robot option MODBUS TCP functions only as a server (master), and therefore, it will exchange I/O input-output data only with customer (slave) devices. Ethernet has not been determined and MODBUS TCP does not guarantee any time. The robot's MODBUS TCP Server will time out and displayed a message: ERROR, if the I/O exchange is not received within the configured timeout, unless the timeouts are disabled. Good network design and topology are crucial for the successful implementation of MODBUS TCP in terms of fast and reliable communication between two devices. Any customer device MODBUS TCP, may be linked to a non-priority connection (a connection that is not marked as priority). Where all available non-priority connections are being used, and another MODBUS TCP customer is trying to establish a non-priority connection, then the oldest non-priority connection in existence will be closed, and the attempt to create a new one will be recorded. This will trigger a PRIO-494 warning alarm. By default, all connections are non-priority ones.

The MODBUS TCP server supports up to two priority connections by indicating an IP address to MODBUS TPC clients. The priority connections can be configured by setting one or both of the following system variables to a valid IP address or hostname:

- $MODBUSTCP.$PRIO1
- $MODBUSTCP.$PRIO2

This means that there is no other MODBUS TCP client device, that can connect to the robot and use a priority connection except the device with the corresponding IP address. The MODBUS TCP server of the industrial robot will never terminate a priority connection without it being closed by the MODBUS TCP client, or in case of a time out.

2.5. Socket programing

The KAREL program language is used in the programing of industrial robots FANUC. It is a combination of high-level languages such as PASCAL and PL/1. KAREL includes structures and models that are common for the high-level languages, as well as functions developed especially for application of robot FANUC. The FANUC robot's messaging option for user socket (port address) Socket Messaging allows using TCP/IP socket messaging in a KAREL environment. Socket Messaging enables data exchange between industrial robots in a networked environment and a remote computer. The combination of the robot's PC-Interface option and the PC-Developed KIT of the computer, is recommended for data exchanged between the robot and a computer with UNIX, LINUX or Windows operational system. Socket Messaging uses TCP/IP protocol for the transfer of unprocessed or original unformatted data in the net.

A typical application of Socket Messaging is a robot that works with a KAREL program that sends process information to a monitoring program on the remote computer. The commands and methods which Socket Messaging uses to transfer data are part of the TCP/IP protocol. Since Socket Messaging supports customer and server tags, applications that require timeout or data formatting commands can also be provided to the customer on the server of the socket messaging link. Socket Messaging is compatible with all network hardware configurations which employ TCP/IP net protocol. Some of these network hardware configurations include Ethernet and serial PPP (Point-to-Point Protocol) connections. Socket messages are an integrated component of KAREL. When socket messaging functions and KAREL utility programs are being used, the syntax is similar to the one of other operations for file reading and saving, except that where messaging functions and other network services are being used, a network connection will have to be established.

3. Remote transmission of information

The remote transmission of information during the operation of robots FANUC through the use of FANUC ROBOGUIDE requires the real robot's controller to be connected to the PC on which FANUC ROBOGUIDE has been installed. The process includes the following steps:

3.1. Server creation

At this stage, it is necessary to create a new project in FANUC ROBOGUIDE with the following installed options: R632 (Karel) and R648 (User Socket Messaging). For the purpose of this document, industrial robot FANUC LR Mate 200iD/7 L has been used. Upon the creation of a server in FANUC ROBOGUIDE one shall use the possibilities of KAREL. The basic commands supported by KAREL for SOCKET MESSAGING are as follows:

- MSG_CONN (string, integer) – a command that indicates the establishment of connection during the transmission of a message. She will return a value for the condition of the remote server which is trying to accept the connection. Should the connection be unsuccessful, the command will return a value showing the emergence of an error in the connection.
- MSG_DISCO (string, integer) – a command used for the closing of a connection. If the connection breaks up without a command, that means that there is an error in either the reading or the recording.
- MSG_PING (string, integer) – utility command used to verify network connections to a remote host. Allows to determine if there is a connection to an external network at a given time, or if a specific host (robot with network adapter enabled) is connected to the network.

With these commands, KAREL enables the server to set up a connection with a remote host in a specific network. A server written in KAREL has been created by modifying a template proposed by FANUC [13].

```
PROGRAM tcp_server
%STACKSIZE = 4000
%NOLOCKGROUP
%NOPAUSE=ERROR+COMMAND+TPENABLE
%ENVIRONMENT uif
%ENVIRONMENT memo
%ENVIRONMENT flbt
%INCLUDE klevkeys
%ENVIRONMENT bynam
%INCLUDE klevkmsk
%ENVIRONMENT fdev
%INCLUDE klevccdf
%ENVIRONMENT kclop
%ENVIRONMENT sysdef
VAR
file_var  : FILE
tmp_int   : INTEGER
tmp_str1 : STRING[128]
tmp_int1 : INTEGER
tmp_str  : STRING[128]
status    : INTEGER
entry     : INTEGER
BEGIN
SET_FILE_ATR(file_var, ATR_IA)
SET_VAR(entry, 192.168.1.2 '*SYSTEM*', '$HOSTS_CFG [3]. $SERVER_PORT', 6008, status)
WRITE TPPROMPT('Connecting.', status, CR)
MSG_CONNECT('S3', status)
WRITE TPPROMPT('Connect Status = ', status, CR)
IF status = 0 THEN
WRITE TPPROMPT('Opening', CR)
FOR tmp_int1 = 1 TO 20 DO
OPEN FILE file_var ('rw', 'S3:')
status = IO_STATUS(file_var)
WRITE TPPROMPT(status, CR)
IF status = 0 THEN
FOR tmp_int = 1 TO 1000 DO
WRITE TPPROMPT ('Reading', CR)
BYTES_AHEAD (file_var, entry, status)
```

```
WRITE TPPROMPT (entry, status, CR)
READ file_var (tmp_str::10)
status = IO_STATUS(file_var)
WRITE TPPROMPT(status, CR)
ENDFOR
CLOSE FILE file_var
ENDIF
ENDFOR
WRITE TPPROMPT('Disconnecting..', CR)
MSG_DISCO('S3:', status)
WRITE TPPROMPT('Done.', CR)
ENDIF
END tcp_server
```

The created server TCP_SERVER.kl written in KAREL, is compiled by using Fanuc ROBOGUIDE, Figure 1.

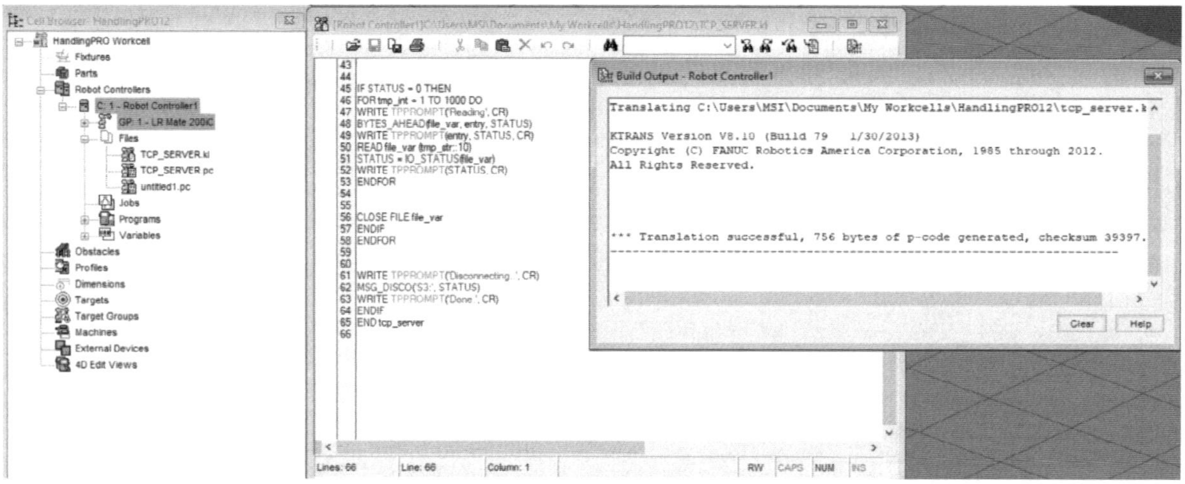

Fig. 1. Compilation of TCP_SERVER.kl in FANUC ROBOGUIDE

By using the virtual TP (Teach Pendant) provided by FANUC ROBOGUIDE, a program has been developed that would connect the server program TCP SERVER.kl Figure 2.

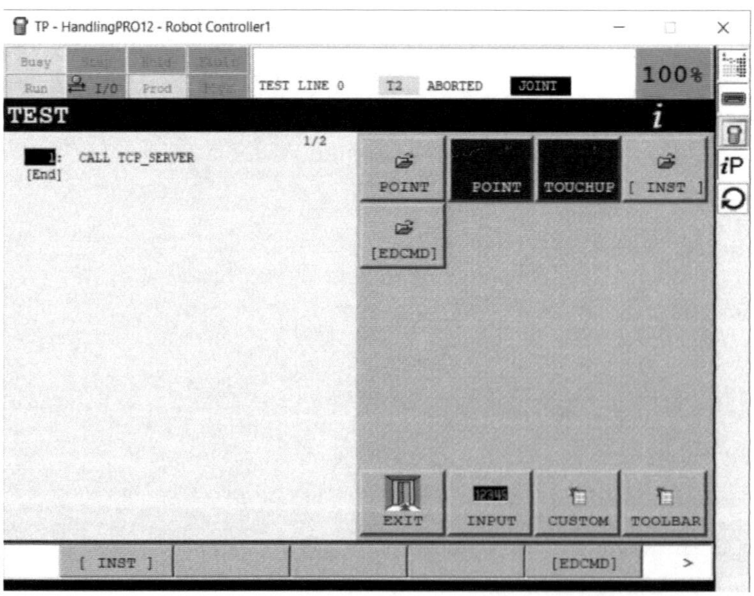

Fig. 2. Development of TEST program in FANUC ROBOGUIDE

3.2. Configuration of Socket Messaging

The next step includes the configuration of Socket Messaging Protocol and the parameters of the variable "$HOSTC_CFC [3]" Figure 3

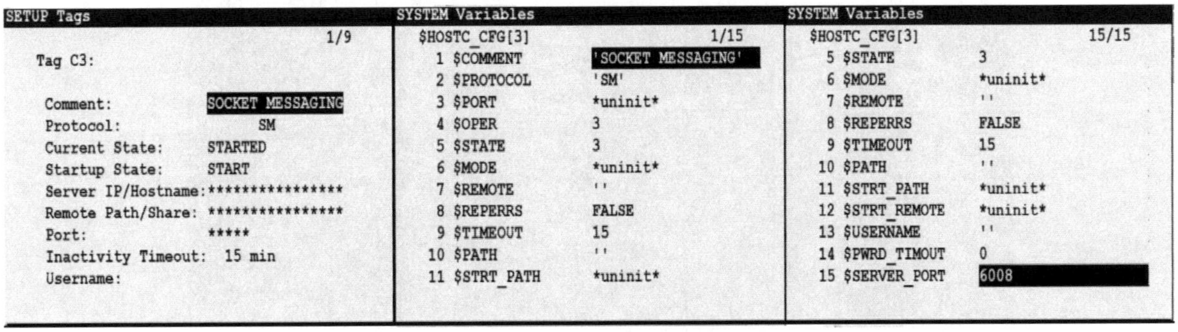

Fig. 3. Configuration of Socket Messaging

Port 6008 has been chosen for the configuration of the protocol for Socket Messaging in System Variable. The port shall be selected after checking for a free port for Socket Messaging. Figure 4 show the host communication setup of IP address 192.168.1.2.

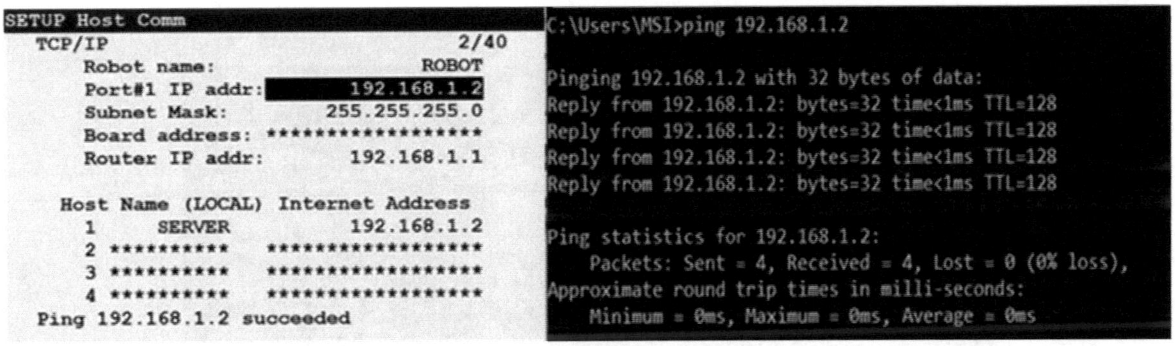

Fig. 4. Host communication setup

The server written in KAREL can be used with a customer program, written in almost any high-level language such as C, C++, C#, JAVA, Python, etc., in which case the Socket Programming syntax shall be observed. When sending messages to an individual network device, TCP can sometimes split the messages into 128-bit fragments. This is because TCP is stream-based and has no concept of a message or frame. Data can be interpreted in this way only by applications that employ TCP connections.

3.3. Create Client socket

At this stage, it is necessary to develop a customer program written in a high-level language. By using the high-level language Python, a customer was created for the purposes of the present document, who was connected to the server at IP address 192.168.1.2.

```
# client.py
import socket
def main():
client_socket = socket.socket(socket.AF_INET, socket.SOCK_STREAM)
client_socket.connect(('192.168.1.3', 6008))
client_socket.sendall(message.encode('utf-8'))
client_socket.close()
message = "Hello World"
print("Message sent.")
    if __name__ == "__main__":
        main()
```

The compilation of the customer client.py program developed through the use of the PyCharm software is shown on Figure 5.

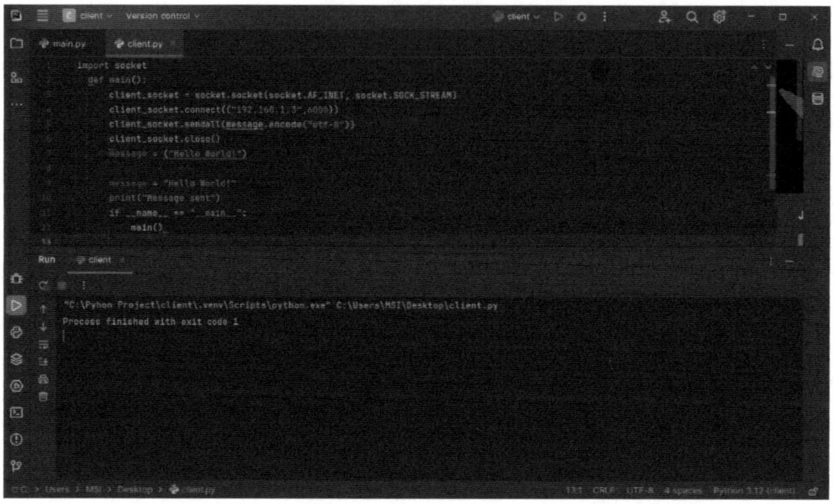

Fig. 5. Compilation of the developed customer program

Figure 6 shows the Ping test result of the developed client program.

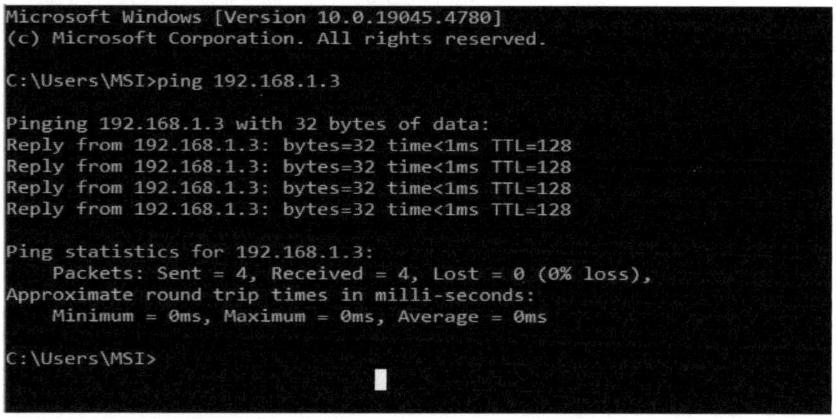

Fig. 6. Ping test result of the developed client program

3.4. Data transfer

At this stage, it is necessary to start the developed client program. If the client and server connect successfully, the program returns a message 1, otherwise 0 Figure 7.

Fig. 7. Client and server successfully connected

Thus, a connection is established between the controller of the industrial robot FANUC LR Mate 200iD/7L and the PC on which FANUC ROBOGUIDE is installed. By modifying the customer's program, one can send messages, generate a particular trajectory of the industrial robot, etc.

4. Conclusion

With the introduction of many new communication protocols, the complexity of configuring industrial robots to work with them also increases. This creates problems when it is necessary to transfer data between different computer systems used in the operation of industrial robots and the controllers of these robots.

In this work, the basic steps necessary to establish a connection between the FANUC industrial robot controller and the PC on which the FANUC ROBOGUIDE system is installed are defined. A server and a client necessary for the implementation of data transfer when operating FANUC robots with using FANUC ROBOGUIDE have been developed. Using them, a test was carried out to establish a connection between the FANUC LR Mate 200iD/7 L industrial robot controller and a PC on which FANUC ROBOGUIDE was installed. The results obtained allow for the work to be continued with the aim of taking advantage of additional possibilities for interaction between the high-level program languages and User Socket Messaging of FANUC. User Socket Messaging is one of the technical solutions in remote programming, avoiding the problems with the secure algorithms of most industrial robot software systems. This will allow sending control commands, generating defined movement trajectories, remote programming, transmission of hashed/encrypted data and control of other functionalities of a given industrial robot.

5. Acknowledgments

The authors wish to thank Scientific and research sector of Technical University of Sofia. This work was supported by Scientific and research sector of Technical University of Sofia Contract No. 241ПД0024-06 Examination the possibilities of programming industrial robots using API.

6. References

[1] Totev, D.; Dimitrova, R. & Dimitrov, S. (2024). Main steps in the process of upgrade of existing systems for automation and control of industrial and manufacturing processes in order to fulfill the requirements of the concept "industry 4.0", International Conference on Electronics, Engineering Physics, and Earth Science EEPES 2023, 21–23 June 2023, Kavala, Greece, AIP Conference Proceedings, Volume 3063, Issue 1, 21 February 2024, Online ISSN 1551-7616, Print ISSN 0094-243X, https://doi.org/10.1063/5.0195885

[2] Vladimirov, B.; Nikolov, S.; & Tsolov S. (2024). Programming Industrial Robots in the Fanuc ROBOGUIDE Environment, Engineering Proceedings 2024, 70, 20., August 2024, https://doi.org/10.3390/engproc2024070020

[3] Pajaziti, A.; Stuja, K.; Gjelaj, A. & Calvin, L.: Design and Optimization of the Welding Robot System for Aluminium Frames using Simulation Software, Proceedings of the 34th DAAAM International Symposium, pp.0040-0049, B. Katalinic (Ed.), Published by DAAAM International, ISBN 978-3-902734-41-9, ISSN 1726-9679, Vienna, Austria DOI: 10.2507/34th.daaam.proceedings.006

[4] Zaharinov, V.; Malakov, I. & Hasansabri, H. (2023). Choosing an Optimal Structural Variant of a Basic Size for a Size Range of Modules, Proceedings of the 34th DAAAM International Symposium, pp.0062-0071, B. Katalinic (Ed.), Published by DAAAM International, ISBN 978-3-902734-41-9, ISSN 1726-9679, Vienna, Austria DOI: 10.2507/34th.daaam.proceedings.009

[5] Chira, A.; Bilu, M. C.; Balasa, R. I. & Ioradche, C. (2023). Enhancing Realism in Robotic Flight Simulators: A Case Study on the Implementation Analysis of a Washout Filter, Proceedings of the 34th DAAAM International Symposium, pp.0193-0202, B. Katalinic (Ed.), Published by DAAAM International, ISBN 978-3-902734-41-9, ISSN 1726-9679, Vienna, Austria DOI: 10.2507/34th.daaam.proceedings.025

[6] Nikolov, S.; Reneta, D.; Malakov, I. & Zaharinov, V. (2023). Using the Quality Function Deployment Method in the Design of Die Casting Cells, Proceedings of the 34th DAAAM International Symposium, pp.0312-0319, B. Katalinic (Ed.), Published by DAAAM International, ISBN 978- 3-902734-41-9, ISSN 1726-9679, Vienna, Austria DOI: 10.2507/34th.daaam.proceedings.041

[7] Fanuc Roboguide, OPERATION MANUAL

[8] Fanuc Robot Series R-30iA/R-30iA Mate/R-30iB/R-30iB Mate/R-30iB Plus/ R-30iB Mate Plus/ R-30iB Compact Plus Controller - EtherNet/IP OPERATION MANUAL, B-82854EN/03

[9] Fanuc Robot Series R-30iA/R-30iB Controller - Profinet I/O OPERATION'S MANUAL, B-82864EN/04

[10] Fanuc Robot Series R-30iA controller Profibus-DP (12M), Operator's manual – MAROCPRDP04071E REV.A – B-82644EN/01

[11] Fanuc Robot Series R-30iA/R-30iA Mate CONTROLLER Modbus TCP Interface, Operator's manual –B-82844EN/01

[12] Siemens, Simatic NET, GSDML Getting Started, Simple instruction for creating a Device Description Data (GSD) File for PROFINET IO, Version 1.3, Date 08/12

[13] FANUC America Corporation, SYSTEM R-30iA, R-30iB and R-30iB Plus Controller, KAREL Reference Manual - MARRC75KR07091E Rev J – 2017

[14] Siemens PG/PC – PROFINET - CP 1604 / CP 1616I. Operating instruction, 01/2020, C70000-G8976-C218-13

DOI: 10.2507/35th.daaam.proceedings.046

INFORMATION ENTROPY BASED DIAGNOSTICS OF MECHANICAL SYSTEMS

Zdenek Neusser

This Publication has to be referred as: Neusser, Z[denek] (2024). Information Entropy Based Diagnostics of Mechanical Systems, Proceedings of the 35th DAAAM International Symposium, pp.0340-0346, B. Katalinic (Ed.), Published by DAAAM International, ISBN 978-3-902734-44-0, ISSN 1726-9679, Vienna, Austria
DOI: 10.2507/35th.daaam.proceedings.046

Abstract

The article deals with the problem of fault diagnosis of mechanical systems. The traditional approach is based on the evaluation of the change of physical parameters of the system based on identification. Such an approach often fails due to various stochastic excitations of the system. This paper describes a different approach based on computational information entropy. Information entropy does not lead directly to the identification of physical parameters, but it can distinguish between the behaviour of the original system and the system with changed physical parameters due to some failure. Another advantage of information entropy is that it provides a single value for detecting changes in the physical parameters of a mechanical system. This approach is particularly advantageous for OMA (operational modal analysis) of elastic mechanical systems where the identification encounters the problem of knowing the excitation.

Keywords: Structural Health monitoring; Predictive maintenance; Non-destructive testing

1. Introduction

Structural health monitoring and fault detection are important tools in the modern approach to predictive maintenance. Structural damage to these structures can cause production losses, cause machinery failure or even injury to operators. There are many applicable methods for detecting and locating structural damage. However, the measurement procedure is usually complex, expensive sensors are required and often only local information about the structure is provided. Natural frequencies, natural modes and associated damping usually represent mechanical properties of the structure. There are several ways to predict and obtain these parameters. The usual way is to perform a modal analysis. The system is excited by some input (usually a hammer blow or a harmonic function with a certain frequency range or white noise). The response of the system is measured at several points, which requires the preparation of an experiment, the experiment itself is time consuming, and data processing follows. Unlike modal analysis, operational modal analysis does not use a specific system input. Excitation is provided by natural and ambient vibrations under its operating conditions (machine motors, external faults, etc.). The analysis requires long-term measurements but does not require expensive measuring equipment. A sufficient level of excitation of the structure is required. Traditional methods are based on mathematical analysis of modal parameters (natural frequencies, mode shapes and modal damping) [1], [2], [3], [4], [5]. These methods assume that structural damage affects the location/shape of modal parameters (usually natural frequencies, which can be easily obtained).

As stated by Salawu [6], eigenfrequency analysis inherently suffers from low sensitivity to change and requires at least a 5% change in the locations of the eigenfrequencies to effectively detect the occurrence of damage. Mode shape analysis shows better sensitivity to damage but is still subject to noise generated by ambient frequency generators (e.g. wind, cars, etc.), leading to many false positives [7]. Mode shapes are also difficult to detect in large structures and are dependent on the number and location of vibration sensors [8]. Modern vibration analysis methods use frequency response function (FRF) usually in conjunction with principal component analysis (PCA), deep principal component analysis (DPCA) [9] using deep learning models and artificial neural networks (ANN) [10], [11], [12], [13], [14]. While PCA-ANN methods are effective in detecting even small structural damage with good accuracy in a noise-free/very low noise environment (less than 2\% [15] or less than 5\% [13] noise), they still fall short of accuracy when significant noise is added to the input data [16]. These methods are also complex and require many adjustments regarding the number of principal components (PCs) of the FRF (or FRF adjustment using cepstrum [17], [18]) that are chosen as input to the ANN, and optimization of the parameters of the ANN itself. State space reconstruction using multi-scale analysis is presented in [19].

This paragraph describes the cross-entropy approach [20], which includes a wide range of methods as presented in [21]. In [22], [23], [24], cross-entropy is applied to the measured signal but with a different state indicator for further evaluation. Eigenmode evaluation with cross-entropy support in [25], [26] performs multi-point measurements. The papers [27], [28], [29] uses cross entropy in data processing with the following optimization. Our research, unlike those papers, uses the behaviour of cross-entropy as a direct indicator of structural change. Our work presents an operational method inspired by modal analysis that seeks frequency change to indicate structural changes. The presented method based on cumulative cross-entropy slope comparison clearly distinguishes between the original and the altered structure, even with very low signal-to-noise ratio. The method is applied to the one mass system with its eigenfrequency alteration.

2. Problem formulation

Typical measurement of mechanical system is usually subject to error due to noise in the signal. Frequency analysis of such a signal has the problem of correctly determining the eigenfrequencies. Let's introduce one-mass one degree of freedom system with known eigenfrequency. The system is excited by random signal. The mechanical system on the Fig. 1. has mass *m*, spring stiffness *k* and external force *F* exciting the system. The excitation force brings disturbance into the system. It is chaotic, noisy signal characterised as white noise. Measured, output signal is mass position. The system has prescribed eigenfrequency ω and mass *m*, the spring stiffness is than obtained using (1).

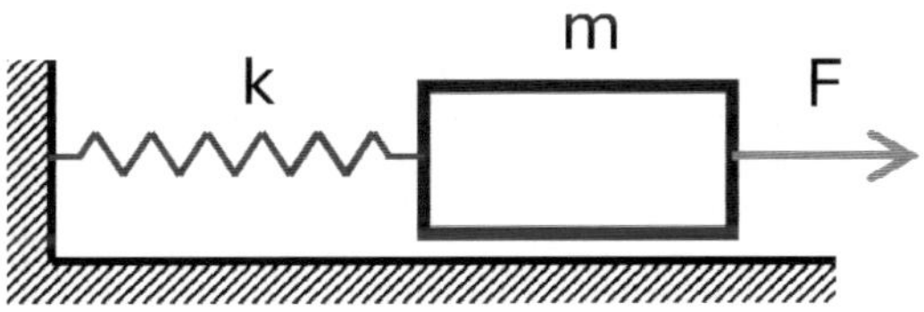

Fig. 1. One-mass model

$$k = m\omega^2 \quad (1)$$

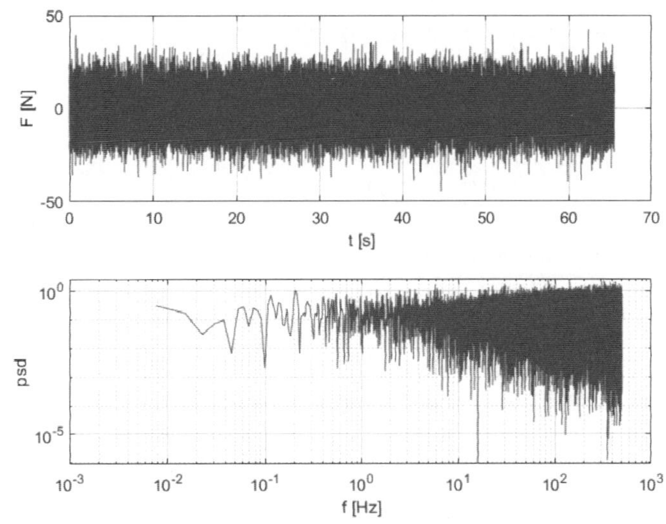

Fig. 2. Input signal - white noise and its frequency spectrum

Example of input signal and its frequency spectrum is presented in the Fig. 2. Such a signal could represent a chaotic excitation of the system, being suitable for operational modal analysis. For mass equal to 1 kg and eigenfrequency at 3 Hz the system response to white noise input signal from Fig. 2. is shown in the Fig. 3.

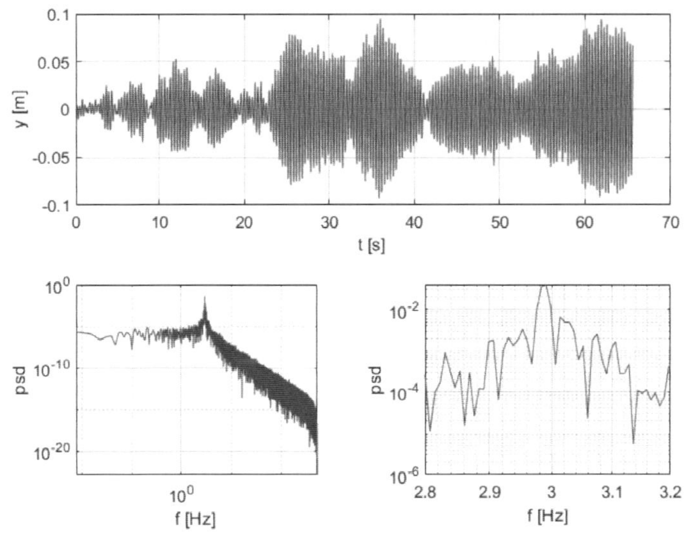

Fig. 3. Signal output - time behaviour and its frequency spectra with detail nearby eigenfrequency

The system output frequency analysis is depicted in the Fig. 3 at the bottom. On the left the overall spectra is shown, on the right the zoom in the surrounding of eigenfrequency is depicted. It is obvious that frequency analysis does not match the correct value. This problem is going to be overcome our entropy based diagnostics approach.

3. Proposed method

The proposed diagnostic method belongs to the group of operational modal analysis methods. The measured data are pre-processed using normalization to emphasis the peak frequencies in the signal, the data are averaged from more measurements and finally compared with reference data using cross entropy function. The whole procedure is shown in the Fig. 4.

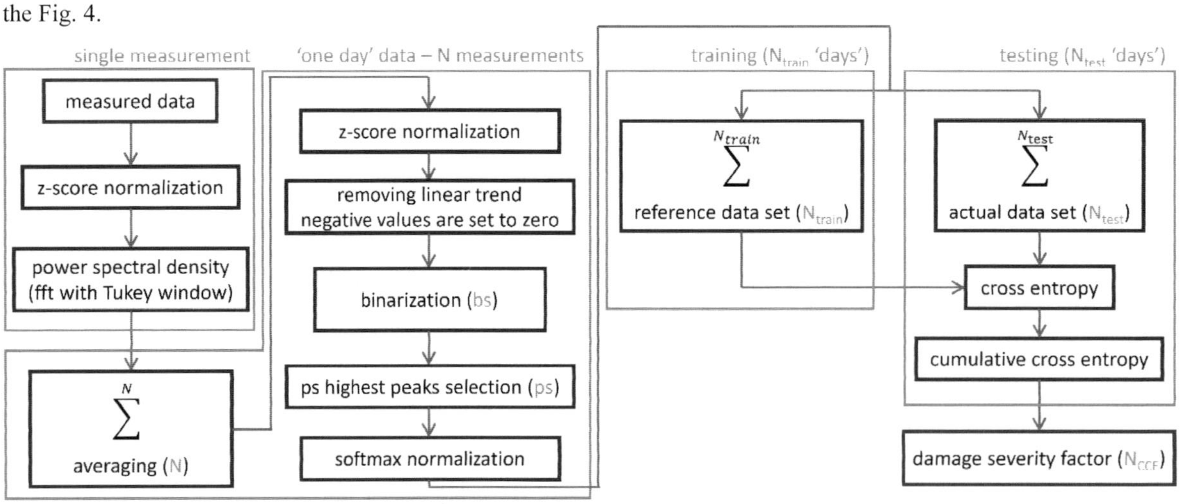

Fig. 4. Proposed method workflow

The individual measured signals are first normalized by z-scores in the time domain, followed by conversion to the frequency domain - power spectral density calculation with window function. This is followed by averaging of N measurements, which could be called "one day" data. The obtained signal is z-score normalized, then the linear trend is removed from the signal and finally all negative values are set to zero. So called binarization is performed in the next step. Every bs adjacent frequencies are averaged into one value – so called bin. Then the ps highest values are taken for further investigation, the rest is zero. The last, but important operation is performed – the softmax normalization, see (2).

$$s = \frac{e^{signal}}{\sum e^{signal}} \quad (2)$$

Variable *signal* in (2) represents the frequency behaviour of peaks in binarized spectra, variable s represents its softmax normalization behaviour. Obtained data could enter two following phases. At first, the training phase, where the reference data set is obtained, created by N_{train} softmax spectra averaged data (a_{ref}). At second follows the testing phase, where the actual data set is created by N_{test} softmax spectra averaged data (a_j). Then the cross-entropy is evaluated.

$$H_j = -\sum_{i=1}^{N_{bin}} a_{ref}(i) \log a_j(i) \quad (3)$$

The cross-entropy value H_j is obtained by summarizing the values in (3) over the number of binarization points N_{bin}. Cumulative cross-entropy behaviour is created by adding the actual cross-entropy value to previous cumulative cross-entropy value. Relative damage severity factor $d\alpha_i$ is evaluated as a comparison of cumulative cross-entropy slopes α_i, with α_1 corresponding to slope of training data, see (4).

$$d\alpha_i = \frac{\alpha_i - \alpha_1}{\alpha_1} \quad (4)$$

4. Method application

The one mass system is subject to white noise excitation. The system has prescribed eigenfrequency equals to 3 Hz. The training is performed and follows the testing phase. At certain point, the eigenfrequency is changed to 3.01 Hz and finally follows the change to 3.02 Hz, imitating the structural changes.

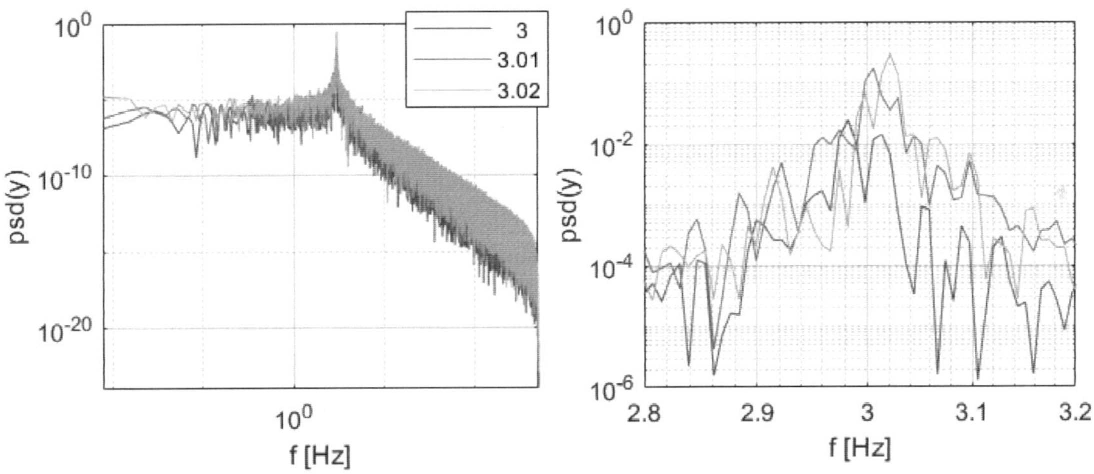

Fig. 5. Frequency response for different system eigenfrequency, four signals averaged

Fig. 5 shows the problem of eigenfrequency recognition. In this case, all signals are averaged from four measurements. The eigenfrequency at 3.02 Hz is recognized clearly, 3 Hz and 3.01 Hz are hard to determine. Approach, we are presenting, uses method of signal averaging together with presented normalization and cross-entropy measure to clearly distinguish between the original system behaviour and altered structure behaviour.

N	bs	ps	N_{train}	N_{test}	N_{CCE}
2	1	10	4	1	1

Table 1. Procedure parameters settings

The procedure settings are presented in the Table 1. There are two signals averaged for so called one day measurement (see Fig. 4.), binarization is unapplied, maximal number of selected peaks is 10, training phase is performed over four 'day' measurements and testing is evaluated at every "day" measurement. Slope is calculated from one adjacent (previous) point.

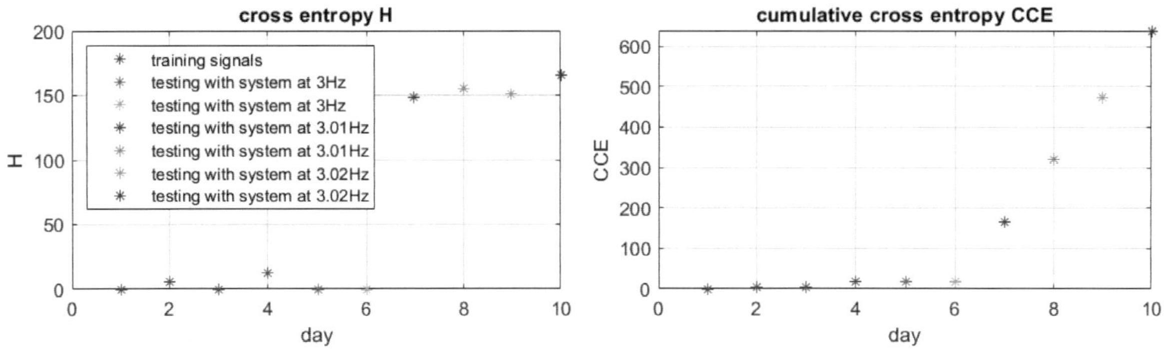

Fig. 6. Cross entropy and cumulative cross entropy behaviour

There is performed 8 training simulations with system 3 Hz eigenfrequency and 12 testing simulations. First four testing simulations are with system eigenfrequency at 3 Hz, next four simulations with system eigenfrequency 3.01 Hz and last four simulations with system eigenfrequency 3.02 Hz. The Fig. 6 shows on the left cross entropy behaviour, at fist for data from training signals, followed by the testing data. Cross entropy with altered system eigenfrequency shows much higher values than original system cross entropy evaluation. On the right of Fig. 6 is presented cumulative cross entropy behaviour. It is obvious, that slope with original system differs considerably in comparison to the slope of system with altered eigenfrequency.

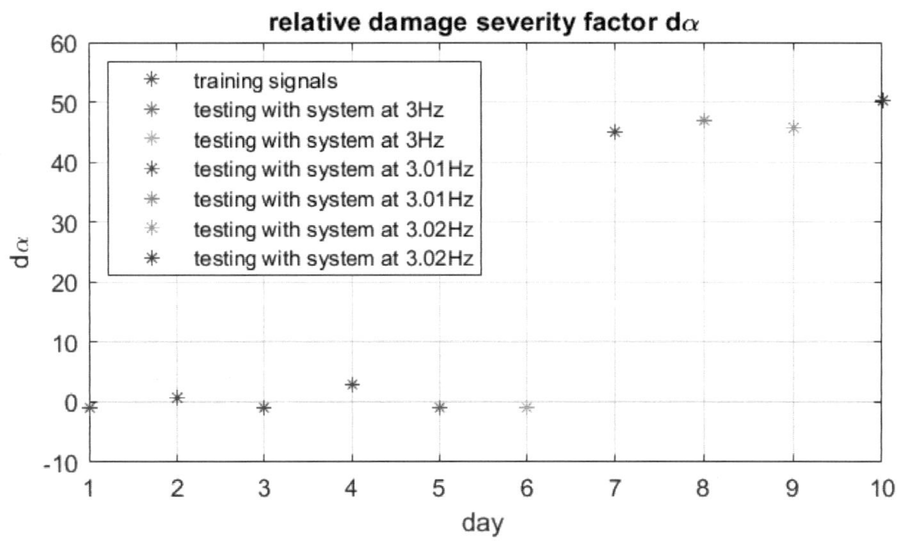

Fig. 7. Relative damage severity factor behaviour

Relative damage severity factor in Fig. 7 is the indicator of mechanical structure condition. There is obvious value difference between the sixth and seventh evaluated measurement "day".

5. Conclusion

In this paper, the problem of fault diagnosis of mechanical systems based on the evaluation of the identified change of physical parameters of the model of the investigated mechanical system was first described. Then, an alternative approach based on computational information entropy was presented and evaluated. This alternative approach allowed to perform successful fault diagnosis. The whole problem was demonstrated on a very simple example of a one-mass system. These results can be interpreted by the hypothesis: the quality of the measured signals does not allow to make a successful identification of the model of the investigated mechanical system. However, the information content in the measured signals is sufficient to detect this change, i.e., the information about the change in the mechanical system is strong enough for fault diagnosis. This hypothesis will be further investigated and validated in future research. The current study is performed on a simple system; application to a more complex structure would require more data to properly evaluate the state of the system.

6. Acknowledgement

This paper was created with the support of the project TN02000010 of National Competence Centre of Mechatronics and Smart Technologies for Mechanical Engineering cofounded by the Technology Agency of the Czech Republic, within the National Centres of Competence Programme.

7. References

[1] Azimi, M.; Eslamlou, A. and Pekcan G. (2020). Data-Driven Structural Health Monitoring and Damage Detection through Deep Learning: State-of-the-Art Review, Sensors, Vol. 20, p. 2778, 10.1002/stc.3100.

[2] Farshidi, R.; Trieu, D.; Park, S. & Freiheit, T. (2010). Non-contact experimental modal analysis using air excitation and a microphone array, Measurement, Vol. 43, No. 6, pp. 755-765, 10.1016/j.measurement.2010.02.004.

[3] Khiem, N. & Toan, L. (2014). A novel method for crack detection in beam-like structures by measurements of natural frequencies, Journal of Sound and Vibration, Vol. 333, No. 18, pp. 4084-4103, 10.1016/j.jsv.2014.04.031.

[4] Sayyad, F. B. & Kumar, B. (2012). Identification of crack location and crack size in a simply supported beam by measurement of natural frequencies, Journal of Vibration and Control, Vol. 18, No. 2, pp. 183-190, 10.1177/1077546310395979.

[5] Li, T.; Hou, R.; Zheng, K.; Zhang, Z. & Liu, B. (2024). Automated method for structural modal identification based on multivariate variational mode decomposition and its applications in damage characteristics of subway tunnels, Engineering Failure Analysis, Vol. 163, p. 108499, 10.1016/j.engfailanal.2024.108499.

[6] Salawu, O. (1997). Detection of structural damage through changes in frequency: a review, Engineering Structures, Vol. 19, No. 9, pp. 718-723, 10.1016/S0141-0296(96)00149-6.

[7] Mehrian, S. Z.; Amrei, S. R.; Maniat, M. & Nowruzpour, S. (2016). Structural health monitoring using optimising algorithms based on flexibility matrix approach and combination of natural frequencies and mode shapes, International Journal of Structural Engineering, Vol. 7, No. 4, p. 398, 10.1504/IJSTRUCTE.2016.079287.

[8] Kim, J.-T.; Ryu, Y.-S.; Cho, H.-M. & Stubbs, N. (2003). Damage identification in beam-type structures: frequency-based method vs mode-shape-based method, Engineering Structures, Vol. 25, No. 1, pp. 57-67, 10.1016/S0141-0296(02)00118-9.

[9] Silva, M.; Santos, A.; Santos, R.; Figueiredo, E.; Sales, C. & Costa, J. C. (2019). Deep principal component analysis: An enhanced approach for structural damage identification, Structural Health Monitoring, Vol. 18, No. 5-6, pp. 1444-1463, 10.1177/1475921718799070.

[10] Shadan, F.; Khoshnoudian, F. & Esfandiari, A. (2016). A frequency response-based structural damage identification using model updating method, Structural Control and Health Monitoring, Vol. 23, No. 2, pp. 286-302, 10.1002/stc.1768.

[11] Bandara, R. P.; Chan, T. H. & Thambiratnam, D. P. (2014). Frequency response function based damage identification using principal component analysis and pattern recognition technique, Engineering Structures, Vol. 66, pp. 116-128, 10.1016/j.engstruct.2014.01.044.

[12] Homaei, F.; Shojaee, S. & Ghodrati Amiri, G. (2016). Multiple-structural damage detection using measured frequency response function, Iranian Journal of Structural Engineering, Vol. 2, No. 1, pp. 13-18.

[13] Padil, K. H.; Bakhary, N.; Abdulkareem, M.; Li, J. & Hao, H. (2020). Non-probabilistic method to consider uncertainties in frequency response function for vibration-based damage detection using Artificial Neural Network, Journal of Sound and Vibration, Vol. 467, p. 115069, 10.1016/j.jsv.2019.115069.

[14] Bandara, R. P.; Chan, T. H. & Thambiratnam, D. P. (2014). Structural damage detection method using frequency response functions, Structural Health Monitoring, Vol. 13, No. 4, pp. 418-429, 10.1177/1475921714522847.

[15] Nguyen, V. D. & Nguyen, N. T. (2015). A Method for Improving the Quality of Collective Knowledge, Proceedings of 7th Asian Conference on Intelligent Information and Database Systems 2015, Bali, Indonesia, March 23–25, pp. 75-84, 10.1007/978-3-319-15702-3_8.

[16] Steinbauer, P.; Neusser, Z.; Bukovsky, I. & Neruda, M. (2019). Lighting Pole Health Monitoring for Predictive Maintenance," Procedia Structural Integrity, Vol. 17, pp. 799-805, 10.1016/j.prostr.2019.08.106.

[17] Dackermann, U.; Smith, W. A.; Alamdari, M. M. & Randall, R. B. (2019). Cepstrum-based damage identification in structures with progressive damage, Structural Health Monitoring, Vol. 18, No. 1, pp. 87-102, 10.1177/1475921718804730.

[18] Dackermann, U.; Smith, W. A. & Randall, R. B. (2014). Damage identification based on response-only measurements using cepstrum analysis and artificial neural networks, Structural Health Monitoring, Vol. 13, No. 4, pp. 430-444, 10.1177/1475921714542890.

[19] Bukovsky, I.; Kinsner, W.; Maly, V. & Krehlik, K. (2011). Multiscale Analysis of False Neighbors for state space reconstruction of complicated systems, 2011 IEEE Workshop On Merging Fields Of Computational Intelligence And Sensor Technology, pp. 65-72, 10.1109/MFCIST.2011.5949517.

[20] Bukovsky, I.; Kinsner, W. & Homma, N. (2019). Learning Entropy as a Learning-Based Information Concept, Entropy, Vol. 21, No. 2, p. 166, 10.3390/e21020166.

[21] Jamin, A. & Humeau-Heurtier, A. (2019). (Multiscale) Cross-Entropy Methods: A Review, Entropy, Vol. 22, No. 1, p. 45, 10.3390/e22010045.

[22] Kumar, A.; Gandhi, C.; Zhou, Y.; Tang, H. & Xiang, J. (2020). Fault diagnosis of rolling element bearing based on symmetric cross entropy of neutrosophic sets, Measurement, Vol. 152, p. 107318, 10.1016/j.measurement.2019.107318.

[23] Chauhan, S.; Singh, M. & Kumar Aggarwal, A. (2021). An effective health indicator for bearing using corrected conditional entropy through diversity-driven multi-parent evolutionary algorithm," Structural Health Monitoring, vol. 20, no. 5, pp. 2525-2539, 9 2021, 10.1177/1475921720962419.

[24] Chauhan, S.; Vashishtha, G.; Kumar, R.; Zimroz, R.; Gupta, M. K. & Kundu, P. (2024). An adaptive feature mode decomposition based on a novel health indicator for bearing fault diagnosis, Measurement, Vol. 226, p. 114191, 10.1016/j.measurement.2024.114191.

[25] Lin, T.-K. & Liang, J.-C. (2015). Application of multi-scale (cross-) sample entropy for structural health monitoring, Smart Materials and Structures, Vol. 24, No. 8, p. 085003, 10.1088/0964-1726/24/8/085003.

[26] Lin, T.-K. & Chien, Y.-H. (2019). Performance Evaluation of an Entropy-Based Structural Health Monitoring System Utilizing Composite Multiscale Cross-Sample Entropy, Entropy, Vol. 21, No. 1, p. 41, 10.3390/e21010041.

[27] Bekker, J. (2013). Multi-Objective Buffer Space Allocation with the Cross-Entropy Method, International Journal of Simulation Modelling, Vol. 12, No. 1, pp. 50-61, 10.2507/IJSIMM12(1)5.228.

[28] Gallimard, L. (2019). Adaptive reduced basis strategy for rare-event simulations, International Journal for Numerical Methods in Engineering, Vol. 120, No. 3, pp. 283-302, 10.1002/nme.6135.

[29] Motwani, A.; Shukla, P. K. & Pawar, M. (2023). Novel framework based on deep learning and cloud analytics for smart patient monitoring and recommendation (SPMR), Journal of Ambient Intelligence and Humanized Computing, Vol. 14, No. 5, pp. 5565-5580, 10.1007/s12652-020-02790-6.

DOI: 10.2507/35th.daaam.proceedings.047

ON THE ROLE OF DISCRETE EVENT SIMULATION IN THE INDUSTRIALISATION OF 3D-PRINTING MANUFACTURING SYSTEMS

Zahra Isania & Giuseppe Casalino

This Publication has to be referred as: Isania, Z[ahra] & Casalino, G[iuseppe] (2024). On the Role of Discrete Event Simulation (DES) in the Industrialization of 3D-printing Manufacturing Systems, Proceedings of the 35th DAAAM International Symposium, pp.0347-0353, B. Katalinic (Ed.), Published by DAAAM International, ISBN 978-3-902734-44-0, ISSN 1726-9679, Vienna, Austria
DOI: 10.2507/35th.daaam.proceedings.047

Abstract

The ability of additive manufacturing (AM) to release designers from the limitations of traditional production is unmatched. This adaptability makes it possible to design components that are strong and perform well while using the least amount of material. Because AM can create designs that are efficient and light, it has become widely used worldwide and has grown into a multibillion-dollar business with a wide range of materials and procedures. However, there are obstacles to overcome in the industrialisation of AM, including maintaining consistent quality, cutting costs, and maximising production efficiency. Discrete Event Simulation (DES) shows up as a vital instrument for conquering these obstacles. In order to improve process quality and encourage wider usage of 3D printing in manufacturing, this article focuses on using DES to simulate and optimise AM's industrialisation.

Keywords: Additive Manufacturing (AM); 3D-printing technology; Industrialisation; Discrete Event Simulation (DES) ; Optimisation

1. Introduction

Additive Manufacturing (AM) often called 3D printing (3DP) has a drastic effect of removing the use of traditional molding techniques. This enables people to design complex geometries where the number of components may be reduced. Using a digital 3D model as a basis, this approach develops components layer by layer and progressively. 3DP has evolved from an original concept to an essential component of modern production thanks to recent developments in materials and procedures [1]. Standards such as ISO/ASTM 52900:2015 are increasingly governing technology as it continues to industrialize [2]. AM has shown to be very valuable in the production of lightweight, integrated structures, especially in the aerospace, medical, and automotive industries. This has become a key area for researchers and engineers in both business and higher learning. These technologies have the enormous potential to change product designs, supply networks, and production methods, making it possible to create more complicated parts more cheaply and efficiently [3].

The need to maximise the utilization of AM facilities is expanding as AM materials and processes continue to advance, especially in terms of cutting the time and cost of 3D printing in industrial production [4]. Due to changing consumer needs and the requirement for quick modifications through flexible production layouts and processes, manufacturing organizations today confront significant hurdles in production planning and management. It may be particularly challenging to increase efficiency in complex production environments, especially when it comes to identifying and eliminating bottlenecks. The dedication of simulation tools, which are powerful and efficient, is essential to the intended end. Those tools are the main ones" that are used to speed up the manufacturing process from start to finish, which includes planning, implementation, and operational management [5].

In the context of 3D printing, the partnership between HP and Siemens at the DFactory in Barcelona is a significant step forward in 3D printing productivity and automation. The DFactory project aspires to streamline the HP Jet Fusion 3D printing production workflow by incorporating Siemens' advanced automation solutions which allow for high-volume and continuous manufacturing with human intervention being minimal. The objective is to accomplish "lights out manufacturing," in which processes are carried out automatically without the need for human oversight—even throughout the dark. In such a setup, all production-related tasks, including material handling, assembly, and quality assurance, are handled by machines, robots, and automated systems, enabling factories to run around the clock. In this way, the labour costs are reduced, the human error is minimised and it is the sustainable production that has more chances to be promoted [6]. Among DES is one of the most used simulation approaches for researching and improving manufacturing processes among other methodologies [7]. The modelling, simulation, optimisation, and visualization of logistics, manufacturing systems, and other operational elements are made easier by DES technologies. Because of its adaptability, DES is very useful for conducting in-depth analyses of system features, identifying possible weak points or bottlenecks in process flows, and enhancing important performance metrics including productivity, material handling costs, energy consumption, and resource usage [8].

For example, Omogbai and Salonitis [9] proposed using DES to simulate and assess lean manufacturing improvement initiatives before their use. Velumani and Tang [10] also demonstrated how to use DES to control production bottlenecks in batch processes and evaluate operational conditions. In addition, Prajapat et al.[11] created a DES model for a repair plant, giving decision-makers a tool to assess different facility layouts and configurations in order to maximise output.

This paper is divided into the following sections: Section 2 serves as an analysis of the existing research on DES. Section 3 looks at the evolution and significance of simulation in AM. Section 4 deals with the use of DES as a simulation tool within this provided context. Section 5 offers an overview of the simulation process that was done in the preceding work. Section 6 explores the role of digital twins, and finally, Section 7 recapitulates the most important findings of the study.

2. State-of-the-Art on DES

For the analysis and optimisation of business processes, simulation modelling presents a large potential. A case in point is the said models can dynamically simulate different parameter values, for instance, by varying arrival rates or service intervals, to help identify the bottlenecks and evaluate the alternatives [12]. Furthermore, graphical representations of processes that may be interactively updated and manipulated to visualize process dynamics are another feature of simulation models. This strategy has shown to be quite successful in resolving issues and enhancing manufacturing system design. Law and McComas [13], have pointed out that modelling and simulation are very helpful in industrial settings. However, multiple objective analyses and optimisation inside these simulations are uncommon. DES technology is developing quickly; many academic articles and new software features are being released every year. Numerous industries, including healthcare patient flows, military tactics, logistics, contact centers, and restaurants, have used DES software and languages. Profit optimisation, which entails determining which options provide the maximum profits over time, is a common goal in DES [14]. Though traditionally the main priority has been income, environmental concerns are becoming more significant. There is an increasing need to address these issues in more detail as long as natural resources are used.

The industrialisation of 3D printing has been greatly affected by the development of DES technology. 3D printing processes have been something manufacturers can see with the aid of advanced DES technology, able to model such aspects as material handling, machine operations, post-processing, and quality control. With this detailed simulation, it is feasible to discover and fix potential problems before the physical production phase, thereby the transition from prototyping to full-scale manufacturing is simpler. Thus, DES is a valuable tool for advancing the development and improvement of 3D printing technology, promoting innovation, and raising industrial productivity.

3. The Evolution and Impact of Simulation in Additive Manufacturing

Since its debut in the late 1980s, AM has seen substantial change. It was first mostly used for fast prototyping, which made it possible for designers to produce physical versions of their designs rapidly and affordably. However as materials, technology, and manufacturing methods continue to progress, AM has evolved into a full manufacturing method that can produce end-use parts for a variety of sectors [15]. Consequently, the fact that AM now boosts industrial productivity through the use of digital systems and automation. Its usage goes beyond rapid prototyping, including both direct production of parts, as well as manufacturing tools for industrial use.

AM possesses distinct benefits which include its ability to produce intricate and lightweight designs with minimal use of materials, making it a desirable option for the production of advanced manufacturing solutions. For instance, AM is commonly used in the aerospace sector for making parts that are lighter as well as stronger than the traditional way of making parts. In addition, in the sector of medicine, AM facilitates the process of a production of bespoke implants and prostheses that are detailedly designed to fit the specific needs of the individual patients [16].

One important aspect of AM is that it produces material properties that can vary from one point to another within the same part. This variation makes it necessary to employ predictive simulation tools for accurate part design, especially for mission-critical applications. Because simulation can simulate and forecast the real behaviour of the manufacturing system, it saves time and money by eliminating the need for expensive trial-and-error trials. This is why simulation is essential to AM industrialisation [17]. With simulation software, models that are dynamic and interactive may be created to mimic the functioning of real production processes. Technology has advanced significantly in the industrial sector recently, making it possible to create intricate and affordable AM components. Enhancing the design of parts for mechanical performance and weight efficiency and modelling the AM process for these components are the two key aims driving this development, which is mostly driven by advancements in equipment and software [18].

It is essential to comprehend and optimise the processes involved in the design and development of 3D-printed parts in order to manufacture high-quality parts. Specific qualities like temperature control and form precision may be attained with the use of efficient simulation approaches [19]. The optimal input values are sought in simulation optimisation in order to minimise resources and improve insights. All aspects of 3D printing, from the melting of raw materials to the final tool path required to make a product, are covered by AM simulation. Throughout the whole printing process, accurate numerical simulations are critical to the success of AM. In order to generate lattice structures, optimise manufacturing processes, create practical designs, ensure performance during usage, and precisely calibrate materials, simulation is essential in several key areas of 3D printing.

Simulation tools are vital for cost-effective AM because they assist in the optimisation of different processes of the technology. By developing and improving lattice structures often utilized in AM, employing optimisation techniques to refine designs, and keeping an eye on phase changes to calibrate materials, they aid in functional design [20]. AM simulation brings several benefits. It increases the likelihood of successful printing, helps in making part design and quality improvements, as well as in preventing print failures. Imparting a full AM process insight allows for production optimisation thus reducing shape changes and increasing reproducibility. In addition, AM simulation is one of the methods used in cost reduction and elimination of the material waste problem, while at the same time, it facilitates the optimisation of designs, thus lowering the production and development costs. Additionally, the simulation also helps to locate the most critical elements of the printing, thus, avoiding the misprint that is very costly [21].

4. The Role of Discrete Event Simulation in AM

Discrete event modelling is best for the development of systems consisting of components that switch states at particular times due to either their own operations or interactions with others. This strategy is particularly applicable when the action of a system can be viewed as a succession of cross-nonoverlapping time intervals, mostly under the influence of some stochastic nature. DES is a system that is mainly composed of entities that are doing some tasks and interacting through activities and events. A system is seen as a collection of linked, attribute-rich entities that carry out tasks and interact with one another through actions and events in a DES model.

Simulation has been proven to be a very useful tool to improve the breakdowns of the manufacturing system and to find solutions for operational challenges. The application of DES in production has been widespread for energy consumption analyses, bottleneck reduction, and line balancing [22]. Although DES has existed since the 1950s, recent advancements in technology have made it possible to, for example, combine DES with virtual reality or to use automatic data generation as input for DES simulations [23]. Research has shown that IoT systems may monitor assembly lines and provide real-time data into DES for analysis and improvement. For instance, Ingemansson et al.[24] investigated how DES might use auto-generated data. The creation of DES systems, which can update their simulations continually using data stored remotely, is one significant accomplishment. Real-time data can be generated by machines equipped with sensors in a smart factory or Industry 4.0 environment, as exemplified in Figure 1. In this setup, the digital manufacturing execution system (DES) is a component of the cyber-physical system (CPS) that creates a real-time digital twin (DT) of the physical objects in the factory. In order to obtain the most recent parameters and produce updated findings, DES has networking capabilities and the ability to make scheduled Internet queries.

A locally run DES can the parameters like standard deviation and averages gotten by the application that is at the cloud service through remote requests. This web application can be the user's interface of a smart production center whereas it is also possible to connect with the intelligent production paradigm. In this case, the data stored in the cloud can be from real-world processes, which are monitored through the Internet of Things (IoT); the cloud-stored data is then analysed to make simulations, so a digital twin is created. In order to facilitate subsequent usage, DES can also transmit its results to the cloud. The purpose of the AR app was to project the digital simulation of things and their operations, which got the input data from the cloud to display the DES processes and results. The behaviour of physical systems, such as interacting entities whose states change, is often studied and simulated using DES [25]. It is frequently used as an auxiliary instrument in production to enable "what-if" encounters in order to check different settings and suggest alternative plans.

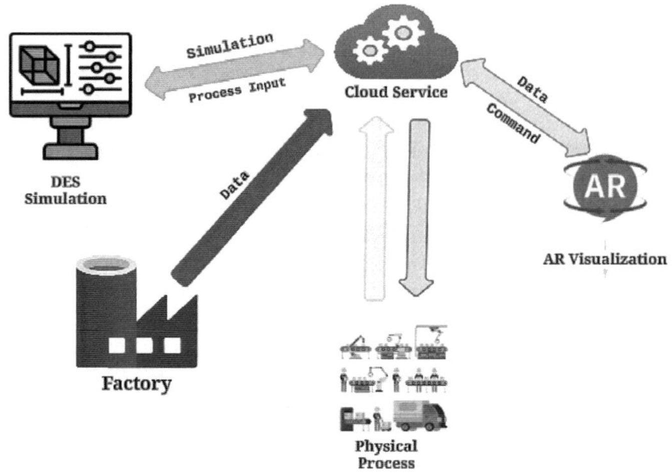

Fig. 1. DES simulation concept from a smart factory.

Many changes are happening in the simulation field nowadays, e.g., the transformation of DES applications from design to operation, and thus, the advantages of simulation in production processes are throughout the whole. A prominent trend is the adoption of hybrid modelling practices and optimisation tools and algorithms [26], along with the improvement of the capabilities of DES and the provision of new value for producers. Although DES does not physically interact with the system, it offers data to change a process based on factors added to the system, such as processes, time, and labour. The combination of DES with other Industry 4.0 technologies such as AR, IoT, and digital twins tore down this barrier by enhancing the application of DES to this zone [27]. Despite its simplicity, DES is a robust and sometimes expensive tool, offering three primary worldviews: activity-oriented, event-oriented, and process-oriented simulations. Although these viewpoints highlight distinct facets of the model, they are all interpretable from one another since they all adhere to basic modelling concepts.

5. Overview of the Simulation Process

A summary of the Selective Laser Sintering (SLS) based AM simulation process was presented in our earlier work (Figure 2 shows the production plan). In order to optimise production, key elements including setup and processing times, cost control, and bottleneck identification were all represented in the simulation. FlexSim software which follows the core principles of Discrete Event Simulation (DES) was used to record the dynamic behaviour of the AM manufacturing setting.

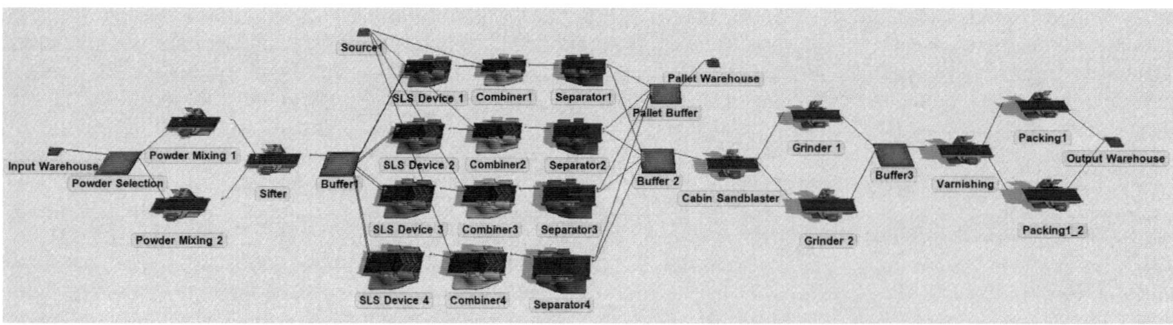

Fig. 2. Simulation model of SLS production line in FlexSim [28]

Important variables were incorporated into the model. For instance, a given number of parts were planned to be prepared concurrently on an SLS machine by taking into account the processing time and setup duration. The investigation also assessed how well machines were being used in terms of strategic configurations to boost productivity, reduce idle time, and enhance performance. The approach comprised simulating the AM process using three main stages—preprocessing, processing, and post-processing—with hypothetical data. A reference model was considered to simulate SLS machine configurations producing the desired number of parts annually, based on operational conditions like the shifts per day and number of working days per year. In order to maximise throughput, our simulation sought to locate idle periods and bottlenecks in the system.

Materials are fed into the system from an input warehouse and mixed in a powder selection buffer before being transported to a sifter to separate grain sizes. Following processing, the parts are arranged so that they depart the machines simultaneously. Due to software restrictions, the plan comprises separator machines for unpacking and combiners to expedite packing, allowing a seamless transition between phases. Buffers are included in the model to gather completed parts and pallets before they are returned to the warehouse for further usage. Parts are given various operations such as sandblasting and grinding to components to enhance their finish and correct any dimensional errors. Every stage is intended to maintain high standards of quality before varnishing and final examinations. Eventually, after being coated for protection and going through the quality control tests, such parts are boxed following rules and sent to the final warehouse for distribution. This methodical technique shows how to integrate efficiency and effectiveness in the AM simulation process in a balanced way while simultaneously improving production efficiency and guaranteeing excellent quality at every stage.

6. The Digital Twins

Modern manufacturing system design is directly impacted by the growing complexity and unpredictability of the present manufacturing and economic environment. More automation, reduced lot sizes, shorter product life cycles, and more product variability are some of the major developments that are being facilitated by intelligent control systems. As a result, suppliers and manufacturers in the discrete manufacturing space are putting more of an emphasis on automation, simulation, and digitization—all of which are critical at the production line and equipment levels. These developments are forcing businesses to incorporate cutting-edge technical solutions and redefine manufacturing methods [29].

Digital Twins (DTs) have become a key element of the changing landscape of digital technology. A virtual model that replicates its physical counterpart in real-time and is used for continuous monitoring, optimisation, and decision-making is called a digital twin [30]. DTs are essential to modern industrial processes because they allow smooth transitions between virtual and physical systems through real-time communication. The use of DTs has been further hastened by the emergence of Industry 4.0, which brings with it technologies such as IoT, Big Data, and predictive maintenance. These tools presently serve as the digital equivalents of Cyber-Physical Systems (CPS), establishing a constant data connection between the digital and real worlds [31].

Digital twins are particularly useful in AM. Industry 4.0 requires the ability to create complex, non-standard structures, which AM makes possible. While AM can speed up production and minimise material waste, improving mechanical characteristics is still difficult and frequently necessitates expensive, time-consuming trial-and-error procedures. Manufacturers may significantly reduce the requirement for physical trials and increase the accuracy of parameter selection by simulating various manufacturing situations using DTs. This integration expedites cuts down on the time needed for product qualification, boosts quality, and optimizes the use of resources [32].

A crucial element of the Digital Twin structure is simulation. Production planning is enhanced, processes are streamlined, and production concepts are validated with the use of techniques like DES [33]. Near-optimal solutions for operational efficiency can be obtained by combining simulation with optimisation methods like genetic algorithms. Nonetheless, there are still issues with gathering high-quality input data and the time needed for comprehensive simulations. A linked data strategy can assist in dealing with these difficulties by automatically extracting data and thus enhancing the connection between the physical and digital systems [34].

Building a Digital Twin for AM is a systemic, collaborative approach. The process is typically divided into three stages: defining the scope of the digital twin, developing the digital master, and creating the digital shadow [35]. This includes defining the elements to be modeled, simulating production systems using DES, and gathering real-time data for traceability and continuous monitoring. From process and design to automation, every step needs teamwork to guarantee that the digital twin can precisely optimise output and decision-making. Even with continuous progress, more work has to be done in areas like data representation, simulation detail, and fortifying the bond between digital and physical systems [36].

7. Conclusion

The industrialisation of 3D printing technology offers a lot of chances for manufacturing innovation and advancement. This study's main finding is that the inefficiencies and bottlenecks in the current AM procedures prevent them from being as effective and scalable as they could be. But it also presents difficulties that need to be resolved to guarantee AM on a large-scale adoption. In order to tackle these obstacles, we utilized Discrete Event Simulation (DES) as a tactical instrument to simulate and evaluate the production process, enabling the pinpointing of significant inefficiencies. DES offers a powerful tool for overcoming these challenges by providing manufacturers with the ability to simulate, analyse, and optimise production processes.

This study has depicted how DES can increase AM production efficiency, decrease expenses and guarantee better quality via relevant case studies and feasible applications as compared to other similar examples. The outcomes show considerable increases in resource allocation, decreased idle periods, and increased production efficiency. Future developments in technology are anticipated to make DES increasingly crucial to AM's industrialization and widespread application in a variety of industries.

Amplitude of AM's capabilities is more likely when DES is combined with cutting-edge technologies like digital twins, AI, and the Internet of Things. To further improve AM operations, future research will concentrate on incorporating real-time data collecting and investigating the possibilities of cutting-edge technologies inside the DES framework. By implementing this technology, manufacturers may find ways to build cheaper, eco-friendly, and more efficient production processes which might lead to the wider use of AM in various industries.

8. References

[1] Atzeni, E. & Salmi, A. (2012). Economics of additive manufacturing for end-usable metal part, The International Journal of Advanced Manufacturing Technology, 62: p. 1147-1155.

[2] Lee, J.; Y. J. An. & C.K. Chua. (2017). Fundamentals and applications of 3D printing for novel materials. Applied materials today,. 7: p. 120-133.

[3] Sykora, J. & L. Kroft, (2023). Additive Manufacturing of Thin-Walled Inconel 718 Brackets for Extreme Conditions. Annals of DAAAM & Proceedings, 34.

[4] Gogate, A. & S, Pande. (2008). Intelligent layout planning for rapid prototyping, International Journal of Production Research, 46(20): p. 5607-5631.

[5] Liu, R. & et al. (2018). A survey on simulation optimization for the manufacturing system operation, International Journal of Modelling and Simulation, 38(2): p. 116-127.

[6] Siemens, The factory of the future for 3D printing (Access date: 19 Sep 2024)

[7] Negahban, A. & J.S. Smith. (2014). Simulation for manufacturing system design and operation: Literature review and analysis, Journal of manufacturing systems, 33(2): p. 241-261.

[8] Huynh, B.H.; H. Akhtar. & W. Li. (2020). Discrete event simulation for manufacturing performance management and optimization: a case study for model factory, in 2020 9th International Conference on Industrial Technology and Management (ICITM), IEEE.

[9] Omogbai, O. & K, Salonitis. (2016). Manufacturing system lean improvement design using discrete event simulation, Procedia CIRP, 57: p. 195-200.

[10] Velumani, S. & H, Tang. (2017). Operations status and bottleneck analysis and improvement of a batch process manufacturing line using discrete event simulation. Procedia Manufacturing, 10: p. 100-111.

[11] Prajapat, N. & et al. (2016). Layout optimization of a repair facility using discrete event simulation. Procedia CIRP, 56: p. 574-579.

[12] Hlupic, V. and S. Robinson. Business process modelling and analysis using discrete-event simulation. in 1998 Winter Simulation Conference. Proceedings (Cat. No. 98CH36274). 1998. IEEE.

[13] Law, A.M. & M.G. McComas. (1998). Simulation of manufacturing systems in 1998 Winter Simulation Conference, Proceedings (Cat. No. 98CH36274). IEEE.

[14] Johansson, B. & et al. (2009). Discrete event simulation to generate requirements specification for sustainable manufacturing systems design, in Proceedings of the 9th Workshop on performance metrics for intelligent systems.

[15] Krishna, R.; M. Manjaiah. & C. Mohan, (2021). Developments in additive manufacturing, in Additive manufacturing, Elsevier. p. 37-62.

[16] Wohlers, T. Wohlers report (2019): 3D printing and additive manufacturing state of the industry. (No Title).

[17] Kumar, S.D. & et al. (2017). Optimization of Thixoforging process parameters of A356 alloy using Taguchi's experimental design and DEFORM Simulation, Materials Today: Proceedings, 4(9): p. 9987-9991.

[18] D'Emilia, G. & et al. (2019). The role of measurement and simulation in additive manufacturing within the frame of Industry 4.0, in 2019 II Workshop on Metrology for Industry 4.0 and IoT (MetroInd4. 0&IoT), IEEE.

[19] Babu, B.; A. Lundbäck. & L.-E. Lindgren. (2019). Simulation of Ti-6Al-4V additive manufacturing using coupled physically based flow stress and metallurgical model, Materials, 12(23): p. 3844.

[20] Mourtzis, D.; M. Doukas. & D. Bernidaki. (2014). Simulation in manufacturing: Review and challenges, Procedia Cirp, 25: p. 213-229.

[21] Sahini, D.K. & et al. (2020). Optimization and simulation of additive manufacturing processes: challenges and opportunities–a review, Additive manufacturing applications for metals and composites, p. 187-209.

[22] Karanjkar, N. & et al. (2018). Digital twin for energy optimization in an SMT-PCB assembly line. in 2018 IEEE international conference on Internet of Things and intelligence system (IOTAIS). IEEE.

[23] Turner, C.J. & et al. (2016). Discrete event simulation and virtual reality use in industry: new opportunities and future trends, IEEE Transactions on Human-Machine Systems, 46(6): p. 882-894.

[24] Ingemansson, A.; T. Ylipää. & G.S. Bolmsjö. (2005). Reducing bottle-necks in a manufacturing system with automatic data collection and discrete-event simulation, Journal of Manufacturing Technology Management, 16(6): p. 615-628.

[25] Zarai, F. & P. Nicopolitidis. (2015). Modeling and simulation of computer networks and systems: Methodologies and applications, Morgan Kaufmann.

[26] Söderberg, R. & et al. (2017). Toward a Digital Twin for real-time geometry assurance in individualized production, CIRP annals, 66(1): p. 137-140.

[27] Schluse, M. & et al. (2018). Experimentable digital twins—Streamlining simulation-based systems engineering for industry 4.0. IEEE Transactions on industrial informatics, 14(4): p. 1722-1731.

[28] Isania, Z.; M.P. Fanti. & G. Casalino. (2024). A Discrete Event Approach to the Simulation of Selective Laser Sintering 3D Printing Large Scale Production, in 3rd International Conference on Visual Pattern Extraction and Recognition for Cultural Heritage Understanding (VIPERC). Bari, Italy.
[29] ElMaraghy, W. & et al. (2012). Complexity in engineering design and manufacturing. CIRP annals. 61(2): p. 793-814.
[30] Tao, F. & et al. (2019). Digital twins and cyber–physical systems toward smart manufacturing and industry 4.0: Correlation and comparison. Engineering. 5(4): p. 653-661.
[31] Morabito, L. & et al. (2021). A discrete event simulation based approach for digital twin implementation. IFAC-PapersOnLine. 54(1): p. 414-419.
[32] Dahmen, U. & J. Rossmann. (2018). Experimentable digital twins for a modeling and simulation-based engineering approach. in 2018 IEEE International Systems Engineering Symposium (ISSE). IEEE.
[33] Kritzinger, W. & et al. (2018). Digital Twin in manufacturing: A categorical literature review and classification. Ifac-PapersOnline. 51(11): p. 1016-1022.
[34] Lidberg, S.; L. Pehrsson. & M. Frantzén. (2018). Applying aggregated line modeling techniques to optimize real world manufacturing systems. Procedia Manufacturing. 25: p. 89-96.
[35] Iriondo, I.R.; A.U. Zearra. & A.K. Igartua. (2020). Discrete Event Simulation Procedure to Build the Production Digital Twin of Highly Automated and Complex Production Systems.
[36] Qin, H. & et al. (2021). Constructing digital twin for smart manufacturing. in 2021 IEEE 24th International Conference on Computer Supported Cooperative Work in Design (CSCWD). IEEE.

DOI: 10.2507/35th.daaam.proceedings.048

ROBOTIZATION PROBLEMS OF ANALYTICAL CONTROL OF THE MELTS CHEMICAL COMPOSITION IN METALLURGY

Aleksey Snopkov & Valentin Pryanichnikov

This Publication has to be referred as: Snopkov, A[leksey] & Pryanichnikov, V[alentin] (2024). Robotization Problems of Analytical Control of the Melts Chemical Composition in Metallurgy, Proceedings of the 35th DAAAM International Symposium, pp.0354-0362, B. Katalinic (Ed.), Published by DAAAM International, ISBN 978-3-902734-44-0, ISSN 1726-9679, Vienna, Austria
DOI: 10.2507/35th.daaam.proceedings.048

Abstract

When smelting steels, in order to maintain quality, it is necessary to carry out analytical control of the alloy before cooling the steel. The control will confirm, that its' chemical composition corresponds to what is expected for this steel grade. It is necessary to carry out analytical control in as little time as possible due to the high costs of maintaining the melt temperature and with maximum accuracy, due to the large influence of the chemical composition of alloys on their further operational characteristics. When working in such conditions, there is a high risk of disruption of the technological process due to the influence of the human factor, not to mention the risks to the health of workers. The need to quickly obtain high-precision laboratory analyses in production conditions led to the creation of automated systems for analytical control of the chemical compositions of metals. In such systems, the analysis of metal samples, previously carried out by trained personnel in laboratory conditions, is performed by automation in the conditions of the melting shop. However, most of these systems analyze metal samples, which are selected and prepared in the process of manual labour. The influence of the human factor is appearing during the procedures for taking a metal sample. It also is destroying the preparation process and analysis; it leads to errors in the operation of automated analytical control systems. The article considers the technology of automation of analytical control of the chemical composition of metal on example of the automated container laboratory "Thermo Techno Engineering", including analyzes of the stages of its work, offers solutions to improve the reliability of such automated systems by robotizing those stages of analysis where the human factor can lead to a malfunction of the automated analytical control system. The proposed solutions will also make it possible to avoid manual input of information and ensure traceability of samples and the technological process at all stages. The work was organized in the frame of the joint project of KIAM, MSTU Stankin and IINET RSUH.

Keywords: analytical control and the human factor in metallurgy; project of a robotic complex for sampling and analysis of melts, automation in analytical production control, algorithms for robotic sampling in metallurgy.

1. Robotization problems of analytical control of the melts chemical compositions in metallurgy

Modern production processes in any field of production cannot be imagined without various steels alloys, the performance characteristics of which are known and predictable for each steel grade. However, such production creates quite extreme conditions for both personnel and devices [1]. Various companies solve the issue of protecting equipment by placing it in containers [2]. However, most of these systems analyze metal samples, which are selected and prepared in the process of manual labour. The influence of the human factor is appearing during the procedures for taking a metal sample.

It also is destroying the preparation process and analysis; it leads to errors in the operation of automated analytical control systems. Suggests method for step-by-step robotization of these processes, justifies the need to develop appropriate algorithms and technical means. When smelting steels, in order to maintain product quality, it is necessary to conduct analytical control of the initial sample of the alloy [13] before cooling the melt (according to the standard [14]), which will confirm, that the chemical composition corresponds to what is expected for this steel grade. Analysis of the chemical composition of metals and alloys is an integral part of many technological processes used in various industries. The study makes it possible to determine the presence of impurities and inclusions in the sample, as well as to predict the performance characteristics of the finished product. It is necessary to carry out analytical control in as little time as possible due to the high costs of maintaining the melt temperature and with maximum accuracy, due to the large influence of the chemical composition of alloys on their further operational characteristics [14].

Steel with an excess content of carbon, nickel, cobalt or any other chemical element in its composition will differ in its strength, hardness, viscosity, abrasive or corrosion resistance or other performance characteristics and will not meet the requirements imposed on the grade of steel, that had to be cast after melting. Moreover, the process of sampling the molten metal and sample preparation is carried out in conditions far from laboratory, sampling takes place near a melting furnace with liquid metal, the temperature of which can be from +1,375 to +1,530 °C, depending on the grade of steel being smelted. In the air environment of foundries, in addition to dust, a large amount of carbon monoxide, carbon dioxide and sulfur dioxide, nitrogen and its oxides, hydrogen and other aggressive substances are released [1]. When working in such conditions, there is a high risk of disruption of the technological process due to the influence of the human factor, not to mention the risk to the health of workers. After sampling and sample preparation, the sample must be cooled, prepared and delivered to the laboratory for spectral analysis. The delivery of samples to the laboratory for analytical control, like other processes, should take place in the shortest possible time due to the high cost of maintaining the melt temperature. Even when using automated sample delivery systems to the laboratory, such as pneumatic transportation systems, the standards for the time for which analytical control of the sample should be performed encourage laboratory staff to work in a hurry, which often leads to analysis errors due to the influence of the human factor. The need to quickly obtain high-precision laboratory analyses in production conditions led to the creation of automated metal analysis systems. In such systems, the analysis of metals, previously carried out by trained personnel in laboratory conditions, is performed by automation in the conditions of the melting shop. This reduces the influence of the human factor, reduces the required number of highly qualified personnel in production, allows for the necessary metal analyses in a smelter and reduces the number of people who must work in conditions harmful to health. To date, the most common solution to this problem is to place an automated analytical control system in production, which is installed in a transportable, protected container with internal condition control, which can monitor the implementation of analysis procedures during the process. Such automated container systems are being developed by such companies as HERZOG [2], Thermo Techno Engineering [3], ARL [4], Flsmidth [16] and others.

2. Review of existing Technologies of automated systems for analytical control of the metal chemical composition

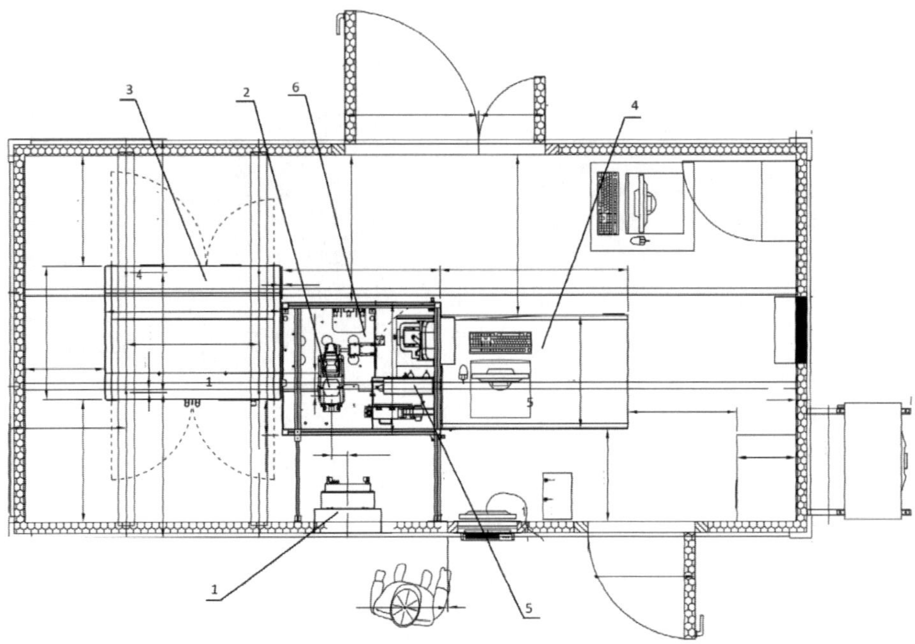

Fig. 1. Generalized composition of the Thermo Techno Engineering automated container laboratory [11].
1 – registration module for samples, 2 – manipulation robot, 3 – sample preparation unit, 4 – spectral analysis unit, 5 – vision system, 6 – sample storage unit.

3. The process of analytical control is the main stages

Fig. 2. Procedure for sampling metal melts for further analysis [8].

3.1. Sampling of metal melts

For fast and high-quality sampling of metal melts (Fig.2.) it is convenient to use disposable submersible samplers in various metallurgical units. After using the sampler, a sample is obtained in the form of a cylinder with the correct geometry: a circle with a diameter of 35 mm or an oval of 35x40mm (depending on the sampler model) and a height of 6, 8, 10 or 12mm (also depending on the model). The samples obtained do not have external and internal defects (pores, shells, cracks, etc.), and have good convergence for further spectral analysis (Fig.3.).

Fig. 3. Sample for spectral analysis after separation from the sampler [9].

3.2. Cooling and pre-sample preparation

Before transferring the sample to the automated container laboratory of analytical control, it is necessary to carry out preliminary sample preparation consisting in cooling the sample to 50 °C and cutting off the sample gate. It is important not to disrupt the technological process of cooling the sample, if the cooling process is disrupted, the sample metal may experience overfilling, which negatively affects the accuracy of the analysis results. It is equally important to observe the technological process of cutting off the sample gate, an incorrectly cut off gate may prevent the capture of the sample by the manipulator of an automated container laboratory (Fig.4).

Fig. 4. A sample for spectral analysis and a cut-off gate after preliminary sample preparation [9].

3.3. Sample registration

When transferring a sample to an automated container laboratory for analytical control, it is necessary to register the sample. When registering a sample, it is important for the operator to observe the temperature characteristics of the recorded sample, since too hot a sample can lead to disturbances in the process of preparing the sample surface for analysis. However, the human factor at this stage is regulated by the sample temperature control system. (Fig.5.)

Fig. 5. Registration module for samples. [10]

3.4. Preparation of the sample for spectral analysis

After registering the sample, the manipulator of the automated container laboratory positions the sample on the vise of the milling machine for subsequent removal of oxide and slag from the sample surface by milling (fig. 6).

Fig. 6. Capture of a sample by a manipulator of an automated container laboratory. [10]

3.5. Visual analysis of the sample with the search for the most suitable sample surfaces for spectral analysis

After milling, the manipulator of the automated container laboratory positions the sample in the working area of the vision system for subsequent assessment of surface quality and search for the most suitable surfaces for spectral analysis of the sample (Fig.7, Fig.8).

Fig. 7. Visual analysis of the sample by the technical vision system [10].

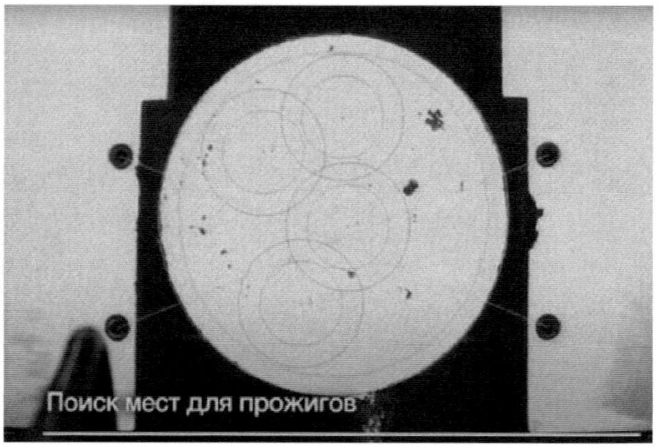

Fig. 8. Selection of the most suitable part of the sample surface by the vision system for further analysis [10].

3.6. Spectral analysis of the sample and archiving of the analysis results

The manipulator positions the sample on an automatic spectrometer stand, in such a way that spectral analysis is performed on the most suitable sample surfaces for analysis. The results of the analysis are displayed on the user interface and archived (Fig.9).

Fig. 9. Spectral analysis of the sample [10].

3.7. Laser marking of the examined sample and sorting

A marker is applied to the sample containing the information necessary for further identification of the sample (Fig.10). The examined and labeled sample is placed in the appropriate compartment of the analyzed samples storage store (Fig.11).

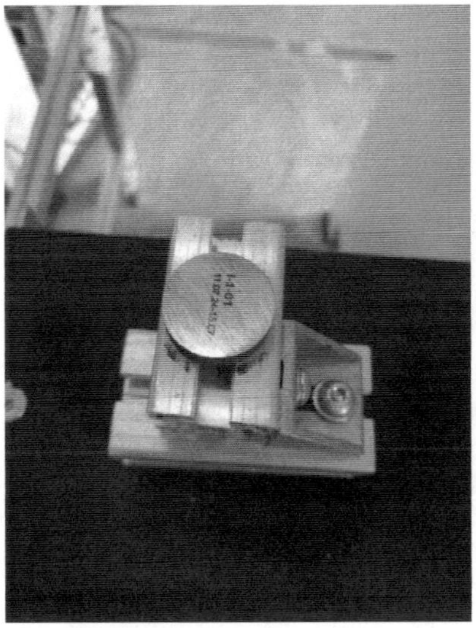

Fig. 10. Labeled sample [10].

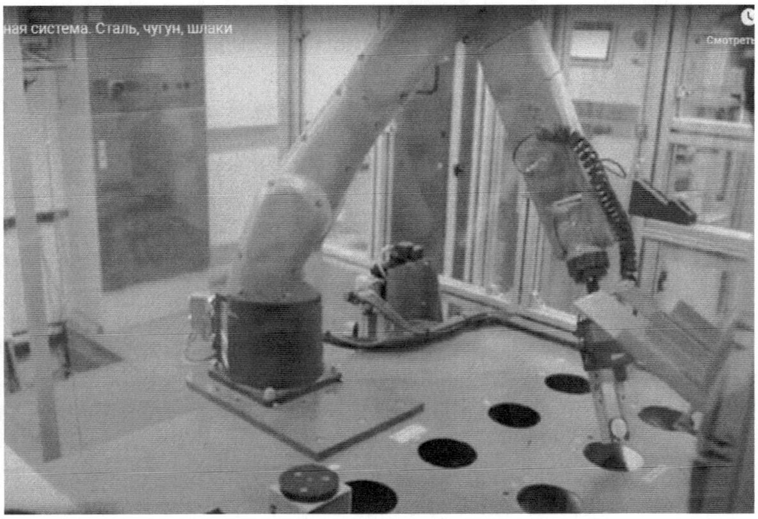

Fig. 11. Sorting of samples after examination [10].

4. Algorithmization of the process of analytical control of melts

The entire process of analytical quality control of metal from sampling to archiving of the analysis results can be represented in the form of several successive stages, some of which are partially or fully automated (fig.12).

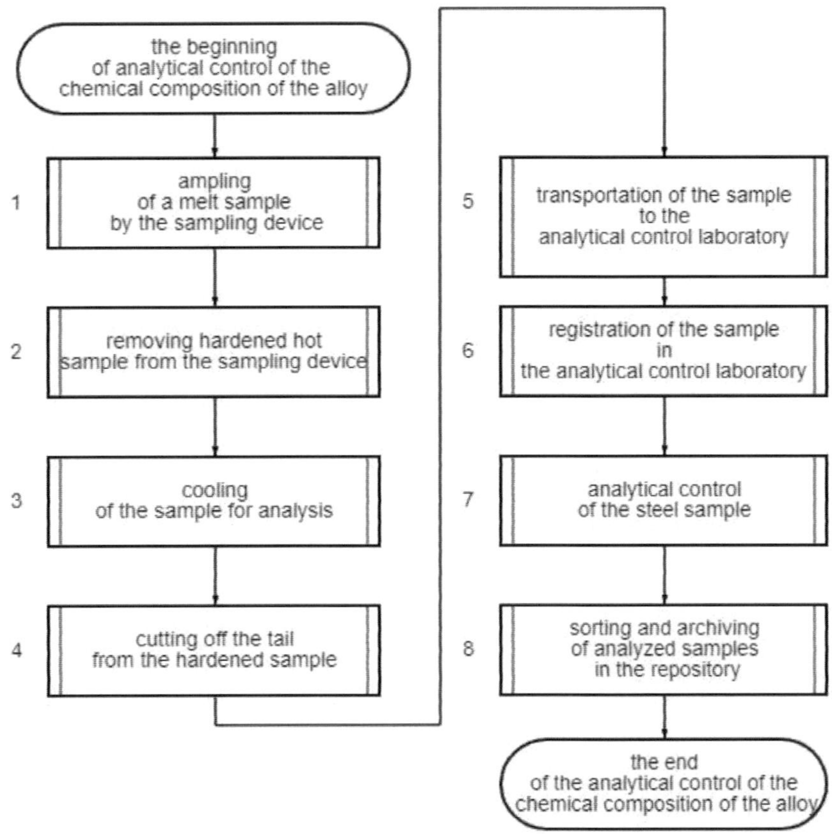

Fig. 12. General algorithm of sample analysis.

Figure 12 shows the sequence of analytical control operations:

Step 1 - The sampling stage is not automated, performed by a person in extreme human conditions;

Step 2 - The stage of sample withdrawal is not automated, performed by a person, often performed with violations due to the influence of the human factor;

Step 3 - The cooling stage of the sample for analysis is not automated, performed by a human, and is often performed with violations due to the influence of the human factor;

Step 4 - The stage of cutting the sample gate is not automated, is performed by a human, is often performed with violations due to the influence of the human factor;

Step 5 - The stage of sample delivery to an automated analytical control laboratory, if necessary, is partially automated through the use of pneumatic delivery systems;

Step 6 - The stage of sample registration is partially automated, the influence of the human factor is excluded by automatic systems controls, the algorithm of which is shown in the block diagram (Fig.13);

Step 7 - The stage of sample analysis in the container laboratory is fully automated by the analytical control system;

Step 8 - The stage of sample sorting and archiving analyzed samples is fully automated.

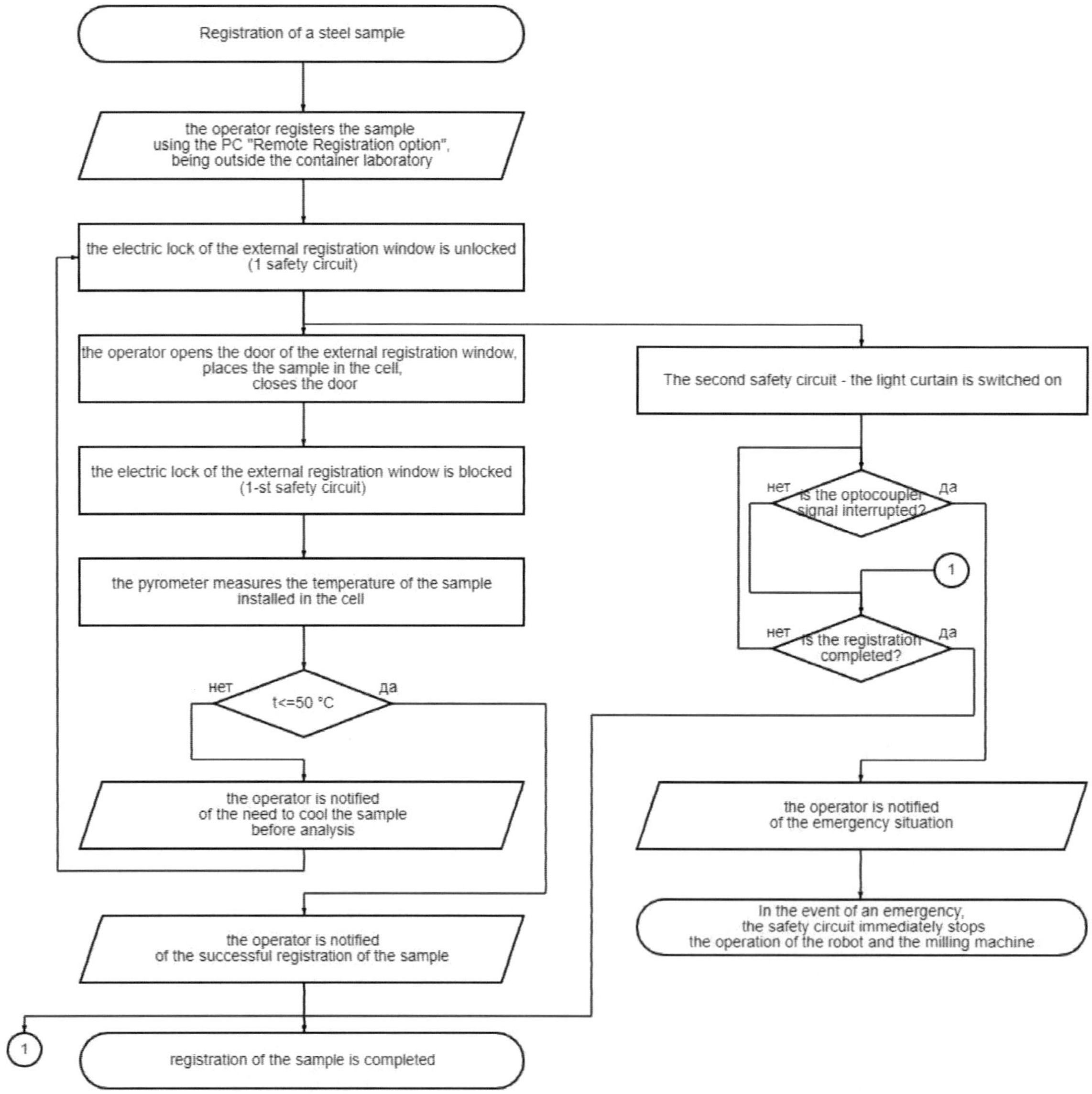

Fig. 13. Flowchart for sample registration in an express laboratory.

5. Conclusion

During the analysis of the stages (steps) of analytical control, it was revealed, that most errors in the analysis of samples occur due to the human factor at the stage of sampling, cooling and stumping of the gate (Fig. 14). Steps 1, 3 and 4 of stages are sufficiently formalized for automation without the use of complex sensor systems, automation of stage 2 will require the use of technical vision and tactile sensing systems to work with objects without clear positioning.

1 2 3 4

Fig. 14. Sampling and sample preparation process.

To increase the reliability of the automated analytical control system, it is necessary to eliminate the influence of the human factor at these stages by automating using a group of mobile and manipulative robots designed to work together in extreme high temperature conditions based on developments [5],[6],[7]. This will make it possible to avoid manual input of information, ensure remote control of samples and the technological process at all stages [7] and increase the accuracy of the results of analytical control of the chemical composition of alloys. The detailed description of the stages of sample analysis presented in the paper is the first step in formalizing this process for its further complete robotization.

6. Literature

[1] Golovina E.I. (2016) Dangerous and harmful factors of foundry production and their influence on the state of the atmosphere in the working area // Bulletin of the Kazan Technological University. (In Russian: Головина Е.И. Опасные и вредные факторы литейного производства и их влияние на состояние атмосферы в рабочей зоне // Вестник Казанского технологического университета).. №23.

[2] Analytical laboratory of Herzog Maschinen Fabrik (Year)

[3] Container Laboratory (Year) https://www.thermotechno.ru/catalog/312/597 / https://www.herzog-maschinenfabrik.de/ru/produkcija/standartnaja-avtomatizacija/metallab/

[4] Analytical equipment of the QUANTOSHELTER container laboratory of ARL (Switzerland) (Year) https://www.krsk.kp.ru/daily/26272/3149953/

[5] V.E.Pryanichnikov, A.A.Aryskin, A.A.Drachkova, K.E.Brundukova, O.I. Vlasova, A.A.Kartseva, Yu.S.Kolesov, A.V.Snopkov, M.D.Solovyova, A.I.Telegin, D.S.Bogotkin, A.N.Tikhomirov, A.V.Fedorov, D.K.Fedotov, V.V.Iastrebov.(2023) Experiments to expand the functionality of mobile service robots. "EXTREME ROBOTICS", A collection of abstracts of the 34th International Scientific and Technical Conference. – St. Petersburg: Central Research Institute of RTK - LLC "Fursov Printing House", (in Russian: В.Е.Пряничников, А.А.Арыскин, А.А.Драчкова, К.Э.Брундукова, О.И. Власова, А.А.Карцева, Ю.С.Колесов, А.В.Снопков, М.Д.Соловьёва, А.И.Телегин, Д.С.Тележкин, А.Н.Тихомиров, А.В.Федоров, Д.К.Федотов, В.В.Ястребов. Эксперименты по расширению функциональных возможностей мобильных сервисных роботов. «ЭКСТРЕМАЛЬНАЯ РОБОТОТЕХНИКА», Сборник тезисов 34 -й Международной научно-технической конференции. – Санкт-Петербург: ЦНИИ РТК - ООО «Типография Фурсова») 2023. – 306 с.: ил., С.34-36, ISBN 978-5-6048032-9-5, https://er.rtc.ru/images/docs/2023/Sbornik-tezisov_ER-2023.pdf

[6] Pryanichnikov, V[alentin]; Mariia, S[oloveva]; Chernyshev, V[adim] & Kalinin, Y[aroslav] (2023). Analysis and Modernization of the Dual Manipulator of Service Robots, Proceedings of the 34th DAAAM International Symposium, pp.0324-0329, B. Katalinic (Ed.), Published by DAAAM International, ISBN 978-3-902734-41-9, ISSN 1726-9679, Vienna, Austria DOI: 10.2507/34th.daaam.proceedings.043

[7] Pryanichnikov V. E., Kuvshinov S. V., Helemendik R. V., Kharin K. V. (2020) Intelligent robotics: scientific and educational technology and building a network of robotariums. In: Strategy for the development of computer reality. Collective monograph / Edited by G.G. Malinetsky, P.A. Vernik, V.V. Ivanov M.:TECHNOSPHERE, (Пряничников В. Е., Кувшинов С. В., Хелемендик Р. В., Харин К. В. Интеллектуальная робототроника:

научно-образовательная технология и построение сети роботариумов. В кн.: Стратегия развития компьютерной реальности. Коллективная монография / Под ред. Г.Г. Малинецкого, П.А. Верника, В.В. Иванова М.: ТЕХНОСФЕРА), 2020. – 360с. С.86-94, ISBN 978-5-94836-575-6
[8] Melt sampling (Year): https://uralsteel.com/media/
[9] Sampling of liquid cast iron and steel using samplers Eurasian devices (2021) https://evrazpribor.ru/otbor_prob_jidkogo_chuguna_i_stali_pri_pomoschi_probootbornikov
[10] Automated system for the preparation and analysis of steel samples, cast iron and slag (Year) https://www.thermotechno.ru/catalog/392/
[11] The generalized composition of the automated container laboratory "Thermo Techno Engineering" (Year) https://www.thermotechno.ru/catalog/397/
[12] Thermotechno (Year) https://www.thermotechno.ru/for_press/756/
[13] GOST R ISO 14284-2009 p3.4 State Standard of Russian Federation (equivalent of ISO Standard) (2019)
[14] GOST R ISO 14284-2009 p3.14 (2019)
[15] GOST 7565-81 (ISO 377-2-89) p1.1 (1981)
[16] Design of FLSmidth express laboratories (Year) https://www.flsmidth.com/en-gb/customer-stories/ese-odin-vladelec-sahty-pozinaet-plody

DOI: 10.2507/35th.daaam.proceedings.049

STATE OF THE ART REVIEW OF SMART FARMING

Marcel Vrdoljak, Mohamed Aburaia, Kemajl Stuja & Ali Aburaia

This Publication has to be referred as: Vrdoljak, M[arcel] (2024). State of the Art Review of Smart Farming, Proceedings of the 35th DAAAM International Symposium, pp.0363-0367, B. Katalinic (Ed.), Published by DAAAM International, ISBN 978-3-902734-44-0, ISSN 1726-9679, Vienna, Austria
DOI: 10.2507/35th.daaam.proceedings.049

Abstract

Smart farming is becoming the future of agriculture by implementing new smart technologies such as robotics, artificial intelligence (AI) and autonomous systems. This integration tries to tackle some of the most prominent challenges like environmental sustainability and resource management. With modern technologies resources can be optimized, crops can be monitored to ensure quality and prevent failures while still minimizing the impact on the environment. The main technologies used in smart farming include autonomous ground and aerial vehicles, IoT sensors and support systems for decision making. The integration of such sophisticated technologies is however challenging, as it includes high costs with many maintenance issues. Overall, smart farming is rising, and the global market values are rising at a significant rate and are paving the way for a sustainable and efficient future for agriculture.

Keywords: Smart farming; Internet of Things; Robotics; Sustainable Agriculture; Precision farming

1. Introduction

Smart farming is an emerging field that uses advanced technologies to be used in agriculture and help improve efficiency, productivity, and sustainability. It allows the use of new and improved tools for a responsible and sustainable use of resources such as water or fertilizer. Farmers can optimize every aspect of their production to ensure a good harvest every time. [1] The new technologies that are being developed are diverse and rapidly changing. Due to this, there are some different terms to describe specific aspects of this field. Smart farming, or agritech, for example, is widely representative of the whole field of technologies and innovations in agriculture.

Precision Farming represents a key component of smart farming. It focuses on the individual and localized management of resources such as water, fertilizer, or pesticides. It allows farmers to individually treat specific parts of a field or animals with all their needs. So, for example, it is possible to water one part of a field more than the others because the crop and the land in the area are drier than in other parts. Similarly, food for stock animals can be adjusted to their specific needs and current health state. Digital farming describes the aspect of gathering and analyzing data for decision-making. This involves all kinds of sensors and IoT devices to gather data all around the farm and monitor any changes to act accordingly. This can minimize the risk of bad crops or spreading infections. Smart farming in general is gathering more traction in recent years due to global challenges such as climate change, scarcity of resources, and a fast-growing population. This can be seen in the economic sector. Between 2020 and 2025, the market value for IoT technologies in agriculture in North America is expected to double, reaching 4 billion US- dollars [2].

Similarly, the global market value of smart farming is projected to reach 33 billion dollars by 2027, a threefold increase since 2021 [3]. This reflects the needs and fast-growing nature of smart farming in recent years and can be observed in figure 1.

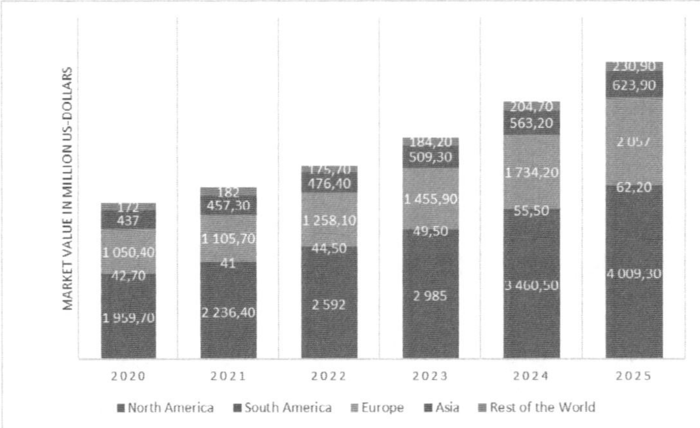

Fig. 1. Market value of IoT technologies integrated in agriculture worldwide by region from 2020 to 2025 Source: [2]

The purpose of this study is to provide an analysis of the currently adopted technologies in smart farming and present their economic impact on the agricultural sector. By exploring a range of technologies and innovations, this study aims to offer an overview of how these technologies are changing old farming practices. Additionally, this study will examine the challenges that are introduced by smart farming.

2. Technologies in Smart Farming

The technologies that are being implemented are widely spread. They are significantly dependent on different factors, one of them being the climate and geographic characteristics of different regions. As countries focus on different crops or different stock animals, the employed technologies and research tend to diverge and focus on different priorities. For instance, Spain has emerged as the leading country in the EU in research and development of robots for wine production. These robots can perform various tasks, such as collecting grapes from grapevines or monitoring the condition of the plants [4]. In contrast, Germany has its priority in harvesting other crops. It has more significant research in autonomous driving, especially with autonomous tractors [5].

Despite all the differences, the technologies used in smart farming can be categorized into two major groups: unmanned ground and aerial vehicles and IoT sensor networks for data collection. Unmanned vehicles play a crucial role in modern farming. Ground vehicles, such as tractors or harvesters, are being used to maximize efficiency and minimize the use of resources [6]. Equipped with modern GPS technology and sensors, they can navigate entire fields accurately. Also, mobile robots are being deployed more often, as research is getting more traction and more funding [4]. Similar, aerial vehicles or drones can observe fields and crops when equipped with cameras. They can cover large fields quickly and precisely, showing data that allows farmers to have a better overview of the fields and identify any issues, making it a competitive technology for IoT networks [7]. However, aerial vehicles are used much less than ground vehicles, as the development didn't start getting traction until 2015, when the word "drone" was created [8]. For comparison, the first research for automatic tractors with GPS tracking was published in 1996 by Stanford University in the USA [9].

Sensor networks are the second big cornerstone of smart farming. As they can provide real-time data about a crop field or various environmental factors, farmers are able to monitor data and act accordingly when needed. Depending on the network, specific crop parts or stock animals can be treated individually to ensure better quality with precision farming [10]. Collected data often gets transmitted through IoT-devices, which enables remote access to all data. This approach leads to better management of resources, allows a greater yield, and reduces the environmental impact. The synergy of these technologies can create an ecosystem for precision farming. With aerial images of the crops and exact real-time data from ground sensors, farmers can evaluate the field conditions and act accordingly. They are also able to react faster to changing conditions or diseases.

3. Examples for application of technologies

One of the earliest examples of smart farming technology is the automated milking machine, first patent and research done in 1987 [11]. Today, these machines not only can milk cows in a steady and unharmful way, they can also collect data about each cow, for instance, their vital values to determine the health state, the yield, and the frequency of each milking procedure. With this data, farmers can adjust feeding practices and react to health issues, which improves productivity and animal welfare [12].

In addition to milking, smart livestock monitoring systems are set to be a critical component in modern animal husbandry. Sensors and IoT-devices can track many different vital values and track the behavior of each individual animal. If unusual data is identified, farmers can react to potential health issues faster than without technology. [13] Agricultural robots have also been pushed by several funded projects in the EU. The European Clearing House for Open Robotics Development (ECHORD++) is an initiative to get researchers and manufacturers together to cooperate [14]. Many projects were funded by this initiative. For instance, the Mobile Agricultural Robot Swarms (MARS) project was creating mobile robots for seeding crops that could work in swarms [15]. Unlike traditional tractors, these robots can minimize soil compaction, which benefits the soil health. Another project was the Ground Robot for Vineyard Monitoring and Protection Project (GRAPE), which aimed to create mobile robots that can autonomously monitor grapevines and harvest them if needed [16].

WaterBee was another EU-funded project that aimed to introduce more technologies for water management. The conservation of water is a critical concern in modern agriculture, especially in very dry climates and regions. WaterBee introduces a network of IoT sensors placed in the soil of a field. They monitor the moisture level in real time and can adjust the watering of the crop depending on the parameters. With this method, specific parts of a field can get watered if needed, and the whole schedule can be adjusted to the real-time needs. This ensures an optimal use of water as a resource. [17]

Artificial intelligence is getting more significant with recent breakthroughs. They are mostly integrated into decision-making components to analyze data and act accordingly. These systems can analyze data from sensors, weather, drones, and all kinds of IoT devices. This ensures an optimal and sustainable use of resources and can even help farmers decide on daily operations and prioritization [18]. As an example, AIs get used as an effective animal repellent combined with ultrasound to protect crops [19].

4. Advantages and challenges

As it offers significant advantages to farmers, the implementation of smart technologies can be challenging and overwhelming. There are high initial costs that need to be carried. Machinery and IoT sensors require a high capital. Small or medium-sized farmers are usually not able to make such an investment. Additionally, after the initial expenses, costs can still be high with regular maintenance, updates, or repair costs [4]. However, there exist many funding programs that can support farmers. In the EU, there exist many different programs and research projects that were discussed in the previous chapter, like ECHORD Plus Plus or WaterBee. Other than that, the EU also has funding projects like the Common Agricultural Policy (CAP), which aims to provide direct development funds for farmers [20]. These funds often require the implementation of smart technologies and digital practices. There is also the Partnership for Agricultural Productivity and Sustainability (EIP-AGRI), which helps to connect farmers with researchers and businesses and provides grants when adopting new smart technologies [21]. Additionally, the European Agricultural Fund for Rural Development (EAFRD) also finances rural projects and farmers and helps to develop new technologies that align with the EU's Farm to Fork Strategy and the Green Deal [22].

While the initial costs can be very high, the long-term savings are considerably higher. The need for manual labor is minimized, and an optimal use of resources is guaranteed. More critical is the ability to create a sustainable agricultural standard. By optimizing the use of all resources, the environmental impact gets minimized and the carbon footprint gets reduced. It also increases the yield of the crops and increases the revenue of the farmers. [23]

Farmers often don't have the necessary skills to employ and use smart farming technologies. A fundamental knowledge about data analytics, machinery, and modern technologies is needed to ensure an efficient use of smart devices [24]. Furthermore, there is currently a lack of standardization across all smart farming technologies. Manufacturers use systems that often don't interact with each other and create inefficiency, as farmers need to work on multiple systems and platforms. This increases the needed knowledge level even more [25]. With the new systems collecting and transmitting large amounts of data, cyber security and data privacy get more important. The fear of a data breach and the misuse of information can hinder many farmers from adopting new technologies. In some regions, there also may be policy constraints that don't allow the implementation of such networks. This includes unclear regulations about the ownership of data. [26]

5. Conclusion

Smart farming shows a shift in agricultural techniques from a traditional way that is highly dependent on manual labor to a modern way with the introduction of sophisticated technologies and innovative solutions. With the gcurrent global challenges of resource shortages, labor shortages, and an unsustainable environmental impact, the integration of smart technologies is getting more crucial. The technologies allow a precise and optimized use of resources, promote more sustainable practices, and reduce the overall costs of farmers. As demonstrated, the use of smart farming technologies can be very diverted and implemented in different ways. Robotic systems are changing tasks like milking, harvesting, or planting to be executed easier and without manpower. Precision farming enables a more specific observation of the farm and introduces a more precise way of managing crops and stock animals. However, these technologies are met with notable challenges. The costs of such systems can be very high, especially for small or medium-scale farmers.

Pairing this with a high degree of complexity and a notable risk of a data breach makes it a very daunting task to implement such systems into a farm. Additionally, the social impact of automation must be considered as a risk of displacement of labor is ensured. Despite all the challenges, it is prognosed that the global market will rise, which reflects the interest in modern technologies in agriculture.

While this study shows the benefits of smart farming technologies, several limitations should be considered. This study focuses more on regions that are technologically advanced and may not be applicable to areas where the infrastructure is not available. Additionally, environmental variability of regions was not considered while discussing the challenges and benefits of smart farming. Unpredictable factors like extreme weather conditions could limit the effectiveness of the technologies and the results. Lastly, while this study showed the numerous benefits of smart farming, it did not account for barriers from a social, economic, or educational point of view that could prevent farmers, especially in rural areas, from implementing new technologies.

In conclusion, smart farming shows to be a promising future for agriculture. As research and development tend to push the boundaries and create new technologies, smart farming will address some of the most crucial global challenges, like global warming. Smart farming will eventually start a new era in agriculture, one with sustainable progress for the future. The next steps in this field will most likely focus more on the improvement of the integration of these technologies to create fully autonomous systems for farmers. Future research will also prioritize using artificial intelligence for decision-making and the use of 5G networks or satellites to solve connectivity issues. The development of affordable and scalable smart farming technologies will be crucial to make them accessible to small farmers and to foster global sustainability in agriculture.

6. References

[1] Singh, G.; Kalra, N.; Yadav, N. & Sharma, A. (2022). Manoj. Smart Agriculture: A Review. Siberian Journal of Life Sciences and Agriculture, vol. 14, no. 6, pp. 423-454.

[2] Marktwert des industriellen Internets der Dinge (IIoT) in der Landwirtschaft weltweit nach Region in den Jahren 2020 bis 2025 (in Millionen US-Dollar) (Graph), Juniper Research, 1. März, 2021. [Online]. Available: https://de-l statista-l com-112m0mpa42e38.han.technikum-wien.at/statistik/daten/studie/1337452/umfrage/landwirtschaftlich-industriell-internet-der-dinge-weltweit-markwert/. Accessed on: 2024-10-09

[3] Prognostizierter Marktwert von Smart Farming weltweit in den Jahren 2021 bis 2027 (in Milliarden US-Dollar) (Graph), BIS Research, 30. Juli, 2023. [Online]. Available: https://de-1statista-1com-112m0mpa42abf.han.technikum-wien.at/statistik/daten/studie/1331861/umfrage/marktwert-smart-farming-prognose-weltweit/. Accessed on: 2024-10-09

[4] Moysiadis, V.; Sarigiannidis, P.; Vitsas, V. and Khelifi, A., Smart farming in Europe. Computer science review, 2021, 39, p.100345

[5] KBVresearch. (2023). Europe Autonomous Tractors Market, KBV-14853

[6] Gonzalez-De-Santos, P.; Fernández, R.; Sepúlveda, D.; Navas, E. & Armada, M. (2020). Unmanned ground vehicles for smart farms, Agron.-Clim. Chang. Food Secur, 6, p.73.

[7] Petkovics, I.; Petkovic, D. & Petkovics, A. (2017). IoT Devices vs. Drones for Data Collection in Agriculture, Chapter 06 in DAAAM International Scientific Book 2017, pp.063-080, B. Katalinic (Ed.), Published by DAAAM International, ISBN 978-3-902734-12-9, ISSN 1726-9687, Vienna, Austria DOI: 10.2507/daaam.scibook.2017.06

[8] Del Cerro, J.; Cruz Ulloa, C.; Barrientos, A. and de León Rivas, J. (2021). Unmanned aerial vehicles in agriculture: A survey. Agronomy, 11(2), p.203.

[9] O'Connor, M.; Bell, T.; Elkaim, G. & Parkinson, B. (1996). Automatic steering of farm vehicles using GPS, Proceedings of the 3rd International Conference on Precision Agriculture, Minneapolis, USA, pp. 767-777

[10] Raj, E.F.I.; Appadurai, M. & Athiappan, K. (2022). Precision farming in modern agriculture, Smart agriculture automation using advanced technologies: Data analytics and machine learning, cloud architecture, automation and IoT (pp. 61-87). Singapore: Springer Singapore.

[11] Noorlander, D.O. & Heckmann, R.A. (1987). Milking Machine Design and Mastitis. Applied Engineering in Agriculture, 3(1), pp.99-103.

[12] Tutkun, M. (2023). Precision dairy farming. Journal of Agricultural, Food and Environmental Sciences, JAFES, 77(1), pp.12-19.

[13] Shrivastav, V.; Yadav, M.; Sharma, A.; Kumar, D.; Sharma, S. & Chauhan, A.S. (2024). IoT and IoE transformations in precision farming agriculture: Sensor based monitoring, Automated irrigation and Livestock monitoring, 2024 IEEE International Students' Conference on Electrical, Electronics and Computer Science (SCEECS) (pp. 1-14). IEEE.

[14] European Clearing House for Open Robotics Development (ECHORD++). (2018). [Online]. Available: https://echord.eu/. Last visit: Accessed on: 2024-10-09

[15] Blender, T.; Buchner, T.; Fernandez, B.; Pichlmaier, B. & Schlegel, C. (2016). Managing a mobile agricultural robot swarm for a seeding task, IECON 2016-42nd Annual Conference of the IEEE Industrial Electronics Society (pp. 6879-6886). IEEE.

[16] Roure, F.; Moreno, G.; Soler, M.; Faconti, D.; Serrano, D.; Astolfi, P.; Bardaro, G.; Gabrielli, A.; Bascetta, L. & Matteucci, M. (2018). GRAPE: ground robot for vineyard monitoring and protection, ROBOT 2017: Third Iberian Robotics Conference: Volume 1 (pp. 249-260). Springer International Publishing.
[17] Project WaterBee. (2010). [Online]. Available: https://www.waterbee.eu. Accessed on: 2024-10-09
[18] Akkem, Y.; Biswas, S.K. & Varanasi, A. (2023). Smart farming using artificial intelligence: A review, Engineering Applications of Artificial Intelligence, 120, p.105899.
[19] Mishra, A. & Yadav, K.K. (2024). Smart Animal Repelling Device: Utilizing IoT and AI for Effective Anti-Adaptive Harmful Animal Deterrence, BIO Web of Conferences (Vol. 82, p. 05014). EDP Sciences.
[20] Common Agricultural Policy (CAP). (1962). [Online]. Available: https://agriculture.ec.europa.eu/common-agricultural-policy/cap-overview_en. Accessed on: 2024-10-09
[21] Partnership for Agricultural Productivity and Sustainability (EIP-AGRI). (2012). [Online]. Available: https://ec.europa.eu/eip/agriculture/en.1.html. Accessed on: 2024-10-09
[22] European Agricultural Fund for Rural Development (EAFRD). (2021). [Online]. Available: https://commission.europa.eu/funding-tenders/find-funding/eu-funding-programmes/european-agricultural-fund-rural-development-eafrd_en. Accessed on: 2024-10-09
[23] Tyagi, M.; Kumar, A.; Guleri, S. & Gaurav, N. (2024). Transforming agriculture: the rise of smart farming, p150-178.
[24] Ahmad, D.; Fatah, F.; Saili, A.; Saili, J.; Hamzah, N.; Nor, R.; Zubaidah, O. & Mohamed, G. (2024). Exploration of the Challenges in Adopting Smart Farming Among Smallholder Farmers: A Qualitative Study, Journal of Advanced Research in Applied Sciences and Engineering Technology
[25] Agarwal, P.; Singh, V.; Saini, G.L. & Panwar, D. (2019). Sustainable Smart-Farming Framework: Smart Farming. 147-173.
[26] Kotal, A., Elluri; L.,Gupta, D.; Mandalapu, V. & Joshi, A. (2023). Privacy-Preserving Data Sharing in Agriculture: Enforcing Policy Rules for Secure and Confidential Data Synthesis.

DOI: 10.2507/35th.daaam.proceedings.050

APPLICATION OF REMEDIAL MEASURES IN COMBINATION WITH ACTIVE THERMAL PROTECTION FOR OPTIMIZING THE INDOOR ENVIRONMENT OF OLDER BUILDINGS

Patrik Stastny

This Publication has to be referred as: Stastny, P[atrik] (2024). Application of Remedial Measures in Combination with Active Thermal Protection for Optimizing the Indoor Environment of Older Buildings, Proceedings of the 35th DAAAM International Symposium, pp.0368-0372, B. Katalinic (Ed.), Published by DAAAM International, ISBN 978-3-902734-44-0, ISSN 1726-9679, Vienna, Austria
DOI: 10.2507/35th.daaam.proceedings.050

Abstract

Damp and unhealthy environments in older buildings that are not listed can be considered as a globally widespread problem. This humidity creates an environment that negatively affects the people who live in this environment and also negatively affects the construction. This problem is also linked to the deterioration of the properties of the construction, such as thermal insulation properties and the like. The paper focuses on a case study on which the deployment of undercutting technology was investigated to reduce the humidity of the construction and its efficiency. Subsequently, it theoretically describes the possible application of active thermal protection (ATP), the implementation of which could significantly improve the internal climate of the building.

Keywords: Rising damp; construction undercutting; rehabilitation; active thermal protection

1. Introduction

The unhealthy environment of constructions, which endangers people living in this environment for a long time, is a problem that was described several decades ago in his publication by Kenwood[1]. It has been a very long time since the publication of said paper before the problem of humidity in buildings has really started to be solved. Today it is clear that many older buildings suffer from this problem. The cause of such humidity can be, for example, condensation of water vapor, transmission, capillary rise and the like. Many of these failures are mainly caused by insufficient building maintenance. However, with rising humidity, this problem can be caused by corrosion and loss of functionality of the original waterproofing layer, a change in the groundwater level, and others. This wetting subsequently has a destructive effect on the constructions and reduces the functionality and lifespan of the attacked constructions. This humidity also contributes to an increase in the thermal conductivity of the construction, which leads to a high consumption of energy used for heating. In this contribution, we will focus mainly on rising humidity and the application of remedial measures on a specific building and its in situ investigation. There are a wide variety of technologies designed to address the problem of rising humidity.

An example can be the division on the basis of which we also divide the technologies in question within the framework of our research. This division was described by Makýš in his publication[2]. It could be considered one of the most comprehensible, as it divides the technologies intended for humidity remediation according to the construction-physical and implementation point of view into seven main groups (Figure 1).

(1) Technologies providing ventilation:
- creation of ventilation ducts
- provision of ventilation by a contact channel
- provision of ventilation through a contact slot
- creation of a pre-wall
- creation of hollow floors

(2) Technology for creating additional impermeable layers:
- newly lined insulation joint
- undercutting of the masonry
- embedding of stainless steel sheets
- laying an additional waterproofing layer

(3) Technology for creating crystalline curtains:
- creation of hydrophobicity curtains
- creation of sealing and hydrophobicity curtains

(4) Technologies using electro-physical principles:
- installation of galvanoosmosis devices
- installation of passive electro-osmosis devices
- installation of active electro-osmosis devices
- installation of wireless dehumidification devices

(5) Technology of heating the structures:
- installation of concealed heating
- installation of microwave drying equipment
- implementation of hot air drying

(6) Additional technologies:
- implementation of waterproofing coatings
- implementation of waterproofing plasters and sealants
- implementation of remediation plasters
- desalination of masonry

(7) Related technologies:
- creation of drainage
- lowering the groundwater level
- creation of vapour-permeable modifications of the surrounding area

Fig. 1. Division of methods intended for the rehabilitation of constructions, according to Makýš, O. [2]

It is clear that there are several divisions, such as the division of these technologies according to the EMERISDA project [3], which divides these technologies into two main groups, or the division according to the author of Witzany [4], who also classifies these technologies into certain groups.

In the paper, we will focus exclusively on the effectiveness of the deployed undercutting technology, the application of which was investigated in situ on a specific case study. This technology consists in the creation of additional impermeable layers and is implemented in successive steps. Masonry undercutting is one of the common technologies used to create a new storage joint for waterproofing in walls [5]. Modern technologies allow the application of undercutting to different types of masonry (stone, brick, mixed) and to walls of different thicknesses. Previous research has confirmed positive results after the application of this technology, both within the author's research [6] and other research carried out outside the Slovak Republic. An example can be an article describing the deployment of cutting technology in the building of the University of Pristina [7]. This technology is also described by Schmidt [8], who describes it on the one hand as radical and financially expensive, but at the cost of speed and high efficiency. However, it is also necessary to focus on the correct implementation, as its incorrect implementation can also lead to the non-functionality of the given technology [9]. Therefore, this technology can be considered globally widespread and functional. Also, research results can positively influence risk management, which can be considered a key factor [10] in the optimization and design of suitable procedures.

2. Methodology

The research methodology was established before the very beginning of the research and is defined by specific procedures that were followed throughout the research. The first step was to inspect the building and determine the degree of wetting in specific parts of the building. These places were marked in the floor plan and the results of the measurements were recorded in a table. The measurements took place in several places and at several heights. In order to examine the moisture in situ, a contact hygrometer BD2 from Dosier Messgeräte, Füssen, was used, but for the objectivity of the results, other devices were also used to determine the degree of humidity of the construction (GANN hygrometer, TESTO 616 hygrometer). With the help of these devices, a set of measurements was carried out by each device and this was subsequently evaluated in order to eliminate as much as possible the inaccuracies of the device's measurements and the measurement conditions. The degree of humidity was determined according to the technical standard ČSN P 73 0610 [11].

Subsequently, a suitable technology was designed to remove rising humidity from the construction. After its implementation, with the passage of time, further measurements were taken at the same places. Based on these measurements, the efficiency of the technology used was evaluated and the current humidity status of the construcion was determined.

3. Characteristics of the object and results of in situ research

The object of the research was a building from 1939, which is not protected as a monument and is used all year round. The object is rectangular in plan. From a constructional point of view, this object is built from bricks with a thickness of 300 and 500 mm. The building is one-story with a carpentry truss and a concrete roof. As described above, undercutting technology was deployed to eliminate the influence of rising humidity. The researched cutting technology was implemented in successive steps and the masonry was cut in sections of 25-30 cm. Subsequently, after each work step, the joint was cleaned and a foil double PE insulation of 2 mm thickness was inserted. Subsequently, the joint was wedged out with plastic wedges and filled with mortar with a low expansion effect. Individual measurement values before and after the application of the given technology are represented in Table 1.

	Height measurements 30 cm	**Height measurement 150 cm**
Average humidity before remediation [%]	11,12	1,40
Average humidity after remediation [%]	3,9	1,23
% difference before and after rehabilitation [%]	64,92	12,14
Masonry type	Brick	

Table 1. Table of the state and percentage loss of damp in the investigated building

From the data presented in the table, it is clear that the object according to the ČSN standard [11] showed a high degree of humidity, which can be defined as wetting at some points because the humidity values exceeded 10%. The table also shows the degree of effectiveness of the technology used. IF we focus at least on the measurement height of 30 cm, a significant decrease in humidity by 64.92% is evident. It should be noted that the measurements took place only a few months after the application of the technology. It is obvious that the reduction of humidity will continue and the humidity will reach the level of 3%, which can be considered as an optimal condition. Regarding measurements at a height of 150 cm, the drop here was low, but the original humidity of 1.40% represented a dry construction and it is likely that the technology may not have had an effect on this part. These measures significantly improved the indoor climate of the facility and also improved the properties of the constructions. It is therefore possible to follow up on this part with the just-mentioned application of active thermal protection (ATP), which will ensure the optimization of the internal environment to the required standards and needs of this object.

4. The possibility of applying ATP on the construction site

After the implementation of renovation interventions on buildings, it is possible to implement active thermal protection (ATP) on perimeter wall constructions. Active thermal protection (ATP) represents a controlled dynamic process specific to building constructions with integrated energy-active components that ensure one or more energy functions in various operating modes of energy systems and heat sources. This proposal could also be a solution for other countries where energy consumption in buildings is addressed [12].

For the purposes of this research, we will for now focus on the theoretical level of the ATP application, as these researches are still ongoing. An example can be a fragment of a solid brick wall, which also defines the constructions under investigation. In order to meet the requirements of static thermal resistance, this construction is insulated with 210 mm thick thermal insulation. In the given construction, the temperature is θm (18.04 °C). Previous research shows [13], [14], [15] that if we apply the ATP system with a thermal insulation thickness of 50 mm, we will reach a temperature of 13.27 °C in the given construction. In ATP systems, it is possible to change the water temperature. This means that if we set the water temperature in the pipeline to 18 °C, this temperature will be equivalent to an insulation thickness of 210 mm. In this way, we can reduce the thickness of the thermal insulation by 160 mm, as shown in picture 2.

5. Conclusion

Previous research applied to previous constructions and also to the current construction proves that the application of an additional waterproofing layer can effectively reduce humidity in the construction. After the implementation of the technology in question, the humidity in the investigated building decreased to an average value of 3.9%, which is a significant decrease in a short period of several months. This downward trend is expected to continue and will be the subject of long-term investigation.

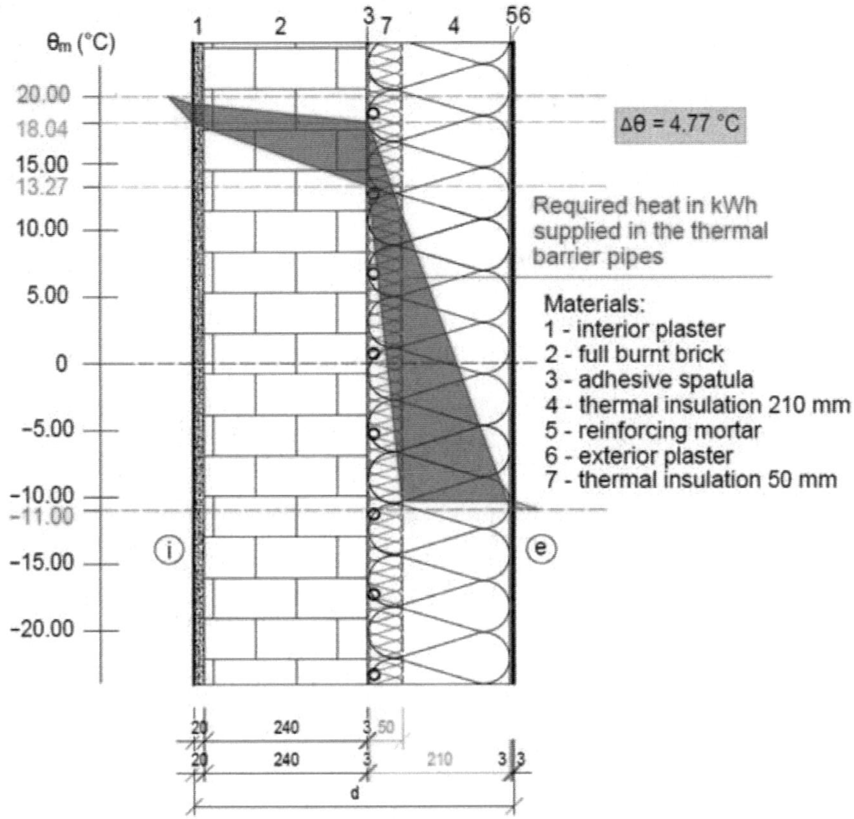

Fig. 2. Fragment of the wall perimeter construction: full burnt brick
θm - the temperature in construction (°C), Δθ - temperature difference (°C), d - construction thickness (mm), e - exterior, i - interior

This will determine the exact degree of effectiveness of this technology on the building in question. It is clear that even at current values, the internal environment of this building has deteriorated significantly. This step created a prerequisite for the application of active thermal protection. As research shows, it can ensure the optimization of the internal climate even with significantly smaller thicknesses of thermal insulation. It will be necessary to follow up on the research in question and also focus on its long-term investigation in order to be able to determine the exact data. These will be acquired at several construction sites. This will provide relevant information about the possibility of applying rehabilitation technologies and the subsequent possibility of implementing active thermal protection. Research determined in this way will provide valuable knowledge about the effectiveness of the investigated technologies and their impact on improving the quality of building interiors, but also on extending the life of the construction. Subsequently, it will be possible to apply this knowledge in the field of protection and maintenance of older buildings.

6. Acknowledgments

Funded by the EU NextGenerationEU through the Recovery and Resilience Plan for Slovakia under the project No. 23-01-03-B

7. References

[1] Kenwood, H. R. (1892). Dampness in and about houses. Public Health, Vol. 5, p. 247–250

[2] Makýš, O. (2023). Technológia Obnovy Budov, Ochrana a Oprava Spodných a Obalových Konštrukcií, SPEKTRUM STU, [Online], ISBN: 978-80-227-4848-3, Bratislava, Slovakia, Available at: https://www.vydavatelstvo.stuba.sk/obchod/technologia-obnovy-budov-ochrana-a-oprava-spodnych-a-obalovych-konstrukcii

[3] EMERISDA (2014). Summary Report on Existing Methods against Rising Damp. D2.1 FINAL Version, [Online], Available at: https://www.emerisda.eu/wp-content/uploads/2014/07/D-2_1.pdf

[4] Witzany, J. (1999). Poruchy a rekonstrukce zděných budov, 1st ed. Technická knižnice autorizovaného inženýra a technika, ISBN 80-902697-5-3, Praha, Czech Republic: Český svaz stavebních inženýrů,

[5] Greško, D., Držka, M., & Adamská, G. (1991). Konštrukcie pozemných stavieb: poruchy a rekonštrukcie stavieb (Vol. 1). Slov. technická univ., Stavebná fakulta, ISBN: 8022704296, Bratislava, Slovakia

[6] Šťastný, P., Gašparík, J. & Makýš, O. (2021). Analysis of moisture and salinity of historical constructions before and after the application of REMEDIATIONS. J. Build. Eng., Vol. 41, p. 102785, DOI: 10.1016/j.jobe.2021.102785

[7] Gavrilović, D. J. (2011). Building waterproofing remediation applying the 'total cut' method. Facta Univ. - Ser. Archit. Civ. Eng., Vol. 9, No. 1, p. 105–118

[8] Schmidt, H. (1999). Historische Bauwerksabdichtungen - Traditionelle und neuzeitliche Maßnahmen zum Schutz gegen Bodenfeuchtigkeit und zur Trockenlegung feuchter Wände. Bautechnik, Vol. 76, No. 7, p. 581–598, DOI: 10.1002/bate.199903790.

[9] Møller, E. B., Olsen, B. (2011). Rising damp, a reoccurring problem in basements: The 9th Nordic Symposium on Building Physics (NSB 2011), 9th Nord. Symp. Build. Phys., Vol. 2, p. 765–772

[10] Kremljak, Z (2011). Risk Management Methods, Chapter 10 in DAAAM International Scientific Book 2011, pp. 119-132, B. Katalinic (Ed.), Published by DAAAM International, ISBN 978-3901509-84-1, ISSN 1726-9687, Vienna, Austria DOI: 10.2507/daaam.scibook.2011.10

[11] ČSN P 73 0610 (2020). Hydroizolace staveb - Sanace vlhkého zdiva - Základní ustanovení [Waterproofing of buildings - The rehabilitation of damp masonry and additional pro-tection of buildings against ground moisture and against atmospheric water - Basic provision]. [Online]. Available at: https://csnonline.agentura-cas.cz/Detailnormy.aspx?k=58539

[12] Kadric, D; Blazevic, R; Zukic, E & Kadric, E (2023). Applying the Design of Experiments to Model the Energy Consumption of Typical Single-Family House in Bosnia and Herzegovina, Proceedings of the 34th DAAAM International Symposium, pp.0219-0228, B. Katalinic (Ed.), Published by DAAAM International, ISBN 978-3-902734-41-9, ISSN 1726-9679, Vienna, Austria DOI: 10.2507/34th.daaam.proceedings.028

[13] Mučková, V. et al. (2023). Analysis of the Dynamic Thermal Barrier in Building Envelopes. Coatings, Vol. 13, No. 3, DOI: 10.3390/coatings13030648

[14] Mučková et al. (2023). Contribution to Active Thermal Protection Research—Part 1 Analysis of Energy Functions by Parametric Study. Energies, Vol. 16, No. 11, DOI: 10.3390/en16114391

[15] Kalús, D. et al. (2023). Contribution to Active Thermal Protection Research—Part 2 Verification by Experimental Measurement. Energies, Vol. 16, No. 12, DOI: 10.3390/en16124595

DOI: 10.2507/35th.daaam.proceedings.051

COMPARING THE SIMULATED AND REAL ENERGY CONSUMPTION OF INDUSTRIAL ROBOTS

Caspar J.F. Conradi, Kemajl Stuja, Mohamed Aburaia & Ali Aburaia

This Publication has to be referred as: Conradi, C[aspar]; Stuja, K[emajl]; Aburaia, M[ohamed] & Aburaia, A[li] (2024). Comparing the Simulated and Real Energy Consumption of Industrial Robots, Proceedings of the 35th DAAAM International Symposium, pp.0373-0377, B. Katalinic (Ed.), Published by DAAAM International, ISBN 978-3-902734-44-0, ISSN 1726-9679, Vienna, Austria
DOI: 10.2507/35th.daaam.proceedings.051

Abstract

Pick and place robot applications in industry are often characterised by high energy consumption. Between 2020 and 2022, electricity costs increased by 150% in several European countries. In order to reduce these costs and prevent global climate change, the energy efficiency of these applications must be optimised. Robot manufacturers provide software tools that can also predict energy consumption. This makes these tools very attractive, even in the early stages of developing these applications. Normally, the energy required to move the robot is predicted. In this case, the energy consumption of the robot controller must also be added. The aim of this work is to compare the total energy consumption of the robot system with the energy consumption of the manipulators (i.e. the simulated energy values of the robot software). To express it more precisely, the ratio of these two major factors is of interest. The ABB software RobotStudio and the robot IRB1100, also from ABB, were used. For compatibility, an ABB EM/S3.16.1 energy meter is connected between the power source and the robot controller, which communicates with the output unit via the KNX bus. The energy consumption of two different motion sequences at five different speeds, ranging from the movement of a single axis to a more complex pick-and-place application, was measured both practically and by simulation. The results of this work show that there is a relationship between the measured values and the simulated energy consumption. In the future, it is planned to extend the test to other robot manufacturers with robots of the same class.

Keywords: Industrial Robots; Automation; RobotStudio; KNX; Energy efficiency

1. Introduction

The design of robotic applications is very important and therefore an essential part of the design [1], [2]. For years, the use of simulation software such as ABB's RobotStudio has improved for the design of such systems [3]. In terms of energy efficiency, the ability of software to estimate power consumption of the robot is of great importance [4] from designing and planning of pick and place applications. Although most robot simulation tools can provide a sufficiently accurate model in many respects, as with all simulations, they can never be a fully accurate representation of a real system [5].

It can therefore be assumed that there is a specific distinction between the real robot application and the simulated one. Since the simulated power usage of the system is one of the most widely used and more important tools of RobotStudio and as there is no reliable data how accurate the simulated energy consumption is, this thesis will examine the difference between the energy consumption calculated by RobotStudio and the real energy consumption of an ABB robot.

2. Problem Statement

Even when companies have no interest in becoming more energy efficient for the sake of environmental or at least economical concerns, the European Union is forcing them more and more with legal provisions to be so [6]. This is thereby how the EU defines the term energy efficiency: The Ratio of output, service, goods or energy yield to energy input [7]. Before taking any steps to improve the energy efficiency of an industrial plant, it is necessary to identify all areas where significant amounts of energy are used. To identify these spots, it is required to determine the energy consumption of each sub system of the process. The easiest way to achieve this is to use energy meters for each system. In practice few or none energy meters are being implemented to evaluate the energy use hence it only can be estimated or in other cases, simulations of the process are conducted [8].

Because of the permanent use of industrial applications for example a robot cell, relatively small modifications in the system can lead to significant changes in the energy efficiency of this application [9]. Therefore, it is necessary that simulations provide a reliable source of accurate data when they are used to design a robot cell in terms of energy consumption. RobotStudio is a widespread program in the industry and also used for research projects [4] to simulate these kind of robot cells. As mentioned one of the most important aspects of optimizing energy efficiency is viewing the simulated power consumption of industrial robots in RobotStudio. But how accurate are these calculations in comparison to the real consumption of a Robot? This thesis is answering this question by investigating the gap of the calculated electricity consumption of a simulated industrial robot and the measured electricity consumption of a real industrial robot.

3. Materials and Methods

This is accomplished by simulating different motion sequences of a robot in RobotStudio and performing the exact same processes with the equal robot in reality where the power is measured by an electricity meter. The main goal of this thesis is to find out if the simulated energy consumption of a industrial robot in RobotStudio is accurate or by how much it deviates from the real consumption. It is not examined how energy can be saved in a robot application or how it is consumed. In the first part of the paper it is explained with which energy meter and how the measurement of the consumed energy of the real ABB IRB1100 robot is performed and how the data from the energy meter is read out. In the second part the data from the different measurements are presented and discussed.

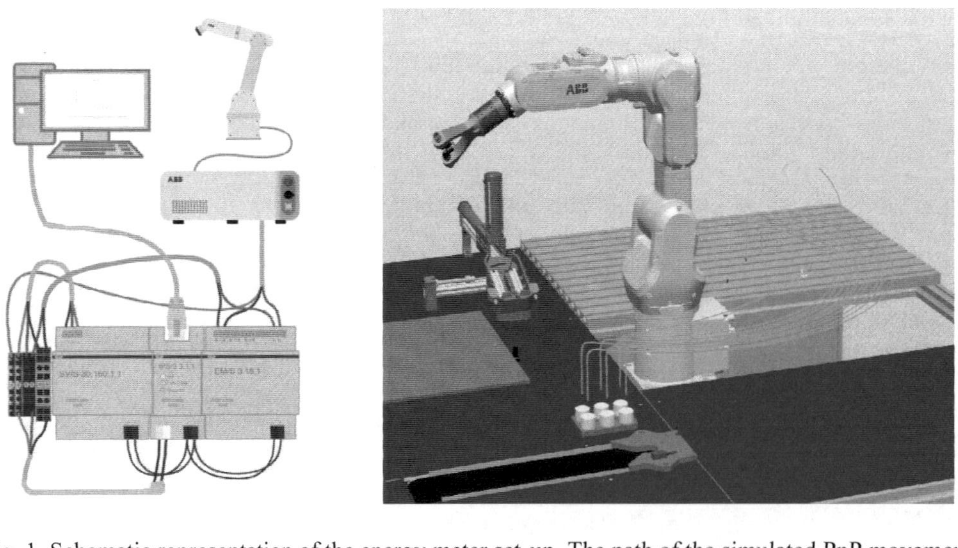

Fig. 1. Schematic representation of the energy meter set-up. The path of the simulated PnP movement

The energy meter that is used to measure the real energy consumption of the robot is the ABB EM/S3.16.1. It is capable of measuring three phase of either 250 V or 450 V alternating current with 16 or 20 Ampere [14]. This is sufficient as the IRB1100 robot that is used is controlled by the Omnicore C30 controller, which has an interrupt fuse for currents in excess of 10A [10]. The connection bus which the EM/S3.16.1 is using is the KNX. KNX was originally developed for home and building control and is the successor to the European Installation Bus (EIB). It is the only open standard worldwide in this field and is recognized in the EU as well as internationally [11].

At the level of the OSI-Model the functionalities of KNX can be assigned to different layers. At the first layer, the physical layer KNX is offering four different alternatives where the TP (Twisted Pair) - which has similarities to the RS-485 standard - is the most used. The three other ones are KNX PL (Power Line), KNX RF (Radio Frequency) and KNX IP (Internet Protocol). The latter one was used in this project where KNX telegrams are wrapped into IP packets so that a conventional LAN network can be used [12].

Therefore, also the ABB IPS/S3.1.1 had to be used to create the connection between the energy meter with the KNX Bus and the PC in the IP Network which was used to read in the measured data. Additionally, as it is shown in the figure 1 the ABB SV/S 30.320.1.1 was used to generate the power for the KNX Bus at 320 mA and monitor the KNX system voltage. The SV/S is supplied with the normal 230 V, 50 Hz voltage source from the electrical enclosure which also provides the 24 V for the IPS/S interface. Apart from these two connections the electrical enclosure is seen as a black box since the only relevant parts where the energy consumption is examined are the ABB Omnicore C90 XT Controller and the ABB IRB 1100 Manipulator as shown in figure 1.

To be able to connect the EM/S over the IPS/S to the computer the XKNX python library was used [13]. After connecting the IPS/S over the LAN with the computer the used python program with the xknx library searches for KNX/IP devices in the network and builds a IP-Tunnel to the corresponding sensor selected by the IP Address. For the first measurements the sensor was not connected directly to the computer but via a wireless access point what lead to a slightly lower scanning rate.

The EM/S is capable of measuring the consumed power in Watt every 0.2 seconds which is enough for slower robot movements at around 250 mm/s but the device reached its limits when the robot speed was increased to more than 1000 mm/s as can be seen further below at the results. The lowest current that can be measured is 25 mA what was in this case accurate enough to measure the power consumption [14].

4. Methods

As mentioned the counterpart of the real movement of the robot is a simulation of the movement in RobotStudio from ABB. To compare the energy consumption of the simulation and the real robot in a useful way it is necessary to ensure that they both are making the exact same movements. To have a better understanding what RobotStudio is measuring, in the beginning simple movements were chosen to determine the differences between simulation and reality. One of the simplest movements that was first picked is the circular path between two points that can be reached by only moving the first axis of the robot by 180 degrees and back. For this kind of movement, the moveJ (J for Joint) command is the most suitable. The forth and back motion of the first axis was performed with three different velocities: 200, 1000 and 5000 mm/s.

The second motion sequence was a little more complex task. At a pick and place (PnP) application six imaginary objects where moved around about 100 degrees as can be seen in figure 2. This was again performed with three different velocities: 2000, 4000 and 7000 mm/s. The PnP offset motion over the yellow cylinders the robot was for safety reasons only moved at 200 mm/s.

In RobotStudio, the energy consumption is computed in the signal analysis and can be exported as a Microsoft Excel file. Different values can be recorded here. For this project only the engine power in Watt and the engine energy in Joule are relevant. The data measured by the EMS was stored by the python program as a .csv file. Each of the three velocities for the two different motion sequences where measured five times to obtain a more accurate average value afterwards. The measurement was started by hand and afterwards the motion sequence was started. It was not practically possible to start both at the same time how RobotStudio is doing it in the simulation.

5. Results

Two pieces of information where looked for. Firstly, how many joules does the robot consume for the entire motion sequence and secondly, how many watts does it consume to move the axis. Since the energy meter is only measuring the current power in watt the joules for the sequence where calculated by numerically integrating the measured watts. This can be seen in equation 1 with t as the current time and W the measured watt. n stands for the current measuring and N for the total number of measurements. For this calculation, it was assumed that the duration of the procedure is the same in both the simulated and the real world.

$$\sum_{n=0}^{N}(t_{n+1} - t_n) \cdot \frac{W_{n+1} + W_n}{2} \quad (1)$$

The watt refers to the average value during the movement. The joules for the whole sequence. The calculated joules for the 180 ° rotation and for the PnP application can be seen in figure 3 and 4. For the simple 180 ° rotation it was also possible to get an average watt consumption for the movement of the first axis. These values are also shown in Figure 3. Calculating an average watt usage was no longer possible for the more complex PnP motions sequence at the higher speeds of 4000 and 7000 mm/s especially because of the positive and negative acceleration of the different axes of the robot. In figure 5 the first half of the motion sequence at 4000 mm/s is illustrated with the measured values and the simulated values on top of them.

6. Discussion

The first thing that can be easily seen is that there is a significant contrast between the measured values of the consumed energy and the simulated ones by RobotStudio. The biggest difference of almost 1100 % more joule for the measurements than for the simulation can be seen in the slow 180 degree movement (figure 3). This huge gap can be explained by the difference of around 138 watt for the rotation movement. Since due to the low speed the rotation takes so long (more that 30 sec back and forth), the gap between the simulation and reality is increasing and increasing every second.

For the higher velocities the difference in the measured watt stays more or less the same (186 W at 1000 mm/s and 177 W at 5000 mm/s) but as the motion sequence is much shorter the gap in the calculated joule is also much smaller. This deviation of around 140 to 180 watt can be seen as an absolute error and could possibly originates that RobotStudio does not take the consumption of the controller or Flex-Pendant into account. Or it is also possible that the simulation has certain other inaccuracies (friction, different masses, other physical resistances) or modelling errors that lead to this deviation. An interesting observation that supports this assumption is that when the robot is in a wait position the simulated power consumption is about 5,14 W but at the same time the energy consumption - the joules - is not rising. Another thing to note in figure 3, is that while the simulated energy increases at higher speeds, the energy calculated from the measured values decreases. In figure 4 for the pick and place sequence the trend of the 180 ° rotation continues. At higher velocities the gap between the simulated and measured values is getting smaller while the simulated energy increases and the measured consumption decreases. A couple of things have to be noted here. Firstly as it shows in figure 5 the sampling rate of the energy meter was probably not high enough to accurately read the robot's consumption. Where the simulated values have clear peaks, the measured ones are more like a rectangular signals. For the highest velocity this was even more decisive. Also for this reason, in order to reduce this constraint, several measurements were carried out what again led to the next challenge to synchronize these different measurements. Since the movement sequence could not be started always at the same time, the first peak was used to overlay the five measurements and calculate the average value. Thirdly, it was observed that the simulated values could also go negative, whereas the measured values couldn't. This is probably due to the robot braking at higher speeds and RobotStudio simulates this differently to how it was measured.

The smaller differences between 4000 mm/s and 7000 mm/s in figure 4 comes likely from the fact that according to the simulation the higher velocities of 7000 mm/s can't be reached in this motion sequence. It was not possible in this project to determine how these speeds correspond to the ones in real world.

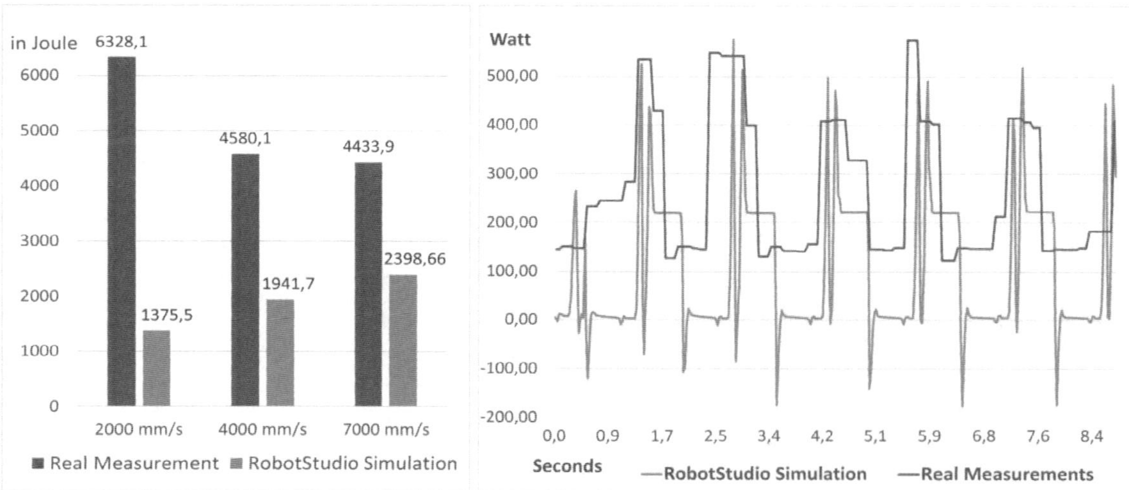

Fig. 2. Measurements of the energy consumption of the PnP application. Figure 5: Illustration of the measured and the simulated power of the PnP application.

7. Summary and Outlook

Pick and Place robot applications Since years the use of simulation software to plan robot application is a standard. RobotStudio from ABB is an example of this software. The tool to visualize the consumed energy of the robot in the simulation is essential to design these applications in an energy efficient manner. To determine how accurate these values are the ABB EM/S3.16.1 energy meter was used to compare them to real measurements from the same robot station. The energy meter was connected to an PC with the KNX IP bus and read out with the xknx python library. Two different motion sequences where examined, one 180 ° rotation and a pick and place application each at three different speeds. Significant differences were observed between the simulated and the measured energy consumption.

As a result of a possible absolute error (which does not increase at higher speeds) between the simulated and measured watts required to move an axis, the difference in energy consumption increased the longer the movement lasted. This observation could be made for the simpler 180 ° rotation and for the more complex Pick and Place sequences. For following projects an energy meter with a higher sampling rate would be necessary to measure high dynamic applications more precisely. It would also be interesting having the possibility to make these kind of comparison between the simulation and the real world with bigger industrial robots where much more energy is needed and also the potential for savings is much higher.

8. References

[1] Carabin, G., Wehrle, E. & Vidoni, R., 2017. A Review on Energy-Saving Optimization Methods for Robotic and Automatic Systems. Robotics, 6(4), 39.

[2] Statista, G., 2022. Index zur Entwicklung der Industriestrompreise in Deutschland bis 2022. (accessed 2022-12-29), https://de.statista.com/statistik/daten/studie/12500/umfrage/ entwicklung-der-industrie-strompreise-in-deutschland-seit

[3] ABB AG, 2023. RobotStudio. (accessed 2023-02-06).

[4] Stuja, K., Katalinic, B., Aburaia, M. & Pajaziti, A., 2021. Design of a Robot Application with Regard to Energy Efficiency. Vol. 1. first: DAAAM International Vienna. http://www.daaam.info/Downloads/Pdfs/ proceedings/proceedings_2021/005.pdf.

[5] Roddeck, W., 2022. Grundprinzipien der Mechatronik: Modellbildung und Simulation mit Bondgraphen. Wiesbaden: Springer Fachmedien Wiesbaden. https://link.springer.com/10.1007/ 978-3-658-38089-2.

[6] Blesl, M. & Kessler, A., 2017. Energieeffizienz in der Industrie. Berlin, Heidelberg: Springer Berlin Heidelberg. http://link.springer.com/10.1007/ 978-3-662-55999-4.

[7] European Union, 2012. Richtlinie 2012/27/EU des Europäischen Parlaments und des Rates zur Energieeffizienz.

[8] Medojevic, M., Medojevic, M., Cosic, I., Lazarevic, M. & Dakic, D., 2017. Determination and Analysis of Energy Efficiency Potential in Socks Manufacturing System. Vol. 1. first: DAAAM International Vienna. http://www.daaam.info/Downloads/Pdfs/ proceedings/proceedings_2017/082.pdf.

[9] Bukata, L., Šůcha, P. & Hanzálek, Z., 2019. Optimizing energy consumption of robotic cells by a Branch & Bound algorithm. Computers & Operations Research, 102, 52–66. https://linkinghub.elsevier.com/ retrieve/pii/S0305054818302533.

[10] ABB AG, 2022. Product Manual - OmniCore C30.

[11] Schnell, G. & Wiedemann, B. eds., 2019. Bussysteme in der Automatisierungs- und Prozesstechnik:Grundlagen, Systeme und Anwendungen der industriellen Kommunikation. Wiesbaden: Springer Fachmedien Wiesbaden. http://link.springer. com/10.1007/978-3-658-23688-5.

[12] Stuckenholz, A., 2020. Basiswissen Energieinformatik: Ein Lehr- und Arbeitsbuch für Studierende und Anwender. Wiesbaden: Springer Fachmedien Wiesbaden. http://link.springer.com/10.1007/ 978-3-658-31809-3.

[13] XKNX, 2023. Asynchronous Python Library for KNX. (accessed 2023-06-22).

[14] ABB AG, 2012. Energy Module EM/S 3.16.1 - Product Manual.

DOI: 10.2507/35th.daaam.proceedings.052

STRATEGY DEVELOPMENT BASED ON INTANGIBLES FOR STARTUPS: CAPACITY BUILDING IN EMERGING EUROPEAN INNOVATION ECOSYSTEMS

Katrin Singer-Coudoux, Henry Nicolai Buxmann & Markus Will

This Publication has to be referred as: Singer-Coudoux, K[atrin]; Nicolai Buxmann, H[enry] & Will, M[arkus] (2024). Strategy Development Based on Intangibles for Startups: Capacity Building in Emerging European Innovation Ecosystems, Proceedings of the 35th DAAAM International Symposium, pp.0378-0386, B. Katalinic (Ed.), Published by DAAAM International, ISBN 978-3-902734-44-0, ISSN 1726-9679, Vienna, Austria
DOI: 10.2507/35th.daaam.proceedings.052

Abstract

The SPREAD2INNO project aims to enhance innovation and strategic development for startups in emerging innovation countries across Europe. As part of its capacity-building activities, the project introduced the Integrated Strategy Development (ISD) method during Innovation Academies held in Bologna, Italy, and Sofia, Bulgaria, in June 2024. The ISD method was presented to assist startups in aligning their business models, managing intangible assets, and developing sustainable competitive advantages in a knowledge-based, digital economy. Through structured workshops, startups learned to analyze their current business strategies, identify growth opportunities, and implement targeted actions for market expansion and business transformation. The paper outlines how ISD was applied as a practical tool to address the strategic needs of startups, with a focus on innovation and business sustainability. To conclude, the method has equipped participants with essential strategic insights, positioning them for sustainable, long-term success in resource-limited environments. The paper highlights key elements of ISD relevant to the project's objectives, including its potential for fostering innovation and competitiveness in Europe's startup ecosystems.

Keywords: Strategy development; Intangibles; Startups; Emerging innovation ecosystems; Capacity building; European innovation; SMEs; Integrated strategy development; Innovation management; Business model

1. Introduction

In today's knowledge-driven economy, the ability to manage intangible assets such as intellectual capital, customer relationships, and organizational culture is increasingly critical to the success of businesses. This is particularly true for startups and SMEs (Small and Medium Enterprises), which often face the challenge of building sustainable competitive advantages in rapidly evolving markets. The traditional focus on tangible resources such as infrastructure and financial capital is no longer sufficient. As the Fraunhofer IPK study on intangible assets in German companies [1], [2] suggests, the knowledge-based economy has led so far that intangible resources have a greater influence on business success than classical material production factors.

In the context of the digital economy, this challenge became even more important as competitive differentials on a globalized market increasingly rely on intangible features rather than "hard" product functionality or price only. Against this background, also small and medium-sized enterprises (SMEs) must increasingly deal with the targeted development of sustainable competitive advantages, which can only be achieved through the specific bundling of intangible assets into core competencies that are difficult to imitate [3]. This paper explores the application of the Integrated Strategy Development (ISD) method as a capacity-building tool within the SPREAD2INNO project, which aims to foster innovation and entrepreneurship in emerging European innovation ecosystems. The project includes innovation academies designed to support startups from various sectors in countries such as Italy and Bulgaria. The ISD method, developed by Fraunhofer IPK, offers a structured approach to strategic development, focusing on the management of intangible assets, business process alignment, and long-term value creation.

In this context, the ISD method was introduced to startups as part of the project's capacity-building activities. This paper examines how ISD can help startups in these ecosystems identify key strategic focus areas, their intangible assets, and develop sustainable business models that enhance competitiveness and innovation capacity. Through this investigation, we aim to highlight the method's practical implications and the early outcomes observed from its application within SPREAD2INNO.

2. Startups in Emerging Innovation Countries

The European innovation ecosystem is facing a pivotal moment as it contends with intensifying global competition [4]. According to the European Union's 2023 European Innovation Scoreboard (EIS), while there has been significant advancement across member states, a substantial gap persists in innovation activities among European countries and regions, especially between southern and eastern Europe, which underperform relative to western and northern Europe [5]. The EIS provides a systematic comparison of research and innovation achievements across EU member states and neighbouring regions, categorizing them into four distinct groups based on their innovation output: Innovation Leaders, Strong Innovators, Moderate Innovators, and Emerging Innovators. Notably, countries classified as Moderate and Emerging Innovators fall below the EU's average innovation score. Specifically, Croatia, Slovakia, Poland, Latvia, Bulgaria, and Romania are designated as Emerging Innovators, whereas Estonia, Slovenia, Czechia, Italy, Spain, Malta, Portugal, Lithuania, Greece, and Hungary are identified as Moderate Innovators.

Bridging this innovation divide is crucial and entails targeted development of educational programs and support systems to inspire young talent and stimulate start-up ecosystems. However, these innovation environment remains riddled with challenges, including disparities in knowledge access, limitations in human capital, financial restrictions, and effective implementation of information technologies [5]. This is supported by the analysis of [6] which states that institutional void, structural gap in innovation networks and resource scarcities are constrains in emerging innovation ecosystems. Hence efficient capacity building is crucial for startups in such environments to conquer the lack of available resources. Further, [7] find that startup growth in the innovation environment of Brazil is strongly influenced by their ability understand customer demands and build scalable business models. Moreover, [8] highlights that survival of startups in incipient entrepreneurial ecosystems is interlinked with their human and organizational capital. Within the SPREAD2INNO project it was thus emphasized to equip startups from emerging and moderate innovation regions with knowledge about efficient capacity building, analysing their own resources base, market demands, and sustainable business models to enable them to develop strong strategic foundations. Therefore, the ISD method was utilized in a workshop setting.

3. SPREAD2INNO Project Overview

The SPREAD2INNO project aims to enhance the innovation ecosystem ins emerging and moderate European countries and to enhance the European innovation ecosystem overall. This goal is pursued through a variety of training activities, including local events and regional academies, which support startups and SMEs in building effective business models and strengthen acceleration and incubation services in less connected regions.

The consortium consists of seven partners across six countries: Fondazione E. Amaldi (Italy) coordinates the project; MINDS & SPARKS (Austria) manages communication and platform development; Fraunhofer IPK (Germany) provides training and knowledge exchange; European Business Angel Network (Belgium) runs acceleration academies; Cleantech Bulgaria oversees ecosystem scaling; and Teseas (Italy) and Patras Science Park (Greece) contribute to implementation. This balanced partnership connects high-innovation regions (Germany, Belgium, Austria) with low-to-moderate innovation areas (Italy, Greece, Bulgaria), fostering equitable ecosystem growth across Europe.

The project's specific objectives are to identify promising startup during local events in Italy, Greece and Bulgaria. provide entrepreneurs with adaptable knowledge through two regional academies and support the commercialization of six promising concepts through comprehensive training. Additionally, SPREAD2INNO connects entities from both developed and less-developed innovation ecosystems, enhancing their capacity, connectivity, and competitiveness. During the six local events, two in each of the countries respectively, were held to determine the four most promising startups at each event. The identified companies had to pitch their business idea and went on to the two SPREAD2INNO academies, held in Bologna, Italy and Sofia, Bulgaria in which they received feasibility studies generated by the consortium focusing on business model development, tech due diligence and funding opportunities.

Further, the startups were offered an opportunity to gain insights through intensive workshops, complemented by one-on-one meetings and roundtable discussions. One of these workshops was conducted by two experts from Fraunhofer IPK which introduced the ISD to enable the startups to increase their capacities. The method was used to guide startups through the process of strategic thinking, particularly emphasizing the management of intangible assets. Finally, at each Academy three winners were identified, and participated in in an intensive seven-day training, providing a unique opportunity to connect with experts from prominent networks and expand their market reach and business scalability. Additionally, each participating company received tailored business coaching, specifically designed to boost their appeal to investors and increase market impact.

4. Introduction to the Integrated Strategy Development (ISD) Method and its application in SPREAD2INNO

As highlighted in the Fraunhofer IPK study on intangible assets in German companies [1], [9] the shift towards a knowledge-based economy has progressed to the point where intangible resources, such as intellectual capital, brand value, and networks, now have a more significant impact on business success than traditional material assets. This trend applies across all sectors and company sizes but is especially critical in the services industry. In today's digital economy, this challenge is even more pronounced, as competitive advantages in a globalized market increasingly depend on intangible features, rather than merely on "hard" product functionality or price. Against this background, SMEs must increasingly focus on developing sustainable competitive advantages by strategically bundling intangible assets into core competencies that are challenging for competitors to replicate. Achieving this requires the development of so-called knowledge-based business models, where strategic business activities are aligned with intangible resources and customer values to establish unique market positions or, at the very least, ensure a viable competitive stance and long-term survival.

The Integrated Strategy Development (ISD) method was introduced within the SPREAD2INNO project as a means of enabling startups and SMEs to systematically integrate intangible assets into their business models to meet the demands of a knowledge-based, digital economy. The ISD method was presented during the SPREAD2INNO Innovation Academies to help participating startups identify their key intangible resources, strategically develop them, and ultimately, foster sustainable competitive advantages.

The ISD method supports the following core management tasks through a structured communication process that utilizes workshop concepts and analytical tools specifically adapted for SMEs:

- **Explication and Documentation of Existing Strategies and Strategic Options:** Startups document and clarify their current business models and strategic alternatives, establishing a clear foundation for strategic planning.
- **Development and Review of Strategic Goals with Respect to Logical Consistency:** Startups formulate strategic goals and evaluate them for coherence, ensuring alignment with their core competencies and available resources.
- **Analysis of Internal and External Influencing Factors, Particularly Intangible Assets, in Relation to Strategic Objectives:** Through ISD, startups assess both internal and external factors, with a special focus on intangible assets, to understand how these elements impact their strategic goals.
- **Derivation and Prioritization of Action Areas, Development Goals, and Measures for Corporate Development in Line with Strategic Goals:** The method guides startups in defining action areas and setting development priorities, establishing clear steps to enhance competitiveness and drive growth.
- **Systematic Alignment of Current and Planned Measures with Strategic Objectives:** ISD ensures that ongoing and planned initiatives align with the company's broader strategic objectives, allowing startups to focus on measures that directly support their growth and sustainability.
- **Continuous Monitoring and Control of Implementation Success:** Startups are equipped to regularly assess the progress of their strategic initiatives, providing the flexibility to adapt their approach as needed.

The ISD method provides flexibility for startups, as it can be used both for comprehensive, continuous business planning and for addressing specific strategic challenges. Within the SPREAD2INNO project, it was presented in a modular manner to meet the diverse needs of the participating startups.

For example, companies can use the method to make targeted strategic decisions such as:

- **Making Specific Strategic Decisions:** For example, ISD enabled startups to answer critical questions such as "Should we expand into a new market segment or launch a new product?" by providing a structured framework for assessing their readiness and strategic fit.
- **Implementing Strategic Decisions Operationally:** The ISD method helped startups determine the specific steps needed to bring strategic decisions to fruition, such as identifying the resources and actions required to successfully launch a new product or service.
- **Ensuring Strategic Alignment of Ongoing and Planned Measures:** Startups were guided to continuously assess their initiatives against their strategic objectives, ensuring that all activities are consistent with the broader goals set within the SPREAD2INNO framework.

The ISD method and its application is presented in more detail the following chapters.

5. Model of Integrated Strategy Development (ISD) for Start-Ups

The structural model of Integrated Strategy Development (cf. Fig. 1) was developed to balance thoroughness with a reduction in complexity [10]. It encapsulates the essential content and process considerations of ISD. In the vertical dimension, the model presents the essential content dimensions of a corporate strategy, referred to as analysis levels in ISD. In the horizontal dimension, the model outlines the strategy development process through key intermediate outcomes, covering both strategy formulation and implementation.

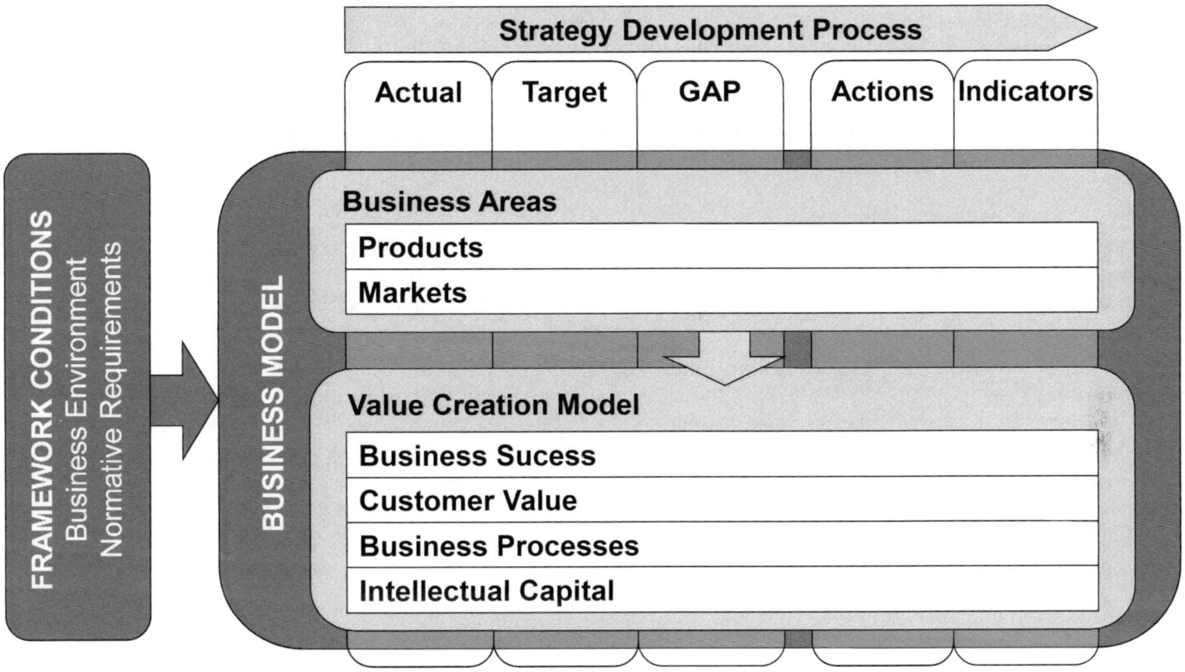

Fig. 1. Structural Model of Integrated Strategy Development
Source: [10]

Framework conditions fundamentally shape corporate development opportunities. On one hand, these include opportunities and risks arising from current developments in the business environment—such as trends in customer and supplier markets, competition, and broader technological, social, and legal trends [11]. On the other hand, normative requirements from company owners and other stakeholders play a significant role. These requirements may encompass overarching goals for the company's development, such as specific profit or growth targets, as well as other general restrictions.

Based on these framework conditions, the business model of each company is refined. The ISD method defines a business model through the following elements: business areas, which represent homogeneous product-market combinations [12]. Consequently, "products" and "markets" form the initial two elements of the business model, summarizing which products or services deliver specific benefits to designated target groups. Additionally, the "architecture of goods and services provision" and the "revenue model" [13] are determined, explaining how the organization leverages resources in business processes to produce and deliver these products/services for the respective target groups, while generating economic returns. Within ISD, this combination is known as the value creation model.

Aligned with ISD's focus on intangible resources, value creation and differentiation are increasingly driven by intangible factors, positioning them as central in the competitive market landscape [11]. Thus, intangible assets are incorporated explicitly within the value creation model from both internal resource and external market perspectives. Business processes act as a link between internal intangible resources—such as intellectual capital—and external customer values. The business success level, in turn, connects the value creation model to the framework conditions (owner and stakeholder requirements, environmental opportunities and risks) and operationalizes the development of the business areas according to the overarching strategic and economic goals.

This model provides startups with a framework for recognizing and developing the drivers of knowledge-based business models, converting them into sustainable strategies. By tying the intangible resource base directly to the corporate strategy, ISD links intellectual capital with the company's specific business functions—its products and markets—through various levels of the value creation model. This approach ensures a practice-oriented and cohesive management concept, where intangible assets are treated not as isolated or additional factors, but as integral to the creation of goods and services and successful market engagement.

Startups present at the SPREAD2INNO academies can apply this approach to build core competencies from unique configurations of intellectual capital that are difficult for competitors to imitate [14]. In the ISD model's dynamic perspective on organizational change, the current state (**ACTUAL**) of the business model is first assessed to determine the startup's starting position. Based on the existing framework conditions, a future target state (**TARGET**) is defined, typically within a 3–5-year horizon. The **GAP**—the strategic difference between the TARGET and ACTUAL states—highlights the strategic objectives and development goals necessary for growth. This GAP analysis helps startups to clarify strategic intentions, assess current capabilities, and identify necessary actions to bridge the gap, providing a structured pathway to growth and resilience. The GAP analysis provides startups with a clear roadmap to bridge current and desired future performance levels. Derived actions are monitored using indicators that allow for tracking progress and ensuring alignment with strategic objectives over time. By aligning strategy formulation and operational implementation, ISD ensures that strategy is not a static plan but an actionable process. This compliance with methodological principles fosters a structured approach for startups, equipping them with a strategy framework that remains adaptive to new market realities. In summary, Integrated Strategy Development (ISD) seeks to answer the following core questions to create a cohesive strategic framework for startups:

Business Areas (Products & Markets):
- What product-market combinations do we engage in, and how does value creation translate into customer benefit?
- What strategies will allow us to grow and expand these areas sustainably?
- Which strategic goals emerge from the desired evolution of our business segments?

External Perspective of the Value Creation Model (Customer Values & Business Success):
- Where do our main competitive advantages lie?
- How can we further differentiate ourselves within our business areas?
- What indicators will confirm we are meeting our strategic goals?

Internal Perspective of the Value Creation Model (Intellectual Capital & Business Processes):
- What are our internal strengths and weaknesses?
- Which business processes and intangible resources require development to achieve our strategic goals?

The following chapter will provide insights on how start-ups can implement the model.

6. Implementation of Integrated Strategy Development (ISD)

The implementation of Integrated Strategy Development (ISD) is structured into three main phases: **Focus**, **Analysis**, and **Change** (cf. Fig. 2).

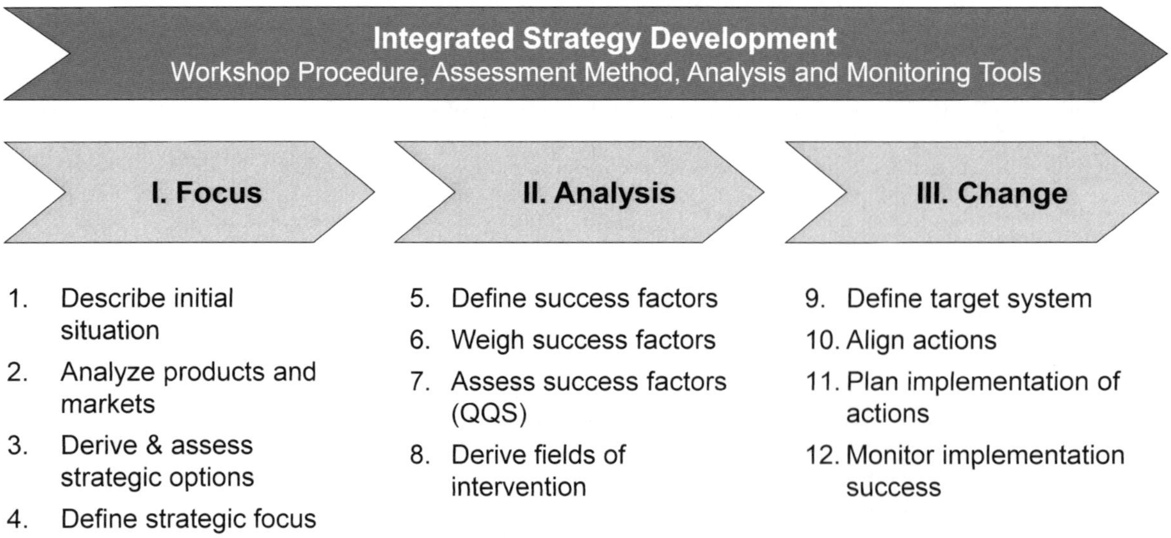

Fig. 2. Procedural Model of Integrated Strategy Development
Source: [10]

Each phase produces actionable intermediate results, which can be applied modularly or as standalone components, depending on specific startup needs and available resources [15].

The ISD approach follows a workshop methodology, adapted from the "Wissensbilanz – Made in Germany" intellectual capital framework, to help startups systematically develop their strategy through facilitated sessions. A trained moderator guides the startup's strategy team—typically comprised of founders and key stakeholders—through each phase to ensure structured and effective implementation.

6.1 Phase I: Focus

The **Focus** phase aims to establish a clear consensus within the startup's strategy team on strategic direction. By analyzing the current business model and assessing the startup's business environment, this phase identifies strategic options that align with the company's goals and market opportunities. For startups, establishing this focus provides immediate guidance for daily decision-making and sets the foundation for long-term objectives [16] The **Focus** phase includes four steps:

- **Step 1:** Describe the initial situation.
- **Step 2:** Analyze products and markets.
- **Step 3:** Derive and assess strategic options.
- **Step 4:** Define the strategic focus.

Using tools like the BCG matrix [17] and the Ansoff Matrix [18], startups can systematically evaluate their product-market fit and explore strategic growth options. These options are mapped within the "Strategic Options Portfolio" (StratOp-Portfolio), which helps startups assess the risks and opportunities of different strategies (cf. Fig. 3).

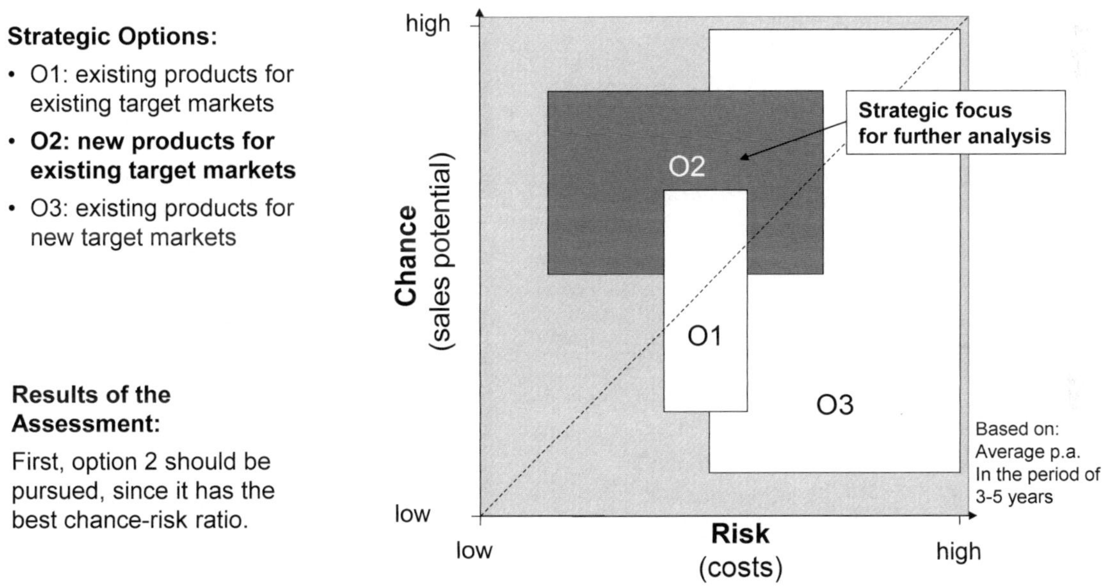

Fig. 3. Strategic Options Portfolio
Source: [10]

In the context of *SPREAD2INNO*, the StratOp-Portfolio helps startups quantify their potential sales growth (best and worst-case scenarios) and evaluate the associated risks. This portfolio enables startups to prioritize strategies with a favorable opportunity-to-risk ratio, allowing them to focus on initiatives most likely to succeed in their specific environment. The final outcome of the **Focus** phase is the identification of a target state (**TARGET**) at the product and market levels, based on the startup's current state (**ACTUAL**). The strategic gap (**GAP**) between **ACTUAL** and **TARGET** clarifies areas for growth and guides the next phase of in-depth analysis.

6.2 Phase II: Analysis

In the **Analysis** phase, the startup's success factors—critical elements that impact strategic goals—are identified and evaluated. This phase results in a strength-weakness profile that pinpoints areas where interventions are needed to achieve the chosen strategic direction. The **Analysis** phase comprises four steps:

- **Step 5:** Define success factors.
- **Step 6:** Weight success factors by importance.
- **Step 7:** Assess success factors (using Quantity, Quality, and Systematic assessment (QQS)).
- **Step 8:** Derive fields of intervention.

Following the QQS framework [19], startups evaluate each success factor's quantity, quality, and systematic relevance to determine a prioritized intervention plan. For example, using a "Management Portfolio" diagram, each factor's improvement potential is plotted to guide discussions on strategic needs. Startups can identify whether they should focus on stabilizing strong areas or improving weaker ones. Prioritizing interventions with the highest significance and improvement potential helps maximize resource efficiency. The resulting intervention fields define specific development goals, which the strategy team can discuss and refine to ensure alignment with the strategic objectives set in the **Focus** phase.

6.3 Phase III: Change

The **Change** phase involves planning and implementing the actions identified in the **Analysis** phase to meet strategic goals. A consistent target system is established to facilitate this, where specific development actions are derived, prioritized, and monitored for impact. The **Change** phase includes four steps:

- **Step 9:** Define a target system.
- **Step 10:** Align actions with the target system.
- **Step 11:** Plan the implementation of actions.
- **Step 12:** Monitor implementation success.

Through impact analysis [15] or simpler scoring systems, startups prioritize actions based on their contribution to key development goals. For example, a workshop can allocate these actions within the target system to visualize their impact. This helps the team reach a shared understanding of each action's scope and objectives before systematically ranking them for implementation. Actions are then organized into strategic programs, each with defined milestones, to create a roadmap that supports ongoing monitoring of their success (cf. Fig. 4).

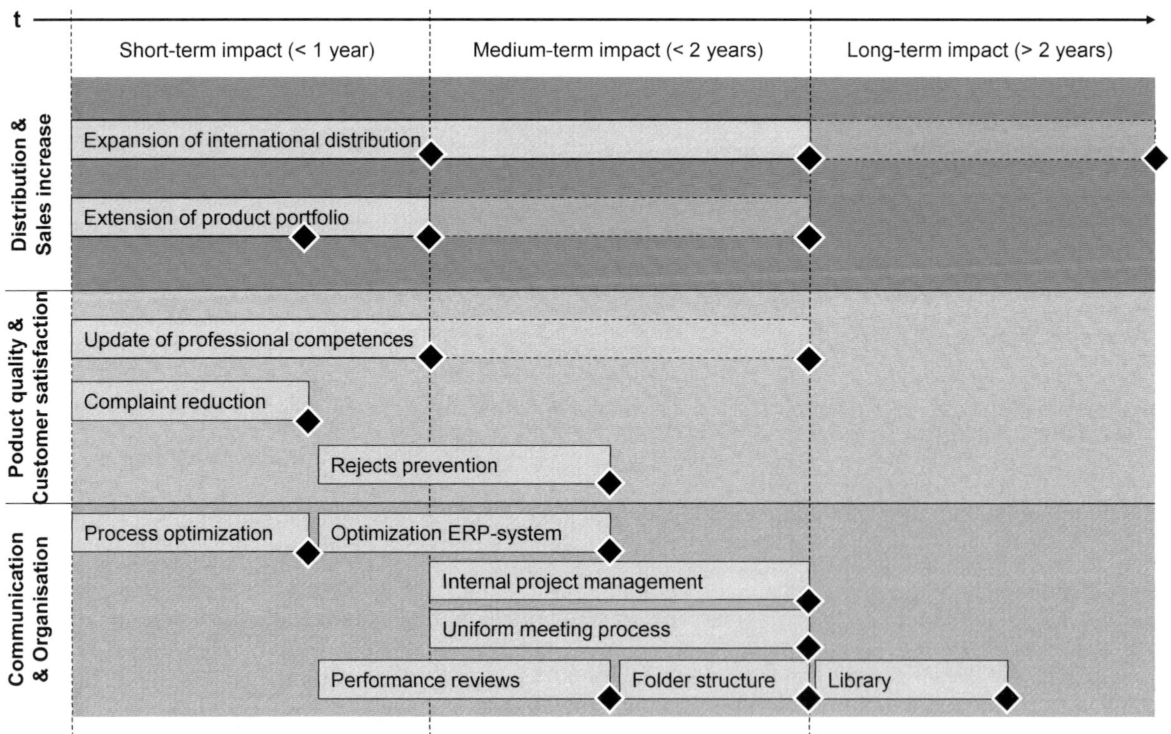

Fig. 4. Exemplary Action Roadmap
Source: [10]

A key output of this phase is an action plan detailing how each prioritized task will be managed. For startups, it is often beneficial to assign responsibility for each action to specific team members or project leads. This assignment provides accountability and enables real-time adjustments based on feedback from performance monitoring. Finally, an indicator-based monitoring system is established to track key performance indicators (KPIs) linked to each development goal. Startups can benefit from this by integrating relevant KPIs into their existing management systems, enhancing their ability to monitor and respond to changes continuously (cf. Fig. 5). This integration, supported by ISD's compatibility with standard tools like the Balanced Scorecard, helps startups manage their growth effectively.

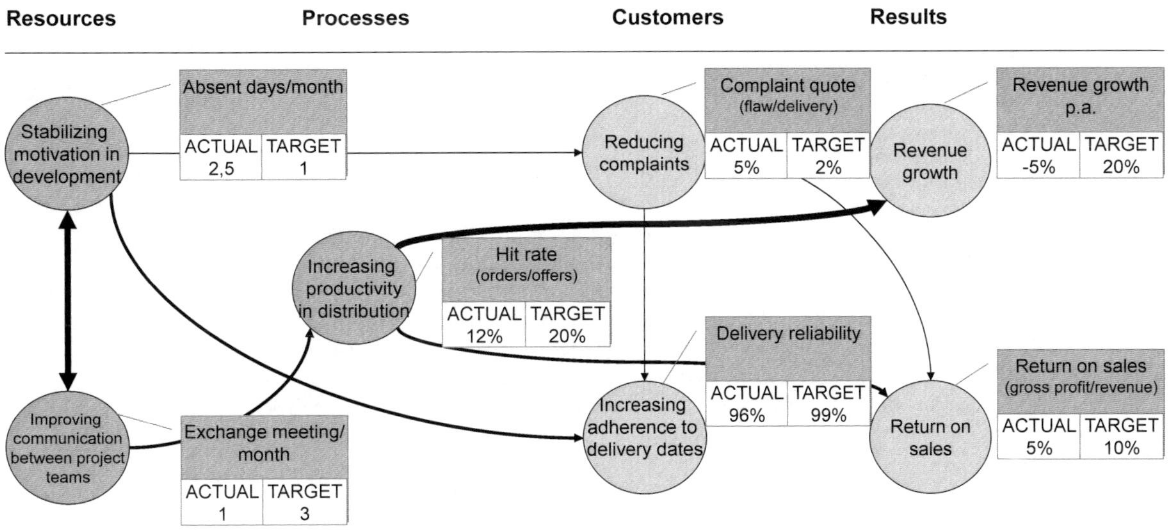

Fig. 5. Exemplary Indicators Allocated in the Target System
Source: [10]

7. Conclusion

This paper underscores the importance of managing intangible assets e.g. intellectual capital, customer relationships, and market understanding as crucial drivers of competitive advantage in today's knowledge-driven economy. Specifically, within the SPREAD2INNO project, the application of the Integrated Strategy Development method has proven to be an effective tool for startups and SMEs in emerging European innovation ecosystems. By guiding these companies in identifying and leveraging their intangible assets, the ISD method fosters strategic development, enabling sustainable business models that can drive long-term growth and competitiveness. The companies were equipped with structured tools for strategic planning, resource alignment, and long-term goal-setting, ultimately supporting their resilience and success in complex markets.

The practical implications of this research highlight the critical need for capacity-building tools that enhance strategic thinking in resource-constrained environments, particularly for startups in moderate and emerging innovation regions within Europe. This research suggests that through structured, knowledge-based frameworks like ISD, startups can bridge innovation gaps and establish robust, competitive positions in their respective markets. By systematically mapping out intangible resources and linking them to strategic business objectives, startups can more effectively identify and leverage their unique competencies, making them better positioned to navigate changing market dynamics and tailor strategic decision-making to their immediate challenge.

Future research could explore the longitudinal impact of the ISD method on startup survival and growth, particularly in diverse economic contexts and across various industries. Additionally, expanding the method to incorporate emerging digital tools and artificial intelligence and assessing their impact could provide new insights into scaling intangible asset management. The findings from this study lay the groundwork for further investigation into strategies that strengthen startup resilience and adaptability, ultimately enhancing regional and global innovation ecosystems.

8. Acknowledgments

The SPREAD2INNO project [project number: 101071977] is funded by the European Union under the call HORIZON-EIE-2021-SCALEUP-01-01. Views and opinions expressed are however those of the author(s) only and do not necessarily reflect those of the European Union or the European Innovation Council and SMEs Executive Agency. Neither the European Union nor the granting authority can be held responsible for them.

9. References

[1] Will M. (2015). Intellectual Capital Statement as a Strategic Management Tool: The European Approach. Ordónez de Pablos P, Edvinsson L: Intellectual Capital in Organizations – Nonfinancial Reports and Accounts.

[2] Alwert K, Bornemann M, Meyer C, Will M, Wuscher S. (2010). Wissensstandort Deutschland - Deutsche Unternehmen auf dem Weg in die wissensbasierte Wirtschaft. Berlin: Fraunhofer IPK.

[3] Will M. (2020). Integrated Strategy Development Based on Intangibles. Ordónez de Pablos & L. Edvinsson (Eds.): Intellectual Capital in the Digital Economy. New York: Routledge.

[4] Abels J, Bieling H-J. (2023). Infrastructures of globalisation. Shifts in global order and Europe's strategic choices. Competition & Change: pp. 516–33, DOI: 10.1177/10245294221130119.

[5] https://research-and-innovation.ec.europa.eu/statistics/performance-indicators/european-innovation-scoreboard_en#documents-and-media. (2023). European Union: European Innovation Scoreboard 2023. Accessed on: 2024-11-11.

[6] Cao Z, Shi X. (2021). A systematic literature review of entrepreneurial ecosystems in advanced and emerging economies. Small Bus Econ: pp. 75–110, DOI: 10.1007/s11187-020-00326-y.

[7] Teixeira EG, Moura GL de, Lopes LFD, Marconatto DAB, Fischmann AA. (2021). The influence of dynamic capabilities on startup growth. RAUSP: pp. 88–108, DOI: 10.1108/RAUSP-08-2019-0176.

[8] Alvarez Salazar J. (2021). Organizational resources and survival of startups firms – a qualitative analysis in the Peruvian context. ARLA: pp. 59–87, DOI: 10.1108/ARLA-04-2020-0080.

[9] Orth R, Wuscher S, Steinhöfel E, Meyer C, Will M, Alwert K, Bornemann M. (2015). Studie Wissensstandort Deutschland - Deutsche Unternehmen auf dem Weg in die wissensbasierte Wirtschaft - Ergebnisse 2014.

[10] Will M. (2012). Strategische Unternehmensentwicklung auf Basis immaterieller Werte in KMU - Eine Methode zur Integration der ressourcen- und marktbasierten Perspektive im Strategieprozess. Stuttgart.

[11] Welge MK, Al-Laham A. (2001). Strategisches Management. Grundlagen – Prozesse – Implementierung. 3rd ed. Wiesbaden.

[12] Hungenberg H. (2008). Strategisches Management in Unternehmen. Ziele - Prozesse - Verfahren. 5th ed. Wiesbaden: Gabler.

[13] Stähler P. (2002). Geschäftsmodelle in der digitalen Ökonomie. Merkmale, Strategien und Auswirkungen. 2nd ed.: Lohmar: Eul.

[14] Prahalad CK, Hamel G. (1990). The Core Competence of the Corporation. Harvard Business Review: pp. 79–91.

[15] Alwert K, Bornemann M, Will M. (2008). Wissensbilanz – Made in Germany. Leitfaden 2.0. Berlin: German Federal Ministry for Economics and Technology Germany.

[16] Will M, Wuscher S. (2014). Leitfaden Strategische Ziele entwickeln. Zusatzmodul zum Leitfaden 2.0 zur Erstellung einer Wissensbilanz. Berlin: Fraunhofer IPK.

[17] Alwert K, Will M. (2014). Leitfaden Maßnahmen managen. Zusatzmodul zum Leitfaden 2.0 zur Erstellung einer Wissensbilanz. Berlin: Fraunhofer IPK.

[18] Ansoff, H, I. (1957). Strategies for diversification. Harvard Business Review: pp. 113–24.

[19] European Commission. (2008). InCaS: Intellectual Capital Statement – Made in Europe: European ICS Guideline.